DRUG-INACTIVATING ENZYMES AND ANTIBIOTIC RESISTANCE

DRUG-INACTIVATING ENZYMES AND ANTIBIOTIC RESISTANCE

2nd International Symposium on
ANTIBIOTIC RESISTANCE
Castle of Smolenice, Czechoslovakia 1974

Editors
S. MITSUHASHI, L. ROSIVAL, V. KRČMÉRY

SPRINGER-VERLAG BERLIN HEIDELBERG GMBH

1975

Originally published by Springer-Verlag Berlin Heidelberg New York 1975
Softcover reprint of the hardcover 1st edition 1975

ISBN 978-3-642-66064-1 ISBN 978-3-642-66062-7 (eBook)
DOI 10.1007/978-3-642-66062-7

Publisher:

SPRINGER-VERLAG BERLIN HEIDELBERG GMBH

2nd INTERNATIONAL SYMPOSIUM on ANTIBIOTIC RESISTANCE: DRUG-INACTIVATING ENZYMES and Other Problems of Resistance to Antibiotics

Castle of Smolenice, Czechoslovakia
5th to 8th June, 1974

Organisers:

THE SLOVAK MEDICAL SOCIETY (on behalf of Czechoslovak Medical Society)

President: Professor R. T. NIEDERLAND, M.D.
Vice-President: Professor J. ŠTEFANOVIČ, M.D.
Secretary General: J. MARIÁNYI, M.D.

CZECHOSLOVAK SOCIETY OF MICROBIOLOGY

President: Professor T. MARTINEC, PhD.
Vice-President: Professor J. ŠTEFANOVIČ, M.D.
Secretary: V. HAVEL, Ph.D.

THE RESEARCH INSTITUTE OF HYGIENE, BRATISLAVA

Director: Professor L. ROSIVAL, M.D.

Chairman of the Symposium:

Professor SUSUMU MITSUHASHI, M.D., Gunma University, Maebashi, Japan

Secretary General of the Symposium:

Professor L. ROSIVAL, M.D.

International Organising Committee:

Dr. JULIAN DAVIES (Madison)
Professor S. MITSUHASHI (Maebashi)
Professor S. M. NAVASHIN (Moscow)
Professor J. ŠTEFANOVIČ (Bratislava)

Organising Secretariat:

J. BARANYAI, M.D.
V. KRČMÉRY, Ph.D.
F. VÝMOLA, M.D.

PREFACE

It was an extraordinary pleasure for me, as the secretary general, to organise, with my collaborators, this Second International Symposium on Antibiotic Resistance in the Castle of Smolenice in Czechoslovakia. We all appreciated many offers all participants must have spent to attend this Symposium and we were glad to do our best to prepare this meeting in a suitable and convenient way.

We gathered in Smolenice Castle after the significant International Congress on Chemotherapy in September 1973 in Athens. We had the chance of getting acquainted, on that world-wide forum, with the latest information about bacterial resistance to antibiotics. It was possible at that Congress to outline main topics of interest in the field of bacterial resistance to antibiotics which then have been placed on the programme of the subsequent Smolenice Symposium. Undoubtedly, enzymatic mechanisms by which bacteria can inactivate older and newer antibacterial drugs and which can be transmitted and spread among bacterial strains, have both medical as well as theoretical priority. Dealing specially with Pseudomonas aeruginosa is highly urgent at present from the clinical, hygienical and genetical point of view.

We have realised that antibiotics have ceased to be "magic bullets" which hit the microorganisms without discrimination. Today we urgently consider the question what should be done in the area of "antibiotic policy" to preserve the effectiveness of antibiotics for the future.

Effort to synthetise new chemotherapeutical substances and their derivatives is undoubtedly justified and it is necessary to support it. But at the same time we should aim our policy against the irrational wasting with antibacterials in our hospital and ambulatory practice in banal illnesses without establishing true and specific causal agents, and for protecting our special intensive care units against problem bacteria of nosocomial origin. We expect this to be one of the leading topics of our next, third Smolenice Symposium in 1976. In this context it is continuously important to reconsider the public-health aspects of antibiotic usage in agriculture and veterinary medicine and to continue in long-term studies of antibiotic residues that are likely to occur in human food as a consequence of their non-medical use. Data should be continuously collected on the dynamics of antibiotic usage in medical and non medical field by means of computers on national as well as international levels not forgetting the monitoring of antibiotic consumption and dynamics of resistance to antibacterials in hospitals, wards as well as in city and country population.

Nevertheless, theoretical problems of bacteria resistant to antibiotics are on advance, and it emerges that bacterial plasmids, of which a high proportion is coding for antibiotic resistance, are in fact potent instruments for genetic manipulation. Thus, all theoretical problems concerning resistance plasmid science must not be put out of our scope, and should be dealt continuously and with highest responsibility. In this connection this volume submits much of a valuable and progressive experimental material of utmost importance.

Thus, papers and discussion at the Second Symposium in Smolenice where scientists from four continents actively participated, focussed the attention on the above-mentioned aspects of international research concerning antibiotic resistance. Antibiotics are now integral part of the human life's chemisation. The future development should not let to turn it against man, but to become one of the key-issues of a happy, healthy life.

The positive significance of the First Smolenice Symposium in 1971 and of the First volume of the Proceedings as well as the positive events of the Second Symposium encourages us to continue in the well-started tradition of both Antibiotic-Resistance Symposia and of editing the mini-library of their Proceedings. The AVICENUM Publishers are doing a great deal of good work in this respect and we are very thankful for their initiatives. Also a fruitful help of SPRINGER VERLAG is appreciated.

The organisers of the Second Symposium also appreciate a financial support of Essex Chemie, Luzerne, Switzerland for expenses of some foreign participants. Beecham Research Laboratories also generously supported necessary expenses of some symposial guests. Essex Chemie financially supported also the edition of these Proceedings.

The editors express their thanks to all contributors for excellent cooperation in setting up their manuscripts which made the edition of these Second Proceedings an exciting and agreeable work.

L. ROSIVAL

CONTENTS

V. Ecology, Epidemiology and Nosocomial Problems of Antibiotic Resistance

(Chairmen: R. E. Dixon and L. Rosival)

WELCOMING ADDRESS

Dear Friends and Colleagues,

I take great pleasure in welcoming you in this country on behalf of the Czechoslovak Society and the International Society of Chemotherapy at the Second International Symposium on Antibiotic Resistance, which, much to my regret, I am unable to attend owing to another long-standing commitment that takes me out of the country.

I am sure that this symposium, to which many of you have travelled a long way to share your findings and exchange views on questions of joint interest, will contribute to the dissemination of knowledge and the progress of research in antibiotic resistance.

I wish your proceedings all success and to all members of the symposium a pleasant stay in our country.

Professor P. Málek, M. D., Dr. Sc.,
President of International Society of Chemotherapy

INTRODUCTORY LECTURES

CONTRIBUTION OF BACTERIAL GENETICS TO CHEMOTHERAPY

S. MITSUHASHI

Department of Microbiology, School of Medicine Gunma University, Maebashi, Japan

It is a real pleasure to give the opening address at this symposium. First of all, I want to express our sincere thanks to the Czechoslovak Medical Society, the Czechoslovak Society of Microbiology of the Academy of Sciences and the Research Institute of Hygiene, Bratislava, which organized successfully this symposium. Special thanks are due to the Director of the Research Institute of Hygiene, Professor L. Rosival, Dr. V. Krčméry, and Secretaries General. Their devotion and expertness played a major role in the success of this Symposium. I feel honored that my duties as a chairman of this symposium invited me from the world of microbial genetics to a closer association with those biochemists, organic chemists and pharmacologists who are actively involved in the study of chemotherapeutic agents.

Since the discovery of prontosil and penicillin, we had lived the drama of the first success of chemotherapy against bacterial infections. We can look back emotionally to the time that is one of the great moments in the long history of medicine. The introduction of antibacterial agents has contributed greatly to improved treatment of infectious diseases. In addition to their extensive use, the progress in preventive medicine and improved sanitary conditions have caused revolutional changes in microorganisms which play important roles in infectious lesions, changing the pattern of infections seen today in the hospital.

TABLE 1

Changes in microorganisms after extensive use of antibacterial agents

1. Population changes in microorganisms which play important roles in infectious lesions.
2. Wide spread of multiple-resistant strains.

Since 1945, large amounts of sulfanilamide (SA) have been used in Japan for the treatment of infections caused by Shigellae. The duration of sulfanilamide effectiveness was only about five years, after which SA-resistant *Shigella* strains appeared rapidly and reached a maximum in which 80–90% of the *Shigella* isolates were resistant to SA. Fortunately, antibiotics such as streptomycin (SM), tetracycline (TC), and chloramphenicol (CM) were introduced in Japan for the treatment of infectious diseases, and they showed striking effectiveness toward *Shigella* infection. However, the incidence of infections caused by Shigellae in Japan did not decrease to any great extent at that time. Following a few cases of the isolation of SM- or TC-resistant strains, multiple-resistant strains of various seroptype were isolated independently in different places in Japan. And the incidence of bacillary dysentery caused by multiple-resistant strains increased enormously and almost explosively since 1956. Especially, the triple (TC.CM.SM)-resistant strains have increased and almost all of them are resistant to sulfanilamide (SA). Such an increase in multiple resistance could not be answered by chromosomal mutation and by epidemic spread of resistant bacteria.

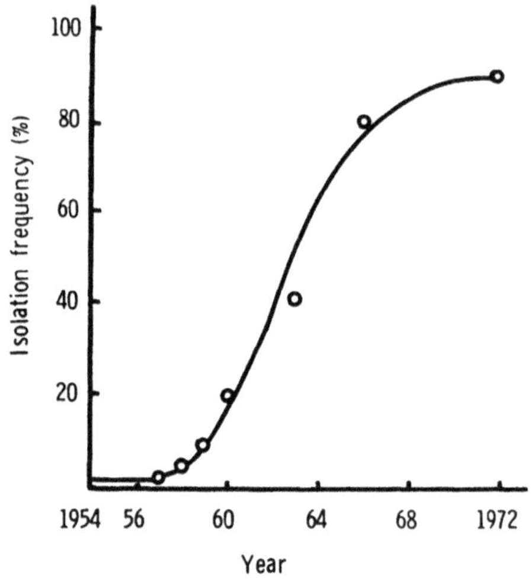

Fig. 1. Yearly changes of the isolation frequency of antibiotic-resistant *Shigella* strains in Japan. Antibiotics include streptomycin (SM), tetracycline (TC) and chloramphenicol (CM).

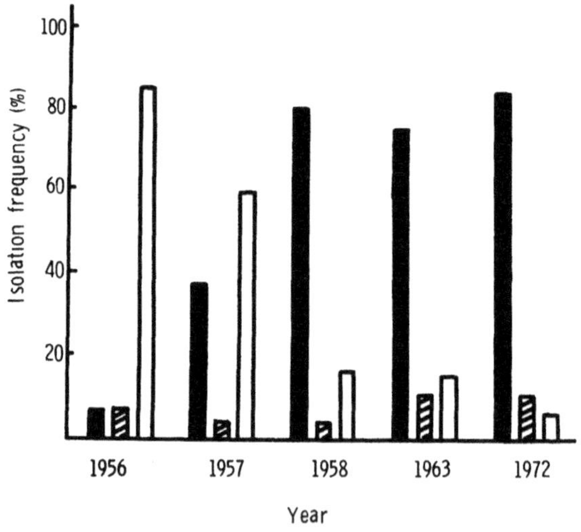

Fig. 2. Yearly changes of resistance patterns among antibiotic-resistant *Shigella* strains. Antibiotics include SM, TC, and CM.

■, triple resistance; ▨, double resistance; and □, single resistance.

TABLE 2

Isolation of antibiotic-resistant Shigella strains in Japan

	Year	1952	1955	1956	1957	1958	1959	1960	1963	1966	1969	1972
	No. of strains examined	5,500	5,327	4,399	4,873	6,563	547	497	4,718	4,292	889	824
	Isolation frequency of resistant strains (%)	0.04	0.09	0.5	2	3.6	8.7	19.5	40.6	79.1	80.9	90.5
Isolation frequency among resistant strains (%)	TC.CM.SM	0	(1)	7	37	80	82	75	75	95.5	90	83.9
	TC.SM	0	0	0	2	1	2	4	2	0.5	1.6	0.4
	CM.SM	0	0	0	2	3	0	5	7	1	4.2	8.4
	TC.CM	0	0	7	0	0	2	1	1	0.1	0.1	1.3
	TC	(1)	0	28	46	8	8	6	13	0.7	1.6	0.8
	SM	(1)	(4)	57	13	8	6	9	2	2.2	2.1	2.4
	CM	0	0	0	0	0	0	0	0	0	0.4	2.7

Number in parenthesis indicates an actual number of isolate. Antibiotics include TC (tetracycline), CM (chloramphenicol) and SM (streptomycin).

Based on these findings, the R factors were discovered and caused big problems in practical medicine and bacterial genetics. The genetic properties of R factor, i.e., especially the wide range of transferability beyond various species of bacteria and an ability to confer multiple-resistance, are of great importance in the medical, hygienic and pharmaceutical sciences.

TABLE 3

Discovery of R factors

Subject	Presentation
Appearence of multiple-resistant *Shigella* strains	1952
Demonstration of *E. coli* with the same resistance pattern as that of *Shigella* from dysenteric patients	1957
Transmission of drug-resistance from *Shigella* to *E. coli*	1959 Nov.
Transmission is not mediated by cell-free extract prepared from multiple-resistant strains	1960 Jan.
Drug-resistance is transmitted regardless of the polarity of F plasmid	1960 Jan.
Drug-resistance is lost spontaneously during storage	1960 March
Drug-resistance is eliminated by treatment with acriflavine	1960 June
Transmission of drug-resistance is interrupted by blendor treatment	1960
The term "R" factor was presented for the transferable drug-resistance property	1960
The range of transmission includes all species of the *Enterobacteriaceae*, *Vibrio comma*, and *P. pestis*	1960, 61, 63

Details, see Review Mitsuhashi (1971).

Similarly, the multiple resistance in staphylococci attracted our attention at that time. Quadruple (TC.SM.SA.PC)-resistant strains were isolated frequently and those resistant to macrolide (Mac) antibiotics in addition to quadruple resistance were also isolated at an increasing rate (see Reviews: Mitsuhashi, 1967; Mitsuhashi et al., 1973). We noticed by chance that Mac, PC and (Mac.PC) resistances were eliminated spontaneously or artificially by treatment with acriflavine (Mitsuhashi et al., 1963; Mitsuhashi, 1967). Transduction experiments could also demonstrate the presence of plasmid carrying the genes responsible for β-lactamase formation (Novick, 1963). Further studies disclosed that various drug-resistance determinants are located extrachromosomally, on a plasmid. The staphylococcal plasmids are not conjugally transmissible but are easily transduced with their lysogenic phages. According to the facts that many plasmids exist in a cell and almost all of staphylococci are lysogenic, the author presented the thought that among staphylococci, lysogenization and transduction of resistance are mainly responsible for the rapid aquisition of resistance, for the wide distribution of multiple-resistant strains and for changes in the phage typing patterns to restricted ones (Mitsuhashi et al., 1965).

As shown in Table 4, there are big differences between R factors and staphylococcal plasmids in genetic and biomolecular properties. Then, I present the term "*r*" instead of R for the non-transferable drug-resistance property such as staphylococcal plasmids.

Epidemiological surveys indicate that the R factors carrying resistance to TC, CM, SM and SA or to various combinations thereof are isolated most frequently. Most of the R factors possessing kanamycin (KM)- or ampicillin (APC)-resistance are isolated

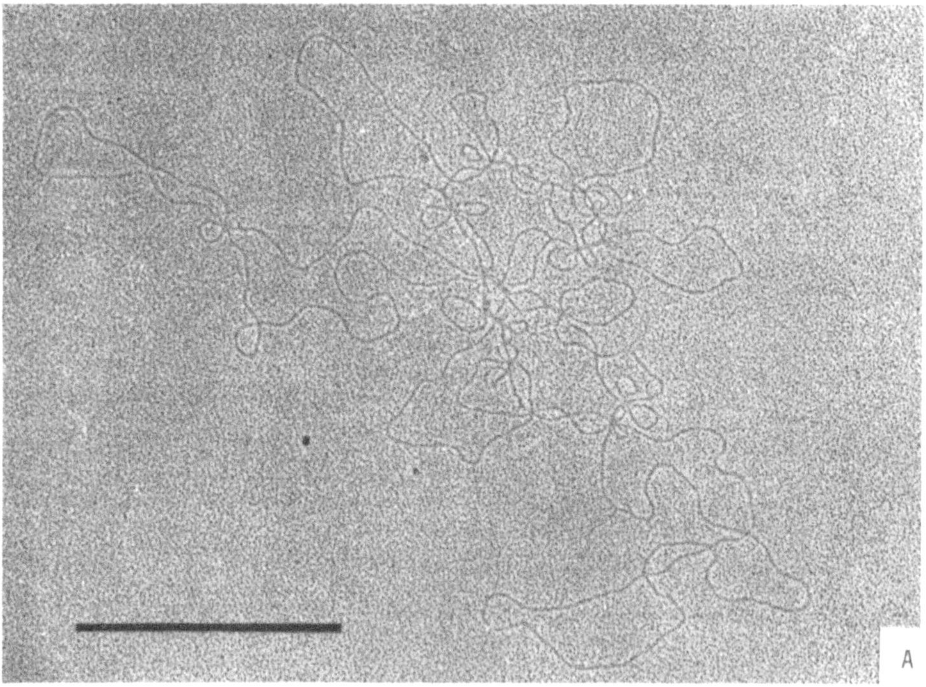

Fig. 3. Electron micrographs from plasmid DNA fractions. Bar represents 1 μm. A, R_{ms201} (TC.CM.SM.SA.APC); B, *r*-St_{ms6} (TC); and C, *r*-St_{ms7} (CM.TC).

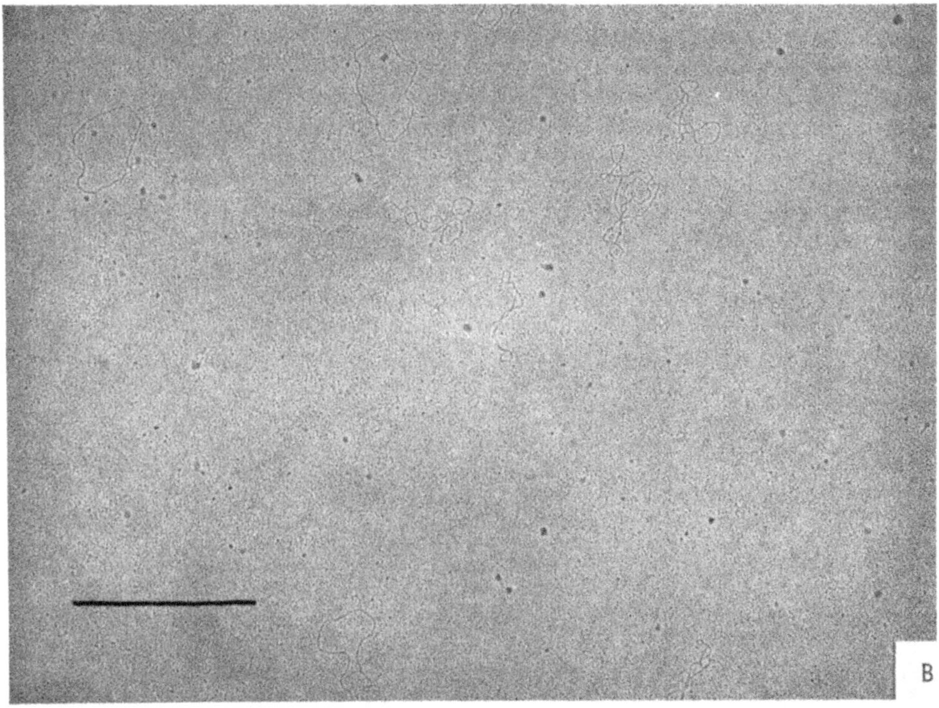
B

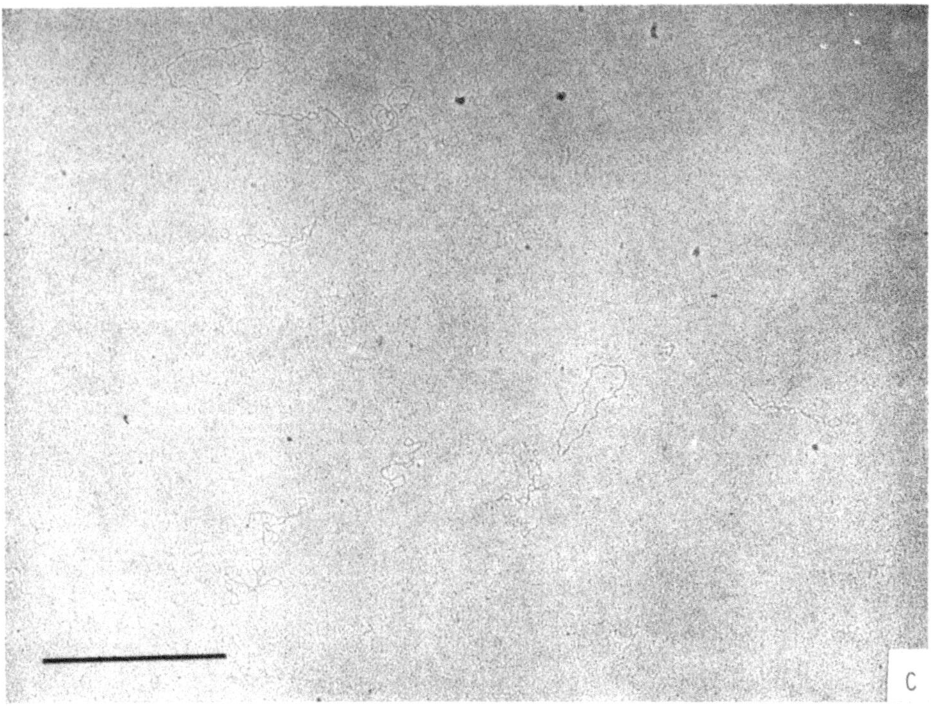
C

together with resistance to the aforementioned four drugs, and the R factors carrying single KM- or APC-resistance are isolated scarcely. Results of survey for resistance type with reference to these four drugs in gram-negative enteric bacteria were reported

TABLE 4

Comparison between R factor and Staphylococcal plasmid

Subject	R factor	Staphylococcal plasmid (*r*-St))
Transferability	+	—
Replication	+	+
Resistance determinant (s)	+	+
DNA size	31—36 μ	1.4—2.7 μ
M.W. (Mdal)	70	3
GC content (%)	52	34
Number of plasmids per chromosome	4—8	4—30

previously (Mitsuhashi et al., 1973) and those in *Shigella* strains are shown in Table 5. Isolation frequency of quadruple (TC.CM.SM.SA) resistance was the highest and followed by single, triple and double resistance in that order. Among triple resistance, the (CM.SM.SA)-resistance type was isolated most frequently. Isolation frequency of (SM.SA)-resistance type was the highest (92.3%) among double-resistant strains and other combinations of double resistance are very few. Single SA-resistance was isolated most frequently (98.1%) and so did scarcely other types of single resistance. According to these results and those reported previously (Mitsuhashi et al., 1973), possible steps of

TABLE 5

Drug-resistance patterns of Shigella strains isolated from 1966 through 1970

Type of resistance		No. of strains isolated	(%) [a]	Isolation frequency [b] of R factor (%)
Quadruple	TC.CM.SM.SA	6224	(100.0)	67.8
Triple	CM.SM.SA	280	(75.9)	88.6
	TC.SM.SA	81	(21.9)	42.0
	TC.CM.SA	8	(2.1)	50.0
Double	SM.SA	133	(92.3)	12.8
	TC.SA	19	(6.2)	78.9
	TC.SM	1	(0.7)	100.0
	CM.SA	1	(0.7)	100.0
Single	SA	1169	(98.1)	2.8
	TC	23	(1.9)	39.1

[a]) Isolation frequency among strains with the same group of resistance type.

[b]) Isolation frequency of R factors from strains possessing the indicated resistance type.

the appearence of multiple resistance in gram-negative enteric bacteria will be postulated as follows: (SA)→ (SM.SA) and (TC.SA)→ (CM.SM.SA) and (TC.SM.SA), and finally (TC.CM.SM.SA). These results can be explained by selection with the practical use of drugs and by biomolecular properties of the genes governing resistance to each drug, probably by homology between them and stability of the recombinants thereof.

Isolation frequency of the (SM.SA)-resistance is high among the double-resistant strains but the frequency of R-factor demonstration therefrom is low. Then, we selected (SM.SA)-resistant bacteria from our stock cultures, whose resistance was not conjugally transferable. It is noteworthy that all of the (SM.SA)-resistant determinants examined so far exist on plasmid (*r* factor) and the mechanism of SM-resistance is due to phosphorylation of the drug similar to that of SM-resistance on R factor. Similarly, the non-transferable (TC.SA)-resistance is located on *r* plasmid, and the TC-resistance mechanism is inducible and due to impermeability of the drug similar to that on R factor. These findings, including staphylococcal plasmid, indicate to us that most of the drug-resistance determinants in naturally occurring resistant strains are located on conjugal(R) or non-conjugal(*r*) plasmids. Wide spread of plasmids in various bacterial species is of great importance in biology and offers a new problem in genetics, and increased knowledge of plasmids will serve to provide clues to gene evolution in addition to chromosomal mutation.

Many scientists have been interested in the structures of a number of antibiotics in addition to natural product structure determination and in complete synthesis of antibiotics. As a result thereof, it has become possible for the biochemist to modify the structure of some antibiotics in a fashion to improve their pharmacologic properties or to render them more effective to resistant bacteria. The possible objectives for modification of antibacterial agents are summarized as follows: (1) broadened microbial spectrum, (2) activity toward resistant microorganisms, (3) greater pharmacologic efficiency, and (4) lessening of toxic effects. In the last thirty year, increased knowledge of chemical structure and of synthesis chemistry have caused a great success in chemical transformation of some of chemotherapeutic agents. You are aware of the successful modification or the total synthesis of some antibiotics, and I want to present some of the examples; (1) total chemical synthesis of chloramphenicol, (2) tetracycline and its analogues, (3) semisynthetic penicillins and cephalosporines, and (4) modification of rifamycin.

TABLE 6

Biochemical mechanisms of bacterial resistance

Drug	Biochemical mechanism	Reference
β-Lactam antibiotics	Hydrolysis of β-lactam ring	
Aminoglycoside antibiotics	Phosphorylation, Acetylation, Adenylation	Umezawa (1974); Benveniste and Davies (1973); Mitsuhashi et al. (1973)
Tetracycline	Impermeability	Izaki et al. (1966); Sompolinski et al. (1970); Inoue et al. (1970)
Synthetic penicillins	Impermeability	
Nalidixic acid	Impermeability	Koga and Mitsuhashi (1974)
Macrolide antibiotics	Decrease in the formation of macrolide-ribosome complex	Saito et al. (1969); Weisblum (1971)

Owing to a collaboration between bacterial geneticists and biochemists, knowledge pertaining to bacterial resistance and its biochemical mechanisms has advanced considerably. We know that the biochemical mechanisms of resistance in naturally occurring resistant strains are different from those of *in vitro* - developed resistant mutants and are generally common in many bacterial strains isolated in different places and from human beings, livestock and cultured fish (Table 6).

As a result thereof, it has become possible for the biochemists to render some chemotherapeutic agents effective against resistant bacteria. Increased knowledge of bacterial plasmid will serve also to provide clues to find out some chemotherapeutic agents which can eliminate plasmid from ist host bacteria or preferentially kill the bacteria carrying plasmid. Fruitful scientific results will be presented at this symposium and The Organizing Committee will edit this symposium with the masterful competence in the coming second volume.

REFERENCES

BENVENISTE, R. and DAVIES, J. (1973). Mechanism of antibiotic resistance in bacteria. Ann. Rev. Biochem. in press.

INOUE, M., HASHIMOTO, H. and MITSUHASHI, S. (1970). Mechanism of tetracycline resistance in strains of *Staphylococcus aureus*. Progress in Antimicrob. and Anticancer Chemoth. - Tokyo, 1969, pp. 433–439.

IZAKI, K., KIUCHI, K. and ARIMA, K. (1966). Specificity and mechanism of tetracycline resistance in a multiple drug-resistant strain of *E. coli*. J. Bacteriol. 9, 628.

MITSUHASHI, S. (1967). Epidemiological and genetical study of drug resistance in *Staphylococcus aureus*. Japan. J. Microbiol. 11, 49.

MITSUHASHI, S. (1971). Transferable drug resistance factor R. University of Tokyo Press. Tokyo; University Park Press. Baltimore and London.

MITSUHASHI, S., MORIMURA, M., KONO, M. and OSHIMA, H. (1963). Elimination of drug resistance of *Staphylococcus aureus* by treatment with acriflavine. J. Bacteriol. 86, 162.

MITSUHASHI, S., OSHIMA, H., KAWARADA, U. and HASHIMOTO, H. (1965). Drug resistance of staphylococci. 1. Transduction of tetracycline resistance with phage lysates obtained from multiple resistant staphylococci. J. Bacteriol. 89, 967.

MITSUHASHI, S., INOUE, M., KAWABE, H., OSHIMA, H. and OKUBO, T. (1973). Genetics and biochemical studies of drug resistance in staphylococci. Microbiology and Immunology, Vol. 1. Staphylococci and Staphylococcal Infections (Karger, Basel 1973), pp. 144–165.

MITSUHASHI, S., HASHIMOTO, H., TANAKA, T. and IYOBE, S. (1974). Genetics and Biochemistry of R factors. Adv. Antimicrob. and Antineoplastic Chemoth.-Athens, 1973, in press.

NOVICK, R. P. (1963). Analysis by transduction of mutations affecting penicillinase formation in *Staphylococcus aureus*. J. Gen. Microbiol. 33, 121.

SAITO, T., HASHIMOTO, H. and MITSUHASHI, S. (1969). Drug resistance of staphylococci. Decrease in the formation of erythromycin-ribosome complex in erythromycin resistant strains. Japan. J. Microbiol. 13, 119.

SOMPOLINSKI, D., ZAIDENZAIG, Y., ZIEGLER-SCHLOMOWITZ, R. and ABRAMOVA, N. (1970). Mechanism of tetracycline resistance in *Staphylococcus aureus*. J. Gen. Microbiol. 62, 351.

UMEZAWA, H. (1974). Biochemical mechanisms of resistance to aminoglycoside antibiotics. Drug action and drug resistance in bacteria. 2. Aminoglycoside antibiotics. University of Tokyo Press. in press.

WEISBLUM, B. (1971). Macrolide resistance in *Staphylococcus aureus*. Drug action and drug resistance in bacteria. 1. Macrolide antibiotics and lincomycin. University of Tokyo Press. Tokyo, pp. 217–238.

GENE EXCHANGE BETWEEN UNRELATED PLASMIDS

N. DATTA
Department of Bacteriology, Royal Postgraduate Medical School,
London, G.B.

The main topic of this symposium is a discussion of the mechanisms by which R factors endow their host bacteria with resistance to antibiotics. My own work has been concerned less with this aspect than with the mode of transfer of R factors from one bacterium to another and with plasmid classification. I shall discuss the distribution of genes determining drug-resistance among the different plasmid classes which my colleagues and I have identified.

The first classification of R factors was into the two classes fi^+ and fi^- (Watanabe *et al.*, 1964) where fi^+ indicated *fertility inhibition*. Thus fi^+ R factors, if introduced into an F^+ or Hfr host, inhibited the production of F pili by the F factor; the class fi^- had no such positive attribute but simply co-existed with the fully-functional F factor (Meynell, Meynell and Datta, 1968).

Watanabe and his colleagues (1964) showed that "superinfection immunity" was demonstrable between two fi^+ R factors or between two fi^- R factors, but not between two R factors where one was fi^+ and the other fi^-. They also showed (Watanabe, 1969) that "superinfection immunity" between plasmids included two phenomena, *surface exclusion* and *incompatibility*. Surface exclusion was ineffective if the second plasmid was introduced by transduction instead of by conjugation, but the two related plasmids were still *incompatible*; i.e. both could not be stably maintained in the same host, just as it was already known that two F′ factors could not co-exist stably (Echols, 1963).

TABLE 1
F Complex: F-piliated Plasmids

Compatibility group	Examples
FI	F, ColV2, R386, R455
FII	R100, R1, R136
FIII	ColB–K98
FIV	R124
FV	Fo lac

Using the criterion of *incompatibility*, my colleagues and I have made a classification of plasmids into groups (Datta, 1974). Plasmids which determine F pili can be subdivided into five different compatibility groups (Table 1). All of them are either fi^+ or, like F itself, lack the pilus repressor. The earliest fi^+ R factors to be identified belonged to group FII; though closely resembling F in the genes which control pilus synthesis and

DNA transfer, they are compatible with F. Group FI includes F itself and also some *fi*⁺ col factors and R factors incompatible with F. The other three groups listed here are each represented, so far, by a single known plasmid, one col factor, one R factor and *Folac*, a plasmid identified in *Samonella typhi* which determines lactose fermentation (Falkow and Baron, 1962). All these are comaptible with one another, with groups FI and FII and with all *fi*⁻ plasmids.

TABLE 2

I Complex I-piliated Plasmids

Compatibility Group	Examples
I α	R64, R144, ColIb-P9, R62
I β	R483
I γ	R621a, ColIb-1M1420
I ω	JR66a
I ζ	R805a

TABLE 3

Classification of fi⁻ Plasmids

Compatibility group	Spectfic Phage	Examples
A	—	RA1
C	—	R40a, R55
A–C	—	R715b, Plac
H	—	R27(TP117), R726
M	—	R446b, R69
N	Ike	N3, R15, R46
O	—	R16, R723
P	PRR1	RP4, R751, R702, R906
T	—	Rts1, R394, R401
W	—	S-a, R7K, R388

Wa have called these groups of plasmids the F complex. Another complex of plasmids (Hedges and Datta, 1973) comprises those which determine I pili (Table 2). These are *fi*⁻. But the I complex by no means includes all *fi*⁻ plasmids. The *fi*⁻ R factors first described by Watanabe *et al.* (1964), N3 and R15, were not I-like, but belonged to a quite distinct group which we subsequently termed N. Table 3 shows more of the compatibility groups we have defined. For some of them, plasmid-specific phages have been isolated.

Classification by compatibility seems to reflect true evolutionary relationships since plasmids of the *same* compatibility group share extensively the same polynucleotide sequences, whereas there is little sharing of sequences between plasmids of *different* compatibility groups (Guerry and Falkow, 1971; Grindley *et al.*, 1973) (Table 4). Exceptions to this rule lie in the F and I complexes, within which there is extensive DNA homology, even between plasmids belonging to different groups (Guerry and Falkow, 1971; Falkow, Guerry; Hedges and Datta, MS in preparation).

These findings suggest that the various compatibility groups have evolved by separate pathways and may be considered, perhaps, as plasmid genera. However, the same antibiotic resistance patterns may be conferred by plasmids of many different groups. Does this indicate convergence of phenotypes in response to environmental selection pressures or does it indicate that different plasmids have acquired identical resistance genes?

If each plasmid group had its own characteristic resistance genes it would be evidence for the former view. An early observation was that chloramphenicol resistance was not conferred by *fi*⁻ R factors (Watanabe *et al.*, 1964). This was true of the N group and I group plasmids. Subsequently, at least one N plasmid has been identified which carries chloramphenicol resistance, but no member of the I complex yet identified does so. Chloramphenicol-resistance is also carried by members of several of the other *fi*⁻ groups listed in Table 3, for example, groups C, W and H. R factor chloramphenicol-resistance is (usually) mediated by the enzyme chloramphenicol acetyltransferase (CAT) about which you will hear more later in this Symposium.

TABLE 4

DNA sequence relationships between Plasmids of different groups

Unlabelled DNA of:		Extent (%) hybridization with ^{3}H-thymine labelled DNA of:				
Plasmid	Group	R1 (FI)	R144 (I)	N3 (N)	RP4 (P)	S-a (W)
F	FI	43	—	—	—	—
R1	FII	100	6	6	9	14
R144	I	8	100	5	<1	2
R64	I	—	78	—	—	—
N3	N	8	2	100	10	10
R15	N	—	—	75	—	—
RP4	P	—	3	4	100	3
S-a	W	4	5	8	4	100
R388	W	—	—	—	—	79

From Falkow *et al.* (1974)

Three classes of CAT have been identified (Foster and Shaw, 1973). Class 1 was specified by F-like R factors, by the unique N plasmid carrying chloramphenicol resistance and by plasmids of group C. Plasmids of group W specified CAT class 2, while for class 3 there was only one example, R387, an R factor which also stands alone as the only known example of its compatibility group, K (Hedges and Datta, 1971). There are two common R factor mediated enzymes which inactivate streptomycin, streptomycin phosphotransferase (SPT) and streptomycin/spectinomycin adenylate synthetase (SAS) (Benveniste, Yamada and Davies, 1970). There is some correlation between the type of streptomycin-resistance and the R factor compatibility group — in general R factors of groups F and N specify SAS while I group plasmids specify SPT, but a few exceptions to the rule have been reported (Hedges, 1972b). R factor β-lactamases are not correlated with compatibility group — genetic information for apparently identical enzymes is found in nearly all known groups (Hedges *et al.*, 1974). The same seems to be true of neomycin phosphotransferase 1, the enzyme most commonly responsible for plasmid-determined resistance to neomycin and kanamycin (J. Davies and R. W. Hedges, unpublished). Thus to summarise (Table 5), there is much sharing of resistance mechanisms between plasmids of different groups.

This in itself is evidence for an exchange of genes between replicons. We have more direct laboratory evidence in some instances.

One example is R62, the only known I group plasmid to carry SAS. R62 has I pili and belongs to group I, but about 25% of its DNA consists of polynucleotide sequences typical of N group plasmids (Guerry, Falkow and Datta, 1974) and it seems clear that it is a naturally-occurring recombinant between an Iα and an N plasmid.

Such recombination is unexpected for two reasons: firstly, selection for recombination cannot be exerted, since the two plasmids are able to co-exist quite stably. (Exposure to two antibiotics simultaneously, resistance to each of which is carried on a different plasmid, exerts powerful selection for recombination between incompatible plasmids, but not between compatible ones.) Secondly, the lack of common DNA sequences between unrelated plasmids should make recombination unlikely. However, there are instances of DNA sequences encoding resistance genes which are transposed very readily from

TABLE 5

Drug-inactivating Enzymes determined by R factors of various groups

Enzyme	Compatibility group
β-lactamase (TEM)	FII, I, N, C, A, T, W, P, X
CAT Class 1	FII, C, (N)
CAT Class 2	W
CAT Class 3	K
SPT	I, (N), „SSu"
SAS	FII, N, W, (I)
NPT1	FII, I, A, N, P

Parentheses indicate that the enzyme is specified by exceptional examples of that group.
"SSu" indicates a class of small, non-conjugative plasmids, very widespread among bacterial genera (P. T. Barth and N. J. Grinter, unpublished).

one replicon to another. One such is the ampicillin resistance of RP4, a P plasmid. When RP4 is introduced into a culture of *Escherichia coli* K12 carrying the I plasmid, R64, it is easy to isolate clones in which R64 has acquired the β-lactamase gene of RP4. This penicillinase gene is transposable also to other plasmids, for example S–a, which belongs to group W. It is also transposable to the chromosome of *E. coli* (Richmond and Sykes, 1972). Acquisition of this gene is accompanied by an increase in molecular weight of about 2×10^6 daltons, more than enough DNA to encode solely the β-lactamase. Upon transposition, the capacity to be transposed is not lost, since this gene or genes can be transposed serially from RP4 to R64 to S–a. It looks as though some specific mechanism is effecting the transposition (Hedges and Jacob, 1974). Another "transposon" includes the gene for neomycin phosphotransferase 2 (NPT2), an enzyme identified by Dr Julian Davies and his colleagues, less common than the NPT1 shown in Table 5. Two observations led us to examine this gene for transposability. Firstly, in a naturally occurring strain of *Shigella sonnei* there were two R factors, one F and one N, which were both transferred by conjugation to *E. coli* K12. The FII plasmid conferred resistance to kanamycin; the N plasmid conferred resistance to ampicillin, streptomycin and tetracycline. In the laboratory, there was spontaneous acquisition of kanamycin resistance by the N plasmid from the F plasmid (Hedges, 1972a). This resistance was associated with NPT2. The second observation, made by Dr Julian Davies, was that a strain of *Klebsiella pneumoniae* carried two R factors, one belonging to the F complex and one to the I complex. Both specified the rather unusual NPT2. My colleague R. W. Hedges (unpublished work) tested the plasmid for ability to transpose kanamycin-resistance to another I group plasmid, R64. Again, the transposition occurred readily and spontaneously. In this instance the weight of the "transposon" was about 10×10^6 daltons.

Yet another example of a high frequency of transposition of resistance genes occurs from the R factor, R483, to the chromsome of *E. coli* K12. R483 specifies resistance to streptomycin and to the synthetic antibacterial, trimethoprim (Hedges and Datta, 1973). (The latter resistance is of particular interest since its mechanism is different from other plasmid resistances: it determines the production of a new, trimethoprim-resistant dihydrofolate reductase (Skold and Widh, 1974)). R483 also determines production of I pili and of colicine Ia. If an incompatible plasmid is introduced into an R483$^+$ culture, the resistance genes for streptomycin and trimethoprim are not always eliminated (Hedges and Datta, 1973), but frequently become incorporated into the *E. coli* chromosome at a position between *dna*A and *ilv* (74.1 min), or, less often, into the incoming plasmid. The transposed DNA segment has a molecular weight of about 7×10^6 daltons and does not carry the genes for I pili or the colicine (P. T. Barth, N. J. Grinter and N. Datta, unpublished).

Thus it appears that exchange of genes is an important feature of plasmid behaviour.

REFERENCES

Benveniste, R., Yamada, T. and Davies, J. (1970). Enzymatic adenylation of streptomycin and spectinomycin by R-factor-resistant *Escherichia coli.* Infection and Immunity **1**, 109–116.

Datta, N., (1974). Epidemiology and classification of plasmids. Microbiol. 1974 (in press).

Echols, H. (1963). Properties of F′strains of *Escherichia coli* superinfected with F-lactose and F-galactose episomes. J. Bacteriol. **85**, 262–268.

Falkow, S. and Baron, L. S. (1962). Episomic element in a strain of *Salmonella typhosa.* J. Bacteriol. **84**, 581–589.

Foster, T. J. and Shaw, W. V. (1973). Chloramphenicol acetyltransferases specified by *fi*$^-$ R factors. Antimicrobial Agents and Chemotherapy **3**, 99–104.

Grindley, N. D. F., Humphreys, G. O. and Anderson, E. S. (1973). Molecular studies of R factor compatibility groups. J. Bacteriol. **115**, 387–398.

Guerry, P. and Falkow, S. (1971). Polynucleotide sequence relationships among some bacterial plasmids. J. Bacteriol. **107**, 372–374.

Guerry, P., Falkow, S. and Datta, N. (1974). R62, a naturally-occurring hybrid plasmid. J. Bacteriol. (in press).

Hedges, R. W. (1972a). Phenotypic characterization of *fi*$^-$ R factors determining the restriction and modification *hsp* II specificity. Molec. and Gen. Genetics **115**, 225–233.

Hedges, R. W. (1972b). Resistance to spectinomycin determined by R factors of various compatibility groups. J. Gen. Microbiol. **72**, 407–409.

Hedges, R. W. and Datta, N. (1971). *fi*$^-$ R factors giving chloramphenicol resistance. Nature (London) **234**, 220–221.

Hedges, R. W. and Datta, N. (1973). Plasmids determining I pili constitute a compatibility complex. J. Gen. Microbiol. **77**, 19–25.

Hedges, R. W., Datta, N., Kontomichalou, P. and Smith, J. T. (1974). Molecular specificities of R factor determined β lactamases: correlation with plasmid compatibilities. J. Bacteriol. **117**, 56–62.

Hedges, R. W. and Jacob, A. E. (1974). Transposition of ampicillin resistance from RP4 to other replicons. Molec. and Gen. Genetics (in press).

Meynell, E., Meynell, G. G. and Datta, N. (1968). Phylogenetic relationships of drug-resistance factors and other transmissible bacterial plasmids. Bacteriol. Rev. **32**, 55–83.

Richmond, M. H. and Sykes, R. B. (1972). The chromosomal integration of a β lactamase gene derived from the P-type R factor RP1 in *Escherichia coli.* Genet. Res. **20**, 231–237.

Skold, O. and Widh, A. (1974). The genetic specification of a trimethoprim resistant dihydrofolate reductase by an R factor mediating resistance to trimethoprim. Proc. Soc. Gen. Microbiol. (in press).

Watanabe, T. (1969). Transferable drug resistance: the nature of the problem. In Bacterial Episomes and Plasmids. G. E. W. Wolstenholme and M. O'Connor (Eds.). Ciba Symposium. Churchill, London p. 81–97.

Watanabe, T., Nishida, H., Ogata, C., Arai, T. and Sato, S. (1964). Episome mediated transfer of drug resistance in Enterobacteriaceae VII. Two types of naturally occurring R factors. J. Bacteriol. **88**, 716–726.

NATION-WIDE SURVEY OF ANTIBIOTIC RESISTANCE BY MEANS OF A COMPUTER ANALYSIS OF 200 000 STRAINS OF PROBLEM BACTERIA ISOLATED IN 1973

V. KRČMÉRY, J. GRUNT, L. ROSIVAL, F. VÝMOLA

Research Institute of Hygiene, Bratislava, and Institute of Hygiene and Epidemiology, Prague, Czechoslovakia

Our computer system of nation-wide surveillance of the dynamics of antibiotic resistance in "Problem bacteria" (1) has been used to evaluate data collected by 18 Public Health Laboratories (PHL) in the Slovak Socialist Republic in 1973. These data comprise the susceptibility or resistance to 21 currently used antibacterials of almost 200,000 bacterial strains belonging to 8 species of clinically important bacteria. Attention was paid also to the analysis of the types of materials sent for bacteriological examination and susceptibility estimation, the data connected with the interest of individual clinical branches to obtain such information as a basis for their medication with antibacterials, and the most frequent species of bacteria examined. The feed-back system comprised a twice-a-year analysis of the results with the chief microbiologists of the PHL involved on the basis of pre-sent computered results, and a regular information service for clinicians provided by these officers in the form of bulletins, seminars etc. In addition, the results were submitted to the corresponding authorities of the Ministry of Health.

A. RESULTS REFLECTING THE WORK OF THE PHL THEMSELVES

1. Comparing the two semesters of 1973, an about 5% increase was registered in the second half of the year in the overall number of strains tested for antibiotic susceptibility. In the second half of 1973, 96,000 strains of "Problem bacteria" were tested for antibiotic (AB) susceptibility compared with 91,000 in the first semester. This may reflect the actual increase in the activities of PHL in antibiotic susceptibility examination.

2. The pediatric wards with pediatric practitioners were the most active suppliers of bacteriological materials for this examination, representing 40% of sensitivity tests (Table 1). The next were internal and pulmonary clinics with 18%. General practitioners ranked third with 10%. In contrast, surgical wards and clinics supplied merely 4% of bacteriological materials for AB susceptibility examination. These data may serve the Public Health authorities to consider the adequacy of individual branches of medical care as regards rational use of the ABs.

3. The most frequent materials sent for bacteriological examination and AB susceptibility tests were swabs from the upper respiratory tract (this may reflect the pediatric practice mentioned above) — 33%, stool (23%) and urine (20%). The remaining five types of materials represented together only 25% of all the materials submitted for bacteriological examination.

TABLE 1

Retrieval of strains investigated by the PHLs in 1973 according to their origin, source of isolation and the species involved

A. Origin of the pathological materials submitted				B. Type of the pathological material examined			C. Bacterial species		
	Clinic or ward or type of medical care	No. of strains	%		No .of strains	%		No. of strains	%
1.	General practitioners Occupational physicians	19 194	10,23	Pus	14 035	7.48	Staph. aureus	64 760	34.52
2.	Internal and pulm. ward	34 365	18.31	Punctate) exsudate	611	0.33	Proteus mirab.	30 498	16.25
3.	Surgical wards	8 255	4.40	Hemocult.	413	0.22	E. coli	61 190	32.61
4.	Gynecol. and Obstetrics	16 410	8.75	Urine	38 336	20.43	E. coli enterop.	2 341	1.25
5.	Pediatric wards + hospitals + + policlinics	73 473	31.16	Urogenit. secretum	5 849	3.12	Salmonella[1]	2 716	1.45
6.	Infectious diseases sections and wards	11 413	6.08	Swabs from Upper R.T.	62 435	33.27	Shigella	5 044	2.69
7.	Dermatovenerol. wards	5 521	2.94	Sputum	19 390	10.33	Klebsiella + + Enterobacter	13 574	7.23
8.	Ear, nose, throat wards	7 866	4.19	Stool	42 005	22.39	Ps. aeruginosa	7 511	4 00
9.	Urological wards	6 883	3.67	Ear swabs	4 560	2.43			
10.	Epidemiological service	4 254	2.27						

[1] non-typhoid

TABLE 2

Resistance of strains belonging to individual species of "Problem bacteria" to antibacterial drugs recorded in 18 PHL of S.S.R. in 1973

	Staph. aureus		Prot. mirabilis		Escher. coli		coli enteropath.		Salmonella		Shigella		Kleb. Enterob.		Pseudom. aeruginosa	
Antibiotics	No. of strains tested	% of resist.	No. of strains tested	% of resist.	No. of strains tested	% of resist.	No. of strains tested	% of resist.	No. of strains tested	% of resist.	No. of strains tested	% of resist.	No of strains tested	% of resist.	No. of strains tested	% of resist.
Penicilin G	63 252	82.41	×	×	×	×	×	×	×	×	×	×	×	×	×	×
Ampicilin	8 038	40.65	28 642	44.76	55 927	58.19	2 201	63.97	2 599	8.11	4 700	38.30	12 449	82.44	6 238	95.55
Carbenicilin	×	×	8 268	36.93	7 510	42.43	466	56.01	247	3.65	318	24.53	1 282	63.11	3 796	36.96
Oxacilin	59 416	11.54	×	×	×	×	×	×	×	×	×	×	×	×	×	×
Streptomycin	13 355	27.57	29 694	30.50	59 038	25.61	2 330	41.42	2 703	2.07	5 022	14.78	13 338	29.46	7 300	39.37
Tetracyclin	61 130	35.56	30 081	92.12	60 596	50.80	2 303	62.44	2 662	14.80	4 982	38.34	13 449	46.94	6 982	83.26
Chloramphen.	63 803	13.69	30 372	41.61	60 925	31.94	2 329	48.43	2 700	2.37	5 016	27.61	13 463	34.93	7 014	78.19
Erytromycin	63 109	25.65	×	×	×	×	×	×	×	×	×	×	×	×	×	×
Spiramycin	42 641	4.97	×	×	×	×	×	×	×	×	×	×	×	×	×	×
Oleandomycin	4 373	12.58	×	×	×	×	×	×	×	×	×	×	×	×	×	×
Kanamycin	56 888	28.24	29 219	35.56	57 935	16.01	2 220	39.50	2 599	0.77	4 835	4.96	13 089	21.72	6 699	78.68
Neomycin	6 091	13.23	8 186	40.49	15 441	16.61	1 033	50.43	1 414	0.99	3 173	4.54	2 284	21.06	2 025	31.36
Gentamycin	4 875	1.04	13 548	2.42	21 502	2.60	802	3.86	763	1.44	2 668	1.91	4 028	1.71	4 802	2.15
Colistin	×	×	24 671	91.64	52 680	12.64	2 195	9.43	2 474	4.40	4 603	3.15	11 067	11.15	7 010	7.23
Linkomycin	33 697	4.77	×	×	×	×	×	×	×	×	×	×	×	×	×	×
Cefaloridin	5 193	13.98	1 232	86.85	1 503	77.77	96	92.71	39	94.87	1 701	47.98	221	64.70	470	93.83
Nalidicin	×	×	12 832	15.64	24 104	12.63	432	40.28	319	3.45	2 320	2.16	3 603	21.04	2 244	90.87
Nitrofurans	×	×	12 403	50.56	24 415	10.49	434	10.14	445	6.52	2 698	4.75	3 606	22.07	2 256	94.77
Sulfonamides	2 317	42.64	11 072	67.62	20 910	49.81	901	56.05	953	34.42	3 154	41.54	2 571	57.45	1 857	84.66
Septrin	677	13.15	7 986	21.39	15 693	9.42	393	3.82	582	2.06	2 237	1.25	2 370	15.40	1 147	85.70
Polymixin	×	×	6 073	94.17	11 217	3.91	707	14.16	472	2.33	1 988	2.01	1 959	2.30	3 142	4.62

4. Species of bacteria most frequently tested for AB susceptibility were *S. aureus* (34%), *E. coli* (32%) and, surprisingly, *Proteus mirabilis* (16%).

5. We compared the percentage of resistant strains (with regard to each drug and each species) among individual PHLs to see the degree of uniformity of susceptibilities, to consider the quality of the laboratory work of each PHL and/or the differences in individual geographic areas.

The homology of the results obtained by individual PHLs was influenced positively by two factors: the introduction of standard (and obligatory) methods of AB susceptibility testing by the Ministry of Health at the very beginning of this study, and semestrial consultations of the computerized results with the chief microbiologists involved.

6. The results obtained in the "main antibiotics" in the first and second half of 1973 were very similar in all PHLs involved. Thus, as a rule, no gross differences can be scored when semestrially pooling the results of a PHL in a given species. More remarkable differences noted will be commented in the following text. Nevertheless, it seems to be superfluous to compute the nation-wide results more frequently than once a year.

7. Special consideration, however, had to be given to the results obtained with two drugs, namely oxacillin in staphylococci, and colistin in *E. coli*, i.e. species-specific drugs. Although, as stated above, instructions had been given as to the uniform reading of results especially in these two drugs, the results obtained varied. Out of 18 PHLs involved, 14 showed less than 10% resistant strains of *S. aureus* to oxacillin, whereas three of them recorded as much as 50 to 67% resistant strains. As regards colistin and *E. coli*, the resistance data obtained in three (different) PHLs considerably differed from the mean values obtained in the remaining laboratories. The mean value for all PHLs is 12.6% of resistant strains, whereas the values obtained in the three PHLs mentioned varied from 33 to 44%.

B. RESULTS OF ANTIBIOTIC RESISTANCE IN INDIVIDUAL SPECIES

The results are summarised and reviewed on a nation-wide scale in Table 2, where total data for the year 1973 are presented. Wherever necessary, a comment will be given concerning semester differences or changes.

1. *Staphylococcus aureus*

In about 64,000 strains it was clearly demonstrated that the classical penicillin seems to be generally ineffective, i.e. that penicillinase plasmids are widely distributed in clinical strains. Also, a further increase in resistance rates to E, oleandomycin and neomycin was noted in the 2nd semester, which in the case of E. reached a warning level of 25% resistant strains. Thus, the so-called *main drugs*, i.e. S, T, C, E and K all show comparatively high percentages of resistant strains, and this was a uniform feature in all PHL concerned.

2. *Proteus mirabilis*

This species seems to be, at present, a typical example of "Problem bacteria" with a rather specific affinity to the urogenital tract. Resistance of about 30,000 strains seems to be absolute to tetracyclines and to colistin, and slightly increased, in the 2nd semester, to A, carbenicillin (Ca) and K. The rates of resistant strains did not change significantly to S, C and gentamicin; the latter drug is extremely effective. In contrast, only 43 strains in all 18 PHL were tested for susceptibility to cephalosporines, which is difficult to understand.

3. *E. coli*

About 60,000 strains of this species were tested in PHL concerned in 1973. The situation in the proportion of strains resistant to individual drugs did not change significantly in the 2nd semester and is rather serious in all the "*main drugs*": S, T, C, K. In contrast, A and Ca improved slightly their position in the 2nd semester and this may be due to a more adequate reading of the results, especially in beta-lactam drugs. The results obtained with colistin have been discussed previously.

4. *E. coli, enteropathogenic serotypes*

The resistance proportions are much more grave than in *E. coli* commune and an increase in resistance rates was observed in the 2nd semester against Ca and gentamicin — here the increase was from 1% to 4% of resistant strains. Again, cephalosporines were not tested either in *E. coli* commune or in enteropathogenic strains.

5. *Salmonellae*

We registered a 60% increase in the numbers of strains of *Salmonella* submitted for antibiotic susceptibility testing. This seems to be a consequence of our insistence on monitoring the antibiotic resistance in all *Salmonellae* when isolated, as R plasmids may be responsible for the increase of multiple drug resistance of *Salmonella observed in* previous studies (2) and in the first semester of this study. All monitored drugs, namely A, T and sulphonamides, and also C, showed a notable increase in resistance rates, e.g. A from 7.4% in the first semester to 8.5% in the second, T from 10.66% to 17.4% of resistant strains in the 2nd semester.

6. *Shigellae*

The increase in the total number of strains tested for resistance to antibacterials was about 100% in the 2nd semester. Resistance to A increased from 35% in the first semester to 40.1% in the second, resistance to T from 37% to 39%. In contrast, resistance to K shows a decrease from 9.5% in the first semester to 2.2% in the 2nd. It is peculiar that a number of *Shigella* strains have been tested for susceptibility to cephalosporine drugs although these are not drugs of choice in this genus. Very good susceptibility of *Shigellae* was noted against septrin.

7. *Klebsiella-Enterobacter*

This is again a typical species of "Problem bacteria" with high levels of resistance to the "*main drugs*": S, T, C, K and even to gentamicin. Notably colistin shows an unfavourable pattern of resistance. Again, cephalosporines were not tested with this species although their position seems to be justified in the treatment of infections caused by *Klebsiella*.

8. *Pseudomonas aeruginosa*

The situation in the resistance of this species, whose significance as a typical nosocomial microorganism increases, is critical and we could note a high level of resistance ratio against Ca, and especially to T, K, C. Gentamicin is highly effective *in vitro* in this country.

Although our study continues with an analysis of multiple resistance data and with an analysis of the dynamics, and developments of relations of individual drugs versus individual species with time, the following preliminary conclusions were drawn from our experience with single drug resistance results:

1. In the near future, we may experience the beginning of therapeutic troubles with some drugs, e.g. with oxacillin in staphylococci, colistin with *E. coli* or *Klebsiella-Enterobacter* etc.

2. We may soon lose, in addition to T, C, E and other "main drugs", further antibacterials, or even their groups, for effective therapy. This concerns not only ampicillin and carbenicillin (although these substances are not in general use in this country) but also aminoglycosides, especially kanamycin.

3. Some drugs are still generally effective; these are gentamicin, cephalosporines, septrin and, in staphylococci, linkomycin. We should regard these drugs as a general chemotherapeutic reserve and elaborate for them a special monitoring system and a special mode of clinical use.

4. We could arbitrarily estimate the *limit of effectiveness* of a drug for general therapeutic use at 30% of resistant strains by *in vitro* tests. Although we are lacking detailed studies correlating *in vitro* testing results with therapeutical results, we have several examples which justify the above conclusion. Even urogenital infections generally show that the outcome of therapy in relation to the *in vitro* testing for drugs is rather pessimistic than optimistic. Complicating factors which interfere in this respect are: nosocomial aspects of persistance and circulation of "hospital strains", the transferable antibiotic resistance and the paucibacillar proportion of resistant cells which might escape the bacteriological reading of *in vitro* test i.e. some sort of inadequacy and low susceptibility of *in vitro* testing in relation to few *resistant cells* which then, however, emerge under therapy pressure. The same is observable in faeces, where only special techniques were able to demonstrate few R^+ bacteria among R^- bulk of bacterial population or among susceptible ones (Wiedemann and Knothe, 1971, Isenberg and Berkmann, 1971). Thus, a drug which shows 30% of resistant strains by simple *in vitro* testing may in fact be ineffective against much more bacterial strains and regarded as falling out of our therapeutic armamentarium.

5. Multiple drug resistance is hidden at present among the numbers shown in Table 1, but undoubtedly plays an immense role in all problem bacteria studied (perhaps with the exception of *Salmonellae)* and this is a further indication that R plasmids are operating on a large scale among the species studied. The data obtained by Bohuš (this Symposium) who followed the R plasmid proportion in the overall resistance within one year in individual PHL materials, are fully convincing in this respect.

This implies that we should test the strains of "problem bacteria" for susceptibility *against much more drugs* than is now usual, which means that testing 7 to 8 discs on a single plate is quite insufficient and we should test at least against 10 antibacterials [(1), see also Isenberg and Berkman, 1971].

6. One minor result should also be mentioned here as it shows the multipurpose service of the system of monitoring used in this study. Many species are tested in the PHLs concerned *in vain*, and this should be avoided. This concerns testing of A against *Ps. aeruginosa*, PNC-G, E or oxacilling against *E. coli*. But one should also consider the obsoleteness of testing PNC-G in *S. aureus*, and T in *P. mirabilis* or *Ps. aeruginosa*, and, surprisingly, K in *Ps. aeruginosa!*

E. S. Anderson pointed out at the 8th Int. Congress of Chemotherapy in Athens, the strong moral impact of a critical and discriminative use of antibacterial drugs and demanded a more scientific approach of the pharmaceutical companies in advertising an excessive use of antibacterial drugs. We agree with this standpoint since we must consider the increase of resistance spectra and the spread of multiple drug resistant bacteria, maybe with R plasmids, to be a *serious social evil* which undermines the overall achievements in the progress of modern medicine and of medical care. However, pharmaceutical companies have been recently showing increased interest in the research and monitoring of antibiotic resistance and in developing new or re-tailored antibacterial substances. We agree with their standpoint that the fast development of resistance accompanying a rather hectic clinical or prophylactic use of new substances is a self-defeating consequence of past developments. But this is only one of the important components of a rational antibiotic policy. It must be based on continuous monitoring of the resistance developments and on a nation-wide strategy and tactics in the use of antibacterials in clinics, general practice and even in veterinary practice or animal husbandry.

REFERENCES

J. GRUNT, V. KRČMÉRY, F. VÝMOLA, Š. BOLCHOVÁ, J. JANOŠEK, J. JEDINÁK, A. POPLUHÁROVÁ and E. ŠTĚPÁNKOVÁ. Nation-wide Survey of Antibiotic Resistance by Means of a Computer. Zbl. Bakt. Hyg., I. Abt. Orig. A224, (1973), 380–390.

JANOUŠKOVÁ, J., V. KRČMÉRY, B. PAULÍKOVÁ, J. JEDINÁK, E. ŠTĚPÁNKOVÁ, D. KADLEČÍK, K. ČERVINKA. Occurrence of multiresistance to antibiotics and of R factors in *Shigella* and *Salmonella* in Czechoslovakia. Čs. Epidem. 21 (1973), 191–198.

B. WIEDEMANN, H. KNOTHE. Epidemiological Investigation of R factor-bearing Enterobacteria in Man and Animal in Germany. Ann. N.Y. Acad. Sci. 182 (1971), 380–383.

H. ISENBERG, J. I. BERKMAN. The role of Drug-resistant and Drug-selected Bacteria in Nosocomial Disease. Ann. N.Y. Acad. Sci. 182 (1971), 52–59.

I.

BETA-LACTAM ANTIBIOTICS AND BETA-LACTAMASES

CHAIRMAN: P. KONTOMICHALOU (GREECE)

OPENING LECTURE

TEM-LIKE PENICILLINASES IN CONTEMPORARY PLASMIDS OF ENTEROBACTERIACEAE AND PSEUDOMONAS. COMPARISON WITH β-LACTAMASES OF PREANTIBIOTIC KLEBSIELLA

POLYXENI KONTOMICHALOU, EFSTATHIA PAPACHRISTOU and G. LEVIS

Athens University School of Medicine, Department of Clinical Therapeutics, "Alexandra" Hospital. Athens, Greece

The existence of genes coding for drug-inactivating enzymes in R plasmids of Enterobacteriaceae was first demonstrated for enzymes hydrolyzing the β-lactam drugs (Datta and Kontomichalou, 1965). The type, which represented two out of the three enzymes of this study, was later established as the predominant episomal β-lactamase in R plasmids of wide taxonomic and geographical origin. Enzymes of this type are known by three terms: as "TEM-like", derived from the designation of one of the penicillinase plasmids of this type first described, as type Ia and Ib according to the terminology of Japanese authors (Sawai et al., 1973) and as type IIIa, according to the classification of Richmond and Sykes (1973).

An important amount of work is being now done in characterizing these enzymes. The enzyme of plasmid R6K (alias RTEM) was first purified in K12 by Datta and Richmond (1966) and later by Melling and Scott (1971) and has been studied by other investigators in crude or in semipurified preparations (Rosselet et al., 1971, Kazmierczak et al., 1973). Further information has been collected from data obtained by purification of other episomal enzymes of the same type, as the enzyme of plasmid R1 (Lindqvist and Nordström, 1970) and the enzymes of R factors R6N14 and RGN823, classified respectively as class Ia and Ib (Yamagishi et al., 1969, Sawai et al., 1970). These studies have revealed certain differences in the properties of these enzymes, which might be termed as a group of TEM-like penicillinases.

Our experience with these types of enzymes is based on a previous study of the β-lactamase specified by randomly taken plasmids, conferring resistance to penicillins originating from the following species: *E. coli, Klebsiella, Citrobacter and Protei*, isolated in our clinical practice from 1963 to 1969 (Kontomichalou et al., 1974). We classified

This work was supported by grants from:

1. The Greek Ministry of Culture and Civilization
2. The Wellcome Trust
3. The Ebirikion Foundation

9 out of the ten episomal β-lactamases of our study as TEM-like on the basis of the identity or close similarity of their properties. This proportion was one of the highest obtained from similar investigations made with plasmids found in Japan (Yamagishi et al., 1969a, and Ooka et al., 1970), England (Jack and Richmond, 1970) or from surveys of broad geographical origin (Jenkins and Drabble, 1971, Hedges et al., 1974).

The universally encountered predominance of this type of β-lactamase in R plasmids is of particular importance since it is in contrast with the variety of types of chromosomally determined β-lactamase in Gram-negative bacteria (Richmond and Sykes, 1973).

The prevailing theory, as to the origin of the genes coding for episomal enzymes, is that they have been picked up by sex factors or silent plasmids, from the chromosome of certain bacterial species (Watanabe, 1971). If this is the case, one could assume that the chromosomal genes for a drug-inactivating enzyme, which became predominant as episomal, might have a higher tendency to be translocated than the chromosomal genes for other enzymes of the same function, or that the properties of that particular enzyme, and the resistance it mediates confer selective advantage to the host of such plasmids.

Contemporary Gram-negative bacteria, producing presumably chromosomal β-lactamases, have been examined to find the species with closer resemblance to the episomal TEM-like enzyme species-specific b-lactamase (Yamagishi et al., 1969a, Dale and Smith, 1971, Sawai et al., 1973). However, contemporary bacteria may have been infected with plasmids which, dissociating in their new hosts, might from then onwards become non-transferable, extrachromosomal genes. For this reason, bacteria either from antibiotics-free areas or from the pre-antibiotics era are more reliable for such studies. We therefore decided to use the oldest collection of bacteria we found in our country, which consisted of clinical Enterobacteriaceae isolated in the early thirties. These strains had been since that time kept unopened, in sealed ampoules with a semisolid medium. Fifty of the strains, labelled *E. coli*, were found according to recent identification to be: 34 *E. coli*, 2 *Klebsiella* and 14 atypical. The two *Klebsiella* strains of the collection were among the few strains resistant to low levels of the β-lactam drugs and the only ones giving a positive β-lactamase test in starch agar plates. Crude extracts of disrupted cells from these two cultures showed a hydrolytic activity towards benzylpenicillin of the order of 10 and 5 units/hr/mg of protein, whereas all other properties of the two enzymes, tested in semipurified preparations, with 20 times higher specific activity, were identical. Repeated attempts to transfer conjugally the resistance and hydrolytic activity to the β-lactam drugs of the two strains to a variety of recipients, including *E. coli* K12, failed. The two enzymes also failed to be released by osmotic shock (Neu, 1968). By these criteria and their low levels or resistance to the β-lactam drugs it was assumed that these β-lactamases were chromosomal.

Tables 1 and 2 give a comparison of the properties of the chromosomal β-lactamases of the pre-antibiotics *Klebsiella* with close of the TEM-like enzymes of the contemporary plasmids. It emerges that the two types of enzymes being compared have many properties in common, i.e., the rates of hydrolysis of 8 penicillins and 6-APA, their molecular weights and the resistance to inhibitors, such as pCMB, Cl^- and EDTA. The rates of hydrolysis of cephalosporins, the electrophoretic mobility and their sensitivity to cloxacillin and temperature differed, the enzymes of the pre-antibiotics *Klebsiella* being less active towards cephalosporins, resistant to cloxacillin, more sensitive to temperature and with positive electrophoretic mobility at pH 6. Despite these differences and considering as more important the similarities existing between the two types of enzymes, we concluded that the pre-antibiotics *Klebsiella* b- actamases could be classified as belonging to the TEM-like group. According to the criteria used by Richmond and Sykes they might be variants of typc IIIa. Dale and Smith (1971) and Sawai et al. (1973) have also found close resemblance between the chromosomal β-lactamases of contemporary *Klebsiella* strains and the episomal TEM-like enzymes.

The penicillinase plasmids, coding for TEM-like enzymes, have increasingly been considered important for the epidemiology of R factors, since it was demonstrated in 1970 that the host range of such plasmids might also include strains of the *Pseudomonadaceae* (Sykes and Richmond, 1970, Fullbroock et al., 1970). The β-lactamases, specified by plasmids transferable intergenerically from *Pseudomonas* highly resistant to carbenicillin,

TABLE 1

*Relative rates of hydrolysis of β-lactam substrates**

Substrate profile	Enzymes coded by: Plasmids in Entero-bacteriaceae**	Enzymes coded by: Pre-antibiotic Klebsiella***
Benzylpenicillin	100	100
Ampicillin	128	156
Carbenicillin	9	7
6-APA	91	91
Phenoxymethylpenicillin	57	95
Phenoxyethylpenicillin	28	63
Phenoxypropylpenicillin	11	21
Cloxacillin	1.8	2
Methicillin	0.8	n.d.****
Cephaloridin	135	20
Cephalothin	13	n.d.
Cephalexin	6.5	n.d.

* Tested with crude preparations with substrates 7 μmoles/ml.
** TEM-like enzymes. Mean from nine plasmids, in various hosts.
*** Mean from strains 205 and 327 isolated in 1932–33.
**** n.d. = non detectable.

TABLE 2

Properties of semipurified episomal and chromosomal β-lactamases

Properties of the enzymes	Enzymes: TEM-like of 9 plasmids	Enzymes: Klebsiella 205 and 327
Molecular weight*	23,000–25,000–26,000	23,000–25,000
Remaining activity		
in pCMB .5mM 15′	75 – 100%	65 – 100%
Cloxacillin 100 μM 15′	27 – 44%	100%
Cl^-.1M 15′	92 – 100%	95 – 100%
EDTA 1mM 15′	90 – 100%	65 – 100%
60°C 15′	50 – 100%	0 – 1.6%
100°C 15′	18 – 83%	0
120°C 15′	2 – 11%	
Electrophoretic mobility**	–	+

* By column chromatography on Sephadex G-75.
** By starch gel electrophoresis at pH 6.0.

have been studied only from strains isolated in two hospitals in England. The presence in these plasmids of genes coding only for typical TEM-like enzymes has been reported by both centres on the basis of studies with semipurified or purified enzymes (Neu 1970). The plasmids described by Witchitz and Chabbert (1972), originating from *Pseudomonas* with gentamicin-inactivating enzymes, also carried resistance determinants to ampicillin. Intergeneric transfer of ampicillin resistance has also been described by Chitkara (1973), but no enzymatic studies have been made the resistance to the β-lactam drugs carried by the plasmids of these authors. Reports on the intergeneric transfer of TEM-like plasmids were consequently until now confined only to strains isolated from England.

TABLE 3

Expression in two hosts of β-lactamase activity and resistance to Carbenicillin, mediated by the Pseudomonas plasmids

Plasmids	*Pseudomonas*-donors		*E. coli* K12 F^- transconjugants	
	β-lactamase specific activity*	Carbenicillin MIC μg/ml**	β-lactamase specific activity*	Carbenicillin MIC μg/ml**
p PK 141	33	64***	1068	4000
p PK 148	965	2000	1183	4000
p PK 170	1652	4000	976	4000
p PK 175	492	4000	775	2000
p PK 192	1500	8000	1500	8000
p PK 184	842	8000		

* Units per mg protein, in crude extracts. Substrate benzylpenicillin 7 μmoles/ml.
** By the tube dilution method, inoculum 10^4 bacteria/ml.
*** MBC = 2,000 μg/ml.

In 1971 we were aware of the high proportion of our clinical isolates of *Pseudomonas* which were highly resistant to carbenicillin. Several strains of this period were kept in egg slopes and the surviving ones, together with a hundred *Pseudomonas*, isolated from November 1972 to February 1974 in our Department, and a group of 24 *Pseudomonas*, isolated at the Athens Accident Hospital, were tested for resistance to carbenicillin and transferability of the resistance to *E. coli* K12. All these *Pseudomonas* strains, a total of 130, were *aeruginosa*. Cumulatively, 13% of the strains were resistant to carbenicillin, criterion for resistance, by the tube-dilution method, being MIC and MBC of >100 μg/ml. None of the *Pseudomonas* of the Accident Hospital was resistant, and the resistance to carbenicillin was thus confined only to our hospital, and this also declined from 60% in 1971 to 8% in 1974. Characteristic of this resistance were the high levels, 1000 to 8000 μg/ml, in 15 out of the 18 resistant strains. Transferability of carbenicillin resistance from *Pseudomonas* to *E. coli* K12 F^- rifampicin resistant was demonstrated in 14 out of the 15 strains highly resistant to carbenicillin. The frequencies of transfer by the membrane method ranged from $10^{-6}-10^{-7}$, with selection made by carbenicillin in either minimal media with lactose or in MacConkey agar with rifampicin. The frequencies of interspecific transfer from F^-K12 to Hfr or F^+ K12 ranged from 10^{-3} to 10^{-6}, showing thus 10 to 10^3 higher frequencies on the interspecies than the intergeneric transfers. The levels of the transferable resistance to carbenicillin were expressed in K12 as high as those found in *Pseudomonas*-donor strains, and in one case they were even considerably higher.

Enzymatic studies have been performed on both the *Pseudomonas*-donor cultures and the F⁻K12 transconjugants. In *Pseudomonas*, the presence of constitutive penicillinase was demonstrated in all 15 strains highly resistant to carbenicillin and in none of a group of sensitive *Pseudomonas* examined. The levels of resistance to carbenicillin were found

TABLE 4

Substrate profiles of constitutive β-lactamases specified by plasmids originating from Pseudomonas aeruginosa

Cultures	Substrates, 7 mM					β-Lactamase
	Benzyl-penicillin	Ampi-cillin	Carbeni-cillin	Cephalo-ridin	Cephalo-thin	
Pseudomonas 148	100	128	10.0	138	10.9	TEM-like
E. coli K12-pPK 148+	100	156	11.7	127	7.9	TEM-like
Pseudomonas 170	100	168	9.5	158	9.8	TEM-like
E. coli K12-pPK 170+	100	152	12.8	164	10.9	TEM-like
Pseudomonas 175	100	106	6.1	139	8.8	TEM-like
E. coli K12-pPK175+	100	138	7.7	133	11.1	TEM-like
Pseudomonas 192	100	148	8.1	122	8.5	TEM-like
E. coli K12-pPK 192+	100	100	8.8	100	6.8	TEM-like
Pseudomonas 184	100	128	5.6	106	9.6	TEM-like

TABLE 5

Properties of a semipurified constitutive enzyme originating from Pseudomonas aeruginosa

Properties of the enzyme	Strain *E. coli* K12 F⁻ p PK 170+
Remaining activity in	
pCMB . 5mM 15′	100%
Cloxacillin 100mM 15′	40%
Cl⁻ . 1M 15′	100%
EDTA 1mM 15′	94%
60°C 15′	100%
100°C 15′	26%
120°C 15′	1.4%
Electrophoretic mobility*	—
Molecular weight**	28,000

* By starch gel electrophoresis at pH 6.0.
** By column chromatography on Sephadex G-75, from one experiment only.

in both the *Pseudomonas*-donor strains and the K12 transconjugants to be directly proportional to the levels of β-lactamase activity of the cultures, towards benzylpenicillin as a substrate per mg of protein in crude extracts (Table 3). An example was the *Pseudomonas* strain number 141, which had been included in the resistant group only because of its high MBC, 2000 μg/ml. For this particular strain only, the resistance to carbenicillin and the enzymatic activity to the β-lactam drugs were expressed in K12 at considerably higher levels than in the *Pseudomonas* donor.

The properties of the enzymes, specified by the plasmids originating from *Pseudomonas* have been tested in respresentative cultures of both *Pseudomonas* and *E. coli* K12 hosts.

We also included the enzyme of the only *Pseudomonas* highly resistant to carbenicillin, strain 184, for which a first attempt for transfer to K12 was negative (Table 4). The substrate profiles of these enzymes, towards five penicillin and cephalosporin derivatives, were for all cultures tested typical of TEM-like enzymes (Table 4). Other properties of one representative enzyme in K12 host were further tested in semipurified preparation, as shown in Table 5. All characters, such as resistance to pCMB, Cl^- and EDTA, sensitivity to cloxacillin, resistance to heat and negative electrophoretic mobility, confirmed the identification of this episomal penicillinase as TEM-like. The molecular weight, tested in one experiment only, was slightly higher than that found in other TEM-like episomal β-lactamases. Experiments to confirm this difference are in progress.

It was thus shown that the high-level resistance to carbenicillin was in all *Pseudomonas* strains, found in our clinical practice, due to the presence of penicillinase plasmids, specifying TEM-like β-lactamases, transferable intergenerically to *E. coli*. The pattern of resistances co-transferred with the penicillinase plasmids from almost all donors included determinants for resistance to the aminoglycosides (gentamicin, tobramycin, kanamycin, streptomycin), to chloramphenicol, tetracyclines and the sulfonamides. As a rule, fragments of this pattern were received by individual clones of the *E. coli* K12 tranconjugants. Two of the penicillinase plasmids were found to belong to fi^+ and the remaining to fi^- class, despite the similarity in their resistance pattern. We therefore assume that the source of these *Pseudomonas* resistance-plasmids might not be unique.

From the epidemiological point of view it was found that the more complex plasmids were gradually built up in nature during the period 1971–1973, the 1971 plasmids carrying only the β-lactamase, chloramphenicol and tetracycline markers.

Although other criteria for characterizing these transmissible genes as extrachromosomal were fulfilled, we observed in the transfer experiments of these plasmids also cases of co-transfer of chromosomal markers. With the donor *Pseudomonas*, transfer of methionine marker rendered the *E. coli* K_{12} met^- recipient prototrophic and another plasmid mobilised the *E. coli* K_{12} F^- chromosome in interspecies transfers (results on this point will be reported in another communication).

Thus the *Pseudomonas* R plasmids encountered in our clinical practice had characteristics in common with those of similar plasmids found in *Pseudomonas* in England (plasmids RPI, RP4 or RP8) in that they specified the synthesis of TEM-like b-lactamases and were intergenerically transferred. Some of them also resemble RPI in that they promote the transfer of chromosomal markers. The number of resistance markers carried by the *Pseudomonas* R plasmids of our study was much greater than that of RP_1, RP_4, or RP_8 and the intergenerically transferred gentamicin-resistant plasmid described by Witchitz and Chabbert (1972). In contrast to all above-listed plasmids the resistance determinants of the *Pseudomonas* plasmids of our study dissociated during or after transfer to K_{12}, which might be consistent with the presence of many independent R plasmids, all of which were intergenerically transferred.

In conclusion, from strains isolated in clinical practice in Greece β-lactamases of the TEM-like group have been found to be specified by genetic elements in the following three locations: the chromosome of the pre-antibiotics *Klebsiella*, the plasmids of Enterobacteriaceae from the sixties and the plasmids detected since 1971 in *Pseudomonas aeruginosa* highly resistant to carbenicillin.

REFERENCES

CHITKARA, Y. K. (1973). Transferable drug resistance in *Pseudomonas aeruginosa* and other gram-negative organisms in burns. Amer. J. Clin. Path. 59:211–215.

DALE, J. W. and SMITH, J. T. (1971). Some relationships between R-factor and chromosomal b-lactamase in gram-negative bacteria. Biochem. J. 123: 507–512.

DATTA, N. and KONTOMICHALOU, P. (1965). Penicillinase synthesis controlled by infectious R-factors in Enterobacteriaceae. Nature (London) 208: 239–241.

DATTA, N. and RICHMOND, M. H. (1966). The purification and properties of a penicillinase whose synthesis is mediated by an R-factor in Escherichia coli. Biochem. J. 98: 204–209.

FULLBROOK, P. D., ELSON, S. W. and SLOCOMBE, B. (1970). R-factor mediated b-lactamase in *Pseudomonas aeruginosa*. Nature, 226: 1054–1056.

HEDGES, R. W., DATTA, N., KONTOMICHALOU, P. and SMITH, J. T. (1974). Molecular specificities of R-factor-determined beta-lactamases: Correlation with plasmid compatibility. J. Bacteriol. 117: 56–62.

JACK, G. W. and RICHMOND, M. H. (1970). A comparative study of eight dinstict b-lactamases synthesized by gram-negative bacteria. J. Gen. Microbiol. 61: 43–61.

JENKINS, P. G. and DRABBLE, W. T. (1971). β-lactamase of R-factors derived from *Shigella* and *Salmonella* strains. J. Bacteriol. 108: 159–169.

KAZMIERCZAK, A., PHILIPPOU, A., CHARDON, H., LABIA, R. et LE GOFFIC, F. (1973). Constantes enzymatiques (Km et Vmax) des b-lactamases mesurées par une méthode micro-acidimétrique couplée à l'ordinateur. Ann. Microbiol. (Inst. Past.) 124 B, 259–268.

KONTOMICHALOU, P., PAPACHRISTOU, E. and LEVIS, G. (1974). R-mediated b-lactamases and episomal resistance to the b-lactam drugs in different bacterial hosts. Antimicrob. Ag. Chemother. In press.

LINDQVIST, C. and NORDSTRÖM, K. (1970). Resistance of *Escherichia coli* to penicillins. VII. Purification and characterization of a penicillinase mediated by the R-factor R_1. J. Bacteriol. 101: 232–239.

MELLING, J. and SCOTT, G. K. (1972): Preparation of gram quantities of a purified R-factor-mediated penicillinase from *Escherichia coli* strain W 3310. Biochem. J. 130: 55–62.

NEU, H. C. (1968). The surface localization of penicillinase in *Escherichia coli* and *Salmonella* typhimurium. Biochem. Biophys. Res. Commun. 32: 258–262.

OOKA, T., HASHIMOTO, H. and MITSUHASHI, S. (1970). Comparison of penicillinase produced by R-factors isolated from ampicillin-resistant gram-negative bacteria. Jap. J. Microbiol. 14: 123–128.

RICHMOND, M. H. and SYKES, R. B. (1973). The b–lactamases of gram-negative bacteria and their possible physiological role, p. 31–88. In A. S. Rose and D. W. Tempest (ed), Advances in Microbial Physiology, vol. 9, Academic Press Inc., London.

ROSSELET, A., KNÜSEL, F. and ZIMMERMANN, W. (1972). Phenotypic expression of b-lactamase activity in different strains carrying the TEM R-factor. Bacterial plasmids and Antibiotic Resistance. Avicenum. Prague, 231–243.

SAWAI, T., TAKAHASHI, K., YAMAGISHI, S. and MITSUHASHI, S. (1970). Variant of penicillinase mediated by an R-factor in *Escherichia colli*. J. Bacteriol. 104: 620–629.

SAWAI, T., YAMAGISHI, S. and MITSUHASHI, S. (1973). Penicillinases of *Klebsiella pneumoniae* and their phylogenetic relationship to penicillinases mediated by R-factors. J. Bacteriol. 115: 1045–1054.

SYKES, R. B. and RICHMOND, H. M. (1970). Intergeneric transfer of a b-lactamase gene between *Ps. aeruginosa* and *E. coli* Nature (London) 226: 952–954.

WATANABE, T. (1971). The origin of R-factors. Ann. N.Y. Acad. Sci. 182: 126–140.

WITCHITZ, J. L. et CHABBERT, Y. A. (1972). Résistance transférable à la gentamicine. II. Transmission et liaisons du caractère de résistance. Ann. Inst. Past. 122: 367–378.

YAMAGISHI, S., SAWAI, T., O'HARA, K., TAKAHASHI, K., and MITSUHASHI, S. (1969a). b-lactamases of R-factor and gram-negative rod bacteria. Proc. 6th Intern. Congr. Chemother. 579–591.

YAMAGISHI, S., O'HARA, K., SAWAI, T. and MITSUHASHI, S. (1969b). The purification and properties of penicillin b-lactamases mediated by transmissible R-factors in *Escherichia coli*. J. Biochem. 66: 11–20.

OPENING LECTURE

R-FACTORS AND INTRINSIC RESISTANCE TO PENICILLINS

M. H. RICHMOND and N. A. C. CURTIS
Department of Bacteriology, University of Bristol, Bristol, G.B.

It has been known for over thirty years that Gram-negative bacteria can be protected against β-lactam antibiotics by the production of an enzyme that destroys these anti-bacterial agents (Abraham and Chain, 1940). For a long time interest centered on the β-lactamases from Gram-positive species, partly because they provided a useful experimental system for the investigation of the fundamental problem of enzyme induction in bacteria (Citri and Pollock, 1966) and partly because the main therapeutic problem posed by these enzymes concerned *Staphylococcus aureus* (Barber, 1962). Since the discovery of the so-called "penicillinase resistant" penicillins (methicillin, cloxacillin, etc.) however, the therapeutic problems associated with β-lactamases have tended to shift to Gram-negative species; and the discovery that penicillin and cephalosporin resistance may be R-factor mediated in these bacteria has intensified this process. Although it was clearly realised that the possession of a β-lactamase often greatly increased the resistance of Gram-negative species to penicillins and cephalosporins, it also seemed likely that Gram-negative bacteria owed some of their resistance to some property (or properties) associated with their surface layers (Knox, 1962). This resistance was known as 'intrinsic resistance', a title that really stressed an ignorance of the molecular basis of the process.

Experiments carried out with mutant Gram-negative bacteria have now shown that resistance to penicillins and cephalosporins is most effective when a β-lactamase is positioned within a barrier which restricts the flow of the antibiotic to the enzyme (Richmond and Sykes, 1973; Richmond and Curtis, 1974). In practice this arrangement is achieved by locating the enzyme in the periplasmic space with the "outer membrane" of the bacterial envelope providing the restriction to access (Fig. 1). Thus the optimum protection is achieved by a careful interaction of both β-lactamase and "intrinsic factors".

Most R-factors that specify resistance to β-lactam antibiotics do so by coding for the synthesis of a β-lactamase (Datta and Kontomichalou, 1965). In these cases the enzymes usually take advantage of the properties of the pre-existing outer membrane of the cell, and the levels of resistance achieved are substantial but rarely very large. There are some R-factors, however, that specify not only the synthesis of a β-lactamase but also seem to modify the surface layers to restrict access even more, and such R-factors give rise to much higher levels of resistance than those which specify β-lactamase synthesis alone (Curtis, Richmond and Stanisich, 1973). This paper will seek to examine some of the changes in properties caused by R-factors of this class. Most attention will be paid to the R-factor RP1, but this plasmid will be contrasted with R1*drd*19 — a plasmid that seems to specify only β-lactamase production. The effects of RP1 on Gram-negative bacteria are certainly complex and the limit of our achievement so far has been to begin to define some of the changes produced to the point at which chemical studies can now be used with some hope of success.

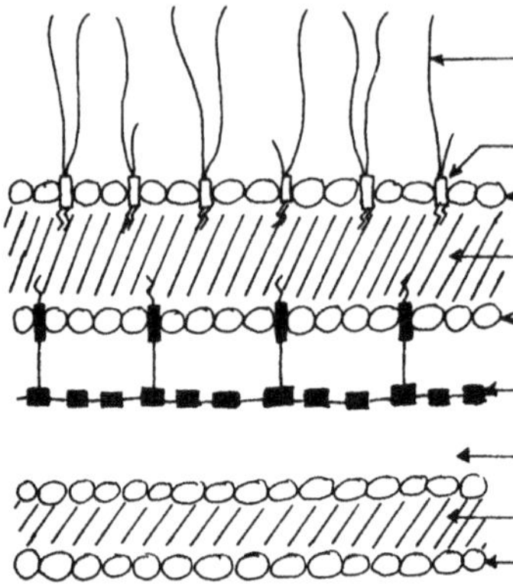

Fig. 1. The structural organisation of the envelope of Gram-negative bacteria (Redrawn from Schnaitman, 1971).

THE EFFECT OF RP1 AND ITS MUTANT DERIVATIVES ON THE RESISTANCE OF *ESCHERICHIA COLI*

The naturally occurring plasmid RP1 specifies the synthesis of Type IIIa β-lactamase (Sykes and Richmond, 1970). Strains that carry the intact plasmid exhibit high levels of resistance to most penicillins, and in some cases (notably carbenicillin) these are higher than can be accounted for satisfactorily solely on the basis of the amount of enzyme activity synthesised (Table 1). The first type of mutant (RP1*amp*1) that has been generated from RP1 by *N*-methyl-*N*-nitro-*N*-nitrosoguanidine (NTG) treatment is characteristic of a class in which the plasmid no longer specifies an enzyme that degrades penicillins — either a β-lactamase or an acylase (Curtis, Richmond and Stanisich, 1973). However *E. coli* organisms carrying this mutant plasmid still show levels of penicillin resistance intermediate between strains which either carry unmutated RP1 on the one hand, or no plasmid at all on the other (Table 1). Moreover this difference is found regardless of whether the plasmid-less *E. coli* strain naturally lacks an R-factor or whether it is obtained by "curing". One may conclude therefore that RP1 carries one or more genes (the *irp* region) that confer some resistance to penicillins in the absence of a β-lactamase or penicillin acylase.

Further treatment of RP1*amp*1 with NTG gave rise to a second class of mutant plasmid (RP1*amp*1 *irp*1) which remained β-lactamase-less but which now no longer specified the intermediate levels of resistance characteristic of RP1*amp*1 (Table 1). It seems likely therefore that RP1*amp*1 *irp*1 has lost the gene(s) specifying intrinsic resistance as well as those coding for β-lactamase.

In contrast to this behaviour of RP1, the R-factor R1*drd*19 loses all ability to give rise to β-lactam resistance when mutated to loss of β-lactamase expression. Since this mutation is reversible at low frequentcy it seems unlikely that it represents a class of mutations in which both the β-lactamase genes *and* those regions responsible for intrinsic resistance have been eliminated by a single point mutation.

TABLE 1

Comparison of the substrate profile of Type IIIa β-lactamase with the MIC values found for E. coli carrying the plasmid RP1

Strain	PenG	Amp	Carb	CER	Tet
	Single cell sensitivities (μg/ml)				
E. coli R^-	20	4	2	2	2
E. coli RP1	10,000	10,000	25,000	8	125
E. coli RP1 *amp*1	625	625	2,500	8	125
E. coli RP1 *amp*1 *irp*1	16	2	1	1	125
Enzyme Type	β-lactamase Substrate Profile				
IIIa	100	180	10	140	—

Abbreviations: PenG, benzyl penicillin; Amp, ampicillin; Carb, carbenicillin; CER, cephaloridine; Tet, tetracycline. Substrate profile values are all related to a rate of benzyl penicillin hydrolysis = 100

EFFECT OF RP1 ON SOME SURFACE PROPERTIES OF *ESCHERICHIA COLI*

Normark and his colleagues have isolated a series of mutants of *Escherichia coli* (the so-called *env*A mutants) in which the mutant bacteria are inordinately sensitive to β-lactam antibiotics and there is much evidence to suggest that these mutants are altered in their surface properties in such a way that these agents reach their target in the cell more easily (Normark, Boman and Matsson, 1969). We have therefore studied the effect of β-lactamase-less mutants of RP1 and of R1 on the penicillin resistance of *env*A mutants of *Escherichia coli* K12. Tables 2 & 3 compare the resistance of *Escherichia coli* *env*A^+ and *E. coli* *env*A^- to penicillins and summarises the effect of various R-factors on them. Whereas the *env*A mutation causes a substantial reduction in the resistance of *E. coli* K12 to penicillins, the presence of RP1*amp*1 repairs the lesion to give MIC values similar to those found in *env*A^+ bacteria. The effect of unmutated plasmids is not easy to test since it is impossible accurately to evaluate the contribution to resistance made by the β-lactamase. Neither RP1*amp*1*irp*1 (Table 2) nor R1*drd*19*amp*2 (Table 3) increase the β-lactam resistance of *env*A mutants. It follows, therefore, that RP1 probably carries a gene (or genes) that can phenotypically repair the *env*A lesion, and that the R-factor determinants responsible for this property are closely linked to, if they are not identical with, those that enhance the intrinsic resistance of naturally-occurring strains. R1 does not seem to carry these determinants.

Another mutant of *Escherichia coli* that is altered in its surface properties is the mutant AS19 (Sekiguchi and Iida, 1967). This strain shows complex phenotypic alterations involving enhanced sensitivity to actinomycin D, to nalidixic acid, to rifampicin and to lysozyme. This last property seems to point conclusively to some change in the surface of the cells since lysozyme cannot normally penetrate to the murein of Gram-negative bacteria. This alteration could also account for sensitivity to antibiotics not normally regarded as penetrating the surface layers of Gram-negative bacteria. Table 4 compares the antibiotic resistance of *Escherichia coli* AS19 with the parental strain and also summarises the effect of RP1, and of mutant derivatives of it, on strain AS19. This strain is markedly more sensitive than the parental strain to benzyl penicillin, ampicillin, carbenicillin, actinomycin D, nalidixic acid and rifampicin. However insertion of RP1 or RP1*amp*1,

TABLE 2

Effect of various RP1 factors on the resistance of Escherichia coli envA

Strain	Single Cell Sensitivity (μg/ml) PenG	Amp	Carb	CER	Neo	Kana	Tet
D21	320	63	2	4	2	2	1
D21 *env*A	32	4	2	2	<1	<1	1
D21 (RP1)	10,000	10,000	25,000	8	32	250	125
D21 *env*A (RP1)	320	625	2,500	4	8	125	63
D21 (RP1 *amp*1)	625	625	1,250	4	32	250	125
D21 *env*A (RP1 *amp*1)	63	63	125	4	16	125	125
D21 (RP1 *amp*1 *irp*1)	320	63	8	4	32	125	250
D21 *env*A (RP1 *amp*1 *irp*1)	32	8	1	2	16	63	125

Abbreviations: PenG, benzyl penicillin; Amp, ampicillin; Carb, carbenicillin, CER, cephaloridine; Neo, neomycin; Kana, kanamycin; Tet, tetracycline. Bracketed entries are plasmid-borne markers.

TABLE 3

Effect of various R1 factors on the resistance of E. coli envA

Strain	PenG	Amp	Carb	CER	Cm
D21	320	63	4	4	6
D21 *env*A	32	4	2	2	2
D21 (R1*drd*19)	1000	1000	2000	4	500
D21 *env*A (R1*drd*19)	320	320	625	4	250
D21 (R1*drd*19 *amp* 2)	320	63	4	4	500
D21 *env*A (R1*drd*19 amp 2)	32	8	1	1	250

Abbreviations: Cm, chloramphenicol; all others as in Table 2.

TABLE 4

Effect od RP1 and some mutant derivatives of RP1 on E. coli AS19

Strain	Single Cell Sensitivities (μg/ml) Rif	Nal	Act	Pen	Amp	Carb	CER	Neo	Kan	Tet
Wild type (*E. coli* B)	25	12.5	16	16	2	2	2	—	—	—
AS19	0.2	0.6	1.6	2	<1	<1	1	1	1	<1
AS19 (RP1)	3.2	1.6	3.2	500	500	1000	4	16	125	63
AS19 (RP1 *amp*1)	3.2	1.6	3.2	32	8	16	1	16	125	63
AS19 (RP1 *amp*1 *irp*1)	0.4	0.6	0.8	2	<1	<1	1	16	125	63
AS19 (R1*drd*19)	0.4	0.6	0.8	63	63	125	1	1	1	<1
AS19 (R1*drd*19 *amp*2)	0.4	0.6	0.8	2	<1	<1	1	1	<1	<1

Abbreviations: Rif, rifampicin; Nal, nalidixic acid; Act, actinomycin D; all others as in Table 2.

but not RP1*amp*1*irp*1, affords some measure of protection against these antibiotics. Where a functional β-lactamase gene is present (e.g. in RP1) resistance to benzyl penicillin and ampicillin is much raised, but even RP1*amp*1 is able to protect strain AS19 against killing by β-lactam compounds to a considerable extent. R1 has no protective effect on AS19 when its β-lactamase gene is inactivated (as in R1*drd*19*amp*2) (Table 4). All variants of RP1 that carry a functional *irp* gene afford some protection to rifampicin, nalidixic acid and actinomycin D, particularly to the first of these.

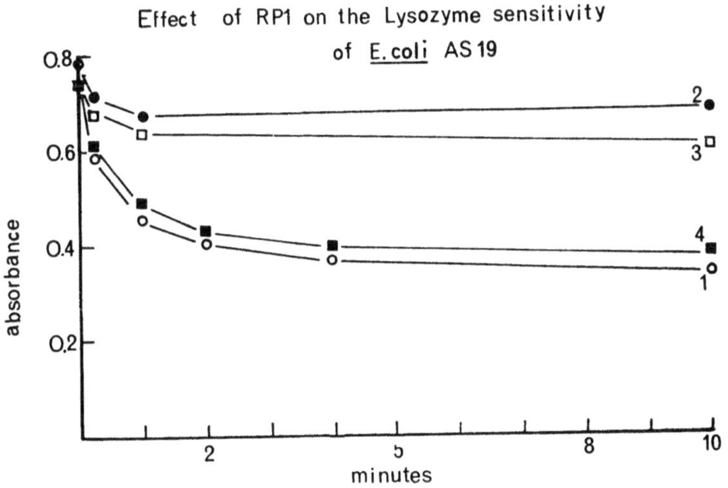

Fig. 2. The effect of various plasmids in protecting *Escherichia coli* AS19 against lysis by lysozyme. 1, AS19 + lysozyme; 2, *E. coli* B + lysozyme; 3, AS19 (RP1) + lysozyme; 4, AS19 (Rp1*amp*1 *irp*1) + lysozyme. *E. coli* B is the parent of strain AS19. *E. coli* AS19 (RP1*amp*1) behaves like *E. coli* AS19 (RP1) when treated with lysozyme. Lysozyme concentration: 10 μg/ml.

Since sensitivity to lysozyme as well as to the antibiotics listed above is part of the phenotype of AS19, it seemed interesting to see whether any plasmids related to RP1 would alter the lysozyme sensitivity of the strain. The results of this experiment are shown in Fig. 2. Both RP1 and RP1*amp*1 almost completely protect AS19 against lysis but RP1*amp*1*irp*1 does not. On the other hand neither R1-19, nor any mutant derivative of it, will protect.

All the experiments with RP1 and its mutant derivatives lead to a consistent picture: loss of the β-lactamase gene has a substantial effect on the penicillin resistance of *E. coli* strains whether unmutated or a member of the AS19 or *env*A mutant series. The major part of this effect is due to the hydrolytic action of the enzyme, a conclusion stressed by the fact that this enzyme confers relatively little protection against cephaloridine, a β-lactam antibiotic against which it is relatively inactive at therapeutic concentrations. However, inactivation of the β-lactamase gene still leave some residual resistance to penicillins and this allows one to infer the presence of a gene or group of genes on RP1 (the *irp* region) which confer intrinsic resistance to penicillins.

Studies on the *env*A and AS19 mutants of *Escherichia coli* show that the *irp* genes have a similar effect in enhancing intrinsic resistance to penicillins, indeed the increase in resistance in these abnormally sensitive mutants is correspondingly great. But the same group of genes, or an extremely closely linked group, also confer some resistance to a number of other antibiotic substances (notably rifampicin, nalidixic acid and actinomycin D) to which strains of *Escherichia coli* are normally insensitive. These results on the properties of plasmids that carry an active *irp* region suggest that the function of this genetic region may be to alter the surface layers of the bacteria to make access by thess compounds to their targets in the cell more difficult. The surface changes, if any, in strain AS19 have not been closely examined, but the abnormally great sensitivity of strain D21*envA* to β-lactam antibiotics is almost certainly due to a freeer degree of access of these antibiotics to their target in the inner membrane of the Gram-negative bacteria. Although there is as yet no confirmatory evidence that RP1 plasmids carrying an active *irp* region do lead to altered surface layers of *Escherichia coli* the fact that this gene, or one closely linked to it on this plasmid, confers lysozyme resistance on strain AS19 is suggestive. The insensitivity of *Escherichia coli* to lysozyme is almost certainly due to the interposition of a barrier between the sensitive murein layers in the wall and the outside of the cell. The sensitivity of AS19 to lysozyme without the addition of EDTA (Hamilton-Miller, 1965) or of other agents known to sensitise the cells, suggests that the protective layer in this strain is missing, and the resistance afforded by RP1 and *irp*$^+$ derivatives of it implies that a protective layer — perhaps not the original one — has been introduced.

Whereas RP1 seems to carry genes giving the properties described above, similar effects are not found with derivatives of R1. Plasmid mediated intrinsic resistance to penicillins is therefore not provided by all R-factors having a β-lactamase gene. This fact may indeed account for an observation which up to this point has been puzzling. RP1 confers very high levels of resistance to carbenicillin on both *Escherichia coli* and Pseudomonas aeruginosa strains (Sykes and Richmond 1970), whereas R1 is much less effective in *E. coli* — the only strain in which it can be tested. The reason for this difference may be the fact that R1 relies for its effect on β-lactamase production within the existing outer membrane of the cells while RP1, on the other hand, specifies an enzyme and also seems to increase the impedence offered by the outer layers of the cells to ingress of the antibiotic. Thus this particular class of R-factors seem to be particularly highly evolved to produce very high levels of resistance by a highly sophisticated interaction of two complementary forms of protection.

As mentioned at the beginning of this article, these experiments merely define the area where one must now look for changes in chemical structure conferred by plasmids of the RP1 class. Such studies should enhance our understanding of just how the cells achieve "intrinsic resistance", a term used for too long now to hide our ignorance.

ACKNOWLEDGEMENTS

The original work described in this article was supported by grants to M.H.R. from the Medical Research Council, the Royal Society and the Smith, Kline & French Foundation. N.A.C.C. was supported by a CAPS studentship awarded jointly by the Science Research Council and Glaxo Research Ltd.

REFERENCES

ABRAHAM, E. P. and CHAIN, E. (1940). An enzyme from bacteria able to destroy penicillin. Nature (Lond.). **146**, 837.

BARBER, M. (1962). Coagulase positive staphylococci resistant to benzyl penicillin, methicillin and other penicillins. In CIBA Foundation Study Group No. 13. "Resistance of Bacteria to the Penicillins". p. 89. Ed. by A. V. S. de Rueck and M. P. Cameron. London: J. & A. Churchill.

Citri, N. and Pollock, M. R. (1966). The biochemistry and function of β-lactamase (penicillinase). Adv. Enzymol., **28**, 237.

Curtis, N. A. C., Richmond, M. H. and Stanisich, V. (1973). R-factor mediated resistance to penicillins which does not involve a β-lactamase. J. gen. Microbiol., **79**, 163.

Datta, N. and Kontamichalou, P. (1965). Penicillinase synthesis controlled by R-factors in Enterobacteriaceae. Nature (Lond.). **208**, 239.

Hamilton-Miller, J. M. T. (1965). Effect of EDTA on bacterial permeability to penicillins. Biochem. Europhys. Res. Commun. **20**, 688.

Knox, R. (1962). Different types of resistance to different penicillins. In CIBA Foundation Study Group No. 13: "Resistance of Bacteria to the Penicillins", p. 76. Ed. by A. V. S. de Rueck & M. P. Camerson. London: J. & A. Churchill.

Normark, S., Boman, H. G. and Matsson, E. (1969). Mutant of *Escherichia coli* with anomalous cell division and ability to decrease chromosomally and episomally mediated resistance to ampicillin and several other antibiotics. J. Bact. **97**, 1334.

Richmond, M. H. and Sykes, R. B. (1973). The β-lactamases of Gram-negative bacteria and their possible physiological role. In Advances in Microbial Physiology, vol. 9, p. 31. Ed. by A. H. Rose and D. Tempest. London & New York: Academic Press.

Richmond, M. H. and Curtis, N. A. C. (1974). The interplay of β-lactamases and intrinsic factors in the resistance of Gram-negative bacteria. Ann. N. Y. Acad. Sci. (in press).

Schnaitman, C. A. (1971). Effect of Ethylenediamine tetraacetic acid, Triton X100, and lysozyme on the Morphology and Chemical composition of Isolated Cell Walls of *Escherichia coli*. J. Bact. **108**, No. 1. 545.

Sekiguchi, M. and Iida, S. (1967). Mutants of *Escherichia coli* permeable to actinomycin. Proc. nat. Acad. Sci., U.S. **58**, 2315.

Sykes, R. B. and Richmond, M. H. (1970). The intergeneric transfer of a β-lactamase gene between *Pseudomonas aeruginosa* and *Escherichia coli*. Nature (Lond.) 226, 952.

MUTATION OF R FACTORS SPECIFYING HYPERSYNTHESIS OF β-LACTAMASE

H. HASHIMOTO, Y. ODAKURA, Y. IKE and S. MITSUHASHI

Department of Microbiology, School of Medicine Gunma University
Maebashi, Japan

We surveyed for ampicillin (APC) resistance in gram-negative enteric bacteria and isolated 330 ampicillin (APC)-resistant strains from 10,800 strains examined (Ooka et al. 1970). One hundred and thirty eight out of 330 strains could transfer their APC resistance to an E. coli K-12 strain. These R^+ strains were classified into 4 groups after examining the level of resistance to 5 penicillins and a cephalosporin (Fig. 1). The major 2 groups are 85% among total isolates. β-Lactamase of these 2 groups showed the same substrate specificity and the difference between the 2 was only in the specific enzyme activity. Sawai et al. (1970) further proved that the purified enzymes isolated from the 2 groups showed the same chemical and serological properties, and they were denoted as type Ib and type Ia penicillinase, respectively. Preliminary experiments disclosed that all of the R^+ strains carrying Ia type ampicillin (*amp*) gene mutated to those showing Ib like penicillinase. This report concerns the mutation of an R factor capable of specifying hyper synthesis of PCase. A part of the results was reported previously (Odakura et al., 1973).

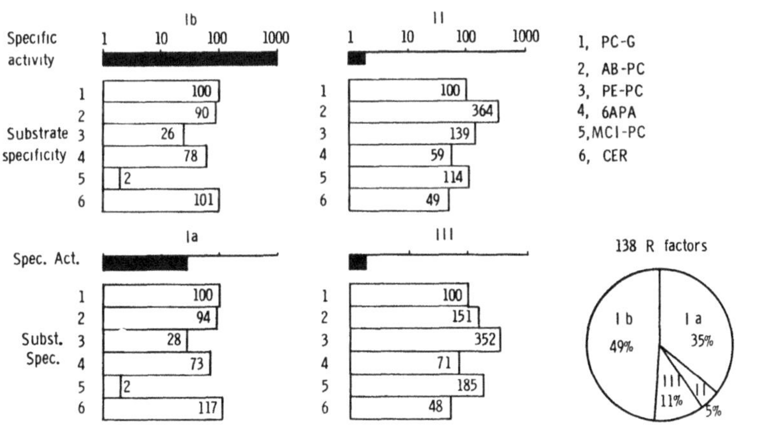

Fig. 1. Classification of R factors specifying ampicillin-resistance. R factors were classified into the 4 groups, after examining the level of drug-resistance in a *E. coli* host to five penicillins and a cephalosporin. The left histograms show the specific activity (unit/mg dry weight) of β-lactamases isolated from the representative strains of each group. PC-G, penicillin-G; APC, ampicillin; PE-PC, phenethicillin; 6-APA, 6-aminopenicillanic acid; MCI-PC, cloxacillin; CER, cephaloridine.

Isolation of a mutant R factor. The R factor used is R_{M201} (Tanaka et al., 1973, Hashimoto et al., 1973), which carries five resistance genes; *amp* (ampicillin, APC), *tet* (tetracycline, TC), *cml* (chloramphenicol, CM), *str* (streptomycin, SM) and *sul* (sulfanilamide, SA). R_{M201} belongs to FII incompatibility group and produces a repressor against F-like pili. The level of APC-resistance of *E. coli* W3630 carrying R_{M201} is 200 μg/ml, but when the culture was plated on a BTB-lactose agar containing 3200 μg/ml of APC, highly APC-resistant colonies grew at a frequency of 10^{-8}. After purification, a mutant was isolated which remained to express a high level of APC-resistance. When the mutant R factor was transferred to a new host by conjugation or transduction, the resultant R^+ recipients showed high resistance to APC. It was concluded that the mutation occurred not on the host chromsome but on the R_{M201}, and the mutant R factor was denoted R_{M201-2}.

TABLE 1

Levels of drug resistance and drug-inactivating activities in various hosts carrying a mutant R_{M201-2}

Host	R factor	Level of resistance				β-lactamase activity (APC)	CATase activity
		TC	CM	SM	APC		
E. coli W3630	R_{M201}	100	200	12.5	100	89,3	0.803
	R_{M201-2}	100	200	12.5	3200	3200	0.750
E. coli ML1410	R_{M201}	50	200	25	200	103.3	0.626
	R_{M201-2}	50	200	25	3200	1665.1	0.720
S. typhimurium LT_2	R_{M201}	50	200	—[a]	200	98.2	0.640
	R_{M201-2}	50	200	—	3200	1328.1	0.670

[a]SM resistance was not determined because of SM resistance of the host strain.

TABLE 2

Comparison of β-lactamases produced by R_{M201} *and its mutant*

Biochemical properties	R_{M201}	R_{M201-2}
Level of APC-resistance (μg/ml)	100	3200
β-lactamase activity (u/mg/protein)	67	888
Substrate specificity for five penicillins and a cephaloridine	similar	
optimal pH	6.5–7.0	6.5–7.0
optimal temperature	45C	45C
Rate of heat inactivation	similar	
Isoelectric point	5.2	5.3

W3630 carrying R factor was harvested by centrifugation and disrupted. The crude enzymes were obtained by centrifugation at 105,000 g for 120min.

β-lactamase coded by a mutant R_{M201-2}. The mutant R factor was transferred to various hosts, and the phenotypes were compared with those of the parent R factor. As shown in Table 1, among these drug-resistance markers, only the ampicillin marker has changed, due to the increased specific activity of penicillinase. The activity of chloramphenicol acetyltransferase remained the same. This indicated that the mutation caused a change of a part of an R genome and not an increase in the copy number of the R_{M201}.

β-Lactamases coded by the mutant and the parent R factor were then compared one another. As shown in Table 2, in quality, we found no difference between th two enzymes. Only the specific activity differred as much as 15 times. By these facts, it was concluded that there was no change in the structural gene responsible for penicillinase formation.

The mutation is thus considered to be a quantitative one, and then two alternative possibilities were assumed: 1) selective amplification of *amp* gene caused the dosage effect on β-lactamase production. 2) the mutation of a promoter or other regulatory gene increased the efficiency of transcription or translation.

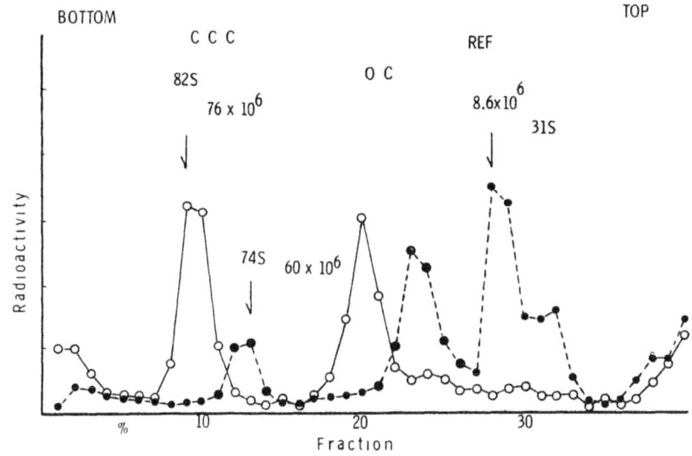

Fig. 2. Neutral sucrose gradient centrifugation of R factor from *E. coli*. Radioactive DNA of ^{3}H-labelled R_{M201} and ^{14}C-labelled R_{M201-2} were collected by pooling satellite fraction obtained by ethidium bromide-CsCl gradient analysis. Each 0.05 ml sample of both R factor DNA was mixed, layered on 4.8 ml of neutral 5 to 20% sucrose gradient, and centrifuged for 60 min at 45 000 rev/min in a RPS65T rotor at 15C. The λdv-1 DNA (Sw_{20}, 31S) was used as a marker.

Determination of the size of the mutant R factor. To determine if the amplified *amp* genes might have been integrated into the R factor, the length of R factor DNA was examined. DNA of W3630 carrying either R_{M201} or R_{M201-2} was differently labelled with ^{3}H or ^{14}C, respectively, and the cells were lysed by Brij58 and Salkosyl after treating them with EDTA and lysozyme (Cohen et al., 1960). The lysate was mixed with ethidium bromide and CsCl, and ultracentrifuged to form a density gradient. The R factor DNA was thus isolated as a satellite band consisting of CCC (covalently closed circular) DNA The amount of CCC DNA of the mutant R factor was as much as 6% of that of the chromosomal DNA, while, in the case of parent R factor, it was 4%. Due to the mutation, 50% increase in the amount of CCC DNA was observed.

The size of each CCC DNA was examined by sucrose gradient analysis and also through electronmicroscope. As shown in Fig. 2, some of the CCC fraction converted to open circular form in each case. But in both fractions, the mutant DNA sedimented faster than parent DNA. $S_{W,20}$ values of the mutant and parent CCC DNA were 82S and 74S, and the molecular weights were calculated as 76×10^6 and 60×10^6 daltons, respectively. Corresponding to these results, electron-microscopic analysis showed that the contour length of mutant DNA was $36\,\mu \pm 0.6\,\mu$, and that of the parent was $31\,\mu \pm 0.6\,\mu$. One of the photographs of the mutant DNA is shown in Fig. 3.

The increase in the size of mutant DNA was thus paralleled to the increase in the specific activity of β-lactamase, and it was assumed that the repeated ampicillin genes were on the mutant R factor, causing dosage effect on enzyme formation. This mutant character was referred to as *amp*-hyper, because of its similarity to the hyper mutation of lactose operon reported by Horiuchi et al., (1963).

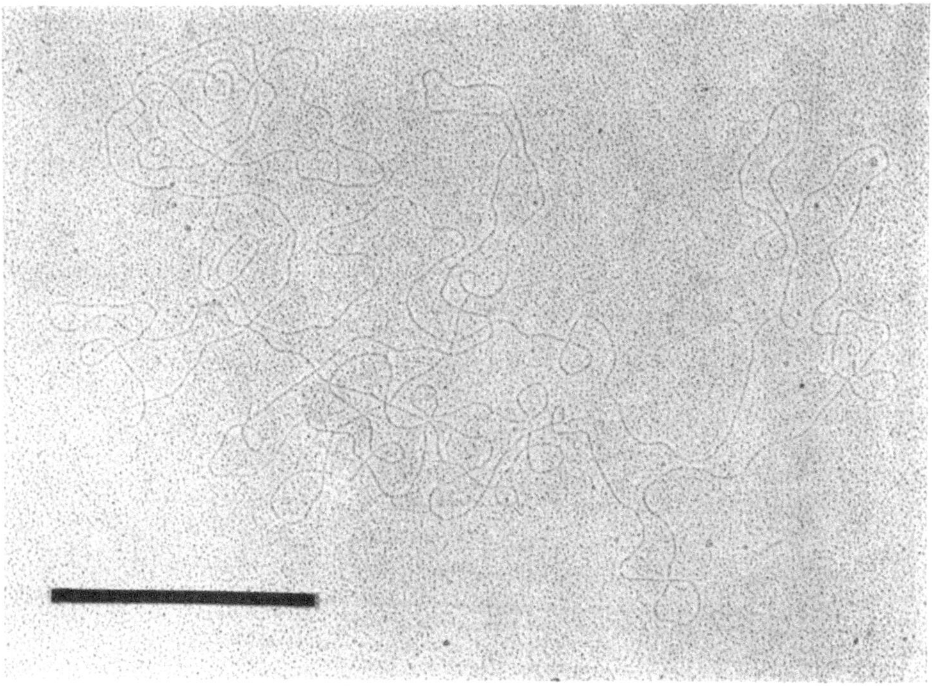

Fig. 3. Electronphotomicrograph of R_{M201-2}DNA. Bar indicate 1μ.

Mechanisms of hyper formation. For the mechanisms of hyper formation, three steps were assumed: 1) dissociation of an *amp* segment from the parent R factor. 2) amplification of the dissociated *amp* gene, and 3) reassociation of the amplified *amp* genes into the parent R factor.

For the first and the third steps, recombination between R genome is a necessary assumption. It is known that recA, not recB or C, highly affects the recombination frequency between R factors (Foster et al., 1971). We used a recA host and examined if hyper mutation would occur. As shown in Table 3, when the parent R_{M201} was carried by the recA host, the mutant colonies grew at a lower frequency on the selective plate containing high concentration of APC, but stable hyper strains were not able to be isolated. Thus it was concluded that recombination is necessary for hyper formation.

For examining the step of *amp* gene amplification, a strain carrying 2 R factors different from each other in their drug-resistance markers and *fi* character was used. The fi^+ R factor carried *tet*, and the fi^- R factor had *str* and *amp*. By selecting cells on plates containing TC and high concentration of APC, only the level of APC-resistance increased, suggesting that hyper mutation occurred. TC-resistance was transduced in each case to see if the *amp* gene might be contransduced. As shown in Table 4, from the parent strain, all of the TC-resistant transductants carried only fi^+R(TC), but from the treated strain, 96% of TC-resistant transductants were also APC-resistant and fi^+.

This result indicates that the *amp* gene was dissociated from the parent fi^- R factor and incorporated to the fi^+ R genome. The high cotransduction frequency suggested that the *amp* gene amplified much.

To visualize the gene amplification, R_{M201}^+ cells were plated on the BTB agar containing 800 μg/ml of APC, and the grown cells were collected directly from the plate,

TABLE 3

Mutation frequency of R_{M201} to mutants with high and stable ampicillin (APC) resistance in either rec⁺ and or rec⁻ strains

Host	APC concn in selective plate (μg/ml)	Survival rate[a]	Mutants showing high and stable APC resistance	
			Isolation frequency[b]	Transmissibility[c]
AB2463 recA	400	2.1×10^{-6}	$<4 \times 10^{-8}$	
	800	1.3×10^{-7}	$<3 \times 10^{-9}$	
	1600	3.8×10^{-8}	$<2 \times 10^{-9}$	
AB1157 *rec*⁺	400	1.1×10^{-4}	1×10^{-4}	+
	800	6.5×10^{-5}	4×10^{-5}	+
	1600	8.9×10^{-7}	5×10^{-7}	+

[a] The number of colonies developed on an APC-containing plate divided by the number of colonies developed on a drug-free plate.

[b] Colonies developed on each selective plate were picked, and each of them was suspended in 0.2ml of nutrient broth. After 2 hr of incubation at 37C, each culture was streaked on BTB agar plate containing the same concentration of APC as that in the first selective plate to confirm whether or not the developed mutants carried stable and high APC resistance. Isolation frequency was expressed by the number of input cells carrying R_{M201} on drug-free plate.

[c] Fifty colonies were randomly picked from each group of mutants and the transmissibility of their APC resistance was examined by the cross-brushing method with W3630.

TABLE 4

Formation of the recombinant between the amp determinant and T-tet factor in a hyper strain

Treatment of donor cells before transduction[a]	Transduction frequency when selected with TC	Genetic properties of TC^r transductants	
		Markers	No. of transductants obtained/examined (%)
TC 25 μg/ml	4.5×10^{-6}	TC^r, fi^+	650/650 (100)
TC 25 μg/ml + APC 1600 μg/ml	5.3×10^{-6}	TC^r, fi^+	61/1560 (4)
		TC^r, APC^r, fi^+	1499/1560 (96)

[a] The donor, W3630 *T-tet*⁺ R_{M91}^+ was obtained by mixed cultivation of W3630 *T-tet*⁺ with ML1410 R_{M91}^+. The resistance pattern of the donor was TC^{r}SM^{r}SA^{r}APC^r. W3630 *T-tet*⁺ R_{M91}^+ in a hetero-R state thus obtained was plated on BTB agar containing both tetracycline and APC or on BTB agar containing only TC. After 18 hr of incubation at 37C, five colonies were randomly picked from each plate and mixed in L broth for the propagation of transducing phage P1.

lyzed, and then CCC DNA was isolated. As shown in Fig. 4, abundant mini circles of 1.5 μ length were observed through electronmicroscope. This result made the present authors examine again carefully the parent DNA samples, and mini circles of similar size were able to be found. Among these mini circles, about one tenth was dimer form. These are considered to be incorporated to the parent DNA and to form one single hyper mutant.

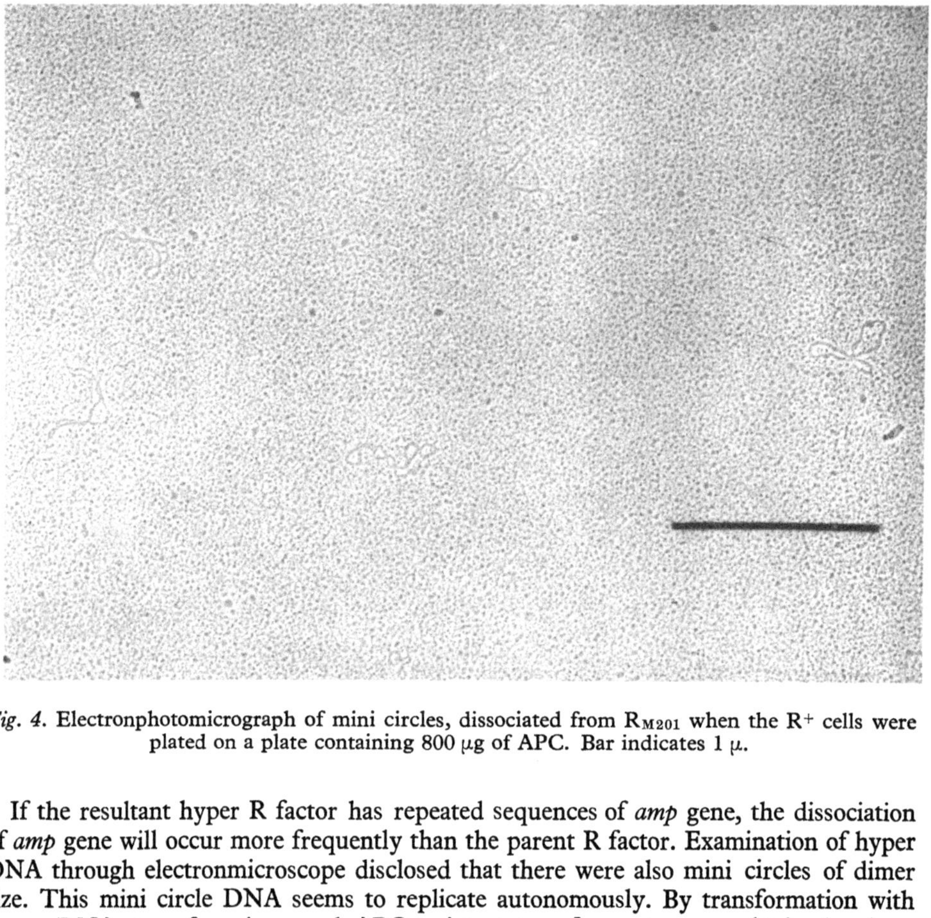

Fig. 4. Electronphotomicrograph of mini circles, dissociated from R_{M201} when the R^+ cells were plated on a plate containing 800 μg of APC. Bar indicates 1 μ.

If the resultant hyper R factor has repeated sequences of *amp* gene, the dissociation of *amp* gene will occur more frequently than the parent R factor. Examination of hyper DNA through electronmicroscope disclosed that there were also mini circles of dimer size. This mini circle DNA seems to replicate autonomously. By transformation with hyper DNA, transfer-minus and APC-resistant transformants were obtained, whose DNA consisted of pure mini circles. The length of the mini circle was the same as that observed with the donor cells.

DISCUSION

From the results described above, it is considered that the R factors coding Ib type penicillinase, may be developed from Ia type R factors by the following way:

1. dissociation of the *amp* gene on the R factor by internal recombination,
2. the newly produced mini plasmid carrying *amp* gene replicates autonomously and the multiple copies of the mini plasmid form dimer or further multimers, and

3. the autonomously replicated *amp* plasmids or its multipliers reassociate sequentially into the parent R genome, forming a longer R factor carrying the repeated sequences of *amp* gene.

If the above assumption is applicable to the natural isolates, there will be various other possibilities to cause a quantitative increase in the activity of β-lactamase. If the dissociated *amp* plasmids replicate stably and form multiple copies as reported Smith et al. (1974), the reassociation of the *amp* genes would not be necessary for explaining the increased β-lactamase activity. We cannot exclude another possibility that much more effective transcription would cause the increased production of β-lactamase activity. Bruenn et al. (1973) reported increased β-galactosidase formation by mutation to highly efficient promoter. The 3 steps assumed by present authors, dissociation, amplification and reassociation, of a part of an R genome is similar to that reported by Rownd et al. (1973) for explaining the phenomenon "transition", the density shift of an R factor when the R^+ cells were incubated with some of drugs. The transition occurred only in *Proteus* and when the transitioned R factor was transferred to *E. coli*, the repeated sequences disappeared. In our case, the amplified *amp* genes were stable even after transduction. This difference may be accountable for by the difference between the concerned genes. The markers *cml-str-sul*, which were responsible for transition were unstable and lost at a high frequency by P1-transduction or spontaneously in *Salmonella*. On the other hand, the ampicillin gene has never been lost spontaneously nor by transduction. The replication locus seems to be linked closely with the *amp* gene and not with *cml-str-sml* clusters. Thus the stability of a dissociated segment as a plasmid seems to be responsible for the effective reassociation.

REFERENCES

BRUENN, I. and HOLLINGSWORTH (1973). A mutant of *E. coli* with a new, highly efficient promoter for lactose operon. Proc. Natl. Acad. Sci. 70, 3697.

COHEN, N. S. and MILLER, C. A. (1969). Multiple molecular species of circular R factor DNA isolated from *Escherichia coli*. Nature (London) 224, 1273.

FOSTER, T. J. and HOWE, T. G. B. (1971). Recombination and complementation between R factors in *Escherichia coli* K-12. Genet. Res. 18, 287.

HASHIMOTO, H., TANAKA, T. and MITSUHASHI, S. (1973). Genetic structure of an R factor conferring ampicillin resistance. Japan. J. Microbiol. 17, 331.

HORIUCHI, T., HORIUCHI, S. and NOVICK, A. (1963). The genetic basis of hyper-synthesis of β-galactosidase. Genetics. 18, 157.

ODAKURA, Y., TANAKA, T., HASHIMOTO, H. and MITSUHASHI, S. (1973). Mutation of R factors capable of specifying hypersynthesis of penicillinase. Antimicrobial Agents and Chemotherapy. 3, 315.

OOKA, T., HASHIMOTO, H. and MITSUHASHI, S. (1970). Comparison of penicillinases produced by R factors isolated from ampicillin-resistant gram-negative bacteria. Japan. J. Microbiol. 14, 123.

ROWND, R., PERLMAN, D., HASHIMOTO, H., MICKEL, S., APPLEBAUM, E. and TAYLOR, D. (1973). Dissociation and reassociation of the transfer factor and resistance determinants of R factors as a mechanism of gene amplification in bacteria. Proceedings of the 6th Miles Symposium on Molecular Biology (Johns Hopkins Univ. Press.) p. 115.

SAWAI, T., TAKAHASHI, K., YAMAGISHI, S. and MITSUHASHI, S. (1970). Variant of penicillinase mediated by an R factor in *Escherichia coli*. J. Bacteriol. 104, 620.

SMITH, R. H., HUMPHREYS, G. O. and ANDERSON, E. S. (1974). Genetic and molecular characterization of some non-transferring plasmids. Molec. gen. Genet. 129, 229.

TANAKA, T., HASHIMOTO, H. and MITSUHASHI, S. (1973). Ampicillin-resistant R factors derived from *Shigella* strains. Japan. J. Microbiol. 17, 323.

IDENTIFICATION OF β LACTAMASES OF PSEUDOMONAS AERUGINOSA BY COMPUTERIZED MICROACIDIMETRY

F. LE GOFFIC and ROGER LABIA
Ecole Normale Supérieure, Laboratoire de Chimie,
Paris, France

One of the most important mechanisms of inactivation of penicillins and cephalosporins is their hydrolysis by β lactamases, which can split selectively the β lactame ring found in all β lactamines.

The methods classically used for the characterization of these enzymes are essentially of two types: the microiodometric method first developed by Perret, then improved by Novick, Sykes and Richmond, and the hydroxylamine technique.

These two techniques do not measure the enzymatic reaction itself but a subsequent one which has to be considered as ideal. Another failure of the first method is that iodine is not very specific for penicilloic acid.

The use of these methods allows, however, a fairly good measurement of the maximal velocity of hydrolysis,but Km, the affinity constant, has sparsely been determined.

With an appropriate equipment it is possible to analyse only one kinetic curve and if the experiment is conducted until completion to obtain the Michaelis Menten constants Km and Vm.

We have developed such a technique by using a computerized pHstat (Labia et *al.*, 1973).

During the enzymatic hydrolysis of β lactamines, a strong carboxylic function appears whereas the amino group liberated is very weak. So pHstat neutralization of penicilloic acid liberated by hydrolysis provides an accurate measurement of the degree of the reaction at any time. Computer analysis of the curve thus obtained from a single experiment gives rise to Km and Vm. These constants are determined starting from the Michaelis Menten equation: $v = \frac{VmS}{Km + S}$

which can be written:

$$v = Vm - Km \frac{V}{S}$$

the computation is a linear regression on v and $\frac{V}{S}$.

The advantages of this new method appear clearly. Its main qualities are its rapidity and its simplicity. Its accuracy is also very acceptable: 3% for Vm and 5% for Km

(Kazmierczak et *al.*, 1973). The method gives also a correlation constant C between a theoretical Michaelis curve and the experimental one obtained. This constant is very important, in particular during the study of bacteria harbouring more than one enzyme.

In summary:

Km: the affinity constant
Vm: the maximum velocity
τ = Km/Vm: the enzymatic stability
C: the correlation constant is now easily available.

$\frac{Km}{Vm}$ which we propose to call: "ENZYMATIC STABILITY" has to be explained (Labia, 1974).

When the concentration of the drug is low (as in the cell), the Michaelis Menten equation: $v = \frac{VmS}{S + Km}$ becomes: $v = \frac{VmS}{Km}$

$$\frac{dS}{dt} = -\frac{Vm}{Km} S$$

$$\frac{dS}{S} = -\frac{Vm}{Km} dt$$

$$\text{Log} \frac{S}{S_0} = -\frac{Vm}{Km} (t - t_0)$$

$$S = S_0 e - \frac{Vm}{Km} t$$

$$S = S_0 e - \frac{t}{\tau}$$

$\frac{Km}{Vm}$ which is homogenous with a time, is also a very important enzymatic constant.

Among the many strains studied in our laboratory, I want to discuss now the case of *Pseudomonas aeruginosa* isolated in "Hospital Cochin" in France. These strains are resistant to penicillin G, ampicillin, carbenicillin, cephaloridin, cephalotin and cephalexin.

The enzymatic constants measured by the microacidimetric method on the crude enzymatic extract are summarized in Tables 1 and 2.

From these results it is clear that these bacteria harbour apparently more than one enzyme: mainly a penicillinase when the culture is performed without induction, the same penicillinase and a cephalosporinase when the bacteria are grown in the presence of penicillin G.

With a better study of these enzymes in view we undertook their purification by affinity chromatography.

Previous experiments have shown that the cephalosporoic acids are inhibitors of cephalosporinases whereas these products have no effect on penicillinases. So a good separation of these two enzymes appeared to be possible by using immobilized cephalosporoic acids.

TABLE 1

Kinetic constants obtained with a crude enzymatic extract obtained from Pseudomonas aeruginosa grown without induction.

Antibiotic	Km	Vm	$\tau = \frac{Km}{Vm}$
Penicillin G	22.9	100	100
Ampicillin	29.8	92	142
Carbenicillin	13.8	8.1	744
Cephaloridin	320	164	852
Cephalotin	24.7	11.7	921
Cephalexin	very slow		

TABLE 2

Kinetic constants obtained with a crude enzymatic extract obtained from Pseudomonas aeruginosa grown with induction with penicillin G.

Antibiotic	Km	Vm	$\tau = \frac{Km}{Vm}$
Penicillin G	15.2	100	100
Ampicillin	27	98	181
Carbenicillin	10.3	7.3	928
Cephaloridin	53	146	237
Cephalotin	21	175	789
Cephalexin	11	9	804

The structure of the biospecific resin we have prepared is shown in Fig. 1, and its efficiency is summarized in Fig. 2.

The chromatography is performed in the cold and is followed by continual determination of the U.V. at 280 nm. The enzymatic activity is monitored with the PHstat. The eluent is a gradient of sodium chloride.

FIG. I

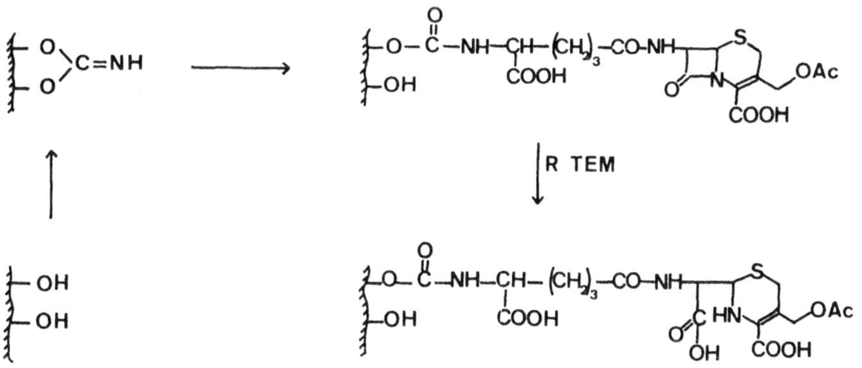

Fig. 1. Structure of the immobilized cephalosporoic acid on indubiose gel.

First a penicillinase activity is found, then on raising the concentration of sodium chloride, the cephalosporinase is eluted where there is a small U.V. absorption at 280 nm.

The fractions eluted first are then put on a column of insoluble ampicillin and purified as we have previously described (Le Goffic F. et *al.*, 1973).

FIG. II

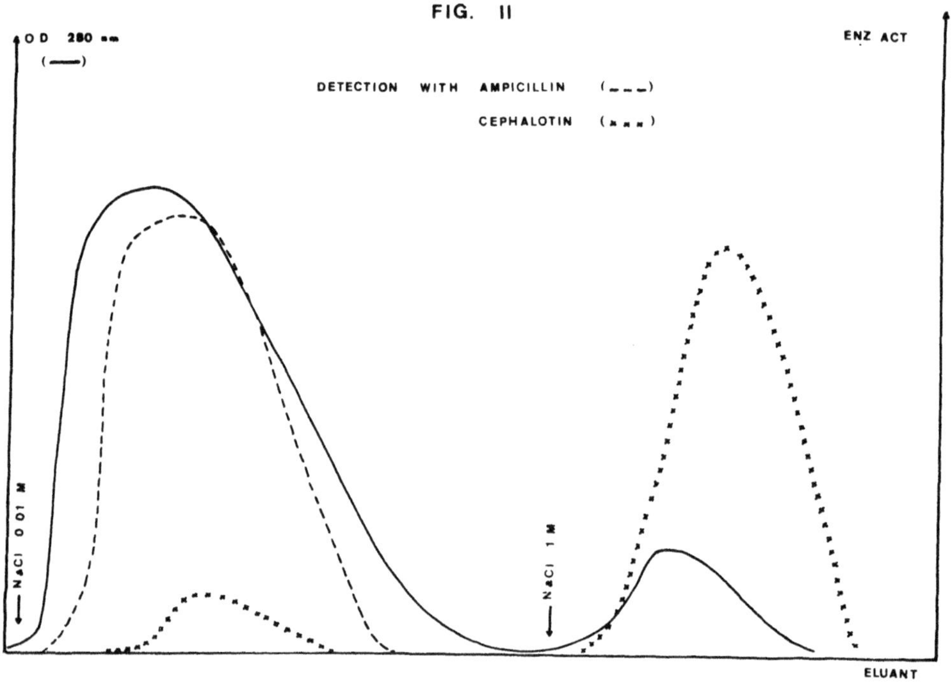

Fig. 2. Elution pattern of penicillinase or cephalosporinase on penicilloic acid – indubiose column: absorbance at 280 nm (———); enzymatic activity: (- - - - -) or (× × ×).

TABLE 3

Kinetic constants obtained with the penicillinase isolated after affinity chromatography. These constants are identical with those obtained with the R TEM penicillinase.

Antibiotic	Km	Vm	$\tau = \frac{Km}{Vm}$
Penicillin G	21	100	100
Ampicillin	29	92	154
Carbenicillin	10	10	524
Cephalotin	230	10	10 000
Cephaloridin	950	170	2 600
Cephalexin	305	65	2 350

Isoelectric point: $p_I = 5.40$.

The kinetic data obtained after these purifications are summarized in Tables 3 and 4.

The penicillinase isolated is identical in all points to the R TEM enzyme which has been extensively studied.

TABLE 4

Kinetic constants obtained with the cephalosporinase isolated by affinity chromatography with immobilized cephalosporoic acid. These constants are identical to those obtained with a cephalosporinase isolated in many different strains in France.

Antibiotic	Km	Vm	$\tau = \frac{Km}{Vm}$
Penicillin G	2,3	100	100
Ampicillin	not hydrolysed		
Carbenicillin	not hydrolysed		
Cephaloridin	51	605	357
Cephalotin	41	890	196
Cephalexin	4.8	40	510

Isoelectric point: $p_i = 9.20$.

The cephalosporinase, which is widespread in France, belong to the group Id in the classification of Richmond.

In conclusion the association of the affinity chromatography technique and the computerized microacidimetric method provide a powerful tool to study β lactamases which we hope to classify soon rationally.

REFERENCES

A. Kazmierczak, A. Philippon, H. Chardon, R. Labia and F. Le Goffic (1973). Ann. Instit. Pasteur, **124 B**, 259–267.

R. Labia, J. Andrillon and F. Le Goffic (1973). FEBS Letters **33**, 42–44.

R. Labia (1974). C.R. Acad. Sc., Série D, in press.

F. Le Goffic, R. Labia and J. Andrillon (1973). Biochimica Biophysica Acta, **315**, 439–442.

OSCILLOPOLAROGRAPHIC DETECTION OF BETA-LACTAMASE ACTIVITY

P. ANTOLÍK

State Institute for Drug Control, Bratislava, Czechoslovakia

Oscillographic polarography was devised by Nobel Prize winner Professor J. Heyrovský (1943) and has been recently applied to biochemical and pharmaceutical problems. Heyrovský (1953) also found this method suitable for purity control of benzylpenicillin. Dušinský (1963) investigated the stability in an alkaline benzylpenicillin solution by this method. The oscillopolarographic behaviour of penicillins was studied in a neutral solution by Dušinský and Antolík (1965).

MATERIALS AND METHODS

The investigators used the international standard of benzylpenicillin 1670 i.u. per mg WHO-National Institute for Medical Research, London, N.W. 7. The standard solution was freshly prepared and kept at 4°C. Phosphate buffer with pH adjusted to 7.0 was used. The experiment was carried out by using the commercial preparation of penicillinase Neutrapen, Riker Lab. containing 400,000 Levy-units per ml. The oscillopolarographic curves were followed by a Polaroscope, Type 576 Křižík, Prague.

The investigations are carried out with 0.2 mg of benzylpenicillin in 5 ml of a phosphate buffer solution. The unattacked penicillins produce only one cathodic incision at Q-value = 0.64 (Fig. 1). After addition of the enzyme, this incision begins to diminish and a second, more positive one is formed belonging to the produced penicilloic acid (Fig. 2) at Q-value = 0.22. At a higher concentration, two other incisions of the degrada-

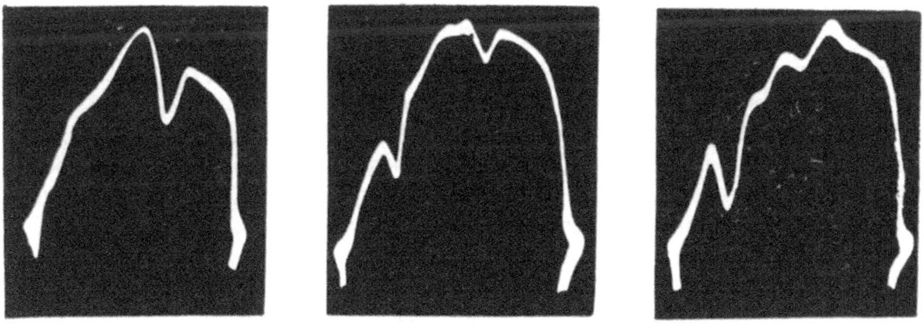

1 2 3

Fig. 1. Oscilopolarogram dE/dt – f/E of 0.2 mg benzylpenicillin/ml in phosphate buffer pH 7.0.
Fig. 2. As Fig. 1. during inactivation by penicillinase.
Fig. 3. As Fig. 1. after complete inactivation by penicillinase.

tion product are formed (Fig. 3). On this basis, it is possible to follow the catalytic enzymatic hydrolysis of penicillins into penicilloic acids. The relation between the depth of the incisions and the concentration of benzylpenicillin and benzylpenicilloic acid is presented in Fig. 4. After addition of penicillinase with a known content of active units of enzyme, we can measure the time of diminution of the incisions of the unattacked penicillins

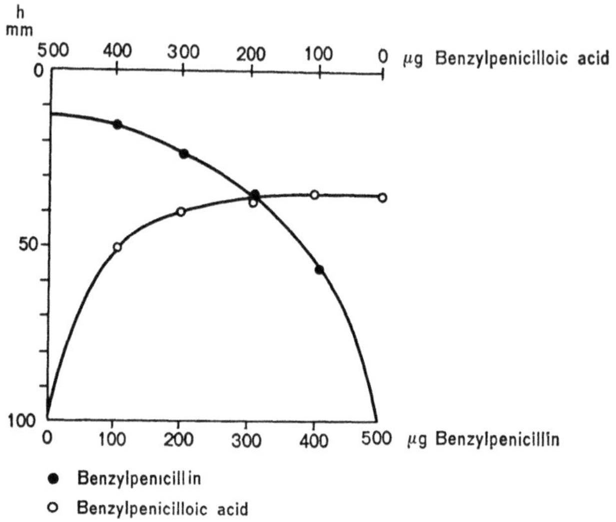

Fig. 4. Relation between the depth of the incisions and the concentration of benzylpenicillin and benzylpenicilloic acid.

up to their total disappearance. Inactivation time ranging from a few seconds to several hours can be measured. This technique allows visual and continuous investigation of enzymatic inactivation of penicillin *in vitro* on the screen of the cathode ray tube. The method was shown to be very specific and easily applicable to routine analyses. We expect that oscillographic polarography will suitably supplement present-day methods of rapid and specific detection of beta-lactamase activity.

SUMMARY

The oscillopolarographic method facilitates very specific and rapid detection of beta-lactamase activity. The assay procedure is based on measuring the rate of the hydrolysis of benzylpenicillin into benzylpenicilloic acid in phosphate buffer pH 7.0. The unattacked benzylpenicillin shows only one incision at Q-value = 0.64 on the curve dE/dt = E. After addition of beta-lactamase, this incision began to diminish and a second, more positive one formed at Q-value = 0.22 belonging to penicilloic acid. This method facilitates visual and continuous investigation of enzymatic inactivation of penicillin.

REFERENCES

DUŠINSKÝ, G. (1963). Chem. zvesti, 17, 265.
DUŠINSKÝ, G., ANTOLÍK, P. (1965). Nature, 206, 196.
HEYROVSKÝ, J. FOREJT, J. (1943). Z. physik, Chem. 193, 77.
HEYROVSKÝ, J. (1953). Chem. Listy 47, 168.

TEMPERATURE-SENSITIVE INDUCTION FOR CEPHALOSPORINASE SYNTHESIS IN *ESCHERICHIA FREUNDII*

T. SAWAI, S. NAKAJIMA, S. YAMAGISHI and S. MITSUHASHI

Faculty of Pharmaceutical Sciences, Chiba University, Chiba, Japan, and Department of Microbiology, School of Medicine, Gunma University, Maebashi, Japan

Cephalosporin β-lactamase (cephalosporinase, CSase) is the most common type of β-lactamase produced by gram-negative rod bacteria (Sawai et al., 1968). The production of CSase is the main mechanism of resistance to β-lactam antibiotics in many gram-negative bacteria, in addition to so-called intrinsic resistance to the drugs known in some gram-negative bacteria such as *Pseudomonas aeruginosa* (Richmond and Sykes, 1973).

There are no established criteria for classifying β-lactamases into two groups, i.e., CSase and penicillinase (PCase), but it is not difficult to classify β-lactamases of gram-negative bacteria into two groups according to their substrate profile. From our previous studies (Sawai et al., 1968, Yaginuma et al., 1973), species-specific β-lactamases of *Aerobacter aerogenes, Arizona, Escherichia freundii, Proteus vulgaris, P. morganii, P. rettgeri, P. inconstans, Ps. aeruginosa* and *Serratia* are known to be of CSase type. On the other hand, species-specific β-lactamases of *Klebsiella pneumoniae* and *P. mirabilis*, and β-lactamases mediated by ampicillin-resistant R-factors are of PCase type. Species-specific β-lactamase of *Escherichia coli* seems to be of an intermediate between the two types. In addition to substrate specificity, onother remarkable difference between those two types of β-lactamase is in the mode of enzyme synthesis. CSases are generally inducible enzymes and β-lactam antibiotics act as the inducer. On the contrary, species-specific PCases and PCases mediated by ampicillin-resistant R-factors are generally constitutive enzymes.

During the past ten years, the gram-negative bacteria resistant to β-lactam antibiotics have become an urgent problem in practical medicine. *E. freundii* strains are generally resistant to β-lactam antibiotics and capable of producing typical CSase without addition of any inducer. We selected *E. freundii* GN346 from our stock cultures because of high CSase activity, and investigated the enzyme properties (Sawai et al., 1970).

CSase activity was determined iodometrically at 30° by the method of Perret (Perret, 1954), except that 0.1M phosphate buffer (pH 6.8) and cephaloridine were used as the reaction buffer and the substrate, respectively, instead of 0.2M phosphate buffer (pH 6.5) and benzylpenicillin. CSase activity was expressed in units, and one unit was defined as the activity of the enzyme that hydrolyze 1 μmole of cephaloridine per hr at 30° in 0.1M phosphate buffer, pH 6.8.

General enzymological properties were examined by using a crude enzyme sample prepared as follow. Bacterial cells at exponential growth phase were suspended in 0.1M phosphate buffer (pH 6.8) to a density of about 5 to 10 mg (dry weight) per ml, and treated for 5 min at 2° with ultrasonic disintegrator. The disrupted-cell suspension was centrifuged at 1,000xg for 30 min at 2°. Dihydrostreptomycin sulfate was added to the supernatant to give about 5%(w/v) concentration for removal of nucleic acids and the precipitate thus formed was removed by centrifugation at 105,000xg for 60 min at 5°. The resulting supernatant was dialyzed against 0.01M phosphate buffer (pH 6.8) and used as a crude enzyme sample. The enzyme sample exhibited 1,430 units of CSase activity per mg protein.

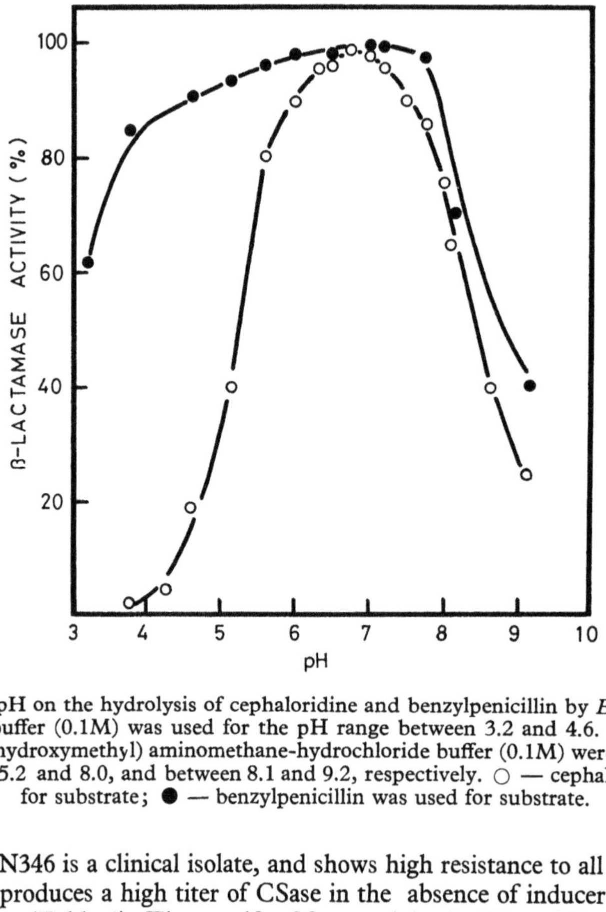

Fig. 1. Effect of pH on the hydrolysis of cephaloridine and benzylpenicillin by *E. freundii* GN346 CSase. Acetate buffer (0.1M) was used for the pH range between 3.2 and 4.6. Phosphate buffer (0.1M) and tris (hydroxymethyl) aminomethane-hydrochloride buffer (0.1M) were used for the pH ranges between 5.2 and 8.0, and between 8.1 and 9.2, respectively. ○ — cephaloridine was used for substrate; ● — benzylpenicillin was used for substrate.

E. freundii GN346 is a clinical isolate, and shows high resistance to all of the β-lactam antibiotics and produces a high titer of CSase in the absence of inducer under a usual growth condition (Table 1). The specific CSase activity is 400 to 900 units per mg of dry weight of bacteria and its value was influenced by bacterial growth age. The highest specific enzyme activity was observed at the late exponential growth phase. The GN346 CSase is cell-bound similar to other β-lactamases of gram-negative bacteria. However, the enzyme does not seem to be "surface enzyme" such as PCases produced by *E. coli* strain harboring ampicillin-resistant R factor, because the GN346 CSase is scarcely released from the cells by the osmotic shock treatment of Neu and Chou (1967). The CSase activity in the disrupted cells was about three times higher than that of the intact cells.

The optimum pH for the GN346 CSase was between 6.8 and 7.0 when cephaloridine was used as a substrate, and a broad optimal pH curve between 3.8 and 7.8 was seen when

TABLE 1

Correlation between β-lactamase activities and levels of resistance to β-lactam antibiotics in E. freundii GN346 and its mutants

Strain	Enzyme activity [b] toward		Drug Resistance (μM) [a]								
	CER	PC-G	7-ACA	CEP-C	P.A.C.	CEG	CEX	CET	CER	PC-G	AB-PC
GN346	880	14.5	1,000	2,000	8,000	500	4,000	4,000	500	8,000	1,000
GN346/16-2	690	16.6	1,000	2,000	4,000	500	4,000	2,000	500	2,000	500
GN346/16-10	392	8.6	1,000	2,000	4,000	250	2,000	1,000	250	2,000	250
GN346/16-9	297	6.5	1,000	2,000	4,000	250	2,000	1,000	250	2,000	250
GN346/1	41.5	3.1	500	1,000	250	63	125	125	32	500	125
GN346/16-24	16.0	< 0.4	500	500	125	32	125	63	16	32	8
GN346/16	12.5	< 0.4	500	250	32	8	32	8	4	32	4
GN346/9	8.4	< 0.4	500	125	16	4	16	8	4	16	4
GN346/8	6.4	< 0.4	500	125	16	4	16	8	4	16	4
GN346/17	< 0.2	< 0.4	500	16	8	2	8	2	2	8	2

[a] Overnight culture in pepton water was diluted 100-fold with pepton water and a loopful of the diluted culture (about 4×10^4 bacteria) was spotted on heart infusion agar plates containing serial two-fold dilutions of each drug. The degree of resistance to each antibiotics was expressed by the maximum concentration (μM) of drug which allowed the visible growth of bacteria after incubation at 37° for 18 hr. Abbreviations: 7-ACA, 7-aminocephalosporanic acid; CEP-C cephalosporin C; P.A.C., 7-phenylacetamido-cephalosporanic acid; CEG, cephaloglycin; CEX, cephalexin; CET, cephalothin; CER, cephaloridine; PC-G, benzylpenicillin; AB-PC, ampicillin.

[b] Enzyme activity was expressed by units per milligram of dry weight of bacteria, and one unit of the enzyme activity was defined as that amount of enzyme activity which hydrolyzed 1 μmole of cephaloridine or benzylpenicillin, respectively, in 1 hr at 30° in 0.1 M phosphate buffer, pH 6.8.

TABLE 2

Effect of iodine on CSase activity

Concentration of iodine (μM)	Activity of CSase (unit)	Inhibition (%)
0	26.0	0
1	22.0	15
2	1.8	93
5	0	100

The enzyme mixture in 7 ml of 0.1M phosphate buffer, pH 6.8, was incubated in the presence of iodine and potassium iodine (ten-fold concentration of iodine) for 2 min at 30°, and then the mixture was quickly titrated with 0.01N sodium thiosulphate to remove free iodine. The residual enzyme activity was assayed.

TABLE 3

Kinetics of hydrolysis of various cephalosporins and penicillins by E. freundii GN346 CSase

Substrate	Vmax[a] (relative)	Km (μM)	Vmax/Km (relative)	Side Chain		
				R_1	R_2	R
7-ACA	0.044	835	0.022	—H	$—OCOCH_3$	——
Cephalosporin C	144	427	150	$—CO(CH_2)_3CHCO_2H$ with NH_2 on CH	$—OCOCH_3$	——
P.A.C.[b]	26.2	49	220	$—COCH_2—$ 	$—OCOCH_3$	——
Cephaloglycin	0.27	9.1	14	—COCH—  with NH_2 on CH	$—OCOCH_3$	——
Cephalexin	33.0	245	59	—COCH—  with NH_2 on CH	—H	——
Cephalothin	18.4	16	500	$—COCH_2$ 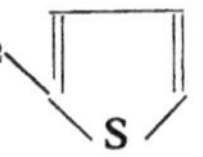	$—OCOCH_3$	——
Cephaloridine	100	455	100	$—COCH_2$ 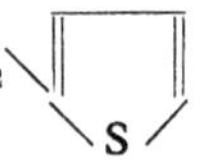	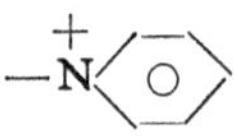	——
Penicillin-G	1.5	5.5	120	——	——	$—COCH_2$ 
Ampicillin	0.073	4.5	8	——	——	—COCH  with NH_2 on CH

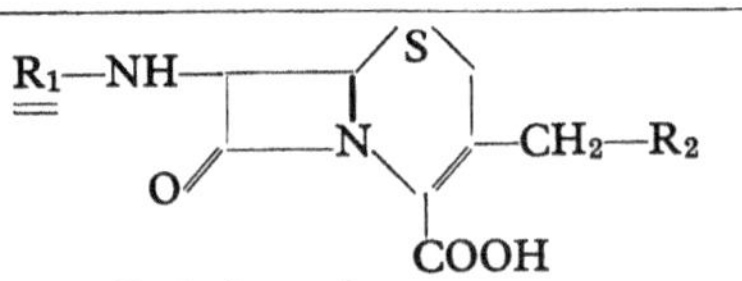

Cephalosporins

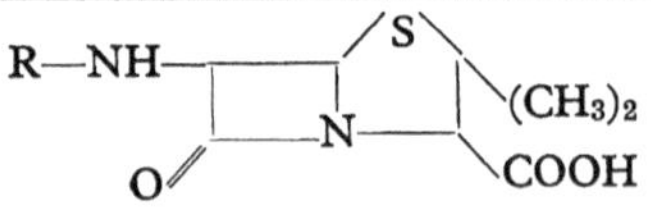

Penicillins

[a] Maximum rate of hydrolysis (Vmax) is expressed in per cent of the hydrolysis of cephaloridine.

[b] 7-Phenylacetamido-cephalosporanic acid.

benzylpenicillin was used as a substrate. The optimum temperature for the enzyme was 44°. One mM of p-chloromercuribenzoate had no effect on the enzyme activity. However, iodine was a powerful inhibitor of the enzyme and its effect was irreversible (Table 2). Table 3 shows the values of the Michaelis constant (Km) and the relative values of the maximum rate of hydrolysis (Vmax) for seven cephalosporins and two penicillins, together

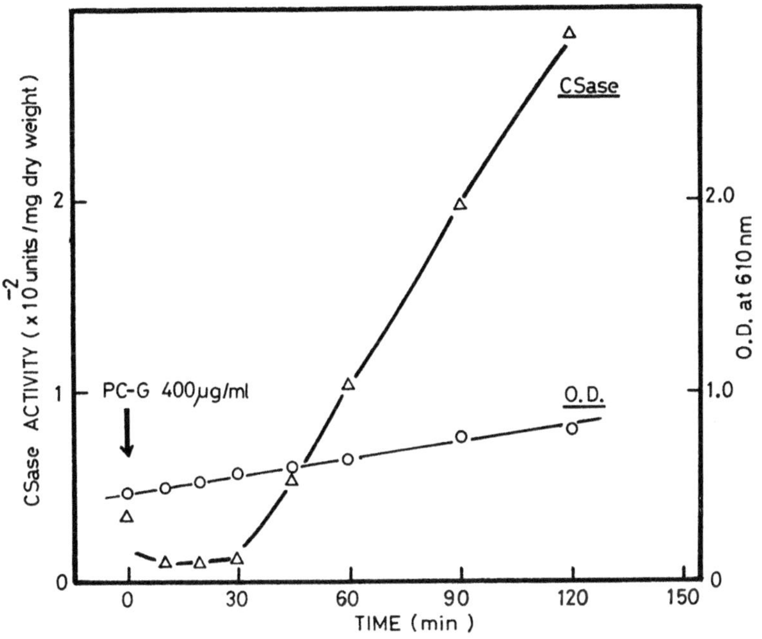

Fig. 2. Induction of CSase synthesis in *E. freundii* GN346 by benzylpenicillin at 20°. Overnight culture of GN346 in nutrient broth at 20° was diluted 10-fold with fresh broth, and incubated with shaking at 20°. When the optical density of the culture at 610 nm reached about 0.5 (in mid exponential growth phase), benzylpenicillin was added to the culture to a final concentration of 400 μg/ml, and then the incubation was continued. After addition of the inducer, a portion of the induced culture was removed at indicated time of incubation for measurement of optical density at 610 nm and assay of CSase activity. The enzyme activity was assayed after the cells were disrupted by ultrasonic disintegrator.

with the physiological efficiency originally proposed by Pollock (1965). Michaelis constants were determined by a modified method of the microiodometric assay (Novick, 1963), and the amount of hydrolyzed substrate was determined photometrically at 620 nm by a standard curve obtained with a known amount of chemically hydrolyzed sample of each substrate.

Role of CSase in bacterial resistance to β-lactam antibiotics. In order to clarify the quantitative correlation between the specific CSase activity and the levels of resistance to β-lactam antibiotics, a series of mutants was prepared from GN346 with the aid of mutagens. Table 1 shows the levels of resistance to nine β-lactam antibiotics and β-lactamase activities in GN346 and its mutants. Strains GN346/8, GN346/9, GN346/16 and GN346/17 were selected after treatment of GN346 with N-methyl-N′-nitro-N-nitrosoguanidine (NTG), according to a modified method described by Adelberg et al., (1965).

Strain GN346/1 was also obtained from GN346 by treatment with ethylmethane sulfonate according to the method of Loveless and Howarth (1959). Strains GN346/16-2, GN346/16-9, GN346/16-10 and GN346/16-24 were the mutants obtained from strain GN346/16 by treatment with NTG. Those mutants were selected from their parent strains by differences in susceptibility to cephaloridine.

The results indicate that the production of the enzyme is closely related to their resistance to β-lactam antibiotics except 7-aminocephalosporanic acid (Table 1). No

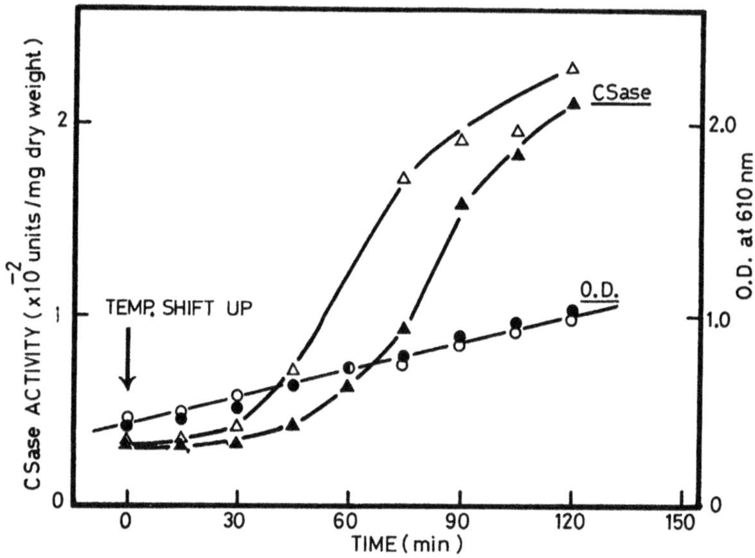

Fig. 3. Induction of the CSase synthesis in *E. freundii* GN346 by shift-up of growth temperature. Overnight culture of GN346 in nutrient broth at 20° was diluted 10-fold with fresh broth, and incubated with shaking at 20°. When the optical density of the culture reached about 0.4 at 610 nm, the growth temperature was shifted from 20° to 37°. Before starting the incubation at 37°, a sample (▲) was prewarmed for 15 min at 37°, and another sample (△) was pretreated for 15 min at 43° and then quickly cooled to 37°. A portion of the culture was removed and assayed its optical density and CSase activity as described in Fig. 2.

detectable β-lactamase activity was found in the disrupted cell-preparation of strain GN346/17. The susceptibility of GN346/17 to β-lactam antibiotics may indicate the intrinsic susceptibility to the drugs without any aid of β-lactamase in the parent strain GN 346.

Temperature-sensitive induction of CSase synthesis

As described above, the CSases produced by gram-negative bacteria in general are inducible enzymes, and the enzyme activity in the cells is very low in the absence of inducer but increases rapidly to a great extent by induction. However, *E. freundii* GN346 and other *E. freundii* strains produce a high titer of CSase in the absence of inducer under usual growth condition. But the production of CSase by *E. freundii* seems to be essentially inducible, because the presence of benzylpenicillin in the growth medium slightly enhances the specific enzyme activity compared with that without inducer. Induction effect of penicillin was not found in the production of the true constitutive β-lactamases in the strains of *E. coli*, *K. pneumoniae*, *P. mirabilis* and the *E. coli* strains harboring ampicilin-resistant R factors.

We examined the effect of growth temperature on the production of GN346 CSase. Specific CSase activities of GN346 grown at various temperatures with or without inducer were compared and the results are shown in Table 4. In the absence of inducer, the specific enzyme activity was markedly affected by growth temperature. The bacterial

TABLE 4

CSase activity in E. freundii GN346 at various growth temperatures

Growth temperature	Concentration of inducer (μg/ml)	CSase activity (unit/mg of bacterial dry weight)
20°	0	<30
	400	430
25°	0	207
	1,600	626
30°	0	787
	3,200	1,660
37°	0	661
	3,200	1,453

The exponential growth phase organisms cultured in nutrient broth at indicated temperature were diluted 5-fold with the fresh broth to give an optical density of about 0.1 at 610 nm. The indicated amount of benzylpenicillin was added to the culture as an inducer. The diluted culture was incubated with shaking at indicated temperature until the optical density of the culture reached 0.5, and the growth was stopped by chilling the culture in an ice-bath. After measured optical density at 610 nm for estimating bacterial growth, the chilled culture was treated with ultrasonic disintegrator and assayed its CSase activity.

cells showed a very low enzyme activity when grown at 20° in the absence of inducer, but the specific enzyme activity increased markedly when grown with inducer at low temperatures. The organisms can grow normally at 20° though its generation time was 140 min. The generation times at 25°, 30° and 37° were 70, 50 and 50 min, respectively. The organisms exhibited higher susceptibility to bacteriocidal action of benzylpenicillin when grow at lower temperatures, and the inducer concentration of 400 μg per ml was the maximum concentration which gave no inhibitory effect on the bacterial growth at 20°. The induced enzyme activity at 20° was found to increase to about 700 units per mg of dry weight of bacteria by further addition of the inducer one hr after the first induction. These facts suggest that the organisms have an ability to produce a large amount of CSase at 20°, and the mode of enzyme production at low growth temperatures is the same as that of typical inducible enzymes. Figure 2 shows the kinetics of induced CSase synthesis in the organisms growing exponentially at 20°. A linear increase in the specific enzyme activity was observed 30 min after addition of the inducer. The lag time, 30 min, is thought to be not longer than the lag time for usual inducible enzymes, when the long generation time of the organisms at 20° is considered.

The results given in Table 4 also indicate that the mode of CSase synthesis in the organisms varies from an usual inducible type to a pseudo-constitutive type with the enhancement of growth temperature. These facts, in addition to the observations mentioned above, suggested that the inducible synthesis of CSase in *E. freundii* strains is governed by a regulatory mechanism which is very susceptible to temperature, i.e., derepressed at higher temperatures than about 25°. Such a possibility was strongly supported by the experiments shown in Fig. 3. When the growth temperature of the organisms growing exponentially at 20° was shifted to 37°, a linear increase of the specific enzyme activity was induced 45 min after temperature shift-up without any chemical inducer. The lag time and the rate of the enzyme synthesis after induction by temperature were almost the same as those observed in the culture induced by benzylpenicillin at 20° (Fig. 2). When chloramphenicol (500 μg/ml) was added to the culture at the same time of temperature shift-up, no increase in the specific enzyme activity was observed during subsequent incubation at 37°.

To know a difference among the CSases synthesized at low and high temperatures, four crude enzyme preparations were extracted from the cells grown at 20 or 37 °C in the presence or absence of the inducer were compared each other in their substrate profiles of six cephalosporins and two penicillins, i.e., cephaloridine, cephalothin, cephalexin, cephaloglycin, cefazolin, cephalosporin C, benzylpenicillin and ampicillin, and no essential differences were observed among those four enzyme preparations in such an enzymological property.

The results demonstrated above suggested that *E. freundii* strain has an inducible system for CSase synthesis which includes a temperature-sensitive repressor. The same temperature-sensitive inducible system was found in other *E. freundii* strains which are highly resistant to cephalosporins and ampicillin, and the unique system is thought to be common in the species. However, we failed to find a similar system in other species of gram-negative bacteria which produce typical CSase.

REFERENCES

ADELBERG, E. A., MANDEL, M. and CHEIN CHING CHEN, G. (1965). Optimal conditions for mutagenesis by N-methyl-N′-nitro-N-nitrosoguanidine in *Escherichia coli* K12. Biochem. Biophys. Res. Commun. **18**, 788.

LOVELESS, A. and HOWARTH, S. (1959). Mutation of bacteria at high levels of survival by ethyl methane sulphonate. Nature (London) **184**, 1780.

NEU, H. C., and CHOU, J. (1967). Release of surface enzymes in Enterobacteriaceae by osmotic shock. J. Bacteriol. **94**, 1934.

NOVICK, R. P. (1963). Analysis by transduction of mutations affecting penicillinase formation in *Staphylococcus aureus*. J. Gen. Microbiol. **33**, 121.

PERRET, C. J. (1954). Iodometric assay of penicillinase. Nature (London) **174**, 1012.

POLLOCK, M. R. (1965). Purification and properties of penicillinase from two strains of *Bacillus licheniformis*: a chemical, physicochemical and physiological comparison. Biochem. J. **94**, 666.

RICHMOND, M. H., and SYKES, R. B. (1973). The β-lactamases of gram-negative bacteria and their possible physiological role. Advances in Microbial Physiology **8**, 31.

SAWAI, T., MITSUHASHI, S., and YAMAGISHI, S. (1968). Drug resistance of enteric bacteria. XIV. Comparison of β-lactamases in gram-negative rod bacteria resistant to α-aminobenzylpenicillin. Jap. J. Microbiol. **12**, 423.

SAWAI, T., MITSUHASHI, S., and YAMAGISHI, S. (1970). Stability of various derivatives of 7-aminocephalosporanic acid and 6-aminopenicillanic acid to cephalosporinases from gram-negative bacteria. Proc. 6th Internat. Congr. Chemotherapy **1**, 410.

YAGINUMA, S., SAWAI, T., ONO, H., YAMAGISHI, S. and MITSUHASHI, S. (1973). Biochemical properties of a cephalosporin β-lactamase from *Pseudomonas aeruginosa*. Jap. J. Microbiol. **17**, 141.

ANTIBIOTIC RESISTANCE AND β-LACTAMASES OF *ESCHERICHIA COLI* BETWEEN 1948 AND 1968

B. SLOCOMBE, R. A. EDMONDSON and R. SUTHERLAND

Beecham Research Laboratories, Chemotherapeutic Research Centre, Betchworth, Surrey, England

It was first reported from Japan in 1959 (Ochiai et al., 1959; Akiba et al., 1960), that multiple antibiotic resistance could be transferred readilly between strains of *E. coli* and *Shigella* and it was shown that R-factor bacteria constituted a high proportion of the antibiotic-resistant isolates of enteric bacteria in Japan at that time. The original report was followed within a fairly short time by other similar reports from Europe, Israel, Taiwan and the U.S.A., (Nakaya, 1960; Datta, 1962; Lebek, 1963; Kabins and Cohen, 1966), and it was evident that transferable antibiotic resistance was common and worldwide in the early 1960's. Little is known however, of the origin or spread of R-factors among bacteria prior to the original report from Japan in 1959 and this study is concerned with the development, in the United Kingdom, of transferable antibiotic resistance among enteropathogenic strains of *E. coli* between 1948 and 1968. (Slocombe and Sutherland, 1973).

A total of 512 strains of *E. coli* was obtained from the Salmonella Reference Laboratory, Colindale, through the courtesy of its former Director, Dr. Joan Taylor. The cultures examined had been isolated from clinical sources throughout the United Kingdom at three different periods of time between 1948 and 1968, namely 1948 to 1951, (112 strains) 1957 to 1960, (200 strains) and 1967 to 1968 (200 strains). Bacteria from these three periods were selected for study as different groups of antibiotics were available during each period. Thus, benzylpenicillin, streptomycin and sulphonamides were widely used between 1948 and 1951, chloramphenicol and the tetracyclines became available in the United Kingdom soon afterwards, and ampicillin was introduced in 1961. Other antibiotics used in 1967 and 1968 for the treatment of infections due to Gram-negative bacteria included cephaloridine and kanamycin, and carbenicillin and gentamicin first became available during this period.

Most of the strains of *E. coli* isolated during the two later periods had been isolated from cases of infantile diarrhoea throughout the country and represented a wide range of serotypes. The origin of the strains isolated between 1948 and 1951 is less certain, but they are known to be of human origin and are thought to be enteropathogenic strains of *E. coli* responsible for infantile gastroenteritis.

Prior to the tests described here, the cultures had been stored in the dark at room temperature on antibiotic-free Dorset egg-agar slopes. The 512 strains were tested for sensitivity to ampicillin, chloramphenicol, streptomycin, tetracyclines and sulphonamides and Minimum Inhibitory Concentrations were determined by serial-dilution tests in agar. Strains of *E. coli* found to be resistant to one or more of the antibacterial agents were tested for ability to transfer resistance to *E. coli* K12/711Nx (Slocombe and Sutherland, 1973).

ANTIBIOTIC RESISTANCE

Results in Fig. 1 compare the sensitivities of the strains of *E. coli* to each of the four antibiotics for the three different time periods.

Ampicillin resistance

Ampicillin inhibited 66% of the 112 strains of *E. coli* isolated between 1948 and 1951 at a concentration of 12.5 μg/ml or less; the remainder (34%) required inhibitory concentrations of from 25 to 250 μg/ml of ampicillin. Of the 200 strains isolated between 1957 and 1960, 94% of the strains were sensitive to ampicillin and the remainder were sensitive to concentrations of 25 to 250 μg/ml of ampicillin. Most (85%) of the strains isolated in 1967 and 1968 were sensitive to ampicillin, but only one-third of the resistant strains were inhibited by 250 μg/ml of ampicillin. The remainder of the resistant strains (10% of the total tested) were insensitive to 1000 μg/ml ampicillin.

Ampicillin-resistant strains destroyed ampicillin during the course of serial-dilution tests showing that resistance was associated with β-lactamase activity.

Streptomycin resistance

In general, the incidence of resistance to streptomycin was similar for the three periods (74% sensitive to 12.5 μg/ml in 1948–1951, 85% sensitive in 1957–1960, and 78% sensi-

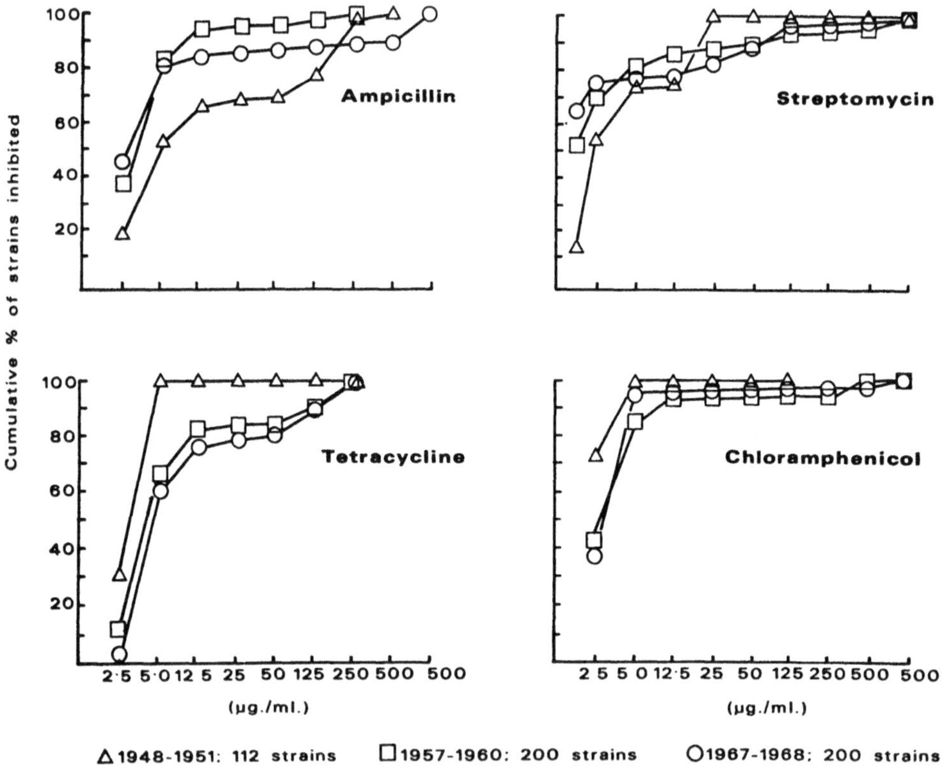

Figure 1. Antibiotic sensitivity of *Escherichia coli* isolated between 1948–1951; 1957–1960; 1967–1968.

tive in 1967–1968). However, whereas all of the resistant strains isolated between 1948 and 1951 were insensitive to 500 μg/ml to streptomycin only a small proportion of the resistant strains isolated in the two later periods were resistant to this concentration of streptomycin.

Chloramphenicol resistance

Only one of the strains of *E. coli* isolated between 1948 and 1951 was resistant to chloramphenicol, and only a small proportion of strains isolated in the two later periods were insensitive to 12.5 μg/ml of chloramphenicol (7.5% in 1957 to 1960 and 5% in 1967 to 1968).

Tetracycline resistance

None of the strains of *E. coli* isolated between 1948 and 1951 was resistant to a concentration of 5 μg/ml of tetracycline but resistance was common among bacteria isolated between 1957 and 1968. Thus, 17.5% of the strains isolated in 1957–1960 and 22% of those isolated in 1967–1968 were resistant to 12.5 μg/ml of tetracycline. The distribution of resistance to tetracycline among these two groups of *E. coli* was very similar.

Sulphonamide resistance

Most (87.5%) of the strains of *E. coli* isolated between 1948 and 1951 were sensitive to 20 μg/ml of sulphamethoxazole. The incidence of sulphonamide-resistant strains was higher in the two later periods when 21% of strains tested were found to be insensitive to 20 μg/ml of sulphamethoxazole.

These results show that resistance to penicillin (or ampicillin) streptomycin and sulphonamides was relatively common between 1948 and 1951, and is presumably a reflection of the selective pressures of the usage of these antibiotics. Resistance to chloramphenicol and tetracycline was rare before 1951, but after their introduction into clinical use resistance to these two antibiotics became more common. It is striking that with the exception of ampicillin resistance, the antibiotic resistance-patterns for the period 1957–1960 were very similar to those of the later period, 1967–1968.

TRANSFERABLE RESISTANCE

Strains shown to be resistant to any of the four antibiotics, or to the sulphonamide, wer tested for their ability to transfer antibiotic resistance to *E. coli* K12/711Nx and the results of these tests are shown in Table 1.

None of the ampicillin-resistant strains isolated between 1948 to 1960 (with ampicillin MIC values of 25–250 μg/ml) was able to transfer ampicillin resistance to the recipient strain, as was the case for strains isolated in 1967–1968 with ampicillin MIC values in the same range. In contrast, all of the ampicillin-resistant strains isolated in 1967–1968 with ampicillin MIC values of greater than 1000 μg/ml transferred ampicillin resistance associated with β-lactamase. Treatment with ethidium bromide had no effect on the resistance to ampicillin of strains that failed to transfer ampicillin resistance but resulted in the elimination of the ampicillin-resistance of the R-factor strains isolated in 1967–1968.

None of the streptomycin-resistant strains of *E. coli* isolated between 1948 and 1951 transferred streptomycin-resistance, but two-thirds of the streptomycin-resistant strains isolated between 1957 and 1968 were capable of transferring streptomycin-resistance to the recipient strain. R-factor mediated streptomycin resistance was eliminated by treatment with ethidium bromide but the streptomycin resistance of strains isolated before 1951 was unaffected by this treatment.

TABLE 1

Transferable antibiotic resistance among Escherichia coli isolated between 1948–1968

Antibiotic	Time of isolation	No. of strains	Resistant strains* number	Resistant strains* % of total	Transferable resistance No. of strains	Transferable resistance % of resistant strains	Transferable resistance % of total strains
Ampicillin	1948–1951	112	38	34.	0	0	0
	1957–1960	200	10	5.0	0	0	0
	1967–1968	200	30	15	21	70	10.5
Chloramphenicol	1948–1951	112	1	1.0	0	0	0
	1957–1960	200	15	7.5	13	87	6.5
	1967–1968	200	10	5.0	10	100	5.0
Streptomycin	1948–1951	112	28	25	0	0	0
	1957–1960	200	28	14	19	68	9.5
	1967–1968	200	44	22	29	66	14.5
Tetracycline	1948–1951	112	0	0	0	0	0
	1957–1960	200	35	17.5	21	60	10.5
	1967–1968	200	44	22	30	68	15
Sulphamethoxazole	1948–1951	112	14**	12.5	3	21	2.7
	1957–1960	200	43**	21.5	17	40	8.5
	1967–1968	200	42**	21	17	40	8.5

* Not inhibited by 12.5 μg/ml of test antibiotic
** Resistant to 200 μg/ml of sulphamethoxazole

Only one strain of *E. coli* isolated in 1948–1951 was resistant to chloramphenicol and this strain failed to transfer its resistance. In contrast, virtually all of the chloramphenicol-resistant strains isolated between 1957 and 1968 transferred chloramphenicol resistance to *E. coli* K12/711Nx. Similarly, tetracycline resistance was readily transferred from resistant strains isolated in this period (60% in 1957–1960 and 68% in 1967–1968).

Three of the fourteen sulphonamide-resistant strains of *E. coli* that were isolated between 1948 and 1951 were found to be capable of transferring sulphonamide resistance of *E. coli* K12/711Nx. These three R-factor strains were isolated in 1951 and probably had a common source. The strains were resistant also to ampicillin, but the ampicillin resistance was not transferred. In the two later periods, 40% of the sulphonamide-resistant strains transferred sulphonamide resistance, usually in association with streptomycin or other resistance.

These results show that R-factor-mediated resistance became prevalent among enteropathogenic strains of *E. coli* in the United Kingdom at some time between 1951 and 1957 although transferable sulphonamide resistance was demonstrated in a small number of strains isolated in 1951. It is of interest, too, that the incidence of R factor-mediated resistance was about as high among antibiotic-resistant strains isolated between 1957–1960 as in strains collected in 1967 and 1968.

Phenotype resistance to β-lactam antibiotics

All ampicillin-resistant strains tested were found to destroy ampicillin in serial-dilution tests to measure Minimum Inhibitory Concentrations of ampicillin, showing that resistance was associated with β-lactamase activity. It was apparent, too, that the ampicillin-resistant bacteria could be separated into two groups on the basis of their resistance to β-lactam antibiotics.

Thus, ampicillin-resistant strains isolated between 1948 and 1968 that had ampicillin MIC values in the range 25–250 μg/ml were sensitive to carbenicillin and to combinations of ampicillin and cloxacillin but were resistant to cephalosporin antibiotics (Table 2). These strains failed to transfer ampicillin resistance, their ampicillin resistance was

TABLE 2
Phenotypic resistance to β-lactam antibiotics

Ampicillin M.I.C. (ug/ml)	Sensitivity to			Resistance	
	Carbenicillin	Ampicillin/ Cloxacillin	Cephalosporins	transferable	effect of curing agents
25—250	S	S	R	—	—
>1000	R	R	variable	+	+

S = Sensitive R = Resistant

unaffected by treatment with ethidium bromide or sodium dodecyl sulphate, and efforts to mobilise or displace the ampicillin resistance by the introduction of plasmids of different compatibilities (RTEM, RA3, S-a, R64, N3, "wild type") were unsuccessful. In contrast, strains isolated in 1967–1968 that were resistant to 1000 μg/ml of ampicillin, were resistant also to carbenicillin and ampicillin/cloxacillin, but were generally more sensitive to the cephalosporins (Table 2). These strains were capable of transferring their ampicillin-resistance and treatment with curing agents resulted in the loss of ampicillin resistance.

It seemed likely, therefore, on the basis of these properties that the strains of *E. coli* that did not transfer ampicillin resistance possessed a chromosomally-mediated β-lactamase that differed from the extrachromosomally-mediated β-lactamase of the R factor strains.

Properties of β-lactamases

Crude, cell-free, β-lactamase preparations were prepared by ultrasonication of selected cultures (Cole, Elson, and Fullbrook, 1972) and these enzyme preparations were tested for electrophoretic mobility and substrate and inhibition profiles. The R-factor strains investigated were transferred to a common host, *E. coli* K12 711/Nx, and the recombinants, enriched by selection in the presence of carbenicillin, were used as the sources of extrachromosomal β-lactamases. For comparison, β-lactamase preparations were also made from the following reference strains:

a) Chromosomally-mediated, *E. coli* D31 (Lindstrom, Boman and Steele, 1970).

b) extrachromosomally-mediated
1. *E. coli R-TEM* (*R6K*) (Datta and Kontomichalou, 1965; Datta and Richmond, 1966.)
2. *E. coli R-GN238* (Egawa, Sawai and Mitsuhashi, 1967.)
3. *E. coli R-1818* (*R46*) (Dale and Smith, 1971).

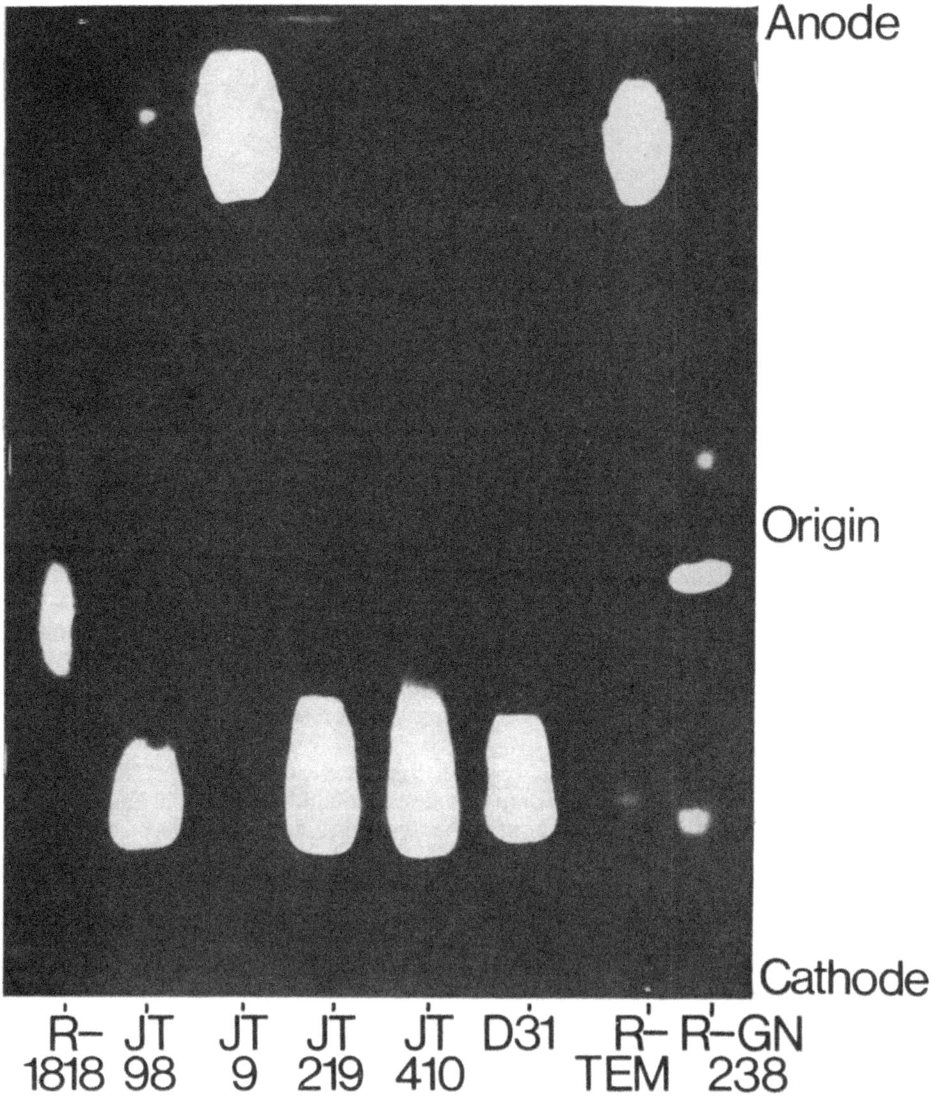

Figure 2. Electrophoretic mobilities of β-lactamases of *E. coli*.

Electrophoretic mobility

The electrophoretic mobilities of the β-lactamases were determined on starch-agar gel (360v, pH 8.5) by a modification of the method described by Jack and Richmond (1970) and results of a typical experiment are illustrated in Fig. 2. The enzyme from the R factor strain of *E. coli* (JT9) moved towards the anode at the same rate as the R-TEM preparation (—2.5 cm/hr) whereas the enzymes from the strains of *E. coli* with non-transferable ampicillin resistance (JT 219, JT 415) migrated towards the cathode (+2.1 cm/hr), slightly faster than the chromosomally-mediated β-lactamase from *E. coli* D31 (+1.8 cm/hr). The reference oxacillin-hydrolising β-lactamases, R-GN238, R-1818, migrated with low positive mobilities (+0.3 cm/hr, +1.1 cm/hr respectively). One strain JT 98, possessed two electrophoretically distinct enzymes, one with positive mobility (+2.0 cm/hr) the other with negative mobility (—2.5 cm/hr).

Substrate and inhibition profiles

Substrate and inhibition profiles of the β-lactamase preparations are shown in Table 3, where the hydroxylamine method described by Dale and Smith (1971) was used for assay of β-lactamase activity. The non-transferable enzymes, JT163 and JT415, resembled the

TABLE 3

Substrate and Inhibition profiles of β-lactamases from ampicillin-resistant E. coli

Enzyme	Relative rates of hydrolysis: Penicillin G	Ampicillin	Carbenicillin	Cephaloridine	Cephalexin	Cloxacillin	Inhibition by: Cloxacillin	Cl⁻ (%)
JT 163	100	0	18	600	32	0	>60	0
JT 415	100	0	0	321	38	0	>60	0
JT 219	100	0	0	335	67	0	>60	0
D-31	100	0	5	350	37	0	>60	0
JT 9	100	87	11	76	0	3	>60	0
JT 20	100	74	13	65	0	5	>60	0
R-TEM	100	86	18	72	0	3	>60	0
R-GN238	100	453	86	24	0	90	—*	60
R-1818	100	150	48	26	0	186	—*	80

* Cloxacillin was a substrate.

chromosomally-mediated β-lactamase, D31, and were more active against the cephalosporins than against ampicillin and carbenicillin. The β-lactamase from the R factor strain JT9, was very similar to that of R-TEM and inactivated the penicillins more readily than the cephalosporins. The oxacillin-hydrolising enzymes, R-GN238 and R-1818 represented a third group of β-lactamases.

One strain, JT98, isolated in 1968, was found to possess both a chromosomally-mediated β-lactamase and an extrachromosomal-mediated β-lactamase, both enzymes having characteristics similar to those described above. This strain was insensitive to both penicillins and cephalosporins. It is perhaps surprising that more strains with both β-lactamases were not found in this series of tests as strains possessing a chromosomal β-lactamase were good recipients of R-factor mediated β-lactamases in transfer experiments.

These results show that the β-lactamases of ampicillin-resistant strains of *E. coli* isolated between 1948 and 1968 fell into two groups according to their biochemical characteristics. Thus, strains that did not transfer ampicillin resistance possessed a β-lactamase similar to the chromosomally-mediated Type Ib enzyme (Richmond and Sykes, 1973) of *E. coli* D31, while the extrachromosomal β-lactamases of R factor strains of *E. coli* were similar to R TEM, a Type IIIa enzyme (Richmond and Sykes, 1973). These facts suggest that the extrachromosomally-mediated β-lactamases of R-factor bacteria were not derived from the chromosomally-mediated β-lactamase found in ampicillin-resistant strains of *E. coli* between 1948 and 1968.

CONCLUSION

This study of a homogenous group of enteropathogenic strains of *E. coli* isolated in the United Kingdom between 1948 and 1968 has shown that transferable antibiotic resistance became prevalent among these bacteria between 1951 and 1957, although a small number of strains with R factor sulphonamide-mediated resistance were isolated before this (in 1951). The incidence of R factor-mediated antibiotic resistance among this group of bacteria was similar in 1957–1960 as it was in 1967–1968 showing that transferable antibiotic resistance was well-established in the United Kingdom at this earlier period of time.

The ampicillin-resistant strains of *E. coli* isolated between 1948 and 1960 did not transfer ampicillin-resistance and appeared to possess a chromosomally-mediated β-lactamase which was quite distinct from the extrachromosomally-mediated β-lactamase of R factor strains of *E. coli* isolated in 1967–1968. R factor mediated β-lactamase woud not appear therefore to have its origin in β-lactamase producing strains of *E. coli*.

ACKNOWLEDGEMENTS

We wish to thank Mr. G. Hill for his skilled technical assistance. We also acknowledge, with thanks, the kind gift of plasmids from Dr. P. Kontomichalou for R6K (= R-TEM), Dr. J. T. Smith for R-GN238 and R-1818 and Dr. Hans Boman for a culture of *E. coli* D31.

REFERENCES

AKIBA, T., KOYAMA, K., ISHIKI, Y., KIMURA, S. and FUKUSHIMA, T. (1960): On the mechanism of the development of multiple drug-resistant doses of *Shigella*. Japan J. Microbiol. **4**, 219 (in Japanese).

COLE, M., ELSON, S. and FULLBROOK, P. D. (1972): Inhibition of the B-lactamases of *Escherichia coli* and *Klebsiella aerogenes* by semi-synthetic Penicillins. Biochem. J. **127**, 295.

DALE, J. W. and SMITH, J. T. (1971): Purification and properties of the B-lactamase specified by the resistance factor R1818 in *Escherichia coli* and *Proteus mirabilis*. Biochem. J. **123**, 493.

DATTA, N. (1962): Transmissible drug resistance in an epidemic strain of *Salmonella typhimurium*. J. Hyg. **60**, 301.

DATTA, N. and KONTOMICHALOU, P. (1965): Penicillinase synthesis controlled by infections R-factors in Enterobacteriaceae. Nature, **208**, 239.

DATTA, N. and RICHMOND, M. H. (1966): The purification and properties of a penicillinase whose synthesis is mediated by an R-factor in *Escherichia coli*. Biochem. J. **98**, 204.

EGAWA, R., SAWAI, T. and MITSUHASHI, S. (1967): Drug resistance of enteric bacteria. XII. Unique substrate specificity of penicillinase produced by R-factor. Japan. J. Microbiol. **11**, 173.

JACK, G. W. and RICHMOND, M. H. (1970): A comparative study of eight distinct B-lactamases synthesized by Gram-negative bacteria. J. Gen. Microbiol. **61**, 43.

KABINS, S. A. and COHEN, S. (1966): Resistance transfer factor in Enterobacteriaceae N. Engl. J. Med. **275**, 248.
LEBEK, G. (1963): Ueber die Enstehung mehrfachresistenter Salmonellen. Ein Experimenteller Beitrag. Zentr. Bakteriol. Parasitenk. Abr. 1, **188**, 494.
LINDSTROM, E. B., BOMAN, H. G. and STEELE, B. (1970): Resistance of *Escherichia coli* to penicillins. VI. Purification and characterisation of the chromosomally-mediated penicillinase present in *amp A* — containing strains. J. Bact. **101**, 218.
NAKAYA, R. (1960): Drug resistance of enteric bacteria. Transmissible drug resistance agent. Rec. Adv. Med. Biol. **1**, 109 (in Japanese).
OCHIAI, K., YAMANAKA, K., KIMURA, K. and SAWADA, O. (1959): Studies on inheritance of drug-resistance between *Shigella* and *Escherichia coli* strains. Nihon Iji Shimpo, **1861**, 34 (in Japanese).
RICHMOND, M. H. and SYKES, R. B. (1973): The β-lactamases of Gram-negative bacteria and their possible physiological role. Adv. Microb. Physiol, **9**, 31.
SLOCOMBE, B. and SUTHERLAND, R. (1973): Transferable antibiotic resistance in enteropathogenic *Escherichia coli* between 1948 and 1968. Antimicrob. Ag. Chemother. **4**, 459.

A CLINICAL ISOLATE OF *ESCHERICHIA COLI* POSSESSING TWO β-LACTAMASES AND AN UNUSUAL RESISTANCE TO β-LACTAM ANTIBIOTICS

S. W. ELSON and B. SLOCOMBE

Beecham Research Laboratories, Chemotherapeutic Research Centre, Betchworth, Surrey, England

In the course of a study of 512 strains of enteropathogenic *E. coli* it became apparent that the ampicillin-resistant strains could be separated into two groups on the basis of their sensitivity to β-lactam antibiotics (Slocombe and Sutherland, 1973). It was observed that strains with Minimum Inhibitory Concentrations (M.I.C.) of 25–250 mg ampicillin/ml were sensitive to carbenicillin but resistant to cephalosporin antibiotics. Transmissable resistance to ampicillin was not demonstrable withthese organisms. In contrast ampicillin-resistant strains that were not inhibited by 1000 μg ampicillin/ml were resistant also to carbenicillin but were relatively sensitive to cephalosporins, and the resistance of these strains was usually transferable. Similar findings have been reported recently by Roupas and Pitton (1974). In the work reported by Slocombe and Sutherland (1973) it was observed that the resistance of these strains to β-lactam antibiotics was associated with drug inactivation as a result of the β-lactamase activities of the organisms.

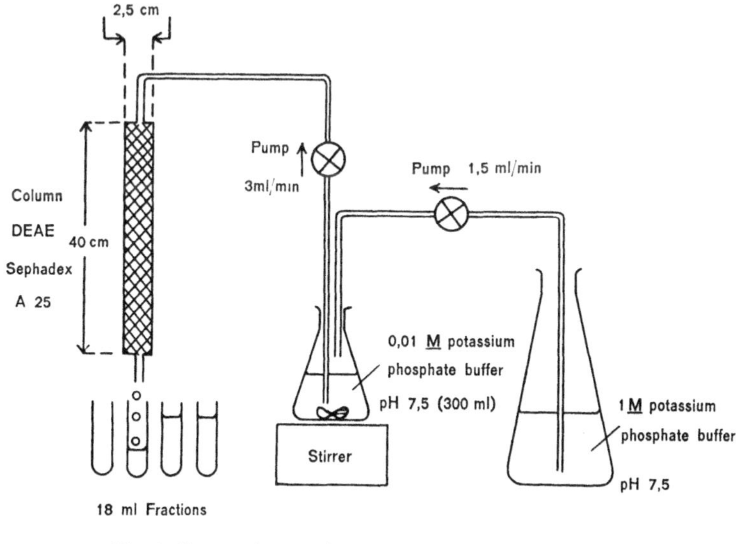

Fig. 1. Ion-exchange chromatography of β-lactamases

It was observed that only one strain of the 512 tested was found to be resistant to penicillins and cephalosporins and the rarity of such strains among *E. coli* was confirmed in subsequent studies (unpublished data).

This report describes the antibiotic resistance and β-lactamase activity of such a strain of *E. coli* (B11, BRL 1983), in comparison with other strains of *E. coli*, and attempts to

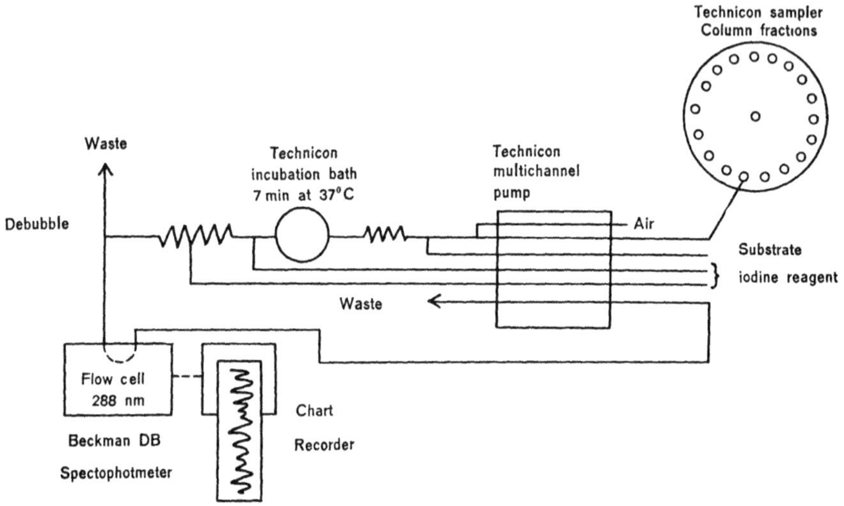

Fig. 2. Automated iodometric assay for measurement of β-lactamase activity (Cole, Elson, Fullbrook, 1972).

Figure 3 Chart Record of Automated Analysis of β-Lactamase Fractions — 1

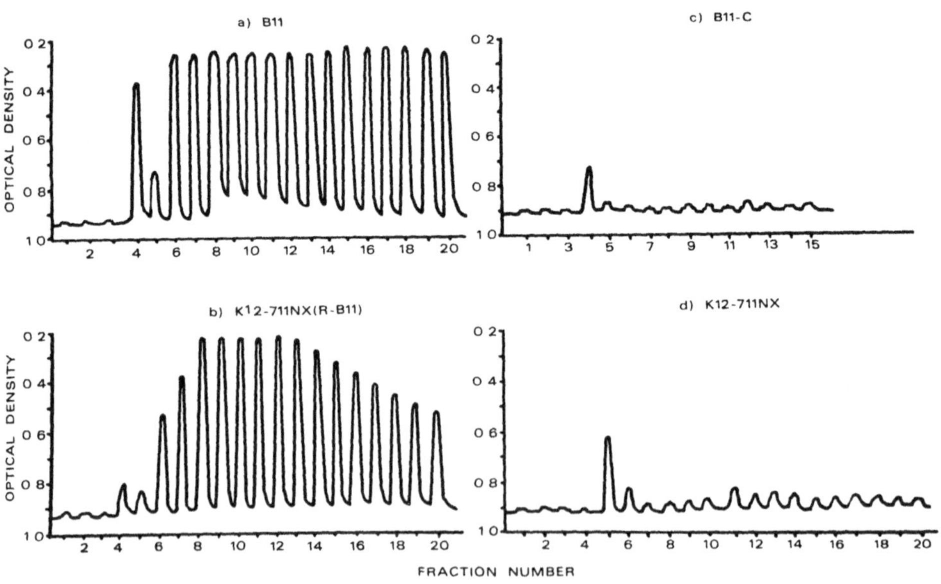

correlate the extent to which resistance to β-lactam antibiotics was associated with β-lactamase activity.

When the supernatant from sonicated late log-phase cells of *E. coli* B11 was eluted from a Sephadex G 100 column with distilled water, two components of β-lactamase activity were obtained. These had different biochemical properties, but the degree of separation was poor. Other grades of Sephadex were no more successful. However, elution through the ion exchange gel, DEAE Sephadex A 25 with a phosphate ion gradient gave good results and this method (Fig. 1) was adopted for studing all the β-lactamases described below. The fractions from the column were assayed for β-lactamase activity by an automated iodometric assay using benzyl penicillin as substrate (Fig. 2).

Two active components were eluted from the column, the first coming off mainly in fraction 4, the remainder in a broad band in fraction 6 onwards (Figure 3a). When the column was eluted with 0.01M buffer only, without an ion gradient, the enzyme in fraction 4 was the only one to elute. This suggested that the first enzyme to elute was not negatively charged, whereas the second enzyme was. Also, starch gel electrophoresis

Figure 4 Chart Record of Automated Analysis of β-Lactamase Fractions —2

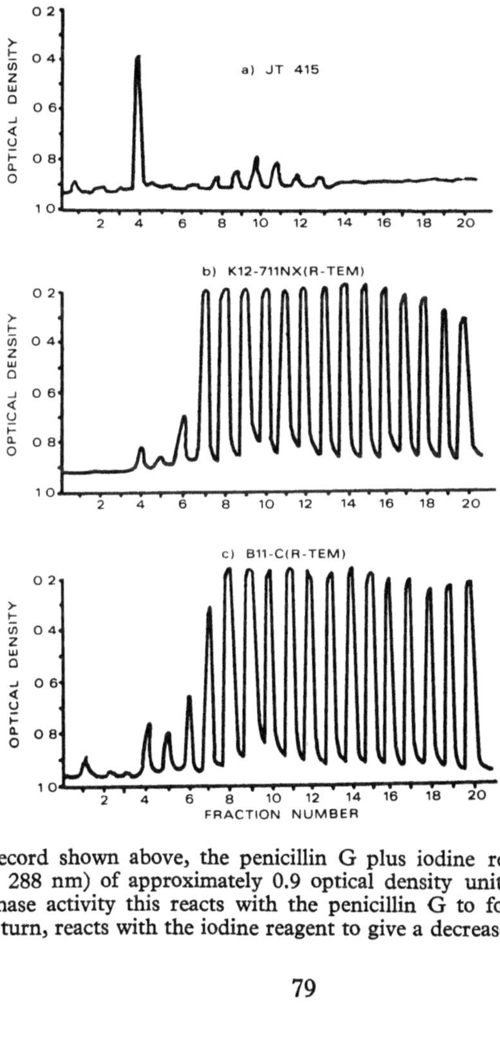

On each chart record shown above, the penicillin G plus iodine reagent mixture has an U/V absorption (– = 288 nm) of approximately 0.9 optical density units. When a column fraction contains β-lactamase activity this reacts with the penicillin G to form benzyl penicilloic acid. This, in turn, reacts with the iodine reagent to give a decrease in U/V absorption.

using the method of Jack and Richmond (1970), of a crude *E. coli* B11 β-lactamase preparation showed the presence of two oppositely charged β-lactamases.

As it was known that *E. coli* possessed transferable resistance to ampicillin, the next stage in the investigation was to see if either or both of the β-lactamases were R-factor mediated. To do this, conjugation experiments were performed with *E. coli* B11 using

PLATE 1

Disc sensitivities of 3 Clinical Strains of E. coli

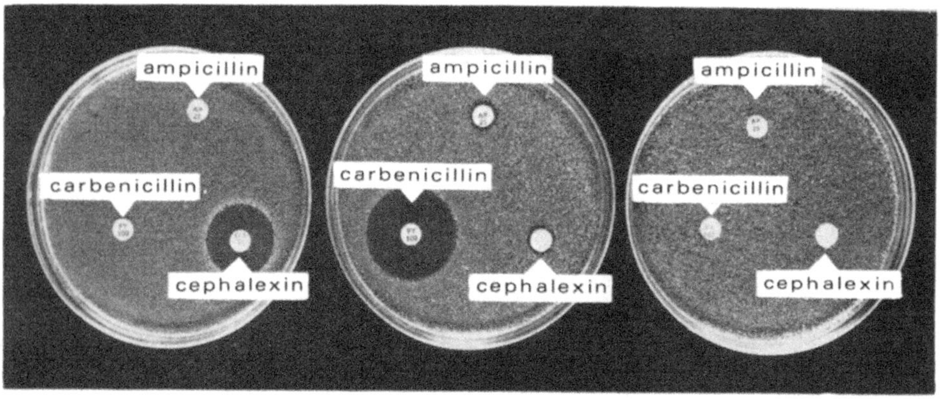

JT4 (*R-factor strain*) *JT415* (*No R-factor*) *B11*

E. coli K12-711Nx, a penicillin sensitive strain, as recipient (Slocombe and Sutherland, 1973). Pure colonies were isolated which were resistant to ampicillin and carbenicillin, but not to cephalexin. This culture, designated *E. coli* K12-711Nx (R-B11) (Plate II), was used to prepare a crude β-lactamase sample which was separated by the ion exchange method described above. Again two components of enzyme activity were obtained, the elution pattern being similar to that of *E. coli* B11 (Fig. 3b).

The results in Table 1 show that the biochemical properties of the negatively charged enzymes from each organism were the same, hydrolysing ampicillin almost as fast as penicillin G, but not hydrolysing carbenicillin, cloxacillin or the cephalosporins. The properties of the β-lactamases eluting in fraction 4 were different in that neither hydrolysed ampicillin and both were inhibited by cloxacillin. They differed from each other in that the enzyme from *E. coli* B11 hydrolysed cephalexin whereas that from *E. coli* K12-711 (R-B11) did not. These findings suggested that the negatively charged enzyme was R-factor mediated. To confirm this, *E. coli* B11 was grown in a medium containing ethidium bromide (Ref. Bouanchaud *et al.*, 1968), in an attempt to eliminate the R-factor. Clones were isolated which were sensitive to carbenicillin, had increased sensitivity to ampicillin but were still resistant to cephalexin (Plate II). Ion exchange chromatography of a β-lactamase preparation from this organism (B11-C) gave enzyme activity in fraction 4 only (Fig. 3c). The properties of this enzyme were the same as those of the corresponding enzyme from *E. coli* B11. From these data it was concluded that in *E. coli* B11, the negatively charged enzyme is mediated by the R-factor.

The genetic information mediating the positively charged enzyme is non-transmissible,

TABLE 1

Kinetic Data for E. coli β-lactamases – I

Culture	[3]Dry Weight	Fraction	[4]Dilution	[1]Relative Initial Velocities:						[2]Inhibition
				Pen G.	Amp.	Carb	Cephthn	Cephx	Clox	Pen G+Clox
B11	56.7 mg	4	1/50	100[5] (2.6)	0	0	ca 10	19	0	100%
		16	1/250	100 (0.6)	93	0	0	0	0	5%
B11 C	49.8 mg	4	1/100	100 (1.2)	0	0	0	26	0	100%
K12-711 Nx (R-B11)	39.5 mg	4	1/20	100 (0.8)	0	0	0	0	0	100%
		17	1/100	100 (0.8)	82	0	0	0	0	17%
K12-711 Nx	169.0 mg	5	1/10	100 (2.6)	3	0	0	0	0	100%

1. Substrate concentrations = 50 μg/ml (ca 140 μM).
2. % Inhibition of hydrolysis of 50 μg/ml Pen G by 5.0 μg/ml cloxacillin.
3. The dry weight of the cells from which the crude β-lactamase that was loaded onto the column was obtained.
4. The final dilution of the fraction in the reaction mixture.
5. Figures in parentheses are actual velocities; μg penicillin G hydrolysed per minute per ml.

TABLE 2

Kinetic Data for E. coli β-lactamases — II

Culture	[3]Dry Weight	Fraction	[4]Dilution	[1]Relative Initial Velocities:						[2]Inhibition
				Pen G.	Amp	Carb	Cephthn	Cephx	Clox	Pen G+Clox
JT 415	93.9 mg	4	1/100	100[5] (1.2)	0	0	0	17	0	100%
K12-711 Nx (R-TEM)	61.8 mg	5	1/10	100 (0.8)	0	0	0	0	0	100%
		17	1/100	100 (1.0)	100	10	0	0	0	7%
B11 C (R-TEM)	34.0 mg	4	1/40	100 (1.4)	0	0	ca 5	25	0	100%
		17	1/50	100 (1.0)	93	13	0	0	0	14%

1. Substrate concentrations = 50 μg/ml (ca 140 μM).
2. % Inhibition of hydrolysis of 50 μg/ml Pen G by 5.0 μg/ml cloxacillin.
3. The dry weight of the cells from which the crude β-lactamase that was loaded onto the column was obtained.
4. The final dilution of the fraction in the reaction mixture.
5. Figures in parentheses are actual velocities: μg penicillin G hydrolysed per minute per ml.

not eliminated by treatment with curing agents and could not be mobilised nor displaced by the introduction of other plasmids. This enzyme is, therefore, probably mediated by the chromosome. The kinetic data for the β-lactamases from *E. coli* B11, *E. coli* K12-711Nx, *E. coli* 711Nx(R-B11) and *E. coli* B11-C where obtained using the automated iodometric assay and are given in Table 1.

PLATE 2

Disc sensitivities of 3 Laboratory Strains of E. coli

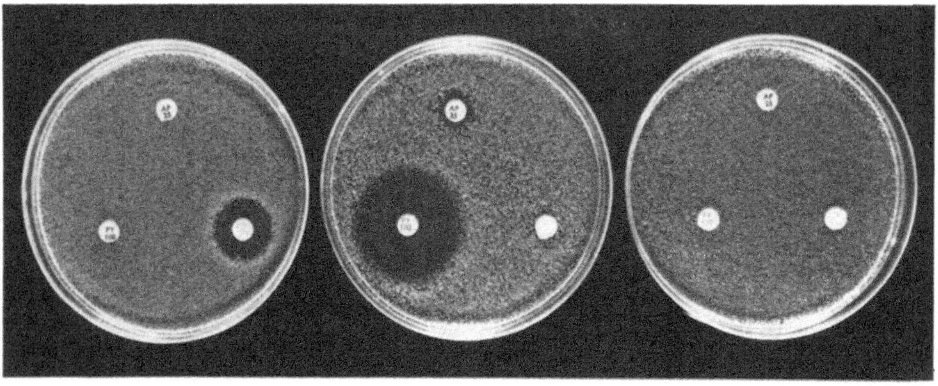

K12-711Nx (R-B11) *B11 C* *B11 C(R-TEM)*

It can be seen by reference to Figure 3d and Table 1 that *E. coli* K12-711Nx produces a β-lactamase that elutes in fraction 5. This is presumably the chromosomally mediated enzyme of *E. coli* K12-711Nx, and is still apparent in the recombinant *E. coli* K12-711Nx (R-B11).

E. coli B11-C showed the same phenotypic resistance as clinical strains of *E. coli* which are resistant to ampicillin but do not possess R-factor mediated β-lactamase. Therefore, we next examined *E. coli* JT 415, which is typical of such clinical strains. When a sonicated cell preparation of *E. coli* JT 415 was eluted through the ion exchange column, one peak of enzyme activity was observed (Fig. 4a). The properties of this enzyme were very similar to those of the enzyme from *E. coli* B11-C (Table 2).

Introduction of the R-factor TEM into *E. coli* B11-C yielded clones that were now resistant to ampicillin, carbenicillin and cephalexin, as was the parent strain *E. coli* B11 (Plate II). However, *E. coli* K12-711Nx(R-TEM) was only resistant to ampicillin and carbenicillin, as was *E. coli* K12-711Nx(R-B11) (Plate II). The data for the β-lactamases of *E. coli* B11-C(R-TEM) and *E. coli* K12-711Nx(R-TEM) are shown in Fig. 4b and 4c, and Table 2.

In summary, the data given above show that *E. coli* B11 produces two distinct β-lactamases. One is probably mediated by the chromosome, and has the ability to hydrolyse cephalexin, which may account for the phenotypic resistance of the organism to this cephalosporin. This enzyme is similar to that found in the clinical isolate *E. coli* JT 415, which has the same phenotype. The other β-lactamase is R-factor mediated. Transfer of the R-factor to a sensitive host causes the exconjugants to become resistant to carbenicillin. This resistance is unexpected, as the R-factor mediated enzyme does not appear to hydrolyse carbenicillin to any great extent. When the R-factor is deleted, the resulting organism is still resistant to ampicillin although much less so than the parent strain. However, the chromosomally mediated enzyme does not readily hydrolyse ampicillin. It is likely, therefore, that other resistance mechanisms are also operating in *E. coli* B11 (e.g. cell wall impermeability to penicillins).

E. coli B11 presumably came into existence when a clinical strain of *E. coli* such as *E. coli* JT 415 received an R-factor. *E. coli* B11 was isolated in 1964 when organisms like *E. coli* JT 415 were probably more prevalent (Slocombe and Sutherland, 1973). As this combination of two enzymes endows *E. coli* B11 with a broad spectrum of resistance to β-lactam antibiotics, it is surprising that in our experince this phenotype is comparatively rare.

We have demonstrated that bacteria possessing the chromosomally mediated β-lactamases act as good recipients for plasmids mediating β-lactamase synthesis. It may be that the plasmids are unstable in these strains and are at a disadvantage in non-selective environments. We have, to date, found only one other strain of *E. coli* which synthesises both enzymes concommitantly.

ACKNOWLEDGEMENT

We wish to thank Miss M. English for expert technical assistance and Drs. G. N. Rolinson and M. Cole for helpful discussion.

REFERENCES

Bouanchaud, D. H., M. R. Scavizzi and Y. A. Chabbert, (1968). J. Gen. Microbiol. **54**: 417–425.
Cole, M., S. W. Elson and P. D. Fullbrook, (1972). Biochem. J. **127**, 295–308.
Jack, G. W. and M. H. Richmond, (1970). J. Gen. Microbiol. **61**, 43–61.
Roupas, A. and J. S. Pitton, (1974). Antimicrobial Agents and Chemotherapy **5** (2) 186–191.
Slocombe, B. and R. Sutherland, (1973). Antimicrobial Agents and Chemotherapy **4** (4) 459–466.

PURIFICATION AND PROPERTIES OF AN UNUSUAL CEPHALOSPORINASE SPECIFIED BY AN R PLASMID IN *ESCHERICHIA COLI*

M. M. BOBROWSKI, and POLYXENI KONTOMICHALOU

Department of Biochemistry, University Medical School, Białystok, Poland and Department of Clinical Therapeutics, University of Athens School of Medicine, Alexandra Hospital, Athens, Greece

One of the best characterized products of R plasmid genes are β-lactamases (penicillin amido-β-lactamhydrolases, E.C.3.5.2.6.), the enzymes that catalyze the hydrolysis of the penicillins, cephalosporins and cephamycins into corresponding biologically inactive compounds. So far, two main types of R plasmid-specified β-lactamases have been recognized and they can be distinguished by their physico-chemical, enzymological and immunological properties (Datta and Kontomichalou, 1965; Yamagishi et al., 1969; Sawai et al., 1970; Dale and Smith, 1971b; Richmond and Sykes, 1973; Rubin and Smith, 1973; Hedges et al., 1974; Bobrowski, 1974).

The enzymes of the first type, usually referred to as the TEM-like β-lactamases (Hedges et al., 1974) or the class III β-lactamases (Richmond and Sykes, 1973), are acidic proteins with a molecular weight of 22,000–26,000 that are active against penicillins as well as cephalosporins. The enzymes of the other type, usually referred to as the oxacillin-hydrolysin β-lactamases (Hedges et al., 1974) or as the class V β-lactamases (Richmond and Sykes, 1973), are basic proteins that hydrolyze primarily penicillins, including methicillin, oxacillin, and cloxacillin. Two enzymes of this type, the R46 and $R_{GN\ 238}$ β-lactamases, were reported to have a molecular weight of about 44,000 and 23,000 respectively (Dale and Smith, 1972).

It should be mentioned that none of the plasmids controlling the synthesis of β-lactamases of either type confers high levels of resistance to cephalosporins, whereas the levels of resistance to ampicillin and other penicillins determined by these plasmids are, as a rule, very high.

Recently, Kontomichalou (1971) described a number of R plasmids governing resistance to β-lactam antibiotics, and one of these plasmids, designated R22K, appeared to be different from the others in that it conferred comparatively high levels of resistance to cephalosporins as compared with penicillins. Furthermore, the author observed that the β-lactamase specified by this R plasmid, unlike the R plasmid-mediated β-lactamases already examined, was predominantly active against cephalosporins, and against cephalothin in particular.

In this study attempts were made to purify the R22K β-lactamase and to determine some of its physico-chemical properties. It is shown, on the basis of the results obtained, that this enzyme represents a novel distinct type of R plasmid-specified β-lactamases.

MATERIALS AND METHODS

Antibiotics and other materials. Benzylpenicillin, phenoxymethylpenicillin, and streptomycin were supplied by the Pharmaceutical Works "Polfa", Warsaw, Poland. Ampicillin, amoxycillin, carbenicillin, ticarcillin, methicillin and cloxacillin were gifts from the Beecham Research Laboratories, Betchworth, U.K. Cephaloridine, cephalothin, cephalexin and cephalosporin C were donated by the Glaxo Laboratories Ltd., Greenford, U.K. Cefazolin was a gift from the Fujisawa Pharmaceutical Co., Osaka, Japan, and cephacetrile was generously supplied by Ciba - Geigy Ltd., Basel, Switzerland.

Sephadex G-75 (superfine grade), Sephadex G-100 (fine grade), CNBr-activated Sepharose 4B and Blue Dextran 2000 were purchased from Pharmacia Fine Chemicals AB, Uppsala, Sweden; CM-cellulose (Whatman CM-23) from Reeve Angel, London, U.K.; "Ampholine" Carrier ampholytes — from LKB-Producter AB; Bromma, Sweden. All ingredients for culture media were obtained from the Difco Laboratories. Other reagents were analytical grade from Serva Fine Biochemicals GmbH, Heidelberg, Germany; Koch-Light Laboratories, Colnbrook, U.K.; or Sigma Chemical Co., St. Louis, U.S.A.

Organism, media and culture methods. All experiments were performed with *Escherichia coli* strain K12 58161 F^- met^- λ^+ into which the R22K plasmid has been transferred by conjugation from a wild type isolate of Proteus mirabilis (Kontomichalou, 1971). The cephalosporin-resistance marker of the R22K plasmids exists as a non-transmissible replicon in the *E. coli* host, and no transfer of the plasmid was possible unless it was mobilized (Naomi Datta — personal communication).

The organism was normally grown in a modified CY medium (Bobrowski et al., 1973). This medium was superior in yields of β-lactamase to any other media tested, including Nutrient Broth, Penassay Broth, Brain Heart infusion, and the original CY medium prepared as described by Novick (1962). Yields of up to 22.5 β-lactamase units per ml of culture, or 5.6 units per mg dry wt. of bacteria, were obtained.

When small amounts of material were required, an overnight culture of the organism was diluted 1: 100 with a fresh medium and incubated with shaking at 37°C for 6–8 hrs. Alternatively, the cultures were incubated without shaking for 24 hrs. The resulting cultures were then centrifuged at 5,000 g for 20 min. and the pellets washed with a 0.9% NaCl solution.

For production of β-lactamase on a larger scale, the bacteria were grown at 37°C in two 10-litre fermentors, each containing 5.5 litres of the medium. The cultures were aerated with 150 litres of air/hr and stirred at 350 rev./min. Foaming was controlled by addition of 5 ml of "Silicone 439" antifoam emulsion. After 5 hrs of incubation the bacteria were collected by centrifugation (5,000 g, 1 hr, 4°C), washed twice with 0.9% NaCl solution and stored at 4°C until required.

Ultrasonic disintegration. The washed cells were resuspended in 0.1-M phosphate buffer, pH 7.0, and ultrasonically treated in a Branson B12 disintegrator at 20 kHz for 3 min. Cell debris was removed by centrifugation, first at 16,000 g for 30 min. at 4°C, and then at 44,000 g for 60 min. at 4°C.

Osmotic shock and spheroplast formation. Osmotic shock was performed according to Neu (1968). Spheroplast formation was carried out as described by Lindstrom et al. (1970).

Enzyme assays. Absolute activities of the β-lactamase preparations against benzylpenicillin, cephalothin and other substrates were determined both by a modification of the hydroxylamine method (Bobrowski and Borowski, 1971) and by the iodometric method of Perret (1954).

For kinetic measurement of the enzyme reaction either the acidimetric indicator method as previously described by Bobrowski and Borowski (1971) or the microiodometric method of Novick (1962), modified by Dale and Smith (1971) was used.

The amount of enzyme required to hydrolyse 1 μmole of benzylpenicillin in 1 hr at pH 7.0 and 30°C was taken as 1 unit (Pollock and Torriani, 1953).

Protein estimations. The protein content of the samples was roughly estimated by measuring the extinction at 280 nm. For more accurate determinations the method of Lowry et al. (1950) was used. Crystalline serum albumin was used as a standard.

Preparation of columns. Sephadex G-75, Sephadex G-100, and GM-cellulose were prepared and packed in columns as recommended by the manufacturers. The adsorbents for affinity chromatography were prepared as follows: Commercial CNBr-activated Sepharose 4 B, 5 g, was suspended in 1-mM HCl solution and washed several times with the same solution. The swollen gel was then resuspended in 20 ml of 0.1-M $NaHCO_3$ solution containing 0.5-M NaCl and either 250 mg of potassium cephalosporin C or 200 mg of ε-aminocapronyl derivative of 6-aminopenicillanic acid was added into the suspension. The reaction mixture was gently stirred at room temperature for 2 hrs and the derivatized Sepharose gel was extensively washed with 0.01-M phosphate buffer, pH 7.0, to remove any excess of uncoupled β-lactam antibiotic, and finally poured into a small column (0.9 × 12 cm).

Determination of molecular weight. The approximate molecular weight of the enzyme was determined by gel filtration on Sephadex G-75 (superfine grade) column (Andrews, 1964). Cytochrome C (mol. wt. 12.400), pancreatic ribonuclease (mol. wt. 13,700), chymotripsinogen A (mol. wt. 25,000), horse-radish peroxidase (mol. wt. 40,100) and ovalbumin (mol. wt. 45,000) were used as internal markers in different runs. The void volume was estimated using Blue Dectran 2000.

Polyacrylamide disc electrophoresis. Analytical disc electrophoresis was performed at pH 8.7 (Tris — glycine buffer) for 2 hrs with 7.5% acrylamide gel to which soluble starch (0.2%, w/v) was added. The voltage used was 120 V, and the current was about 5 mA per tube.

Protein bands were located with 1% Amido Black in 7% (w/v) acetic acid. The enzyme band was located by the staining method described by Ross and Boulton (1973).

Thin layer isoelectrophocusing. Glass plates (9 × 10 cm) covered with 1-mm layer of 5% polyacrylamide gel containing Ampholine carrier ampholytes (pH range 3–10) at a final concentration of 2% were prepared according to Vesterberg (1973). Platinum electrodes were connected to the gel surface through filter paper soaked in 1-M H_3PO_4 solution (anode) or in 1-M NaOH solution (Cathode), and then 10 μl samples of 0.2% β-lactamase solution were applied onto the gel at two different points close to each electrode. The isoelectrofocusing was performed at 4°C in an electrophoretic chamber, first for 1 hr at a voltage increasing from 300–600 V, and then for another 7 hrs at a constant voltage of 600 V. The current was about 80 mA at the start, and about 10 mA at the end of focusing. After the run, the pH gradient was determined at points 10 mm, apart along the gel and the plates were overlayed with 1% agarose gel incorporating 0.1-M phosphate buffer, pH 7.0; benzylpenicillin, 20 mM; soluble starch, 0.2 % (w/v) and Lugol's solution, to detect the β-lactamase band.

CELLULAR LOCATION OF THE ENZYME

Preliminary experiments showed that the R22K β-lactamase, when harbouring in *E. coli* K12 host, is entirely cell-bound, and practically no enzyme could be detected in the culture supernatants. In an attempt to obtain more detailed information as to whether the enzyme is located in, or on, the bacterial cell, the organisms were subjected to an osmotic shock procedure and were also spheroplasted by EDTA — lysosyme treatment. The release of a cell-bound enzyme into the surrounding medium by either of these procedures is regarded as evidence of the surface location of its activity, whereas failure to liberate the enzyme suggests that it is situated in the cytoplasm (Neu, 1968; Lindstrom et al., 1970; Smith and Wyatt, 1974).

It was established that the R22K enzyme, unlike the TEM-like β-lactamases, was only partly liberated by the osmotic shock procedure, and not more than 50% of the enzymic activity was found in the post-shock fluids. Nevertheless, it is believed that this enzyme is located at the cell envelope, presumably in the so-called "periplasmic space", since it was found to be almost completely released by the EDTA — lysozyme treatment.

PURIFICATION OF THE ENZYME

Two different procedures were used to obtain purified preparations of the R22K β-lactamase. The results of the first purification procedure are summarized in Table 1.

TABLE 1

Purification of R-22K β-lactamase (Procedure A)

Step	Total activity (units × 10^3)	Total protein (mg)	Specific activity (units/mg)	Recovery (%)	Purification index
1. Ultrasonic disintegration; centrifugation	140.4	15,600	9.0	100.0	1.0
2. Streptomycin sulphate treatment; dialysis	99.5	8,500	11.7	70.5	1.3
3. CM-cellulose column	68.6	330	209.0	48.5	23.2
4. Sephadex G-100 column	49.4	51	970.0	35.0	107.7

The washed cells, 200 g (wet weight), were resuspended in 400 ml of 0.1-M phosphate buffer, pH 7.0, and the organisms were then subjected to ultrasonic treatment in a Branson sonifier for 3 min. Cell debris was removed by centrifugation, and the supernatant fluid was treated with streptomycin sulphate (50 mg/ml) to separate nucleic acids. After standing at 0°C for 6 hrs the precipitate formed was centrifuged at 8,000 g for 10 min. at 4°C. The supernatant was dialyzed three times, each time for 12 hrs, against 5 litres of 5-mM phosphate buffer, pH 6.0. A slight precipitate formed during this step was removed by centrifugation at 4,800 g for 15 min. The resulting crude enzyme solution was applied to a CM-cellulose column (3,2 × 34 cm) previously equilibrated with 0.01-M phosphate buffer, pH 6.0. The column was washed with 750 ml of the same buffer to remove any unadsorbed material. A combined pH — salt concentration gradient formed from 360 ml each of the 0.01-M phosphate buffer pH 6.0, and a 0.1-M phosphate buffer, pH 7.0, was then applied, and 18-ml fractions were collected at a flow rate of about 30 ml/cm²/hr. Six fractions containing β-lactamase at the highest specific activity were pooled, dialyzed against 5 litre of water and freeze-dried. The material was reconstituted to 10 ml and the sample was loaded into a Sephadex G-100 column (2,6 × 35 cm) which had been preequilibrated against a 0.1-M phosphate buffer, pH 7.0. The column was eluted with the same buffer at a flow rate of 5 ml/cm²/ml, and the fractions (5 ml) were collected. Eight fractions containing the enzyme were pooled, dialyzed overnight against a 5-mM phosphate buffer, pH 7.0, and the non-diffusible material freeze-dried. The specific activity of the preparation obtained after this stage was 970 units/mg protein.

Although over 100-fold purification of the β-lactamase was achieved using this procedure, the preparation was still impure and gave several bands on analytical polyacrylamide electrophoresis. Attempts to purify it further were unsuccessful, because the enzyme, which is very unstable under various conditions, underwent partial denaturation.

Another purification procedure involved release of the enzyme by spheroplast formation, dialysis, and affinity chromatography on derivatized agarose gel.

The cells obtained from 1-litre culture of *E. coli* K12 strain carrying the R22K plasmid were suspended in 50 ml of a 0.03-M Tris-HCl buffer, pH 7.8, containing 0.5-M sucrose and 1-mM EDTA, and then lysozyme was added to a concentration of 40 μg/ml. The suspension was incubated at room temperature for 10 min. and the spheroplasts formed was removed by centrifugation at 11,000 g for 20 min. at 4°C. The pellet was resuspended in 50 ml of Tris — sucrose — EDTA solution, and the operation was repeated. The supernatants obtained after the first and second treatments were combined and dialyzed overnight against a 0.01-M phosphate buffer, pH 7.0. A slight precipitate formed during this step was removed by centrifugation at 16,000 g for 15 min. at 4°C. It is noteworthy that the specific activities of the preparations after this stage were from 36 to 52 units/mg protein, i.e. they were about 3 to 5 times higher than that of the crude enzyme preparation obtained after ultrasonic treatment.

The enzyme solution was then passed through the derivatized Sepharose column, the column was washed with 50 ml of 0.01-M phosphate buffer, and any material adsorbed was eluted with a 0.1-M phophate buffer, pH 7.0, containing 1-M NaCl.

Of the two adsorbents used for affinity chromatography much better results were obtained with the Sepharose gel to which 6-(ε-aminocapronylamido)-penicillanic acid was coupled. About 90% of the β-lactamase activity applied to the column was adsorbed at the first step of the procedure, and the maximal specific activity in fractions eluted from the column at the final step was about 2,600 units/mg protein. The amount of enzyme protein recovered after affinity chromatography was, however, insufficient to enable any evaluation of purity. Concentration of the pooled eluates was avoided because of the risk of denaturation of the enzyme.

With the cephalosporin C — Sepharose gel matrix, which were alternatively used, less than 10% of the β-lactamase sample applied into the column was adsorbed. The reason for this poor adsorption was probably that the affinity of the coupled ligand for the R22K β-lactamase active centre was too low, or that a partial hydrolysis of the ligand occurred while the chromatography still proceeded.

PROPERTIES OF THE ENZYME

Physicochemical properties. The molecular weight of the enzyme was calculated from the elution volumes, which were determined in two independent runs on Dephadex G-75 column (2,6 × 35 cm). The value of 32,050 ± 1050 was obtained, and this is higher than those reported for most R plasmid-mediated lactamases, except for the R46 enzyme (Dale and Smith, 1971; Bobrowski, 1974).

During polyacrylamide disc electrophoresis, which was performed at pH 8.7, the R22K β-lactamase moved slowly toward the anode. Preliminary data obtained by using the thin layer isoelectrofocusing technique indicated that the isolelectric point of the enzyme is about 7.5.

The pH — activity curve of the enzyme (Fig. 1) was determined over a range of pH values from 4.4 to 9.0 using 4-fold diluted Mc Ilvaine's citrate — phosphate or 0.05-M borate buffer mixtures. The optimum pH was 7.6, with the 90% limits of activity being

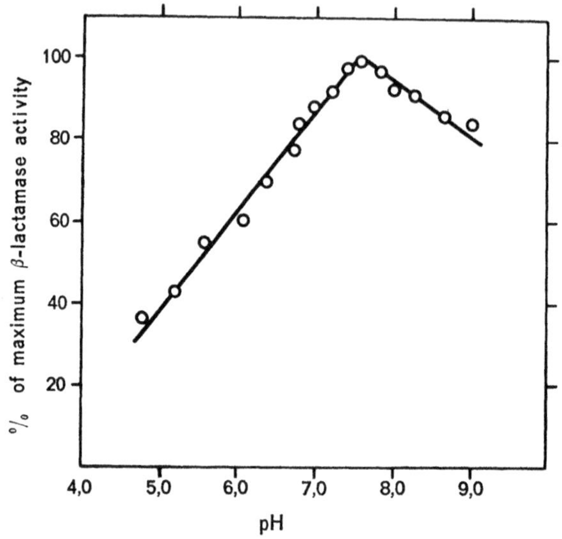

Fig. 1. Effect of pH on the activity of R22K β-lactamase

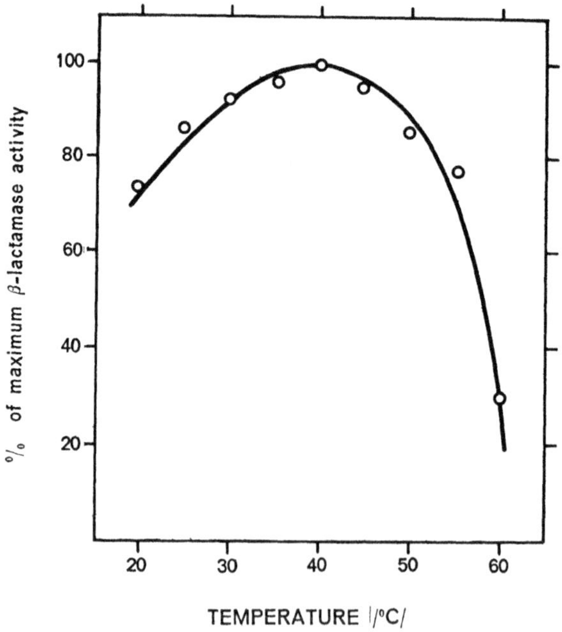

Fig. 2. Effect of temperature on the activity of R22K lactamase

7.1–8.3. The activity decreased more sharply at the acidic pH values, and practically no activity could be detected at pH 4.4.

The temperature — activity dependence of the R22K enzyme (Fig. 2) showed a diffuse optimum at about 40°C and a steep decline of activity at temperatures above 50°C.

The rate of heat inactivation of the enzyme was examined by measuring the activity of enzyme solutions heated in a water bath at 60°C for various time periods and then

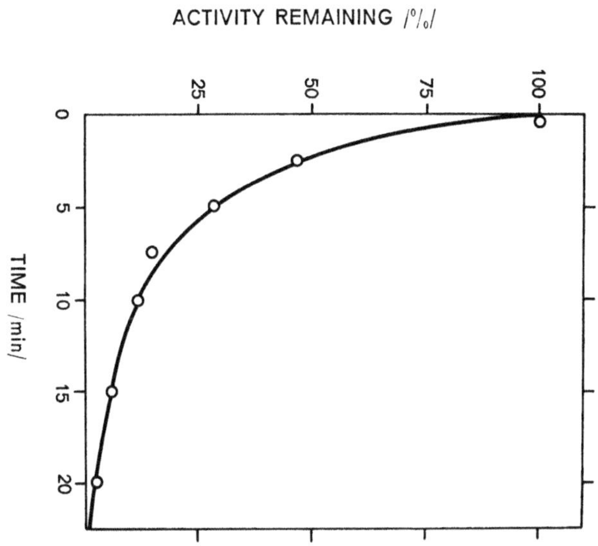

Fig. 3. Rate of heat inactivation of R22K β-lactamase at 60°C

TABLE 2

Substrate Profile and Apparent Affinity Constants of R22K β-lactamase

Substrate	Maximal rate of hydrolysis (relative)	Apparent affinity constant
Benzylpenicillin	100	6.5×10^{-5}M
Phenoxymethylpenicillin	40	
Ampicillin	2.8	
Amoxycillin	1.9	
Carbenicillin	0.8	1.7×10^{-7}M*)
Ticarcillin (BRL-2288)	0.7	
Methicillin	0.8	2.0×10^{-6}M*)
Cloxacillin	0.6	7.5×10^{-9}M*)
Cephalosporin C	350	4.4×10^{-4}M
Cephaloridine	285	5.9×10^{-4}M
Cephalothin	450	2.0×10^{-4}M
Cephacetrile	1050	6.0×10^{-3}M
Cefazolin	1150	6.3×10^{-3}M
Cephalexin	35	

*) Ki values

assayed at 30°C. As shown in Fig. 3, the treatment for 2.5 min. resulted in an over 50% loss of the original activity, and heating for 20 min. was sufficient to inactivate the enzyme almost completely. A comparison of these results with those reported for other R plasmid specified β-lactamases (Yamagishi et al., 1969; Rubin and Smith, 1973) indicated that the R22K enzyme is much more labile than the TEM-like β-lactamases, being, however, a little less labile than the oxacillin-hydrolysing enzymes.

Substrate specificity. The relative rates of hydrolysis of various substrates by the R22K enzyme are given in Tab. 2, together with the kinetic constants for some penicillins and cephalosporins. The Michaelis constants were obtained from the Lineweaver-Burk (1934) plots, and the inhibition constants were derived by the method of Dixon (1953).

It can be seen that all the cephalosporins, except cephalexin, were hydrolyzed at much higher rates than the penicillins. The affinity of the latter for the R22K active centre, however, was found to be higher by one to even six orders of magnitude than the affinity for cephalosporins substrates.

The highest maximal rates of hydrolysis were observed for cefazolin, cephacetrile, and cephalothin. It should be noted, however, that the K– value for cephalothin was about 30 times lower than the respective values for cefazolin and cephacetrile. Therefore, at low substrate concentrations the efficiency of the enzyme in hydrolyzing cephalothin was much greater than in hydrolyzing cefazolin or cephacetrile.

TABLE 3

Effect of Various Inhibitors on the Activity of R22K β-lactamase

Inhibitor	Concentration (mM)	Inhibition (%)
Iodine	0.01	87
	0.005	11
Tetranitromethane	1.0	85
	0.1	36
Difluorodinitrobenzene	1.0	10
p-Chloromercuribenzoate	2.0	27
	0.5	—*)
Iodoacetamide	1.0	—
N-Ethylmaleimide	1.0	—
Phenylmethylsulfonylfluoride	1.0	—
Ethylenediaminotetraacetate	1.0	—

*) no inhibition at given concentration

Ampicillin, carbenicillin, and their analogues, amoxycillin and ticarcillin, were very poor substrates for the enzyme, and they were highly resistant to hydrolysis, as were methicillin and cloxacillin. The latter penicillin derivatives were found to be very strong inhibitors of the hydrolysis of cephalothin by the R22K β-lactamase. A comparison of the K_i values determined for cloxacillin and methicillin leads to the conclusion that the former is able to inhibit the enzyme at much lower concentrations. This contrasts with the situation observed with the TEM-like-lactamases which, as a rule, are more sensitive to inhibition by methicillin (Bobrowski, 1974).

Effect of inhibitors. Various organic and inorganic compounds were examined for their possible inhibitory effect on the activity of R22K β-lactamase. The results are presented in Table 3.

An about 85% inhibition of the β-lactamase activity was observed when the enzyme was exposed to iodine and tetranitromethane used at concentrations of 10 μM and 1mM respectively. Both these inhibitors are known to react with tyrosyl residue in protein molecules. It is therefore suggested that this residue constitutes a part of the catalytical active centre of the R22K β-lactamase.

Other inhibitors tested, including p-chloromercuribenzoate, had very little, or no, effect on the enzyme activity.

CONCLUSIONS

The R22K β-lactamase is unique among R plasmid determined β-lactamases so far examined. It could be easily distinguished from them by its substrate profile, electrophoretic behaviour and molecular weight. However, the R22K enzyme resembles chromosomally-mediated β-lactamases of *E. coli*, in particular those which have been characterized by Lindstrom et al. (1970) and by Dale and Smith (1971). Direct comparison will show the extent of the similarity or possible identity of the R22K enzyme with the *E. coli* chromosomal β-lactamase.

REFERENCES

ANDREWS, P. (1964). Estimation of the molecular weights of proteins by Sephadex gel-filtration. Biochem. J., **91**, 222–233.

BOBROWSKI, M. M. (1974). Bacterial beta-lactamases and their significance in resistance to penicillins and cephalosporins. Post. Hig. Med. Dośw. (Wrocław) (in press).

BOBROWSKI, M. M., and BOROWSKI, E. (1971). Interaction between carbenicillin and β-lactamases from Gram-negative bacteria. J. Gen. Microbiol., **68**, 263–272.

BOBROWSKI, M. M., GEBICKA, K., and BOROWSKI, J. (1973). Combined action of β-lactam antibiotics against Gram-negative bacilli. J. Hyg. Epidemiol. Microbiol. Immunol. (Prague) **17**, 129–140.

DALE, J. W., and SMITH, T. J. (1971a). The purification and properties of the β-lactamase specified by the resistance factor R-1818 in *Escherichia coli* and *Proteus mirabilis* Biochem J.,. **123**, 493–500.

DALE, J. W., and SMITH, J. T. (1971b). Some relationships between R factor and chromosomal β-lactamase in Gram-negative bacteria. Biochem. J., **123**, 507–512.

DALE, J. W., and SMITH, J. T. (1972). A direct comparison of two unusual R factor-mediated β-lactamases. Biochem. J., **128**, 173–174.

DATTA, N., and KONTOMICHALOU, P. (1965). Penicillinase synthesis controlled by infectious R-factors in Enterobacteriaceae. Nature, **208**, 239–241.

DIXON, M. (1953). The determination of enzyme inhibitor constants. Biochem. J., **55**, 161–170.

HEDGES, R. W., DATTA, N., KONTOMICHALOU, P., and SMITH, J. T. (1974). Molecular specificities of R factor determined β-lactamases: Correlation with plasmid compatibilities. J. Bacteriol., **114**, 56–62.

KONTOMICHALOU, P. (1971). R-factor carrying genes controlling resistance to the penicillins. Habilitation thesis. University of Athens.

LINDSTROM, E. B., BOMAN, H. G., and STEELE, B. B. (1970). Resistance of *Escherichia coli* to penicillins. VI. Purification and characterization of the chromosomally-mediated penicillinase present in amp A – containing strains. J. Bacteriol. **101**, 218–231.

LINEWEAVER, H., and BURK, D. (1934). The determination of enzyme dissociation constants. J. Amer. Chem. Soc. 56, 658.

LOWRY, O. H., ROSEBROUGH, N. J., FARR, A. L., and RANDALL, R. J. (1951). Protein measurement with the Folin phenol reagent. J. Biol. Chem., **193**, 265.

NEU, H. C. (1968). The surface localization of penicillinase in *Escherichia coli* and *Salmonella typhimurium*. Biochem. Biophys. Res. Comm. **32**, 258–263.

NOVICK, R. P. (1962). Micro-iodometric assay for penicillinase. Biochem. J. **83**, 236–240.

NOVICK, R. P. (1963). Analysis by transduction of mutations affecting penicillinase formation in *Staphylococcus aureus*. J. Gen. Microbiol. **33**, 121–136.
PERRET, C. J. (1954). Iodometric assay of penicillinase. Nature, **174**, 1012–1013.
POLLOCK, M. R., and TORRIANI, A. M. (1953). Purification et caracteristiques physicochimiques de la penicillinase de *Bacillus cereus*. C.R. Acad. Sci. (Paris) **237**, 276.
RICHMOND, M. R., and SYKES, R. B. (1973). The β-lactamases of Gram-negative bacteria and their possible physiological role. Advan. Microb. Physiol. 2, 31–88.
ROSS, G. W., and BOULTON, M. G. (1973). Purification of β-lactamases on QAE-Sephadex. Biochim. Biophys. Acta, **309**, 430–439.
RUBIN, F. A., and SMITH, D. H. (1973). Characterization of R factor β-lactamases by the acidimetric method. Antimicrob. Ag. Chemother., 3, 68–73.
SAWAI, T., TAKAHASHI, K., YAMAGISHI, S., and MITSUHASHI, S. (1970). Variant of penicillinase mediated by an R factor in *Escherichia coli*. J. Bacteriol. **104**, 620–629.
SMITH, J. T., and WYATT, J. M. (1974). Relation of R factor and chromosomal β-lactamase with the periplasmic space. J. Bacteriol. **117**, 931–939.
VESTERBERG, O. (1973). Isoelectric focusing of proteins in thin layers of polyacrylamide gel. Science Tools, The LKB Instrument Journal, **20**, 22–29.
YAMAGISHI, S., O'HARA, K., SAWAI, T., and MITSUHASHI, S. (1969). The purification and properties of penicillin β-lactamase mediated by transmissible R factors in *Escherichia coli*. J. Biochem. (Tokyo), **66**, 11–20.

COMPARISON OF THE ANTIBACTERIAL ACTIVITY OF BETA-LACTAM ANTIBIOTICS AND THEIR RESISTANCE TO HYDROLYSIS BY BETA-LACTAMASES

H. C. NEU

Division of Infectious Diseases, Department of Medicine, Columbia University, New York, U.S.A.

The clinical usefulness of both oral and parenterally administered cephalosporin antibiotics is well established. The various pharmacological problems associated with the clinical use of cephalosporins, such as a cephalothin and cephaloridine, have resulted in the development of new agents which differ in spectrum against both gram-positive and gram-negative bacteria. We have been interested in the role that beta-lactamases and type of beta-lactamase may contribute to the resistance of enteric bacteria to cephalosporins. We have compared the activity of new cephalosporins against a number of clinical isolates and investigated with certain strains the contribution of beta-lactamase to resistance.

MATERIALS AND METHODS

Sodium cephalothin, cephaloridine, cefazolin, cephalexin, cefamandole, cephanone were gifts from Eli Lilly. Cephacetrile was a gift of Ciba-Geigy and cepharim a gift of Bristol Laboratories. Carbenicillin, ticarcillin, and ampicillin were a gift of Beecham-Massengill Pharmaceuticals. Bacterial cultures were clinical isolates from patients hospitalized at the Columbia-Presbyterian Medical Center. Minimal inhibitory concentration (MIC) values were determined by an agar-dilution procedure utilizing 10^4 colony forming units (CFU) inoculated with a replicating device onto Mueller-Hinton agar. MIC values were determined after overnight incubation at 35C. Standardized disc susceptibilities were performed using the Bauer technique (Bauer et al., 1966).

Assays of beta-lactamase were performed using either a microiodometric technique (Neu et al. Winshell, 1970), or a spectrophotometric assay (Neu, 1971). Assays of destruction of antibiotic were performed using *Bacillus subtilis* as a test organism in a punched well assay method (Neu, 1971). Beta-lactamases used were purified by previously described methods or by use of sonic extracts of bacteria which had been subjected to centrifugation to remove cellular debris.

RESULTS AND DISCUSSION

Table 1 illustrates the MIC values obtained with eight cephalosporins tested against *E. coli*. Differences can be noted in the MIC values at low concentrations. Cefazolin, cephanone, and cefamandole are much more active against *E. coli* at a concentration of

TABLE 1

Minimum Inhibitory Concentrations of Various Cephalosporins against Escherichia coli

Antibiotic	MIC (μg/ml) *0.8*	*1.6*	*3.1*	*6.2* % Sensitive	*12.5*	*25*	*50*
Cefoxitin		4	12	67	100		
Cephaloridine	4	12	21	87			100
Cephalexin			4	8	95	100	
Cephacetrile			4	33	83	100	
Cephalothin	4		17		62	95	100
Cefazolin	4	8	87	95		100	
Cephanone	8	28	95			100	
Cefamandole	23	62	89		100		

TABLE 2

Hydrolysis of Cephalosporins by a Purified E. coli (TEM type) Beta-Lactamase

Substrate	*μMole/min* hydrolyzed*
Cephalothin	4.1
Cephaloridine	18.4
Cephaloglycin	3.5
Cephalexin	0.3
Cephanone	2.8
Cephacetrile	2.2
Cefoxitin	0.2
Cefamandole	0.1

* Equimolar solutions of cephalosporins were prepared in 0.05M, pH 7.0 potassium phosphate buffer. Change in extinction at 255 A° was followed and μ Mole hydrolyzed per minute.

3.1 μg/ml, but at 12.5 μg/ml essentially all of the *E. coli* are inhibited. Many of these strains were resistant to ampicillin and possessed Richmond type III (TEM) beta-lactamases. We thus investigated whether the differences that are seen could be due to the beta-lactamase. Table 2 shows the comparison of the hydrolysis of the cephalosporins by the *E. coli* beta-lactamase. The rate of hydrolysis of cephaloridine by this enzyme is equal to the rate of hydrolysis of ampicillin. There is a suggested effect in that cephaloridine and cephalothin, which have higher MIC values, are better substrates for the enzyme, but the cephanone which is highly active against *E. coli* is also a good substrate and cephalexin, which is one of the poorest substrates, has much higher MIC values than does cephanone.

We studied another group of *E. coli* strains that were sensitive to carbenicillin and ticarcillin, but resistant to cephalosporins such as cephalothin. Table 3 shows that the relative rate of hydrolysis of the cephalosporin or penicillin does not correlate closely with the MIC. For example, the rate of hydrolysis was 0.1 for both cefoxitin and cephanone, but the MIC was 17 times greater for cefoxitin. These results suggest that it is not possible to directly correlate inhibitory levels of cephalosporins with their beta-lactamase activity. We further investigated whether the location in the cell of the beta-lactamase had an effect on resistance (Neu, 1969). EDTA treated or disrupted cells did not alter activity against the cephalosporins.

Klebsiella characteristically are highly susceptible to inhibition by cephalosporins. All of the strains in Table 4 are ampicillin and carbenicillin resistant with MIC values in excess of 200 μg/ml, but the strains are readily inhbited by cephaloridine which is hydrolyzed by some of the strains. Indeed several strains degraded up to 50% of cephaloridine but were sensitive to the antibiotic. Cephalexin and cephacetrile have relatively higher MIC's than do the other agents, but they are not destroyed to any degree by the klebsiella when incubated with either intact or sonically disrupted organisms. Cephalothin is active against klebsiella as are cefoxitin and cefamandole which are the most resistant to hydrolysis. We were unable to find strains of *K. pneumoniae* resistant to cefoxitin or cefamandole. However, Onishi et al. (1974) did have six strains which were resistant to cephalothin and cefoxitin and did not hydrolyze either drug.

TABLE 3

Comparison of MIC Values of a Cephalosporin Resistant E. coli with Hydrolysis of Antibiotics

Antibiotic	*MIC* μg/ml	Relative *Hydrolysis*
Penicillin	2000	33.0
Ampicillin	125	2.6
Carbenicillin	25	3.5
Cephalothin	2000	1.6
Cephaloridine	2000	8.0
Cephanone	6	0.1
Cefamandole	50	0.2
Cefoxitin	100	0.1

TABLE 4

MIC Values of Cephalosporins Against Klebsiella

	MIC (μg/ml)				
Antibiotic	*1.6*	*3.2*	*6.2*	*12.5*	*25*
	% Sensitive				
Cefoxitin	9	85	100		
Cephaloridine	13	52	100		
Cephalexin	0	4	8	56	95
Cephacetrile	0	12	47	95	
Cephalothin	32	95			
Cefazolin	21	100			
Cephanone	28	92	96	100	
Cefamandole	88	88	100		

Comparison of the activity of the cephalosporins against *Proteus mirabilis* strains, Table 5, do not show any significant differences among the agents except for the higher activity of cefamandole. Of these strains only three produced beta-lactamases which were of the TEM type due to R-factors. The MICs against these three strains were the same as against strains without beta-lactamase production.

A large number of salmonellae strains which were ampicillin resistant were investigated for resistance to cephalosporins. Table 6 demonstrates that all are very sensitive to the agents with little difference in the MIC values. Subsequently, six salmonella strains which were resistant to cephalothin were found. None of the strains transferred cephalosporin resistance to a suitable *E. coli* recipient although resistance to ampicillin and carbenicillin was transferred. All of the strains were resistant to cefamandole, cefazolin and cephalexin, but sensitive to cefoxitin and cephanone. This resistance correlated with their ability to hydrolyze these antibiotics.

TABLE 5

MIC Values of Cephalosporins against Proteus Mirabilis

	MIC (μg/ml)				
Antibiotic	*1.6*	*3.1*	*6.2*	*12.5*	*25*
	% Sensitive				
Cefoxitin	4	17	65	82	87
Cephalexin	4	4	14	81	86
Cephalothin	20	28	71	84	88
Cefazolin	4	52	67	81	90
Cefamandole	77	82	87	87	91
Cephanone	4	21	57	77	90

TABLE 6

MIC of Cephalosporins Against Salmonella

	MIC (μg/ml)			
Antibiotic	*0.4*	*0.8*	*1.6*	*3.2*
	% Sensitive			
Cefoxitin		12	87	100
Cephalexin		4	17	100
Cephalothin		12	83	100
Cefazolin		96	100	
Cefamandole	96	100		
Cephanone	87			100

The other organisms investigated in this study were indole-positive Proteae, Citrobacter, Serratia, Enterobacters, and Herelleae. Table 7 gives a comparison of the activity of penicillins and cephalosporins against these organisms. The data could suggest that resistance to beta-lactamase activity contributes to the activity of the newer cephalosporin agents cefoxitin and cefamandole. However, cephalexin is also a poor substrate for the enzymes produced by these bacteria. Carbenicillin is a poor substrate for the enzymes, but is a very effective inducer of beta-lactamases in the indole positive Proteae, Providence and Enterobacter strains. One of the *Proteus morganii* strains had a cefamandole MIC of 50 μg/ml, hydrolyzed 8.6 μM/min of cefamandole, but the cephalothin MIC was 1000 μg/ml and only 0.63 μM/min was hydrolyzed. Furthermore, a number of Enterobacter and Serratia strains from this group which were resistant to cefoxitin were unable to destroy the antibiotic in the presence of intact cells or sonically disrupted cells. Cefoxitin acts as a competitive inhibitor of the hydrolysis of penicillins so that it binds to the beta-lactamase although it is not destroyed.

Sensitivity to carbenicillin in the case of Enterobacter and Serratia seems to indicate that a strain may be susceptible to cefoxitin and to cefamandole. In most instances the organisms do not hydrolyze cefoxitin. In contrast with cefamandole it appears that growth resumes when the antibiotic is hydrolyzed by beta-lactamase released from lysed cells or following induction of the enzyme. For example, a strain of Serratia in the first three hours of incubation with cefamandole falls from 1.5×10^6 CFU to 3×10^5 CFU, but in the next three hours as 75% of the cefamandole is destroyed the CFU increases to 3×10^8. A similar situation occurs with a resistant *P. morganii*.

These studies indicate that knowledge of the beta-lactamase activity of a particular strain of bacteria do not always predict the activity of a particular cephalosporin against the organism. In the case of *E. coli*, *P. mirabilis*, Klebsiella and Salmonella, there is

TABLE 7

Comparison of Activity of Agents Against Cephalothin Resistant Enteric Bacteria

Organism	*Ampicillin*	*Carbenicillin*	*Cephathin*	*Cephalexin*	*Cephacetrile*	*Cephanone*	*Cefoxitin*	*Cefamandole*
	Antibiotic % Sensitive Strains							
P. morganii	0	90	0	0	0	14	70	60
P. rettgeri	0	70	0	0	0	14	60	60
P. vulgaris	0	80	0	0	0	0	85	90
Providence	0	80	0	0	0	10	50	50
Citrobacter	16	74	5	5	5	20	31	40
Herellea	6	80	5	5	5	25	50	12
Enterobacter	3	75	0	0	0	40	33	43
Serratia	0	40	0	0	0	10	35	23

TABLE 8

Antibiotic Susceptibility of Clinical Isolates and Capacity to Degrade Antibiotics

Organism	*Ampicillin*	*Carbenicillin*	*Cephaloridine*	*Cephalothin*	*Cefoxitin*
	Susceptibility *Percent Antibiotic Hydrolyzed*				
E. coli	*R* 100	*R* 10	*S* 100	*R* 10	*S* 0
E. coli	*R* 20	*S* 2	*R* 80	*R* 100	*R* 0
E. cloacae	*R* 0	*R* 0	*R* 0	*R* 0	*S* 0
E. cloacae	*R* 70	*R* 10	*R* 100	*R* 65	*R* 0
S. marcescens	*S* 0	*S* 0	*R* 0	*R* 5	*S* 0
P. morganii	*R* 100	*R* 100	*R* 100	*R* 100	*R* 20
P. rettgeri	*R* 20	*R* 10	*R* 20	*R* 20	*R* 5

fairly good correlation of beta-lactamase activity and sensitivity. With Enterobacter and Serratia this is not seen (Table 8). Previous studies which have shown the lowering of MIC values in the presence of ethylenediaminetetraacetic acid suggest permeability is important (Hamilton-Miller, 1965, Winshell a. Neu, 1970). We have not seen this to be true with the strains of *P. morganii* and Enterobacter which are resistant to cefoxitin but fail to hydrolyze it. Studies by our group of cell wall lipid in strains of Serratia that differ in sensitivity to carbenicillin and cefoxitin also fail to show any significant differences (Winshell a. Neu, 1974). Table 8 demonstrates clearly the problems of predicting the susceptibility patterns of organisms, since in several of the organisms sensitivity would have been predicted on the basis of the beta-lactamase activity known to be present. It is clear that further study of the "so-called" intrinsic resistance of these organisms is necessary before we can suggest molecular changes in penicillins or cephalosporins which will increase their range of activity.

REFERENCES

BAUER, A. W., KIRBY, W. M. M., SHERRIS, J. C. and TURCK, M. (1966). Antibiotic susceptibility testing by a standardized single disc method. Amer. J. Clin. Path. 45, 493.

HAMILTON-MILLER, J. M. T. (1965). Effect of EDTA upon bacterial permeability to benzylpenicillin. Biochem. Biophys. Res. Comm. 20, 688.

NEU, H. C. (1969). Effect of B-lactamase location in *Escherichia coli* on penicillin synergy. Appl. Microbiol. 17, 783.

NEU, H. C. (1971). B-lactamase production by *Pseudomonas aeruginosa*. Antimicrob. Ag. Chemother. 1970, p. 534.

NEU, H. C. and WINSHELL, E. B. (1970). Purification and characterization of penicillinases from *Salmonella typhimurium* and *Escherichia coli*. Arch. Biochem. Biophys. 139, 278.

ONISHI, H. R., DAOUST, D. R., ZIMMERMAN, S. B., HENDLIN, D. and STAPLEY, E. O. (1974). Cefoxitin, a semisynthetic cephamycin antibiotic: Resistance to beta-lactamase inactivation. Antimicrob. Ag. Chemother. 5, 38.

RICHMOND, M. H., JACK, G. W. and SYKES, R. B. (1971). The B-lactamases of gram-negative bacteria including Pseudomonads. Ann. N.Y. Acad. Sci. 182, 243.

WINSHELL, E. B. and NEU, H. C. (1970). Relation of lipopolysaccharide and fatty acid ester release to the ethylenediaminetetraacetic acid, alteration of permeability in Enterobacteriaceae. J. Bacteriol. 102, 537.

WINSHELL, E. B. and NEU, H. C. (1974). Relation of cell wall lipid content of *Serratia marcescens* to resistance to antimicrobial agents. Antimicrob. Ag. Chemother. 6, 73.

II.

AMINOGLYCOSIDE ANTIBIOTICS AND ENZYMES INVOLVED IN THEIR INACTIVATION

CHAIRMEN: H. UMEZAWA (JAPAN) and J. DAVIES (U.S.A.)

OPENING LECTURE

HISTORY AND DEVELOPMENT OF BIOCHEMICAL MECHANISM OF RESISTANCE TO AMINOGLYCOSIDIC ANTIBIOTICS AND THEIR ACTIVE DERIVATIVES

H. UMEZAWA

Institute of Microbial Chemistry, Kamiosaki, Shinagawa-ku, Tokyo, and Department of Antibiotics, National Institute of Health Kamiosaki, Shinagawa-ku, Tokyo

At present, in addition to searching for new and useful antibiotics, chemical transformations of known antibiotics to more useful agents are under study, and the biochemical mechanisms of resistance provide a rational approach to the development of effective agents useful in treating resistant infections.

In 1961, Miyamura, Department of Bacteriology, School of Medicine, Niigata University, reported that resistant dysentery bacilli isolated from patients inactivated chloramphenicol, and in 1965, Okamoto and Suzuki, Department of Chemistry, National Institute of Health, Tokyo, found that *E. coli* carrying R factor produces an intracellular enzyme which acetylates and inactivates chloramphenicol in the presence of acetyl CoA. They also suggested the presence of another enzyme which transfers acetyl group to kanamycin from acetyl CoA. I isolated the reaction product of the enzyme of this strain (*E. coli* K12 R5) and determined its structure to be 6′-N-acetylkanamycin A (Umezawa, H., Okanishi, Utahara et al., 1967). This strain was resistant to kanamycins A and B, but not to kanamycin C. I studied the other strain (*E. coli* K12 ML-1629) which was obtained from Mitsuhashi and which was resistant to all kanamycins, and found that the enzyme of this strain transfers phosphate group of adenosine triphosphate to 3′-hydroxyl group of kanamycins (Umezawa, H., Okanishi, Kondo et al., 1967). I found also the presence of similar phosphatotransferases in resistant staphylococci (Doi, Ogura et al., 1968) and *Pseudomonas aeruginosa* (Doi, Miyamoto et al., 1968; Umezawa, H., Doi et al., 1968). Thus, the study of biochemical mechanisms of resistance to aminoglycosidic antibiotics was initiated in my laboratories, and we determined the structures of most of the reaction products of enzymes involved in resistance mechanisms.

The structures of the inactivated antibiotics thus determined suggested the structures of the derivatives which would not undergo the reactions with these enzymes and which would therefore remain active in inhibiting growth of resistant organisms. Compounds with these structures were synthesized and proved the direct relationship of the enzymes to the resistance. From these observations S. Umezawa and I were able to develop derivatives of kanamycins, ribostamycin and butirosins which are useful in treating resistant infections.

In this paper, I will review our studies on the biochemical mechanisms of resistance, development of active derivatives and methods of isolating the inactivated antibiotics and of determining their structures.

ISOLATION OF REACTION PRODUCTS OF ENZYMES INVOLVED IN MECHANISMS OF RESISTANCE

The enzymes which inactivate aminoglycosidic antibiotics are obtained in the supernatant of 100,000 g centrifugation of disrupted cells (Okanishi et al., 1967 and 1968) or in the osmotic shockate of resistant bacteria (Benveniste et al., 1971). These enzymes catalyse the reactions of aminoglycosidic antibiotics with adenosine triphosphate or with acetyl CoA, yielding phosphorylated, adenylated or acetylated products and these products are extracted and purified by column chromatography, thin layer chromatography or by electrophoresis. Recently, I found that, if the enzymes are immobilized, the reaction products are easily purified (Umezawa, H., Matsuhashi et al., 1974).

TABLE 1

Immobilized Enzymes and Their Activities

CBS = dried CNBr-Sepharose 4B CBC = dried CNBr-cellulose
pr. = protein activity = μmoles/g in the wet state/hr

Prepared from	Enzyme and activity
S-100 (1 ml, 16 mg pr.) of *E. coli* K12 J5 R11-2; 125 mg of CBS	Phosphotransferase I 9.1 μmoles of lividomycin A/g/hr 4.2 μmoles of kanamycin A/g/hr
S-100 (1 ml, 23 mg pr.) of *P. aeruginosa* TI-13a 125 mg of CBS	Phosphotransferase I 3.0 μmoles of lividomycin A/g/hr 1.7 μmoles of kanamycin A/g/hr Streptomycin phosphotransferase 0.12 μmole of streptomycin/g/hr
S-100 (1 ml, 70 mg pr.) of *E. coli* JR66/W677; 125 mg of CBS	Phosphotransferase II 33 μmoles of butirosin A/g/hr 20 μmoles of kanamycin A/g/hr Streptomycin phosphotransferase 1.6 μmoles of streptomycin/g/hr
S-100 (10 ml, 700 mg pr.) of *E. coli* JR66/W677; 10 g of CBC	Phosphotransferase II 1.2 μmoles of butirosin A/g/hr 0.9 μmole of kanamycin A/g/hr Streptomycin phosphotransferase 0.1 μmole of streptomycin/g/hr

Growing cells of *E. coli* K12 J5 R11-2, *E. coli* JR66/W677 and *P. aeruginosa* TI-13 were disrupted and the 100,000 g supernatant was dialysed against 20 mM phosphate buffer, pH 7.2. The enzymes in the dialysed solution were coupled with cyanogen bromide-activated Sepharose 4B or with cyanogen bromide-activated cellulose under stirring at 4°C for 16 hours and the solids with bound enzymes were washed with 1.0 M monoethanolamine, 0.1 M acetate buffer (pH 4.0) in 1.0 M sodium chloride and 0.1 M borate buffer (pH 8.0) in 1.0 M sodium chloride, successively. When 1 ml of the supernatant was coupled with 125 mg of cyanogen bromide-activated Sepharose 4B, the immobilized enzyme weighing from 1,300 mg to 1,870 mg in the wet state was obtained. The activities of the immobilized enzymes are shown in Table 1. Thus, kanamycin-neomycin phosphotransferases I and II and streptomycin phosphotransferase were immobilized, and the activities of these enzymes were more stable in the immobilized states than in the solutions.

Continuous enzyme reaction can be carried out on a column of an immobilized enzyme. Kanamycin-neomycin phosphotransferase II-Sepharose 4B (22 μmoles of kanamycin A/wet g/hr) prepared as described above was packed into a column (0.6 × 4.4 cm) and a solution (250 ml) which contained kanamycin B (121 mg, 250 μmoles), disodium adenosine triphosphate trihydrate (756 mg, 1 mmole), magnesium acetate (2.5 mmoles) and potassium chloride (15 mmoles) in 20 mM phosphate buffer, pH 7.8, was passed through the column at the rate of 4 ml/hr at 37°C. In this way, kanamycin B was completely phosphorylated and the phosphorylated product (115 mg) was isolated by adsorption on a column of Amberlite CG-50 resin in NH_4 form and elution with 0.1 N ammonia. Dihydrostreptomycin 3″-phosphate was easily crystallized after the phosphorylation on streptomycin phosphotransferase-Sepharose 4B column followed by adsorption on a column of Amberlite CG-50 resin in partial ammonium form (NH_4:H, 7:3) and elution with 0.5 N aqueous ammonia.

DETERMINATION OF STRUCTURES OF PRODUCTS OF THE ENZYME REACTIONS

The structure of kanamycin A was determined by chemical methods and also by X-ray analysis (Koyama et al., 1968). Following this X-ray structure, the structures of kanamycins, ribostamycin, butirosins, lividomycins and gentamicins are shown in Fig. 1.

Enzyme reactions involved in biochemical mechanisms of resistance yield 3′-phosphate of kanamycins, ribostamycin, neamine, paromamine (Kondo, Okanishi et al., 1968; Maeda et al., 1968) and butirosins, 5″-phosphate of lividomycins (Kondo, Yamamoto et al., 1972) and 3′,4′-dideoxyribostamycin, 2″-adenylyl products of kanamycins, 3′,4′-dideoxykanamycin B (Yagisawa et al., 1971; Yagisawa, Naganawa et al., 1972a) and gentamicins (Fig. 1–2), 6′-N-acetylated products of kanamycins, 3′,4′-dideoxykanamycin B, ribostamycin and neamine, 3-N-acetylated products of gentamicins (Brzezinska et al., 1972) and 2′-N-acetylated products of sisomicin (Chevereau et al., 1974) and 3′-deoxy-paromamine (Yamaguchi et al., 1974), 3″-0-adenylyl (Umezawa, H., Takasawa et al., 1968; Takasawa et al., 1968) or 3″-0-phosphoryl (Ozanne et al., 1969) product of streptomycin and dihydrostreptomycin (Fig. 2).

Structures of the inactivated antibiotics can be determined chemically, but as reported by us, with the exception of 3-N-acetylgentamicin C, can be determined solely by pmr with application of double or triple resonance technique. Signals of protons on C-1′ and C-1″ of kanamycin A in pmr spectrum can be easily identified, because they appeared at the lowest field comparing with those of the other protons. Irradiating these protons followed by irradiation of protons on C-2′ or C-2″, protons on C-3′ and C-3″ can be identified, because C-3″ binds to the amino groups and the proton on C-3″ should appear in a significantly higher field than that on C-3′. Thus, the signal of 3′-H of kanamycin A is shown to be located at δ 4.25 and this signal shifts by about 0.45 ppm to lower field in kanamycin A 3′-phosphate (Naganawa, Kondo et al., 1971). The structures of ribostamycin 3′-phosphate and butirosin A 3′-phosphate can be determined by a similar method based on the lower shift of 3′-H in the inactivated products. The structure of lividomycin A 5″-phosphate was also shown by pmr (Kondo, Yamamoto et al., 1972). In this case,

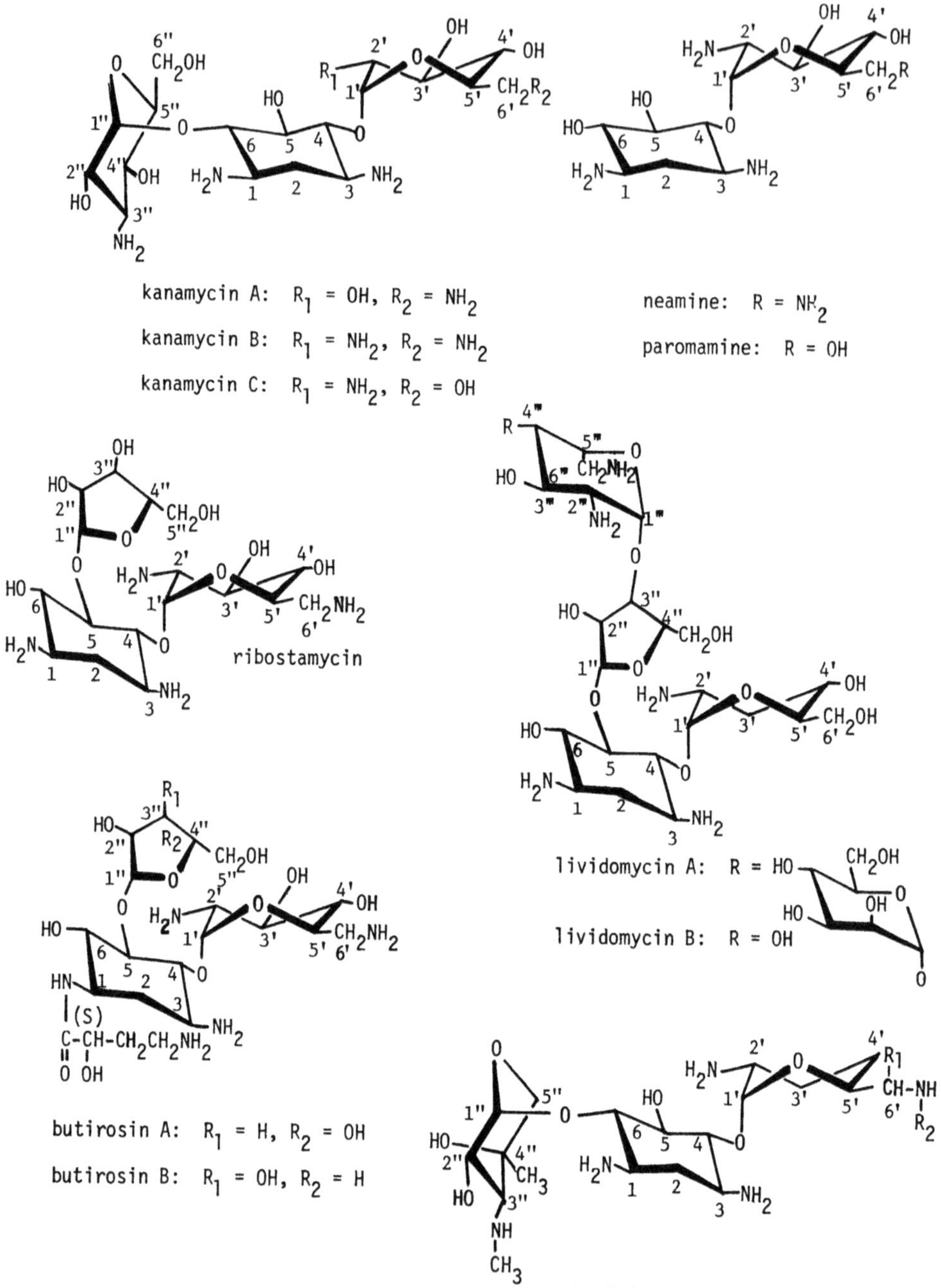
kanamycin A: R1 = OH, R2 = NH2
kanamycin B: R1 = NH2, R2 = NH2
kanamycin C: R1 = NH2, R2 = OH
neamine: R = NH2
paromamine: R = OH
ribostamycin
lividomycin A: R = HO-
lividomycin B: R = OH
butirosin A: R1 = H, R2 = OH
butirosin B: R1 = OH, R2 = H
gentamicin C1a: R1, R2 = H
gentamicin C1 : R1, R2 = CH3
gentamicin C2 : R1 = CH3, R2 = H

Fig. 1

5″-H signal shifts to lower field (δ 4.6–4.7) in the inactivated product. Lividomycin A 5″-phosphate was also confirmed by chemical synthesis (Yamamoto, Kondo et al., 1972a).

The structure of 3′,4′-dideoxykanamycin B 2″-adenylate can be also determined by pmr (Naganawa, Yagisawa et al., 1971). The adenylyl moiety can be shown by the presence of aromatic protons in such low fields as δ 8.63 and δ 8.85 and six protons of

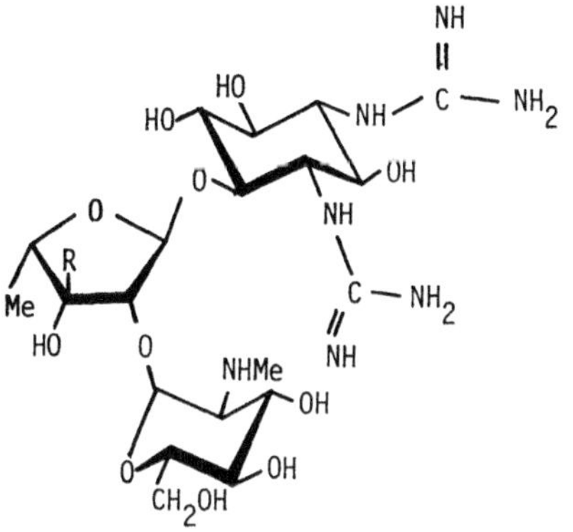

streptomycin: R = CHO

dihydrostreptomycin: R = CH_2OH

Fig. 2

ribose moiety. Similary as in the case of kanamycin A 3′-phosphate, by application of double resonance technique, it can be shown that the 2″-H signal of inactivated 3′,4′-dideoxykanamycin B is shifted by 0.34 ppm to lower field compared with 3′,4′-dideoxykanamycin B.

Application of pmr to structure determination of 6′-N-acetyl derivatives has been described on 6′-N-acetyl-3′,4′-dideoxykanamycin B (Yagisawa, Nagasawa et al., 1972b) and 6′-N-acetylribostamycin and 6′-N-acetylkanamycin A (Yamamoto, Yagisawa et al., 1972). N-Acetyl signal is seen in δ 2.43 and the signal of the C-6′ methylene protons at δ 3.15 in 3′,4′-dideoxykanamycin B shifts to δ 3.7 in the 6′-N-acetyl product.

Structures of streptomycin 3″-phosphate, dihydrostreptomycin 3″-phosphate or streptomycin 3″-adenylate can be determined also pmr (Naganawa, Kondo et al., 1971). The signal of 3″-H at δ 4.43 in the inactivated products of streptomycin or dihydrostreptomycin shifted to a lower field by about 0.4 ppm compared with streptomycin or dihydrostreptomycin.

Mass spectrometry of gentamicin C and 3′,4′-dideoxykanamycin B gives their molecular peaks and as first described by Brzezinska et al., 1972, mass spectrometry of gentamicin C provides the information for assigning an acetyl group to either one of the amino groups of the 2-deoxystreptamine moiety. As we reported (Umezawa, H., Yagisawa et al., 1973), 3-N-acetylgentamicin C_1 can be identified by isolating levorotatory 1-N-ethoxycarbonyl-2-deoxystreptamine after hydrolysis of tetra-N-ethoxycarbonyl derivative of 3-N-acetylgentamicin C_1.

Thus, at the present time, the structures of the products of the enzyme reactions involved in the mechanism of resistance can be rapidly determined by physical methods.

KANAMYCIN-NEOMYCIN PHOSPHOTRANSFERASES I AND II

Both kanamycin-neomycin phosphotransferases I and II transfer the terminal phosphate of adenosine triphosphate to the 3′-hydroxyl of kanamycins, ribostamycin, neamine, paromamine, probably 3′-hydroxyl of neomycins and paromomycins. Phosphotransferase I was purified by affinity chromatography carrying lividomycin A as the functional group, and we found that this enzyme also phosphorylates the 5″-hydroxyl group of ribose moiety of lividomycins and 3′,4′-dideoxyribostamycin (Umezawa, H., Yamamoto et al., 1973). Phosphorylation of lividomycin A by this enzyme is competitively inhibited by kanamycins. If ribostamycin is used as the substrate, this enzyme phosphorylates the 3′-hydroxyl group but not the 5″-hydroxyl group. These results suggest that in this enzyme reaction the terminal phosphorous atom of adenosine triphosphate is located enough close to both 3′-hydroxyl and 5″-hydroxyl groups to react with these groups, but more close to the former. Therefore it is considered that in the presence of a 3′-hydroxyl group, the reaction with this hydroxyl group is preffered than with the 5″-hydroxyl group. The modification of the l-amino group of deoxystreptamine moiety is thought to weaken the binding of the substrate antibiotic with this enzyme, because butirosins do not undergo this enzyme reaction.

Kanamycin-neomycin phosphotransferase II also phosphorylates the 3′-hydroxyl group of butirosins (Yagisawa, Yamamoto et al., 1972). It does not phosphorylate the 5″-hydroxyl group of lividomycins and 3′,4′-dideoxyribostamycin. It suggests that in this enzyme reaction the terminal phosphorous atom of adenosine triphosphate is located near the 3′-hydroxyl group but far from the 5″-hydroxyl group. Recently, it was found that the phosphotransferase II does not so actively phosphorylate 3′-hydroxyl group of 4′-deoxykanamycin A as that of kanamycin A (Umezawa, S. et al., 1974), suggesting that the 4′-hydroxyl group of kanamycins, ribostamycin or butirosins is involved in binding with this enzyme. On this point also, phosphotransferase II is different from I. The phosphotransferase I phosphorylates the 3′-hydroxyl group of 4′-deoxykanamycin A similarly as that of kanamycin A.

Considering the location of the terminal phosphorous atom in these enzyme reactions, it is postulated that there might be another enzyme which catalyses the transfer of the phosphate to the 5″-hydroxyl group of ribostamycin but not to the 3′-hydroxyl group. Kida et al. (1974) reported the presence of an enzyme in *Pseudomonas* which transfers the phosphate group to the 5″-hydroxyl group of ribostamycin.

Kanamycin-neomycin phosphotransferase I described above was obtained from *E. coli* carrying R factors such as *E. coli* K12 J5 R11-2, *E. coli* K12 ML-1629 and *E. coli* K12 ML-1410 R81 and the phosphotransferase II was obtained from *E. coli* JR66/W677. We recently studied the types of kanamycin-neomycin phosphotransferases in *P. aeruginosa*. We then found that a strain, *P. aeruginosa* TI-13 which had been obtained from Mitsuhashi about 4 years ago and which is designated TI-13b in this paper, produces type II enzyme. The same strain which was obtained from Mitsuhashi a year ago and which is designated TI-13a in this paper produces type I enzyme. Another strain, *P. aeruginosa* H-9, produces type II, and the other, B-13 produces both types I and II. These enzymes are similar to those of *E. coli* carrying R factor in relation to the phosphorylation of lividomycin A: Type I phosphorylates this antibiotic but not II. However, the lowest concentrations of sodium chloride which are necessary to elute these enzymes from affinity columns vary depending on the sources of these enzymes. As described above, types I and II which have similar substrate specificities as those of kanamycin-neomycin phosphotransferases I and II of *E. coli* carrying R factor are present in *P. aeruginosa*, but it seems that these enzymes in *P. aeruginosa* differ from each other in their entire structures depending on the sources.

DEVELOPMENT OF ACTIVE DERIVATIVES BASED ON MECHANISMS OF RESISTANCE

Based on mechanisms of resistance, it is considered that modification or removal of the group which undergoes the enzyme reaction gives a compound which is resistant to the enzyme reaction and which might be active in inhibiting resistant organisms. Thus, 3′-O-methylkanamycin (Umezawa, H., Tsuchiya et al., 1972) and 3′-O-methylneamine (Umezawa, S., Jikihara et al., 1972) were synthesized, but these derivatives had no antibacterial activity. Methyl group on 3′-oxygen seems to cause steric hindrance to binding of these antibiotics with bacterial ribosomes. Contrarily, the removal of the 3′-hydroxyl group gave active compounds: 3′-deoxy derivatives, 3′-deoxykanamycin A (Umezawa, S., Tsuchiya, Muto et al., 1971), 3′-deoxykanamycin B (Takagi et al., 1973), 3′-deoxyribostamycin (Ikeda et al., 1973a), and 3′,4′-dideoxy derivatives, 3′,4′-dideoxykanamycin B (Umezawa, H., Umezawa, S. et al., 1971), 3′,4′-dideoxy-6-*N*-methylkanamycin B (Umezawa, H., Nishimura et al., 1972), 3′,4′-dideoxyribostamycin (Umezawa, S., Tsuchiya et al., 1972), 3′,4′-dideoxybutirosin B (Ikeda, et al., 1973b), 3′,4′-dideoxyneamine (Umezawa, S., Tsuchiya, Jikihara et al., 1971) exhibit antibacterial activity and inhibit resistant organisms which produce kanamycin-neomycin phosphotransferases. As reported by Daniels (1974), removal of the 2″-hydroxyl group from gentamicin C_2 gives the derivative which is resistant to gentamicin-kanamycin nucleotidyl-transferase and inhibits the resistant organisms. However, removal of the 5″-hydroxyl group from lividomycins A and B reduced the antibacterial activity strongly (Umezawa, S., Watanabe et al., 1972). Substitution of this hydroxyl with an amino group also reduced the activity (Yamamoto et al., 1972b). 5″-Amino-5″-deoxylividomycin A had only very weak antibacterial activity. Among lividomycin derivatives, 6′-amino-6′-deoxylividomycin B (3′-deoxyneomycin B) showed the most interesting activity, inhibiting some resistant strains of *P. aeruginosa* (Watanabe et al., 1973b).

In an enzyme reaction which catalyses the reaction of an aminoglycosidic antibiotic with adenosine triphosphate or acetyl CoA, these substrates should bind with the enzyme and the binding should bring the reactive groups close to each other. Thus, in the case of kanamycin-neomycin phosphotransferases, the 3′-hydroxyl group should be brought close to the terminal phosphorous atom of adenosine triphosphate. Some hydroxyl groups or some amino groups of the antibiotic should be involved in binding the substrate antibiotic with the enzyme (Doi et al., 1969; Umezawa, H., 1970). Then, modification of such a group is thought to give a compound which does not undergo the enzym reaction and inhibits the resistant organisms. As already described, butirosins in which the 1-amino group of 2-deoxystreptamine moiety is modified are resistant to kanamycin-neomycin phosphotransferase I. Based on this reasoning, 1-N-[(S)-4-amino-2-hydroxybutyryl] kanamycin A which was synthesized by Kawaguchi et al., 1972, 1-N-(D-isoseryl)-kanamycin A, 1-N-(L-isoseryl)kanamycin A, 1-N-(DL-isoseryl)kanamycin A, 1-N--(DL-isoseryl)kanamycin B, 1-N-(DL-isoseryl)-3′,4′-dideoxykanamycin B synthesized by Kondo et al., 1974, and (S)-4-amino-2-hydroxybutyryl derivatives of kanamycin B, 3′,4′-dideoxykanamycin B (Kondo et al., 1973a and 1973b), 3′,4′-dideoxyneamine, (Umezawa, S., Ikeda et al., 1973) and lividomycin A, (Watanabe et al., 1973a) all inhibit the resistant organisms of the phosphotransferase I.

Modification of a group close to the group which binds to the enzyme can cause a steric hindrance of the enzyme reaction and can give an active derivative. The molecular model of kanamycins suggests that the 1-amino group of 2-deoxystreptamine moiety is located close to the 2″-hydroxyl group which undergoes the reaction of gentamicin-kanamycin nucleotidyltransferase. Such compounds as 1-N-4-amino-2-hydroxybutyryl and 1-N-isoseryl derivatives of kanamycins and 3′,4′-dideoxykanamycin B described above are resistant to gentamicin-kanamycin nucleotidyltransferase and inhibit the resistant organisms.

We synthesized 6′-N-methyl derivatives of kanamycin A and 3′,4′-dideoxykanamycin B (Umezawa, H., Nishimura et al., 1972), but this modification was not sufficient to overcome the resistance. Methyl on 6′-carbon is suggested by gentamicin C_1 to be more efficient. Among the active derivatives described above, 3′,4′-dideoxykanamycin B and BBK-8 have been clinically studied in a large scale, and their usefulness in treating infections has been confirmed.

The parallel relation between resistance to the enzymes and activity against the resistant strains is strong proof for the involvement of the enzymes in resistance. However, an exceptional case of unsolved relationships should be noted here. If kanamycin-neomycin phosphotransferase II is added to the reaction mixture in large quantitities, then the 3′-hydroxyl group of 1-N-[(S)-4-amino-2-hydroxybutyryl] kanamycin A is phosphorylated. Moreover, the 3′-hydroxyl group of 1-N-[(S)-4-amino-2-hydroxybutyryl] kanamycin B is phosphorylated almost identically to kanamycin B by phosphotransferase II, and both of these derivatives strongly inhibit the resistant strain, *E. coli* JR66/W677, which produces this enzyme. This exceptional case is now under study.

SUMMARY

The author who initiated the study of biochemical mechanisms of resistance to aminoglycosidic antibiotics reviewed his studies on methods of isolating reaction products of enzymes involved in resistance, on physical methods of elucidating the structures of the reaction products, on kanamycin-neomycin phosphotransferases in *E. coli* carrying R factor and in *P. aeruginosa* and on development of derivatives active in inhibiting resistant organisms. The enzymes were immobilized by coupling with CNBr-activated Sepharose 4B or with CNBr-activated cellulose, and were maintained in a more active state when thus immobilized than in solution. The immobilized enzymes catalysed the reaction and the reaction products, for instance, kanamycin B 3′-phosphate and dihydrostreptomycin 3″-phosphate, were easily purified. The structures of most of the reaction products were determined solely by pmr with application of double resonance technique. The structures thus determined were reviewed. Among the enzymes involved in resistance mechanisms, kanamycin-neomycin phosphotransferases I and II in *E. coli* carrying R factor and in *P. aeruginosa* were described in more detail: I phosphorylates 3′-hydroxyl of kanamycins, 4′-deoxykanamycin and ribostamycin, 5″-hydroxyl group of lividomycins and 3′,4′-dideoxyribostamycin but not 3′-hydroxyl of butirosins; II phosphorylates 3′-hydroxyl of kanamycins, ribostamycin, butirosins, but not 5″-hydroxyl of lividomycins and 3′,4′-dideoxyribostamycin. The phosphorylation of the 3′-hydroxyl of 4′-deoxykanamycin by II is very weak. These enzymes were purified by affinity chromatography. Active compound were developed based on the enzymatic mechanisms of resistance, by removing the group which undergoes the enzyme reaction, by modifying the group which binds with the enzyme or by modifying the group close to the group in binding with the enzyme. The compounds thus synthesized indicated that synthesis based on the biochemical mechanisms of resistance is a rational approach to the development of agents useful in the treatment of resistant infections.

REFERENCES

BENVENISTE, R. and DAVIES, J. (1971). R-factor mediated gentamicin resistance. A new enzyme which modifies aminoglycoside antibiotics. FEBS Lett. 14, 293.

BRZEZINSKA, M., BENVENISTE, R., DAVIES, J., DANIELS, P. J. L. and WEINSTEIN, J. (1972). Gentamicin resistance in strains of *Pseudomonas aeruginosa* mediated by enzymatic N-acetylation of the deoxystreptamine moiety. Biochemistry 11, 761.

CHEVEREAU, M., DANIELS, P. J. L., DAVIES, J. and LEGOFFIC, F. (1974). Aminoglycoside resistance in bacteria mediated by gentamicin acetyltransferase II, an enzyme modifying the 2′-amino group of aminoglycoside antibiotics. Biochemistry 13, 598.

DANIELS, P. J. L., WEINSTEIN, J., TKACH, R. W. and MORTON, J. (1974). Gentamicin derivatives modified at the 2″-position. The preparation of 2″-epi-gentamicin C_1 and 2″-deoxygentamicin C_2. J. Antibiotics 27, 150.

DOI, O., KONDO, S., TANAKA, N. and UMEZAWA, H. (1969). Purification and properties of kanamycin phosphorylating enzyme from *Pseudomonas aeruginosa*. J. Antibiotics 22, 273.

DOI, O., MIYAMOTO, M., TANAKA, N. and UMEZAWA, H. (1968). Inactivation and phosphorylation of kanamycin by drug-resistant *Staphylococcus aureus*. Appl. Microbiol. 16, 1282.

DOI, O., OGURA, M., TANAKA, N. and UMEZAWA, H. (1968). Inactivation of kanamycin, neomycin, and streptomycin by enzymes obtained in cells of *Pseudomonas aeruginosa*. Appl. Microbiol. 16, 1276.

IKEDA, D., TSUCHIYA, T., UMEZAWA, S. and UMEZAWA, H. (1973a). Synthesis of 3′-deoxyribostamycin. J. Antibiotics 26, 799.

IKEDA, D., TSUCHIYA, T., UMEZAWA, S., UMEZAWA, H. and HAMADA, M. (1973b). Synthesis of 3′, 4′-dideoxybutirosin B. J. Antibiotics 26, 307.

KAWAGUCHI, H., NAITO, T., NAKAGAWA, S. and FUJISAWA, K. (1972). BB-K8, a new semisynthetic aminoglycoside antibiotic. J. Antibiotics 25, 695.

KIDA, M., IGARASHI, S., OKUTANI, T., ASAKO, T., HIRAGA, K. and MITSUHASHI, S. (1974). Selective phosphorylation of the 5″-hydroxy group of ribostamycin by a new enzyme from *Pseudomonas aeruginosa*. Antimicrob. Ag. Chemother. 5, 92.

KONDO, S., IINUMA, K., HAMADA, M., MAEDA, K. and UMEZAWA, H. (1974). Syntheses of isoseryl derivatives of kanamycins and their antibacterial activities. J. Antibiotics 27, 90.

KONDO, S., IINUMA, K., YAMAMOTO, H., IKEDA, Y., MAEDA, K. and UMEZAWA, H. (1973a). Syntheses of (S)-4-amino-2-hydroxybutyryl derivatives of 3′, 4′-dideoxykanamycin B and their antibacterial activities. J. Antibiotics 26, 705.

KONDO, S., IINUMA, K., YAMAMOTO, H., MAEDA, K. and UMEZAWA, H. (1973b). Syntheses of 1-N- [(S)-4-amino-2-hydroxybutyryl] -kanamycin B and 3′-4′-dideoxykanamycin B active against kanamycin-resistant bacteria. J. Antibiotics 26, 412.

KONDO, S., OKANISHI, M., UTAHARA, R., MAEDA, K. and UMEZAWA, H. (1968). Isolation of kanamycin and paromamine inactivated by *E. coli* carrying R factor. J. Antibiotics 21, 22.

KONDO, S., YAMAMOTO, H., NAGANAWA, H., UMEZAWA, H. and MITSUHASHI, S. (1972). Isolation and characterization of lividomycin A inactivated by *Ps. aeruginosa* and *E. coli* carrying R factor. J. Antibiotics 25, 483.

KOYAMA, G., IITAKA, Y., MAEDA, K. and UMEZAWA, H. (1968). The crystal structure of kanamycin. Tetrahedron Letters 15, 1875.

MAEDA, K., KONDO, S., OKANISHI, M., UTAHARA, R. and UMEZAWA, H. (1968). Isolation of paromamine inactivated by *Pseudomonas aeruginosa*. J. Antibiotics 21, 458.

MIYAMURA, S. (1961). Dysentery bacilli and its relation to the resistance. Nihon Saikingaku Zasshi 16, 115.

NAGANAWA, H., KONDO, S., MAEDA, K. and UMEZAWA, H. (1971). Structure determination of enzymatically phosphorylated products of aminoglycosidic antibiotics by proton magnetic resonance. J. Antibiotics 24, 823.

NAGANAWA, H., YAGISAWA, M., KONDO, S., TAKEUCHI, T. and UMEZAWA, H. (1971). Structure determination of an enzymatic inactivation product of 3′,4′-dideoxy-kanamycin B. J. Antibiotics 24, 913.

OKAMOTO, S. and SUZUKI, Y. (1965). Chloramphenicol-, dihydrostreptomycin- and kanamycin-inactivating enzymes from multiple drug-resistant *Escherichia coli* carrying episome 'R'. Nature 208, 1301.

OKANISHI, M., KONDO, S., SUZUKI, Y., OKAMOTO, S. and UMEZAWA, H. (1967). Studies on inactivation of kanamycin and resistances of *E. coli*. J. Antibiotics 20, 132.

OKANISHI, M., KONDO, S., UTAHARA, R., MAEDA, K. and UMEZAWA, H. (1968). Phosphorylation and inactivation of aminoglycosidic antibiotics by *E. coli* carrying R factor. J. Antibiotics 21, 13.

OZANNE, B., BENVENISTE, R., TIPPER, D. and DAVIES, J. (1969). Aminoglycoside antibiotics: Inactivation by phosphorylation in *Escherichia coli* carrying R factors. J. Bacteriol. 100, 1144.
TAKAGI, Y., MIYAKE, T., TSUCHIYA, T. and UMEZAWA, S. (1973). Synthesis of 3′-deoxykanamycin B. J. Antibiotics 26, 403.
TAKASAWA, S., UTAHARA, R., OKANISHI, M., MAEDA, K. and UMEZAWA, H. (1968). Studies on adenylylstreptomycin, a product of streptomycin inactivation by *E. coli* carrying the R factor. J. Antibiotics 21, 477.
UMEZAWA, H. (1970). Mechanism of inactivation of aminoglycosidic antibiotics by enzymes of resistant organisms of clinical origin. Progress in Antimicrobial and Anticancer Chemotherapy (Proceedings of the 6th International Congress of Chemotherapy), II, 566, University of Tokyo Press.
UMEZAWA, H., DOI, O., OGURA, M., KONDO, S. and TANAKA, N. (1968). Phosphorylation and inactivation of kanamycin by *Pseudomonas aeruginosa*. J. Antibiotics 21, 154.
UMEZAWA, H., MATSUHASHI, Y., YAGISAWA, M., YAMAMOTO, H. and TAKEUCHI, T. (1974). Immobilization of phosphotransferases obtained from resistant bacteria. J. Antibiotics 27, 358.
UMEZAWA, H., NISHIMURA, Y., TSUCHIYA, T. and UMEZAWA, S. (1972). Syntheses of 6′-N-methylkanamycin and 3′,4′-dideoxy-6′-N-methylkanamycin B active against resistant strains having 6′-N-acetylating enzyme. J. Antibiotics 25, 743.
UMEZAWA, H., OKANISHI, M., KONDO, S., HAMANA, K., UTAHARA, R., MAEDA, K. and MITSUHASHI, S. (1967). Phosphorylative inactivation of aminoglycosidic antibiotics by *Escherichia coli* carrying R factor. Science 157, 1559.
UMEZAWA, H., OKANISHI, M., UTAHARA, R., MAEDA, K. and KONDO, S. (1967). Isolation and structure of kanamycin inactivated by a cell free system of kanamycin-resistant *E. coli*. J. Antibiotics 20, 136.
UMEZAWA, H., TAKASAWA, S., OKANISHI, M. and UTAHARA, R. (1968). Adenylylstreptomycin, a product of streptomycin inactivated by *E. coli* carrying R factor. J. Antibiotics 21, 81.
UMEZAWA, H., TSUCHIYA, T., MUTO, R. and UMEZAWA, S. (1972). Studies on aminosugars. XXIX. The synthesis of 3′-0-methylkanamycin. Bull. Chem. Soc. Japan 45, 2842.
UMEZAWA, H., UMEZAWA, S., TSUCHIYA, T. and OKAZAKI, Y. (1971). 3′,4′-Dideoxykanamycin B active against kanamycin-resistant *Escherichia coli* and *Pseudomonas aeruginosa*. J. Antibiotics 24, 485.
UMEZAWA, H., YAMAMOTO, H., YAGISAWA, M., KONDO, S., TAKEUCHI, T. and CHABERT, Y. A. (1973). Kanamycin phosphotransferase I: Mechanism of cross resistance between kanamycin and lividomycin. J. Antibiotics 26, 407.
UMEZAWA, H., YAGISAWA, M., MATSUHASHI, Y., NAGANAWA, H., YAMAMOTO, H., KONDO, S., TAKEUCHI, T. and CHABBERT, Y. A. (1973). Gentamicin acetyltransferase in *Escherichia coli* carrying R factor. J. Antibiotics 26, 612.
UMEZAWA, S., IKEDA, D., TSUCHIYA, T. and UMEZAWA, H. (1973). Synthesis of 1-N-((S)-4-amino-2-hydroxybutyryl)-3′,4′-dideoxyneamine. J. Antibiotics 26, 304.
UMEZAWA, S., JIKIHARA, T., TSUCHIYA, T. and UMEZAWA, H. (1972). Syntheses of 3′- and 4′-0-methylneamine. J. Antibiotics 25, 322.
UMEZAWA, S., NISHIMURA, Y., HATA, Y., TSUCHIYA, T., YAGISAWA, M. and UMEZAWA, H. (1974). Synthesis of 4′-deoxykanamycin and its resistance to kanamycin phosphotransferase II. J. Antibiotics 27, in press.
UMEZAWA, S., TSUCHIYA, T., IKEDA, D. and UMEZAWA, H. (1972). Syntheses of 3′,4′-dideoxy and 3′,4′,5″-trideoxyribostamycin active against kanamycin-resistant *E. coli* and *P. aeruginosa*. J. Antibiotics 25, 613.
UMEZAWA, S., TSUCHIYA, T., JIKIHARA, T. and UMEZAWA, H. (1971). Synthesis of 3′,4′-dideoxyneamine active against kanamycin-resistant *E. coli* and *P. aeruginosa*. J. Antibiotics 24, 711.
UMEZAWA, S., TSUCHIYA, T., MUTO, R., NISHIMOTO, Y. and UMEZAWA, H. (1971). Synthesis of 3′-deoxykanamycin effective against kanamycin resistant *E. coli* and *Ps. aeruginosa*. J. Antibiotics 24, 274.
UMEZAWA, S., WATANABE, I., TSUCHIYA, T., UMEZAWA, H. and HAMADA, M. (1972). Synthesis of 5″-deoxylividomycin B. J. Antibiotics 25, 617.
WATANABE, I., TSUCHIYA, T., UMEZAWA, S. and UMEZAWA, H. (1973a). Synthesis of 1-N-((S)-4-amino-2-hydroxybutyryl) lividomycin A. J. Antibiotics 26, 310.
WATANABE, I., TSUCHIYA, T., UMEZAWA, S. and UMEZAWA, H. (1973b). Syntheses of 6′-amino-6′-deoxylividomycin B and 6′-deoxy-6′-methylamino- and 6′-deoxy-6′-(2-hydroxyethylamino)-lividomycin B. J. Antibiotics 26, 802.
YAGISAWA, M., NAGANAWA, H., KONDO, S., HAMADA, M., TAKEUCHI, T. and UMEZAWA, H. (1971). Adenylyldideoxykanamycin B, a product of the inactivation of dideoxykanamycin B by E. coli carrying R factor. J. Antibiotics 24, 911.

Yagisawa, M., Naganawa, H., Kondo, S., Takeuchi, T. and Umezawa, H. (1972a). Inactivation of 3′,4′-dideoxykanamycin B by an enzyme solution of resistant *E. coli* and isolation of 3′,4′-dideoxykanamycin B 2″-guanylate and 2″-inosinate. J. Antibiotics 25, 492.

Yagisawa, M., Naganawa, H., Kondo, S., Takeuchi, T. and Umezawa, H. (1972b). 6′-N-acetylation of 3′,4′-dideoxykanamycin B by an enzyme in a resistant strain of *Ps. aeruginosa.* J. Antibiotics 25, 495.

Yagisawa, M., Yamamoto, H., Naganawa, H., Kondo, S., Takeuchi, T. and Umezawa, H. (1972). A new enzyme in *Escherichia coli* carrying R factor phosphorylating 3'-hydroxyl of butirosin A, kanamycin, neamine and ribostamycin. J. Antibiotics 25, 748.

Yamaguchi, M., Mitsuhashi, S., Kobayashi, F. and Zenda, H. (1974). A 2′-N-acetylating enzyme of aminoglycosides. J. Antibiotics 27, in press.

Yamamoto, H., Kondo, S., Maeda, K. and Umezawa, H. (1972a). Synthesis of lividomycin A 5″-phosphate, an enzymatically inactivated lividomycin A. J. Antibiotics 25, 485.

Yamamoto, H., Kondo, S., Maeda, K. and Umezawa, H. (1972b). Syntheses of 5″-deoxylividomycin A and its amino derivative. J. Antibiotics 25, 487.

Yamamoto, H., Yagisawa, M., Naganawa, H., Kondo, S., Takeuchi, T. and Umezawa, H. (1972). Kanamycin 6′-acetate and ribostamycin 6′-acetate, enzymatically inactivated products by *Ps. aeruginosa.* J. Antibiotics 25, 746.

PROPOSAL FOR A RATIONAL NOMENCLATURE FOR PHENOTYPE, GENOTYPE AND AMINOGLYCOSIDE-AMINOCYCLITOL MODIFYING ENZYMES

S. MITSUHASHI

Department of Microbiology, School of Medicine, Gunma University, Maebashi, Japan

I was asked by the Plasmid Nomenclature Group to propose a rational nomenclature for phenotype and genotype of aminoglycoside antibiotics resistance in bacteria and for aminoglycoside-aminocyclitol modifying enzymes. After discussion of the fundamental problems of aminoglycoside antibiotics resistance in bacteria with Dr. H. Umezawa, Institute of Microbial Chemistry, Tokyo and with Dr. J. Davies, Institute of Molecular Biology, University of Wisconsin, we propose the following nomenclature for phenotype and genotype of aminoglycoside-aminocyclitol modifying enzymes, which are mainly responsible for bacterial resistance to the drugs.

1. Aminoglycoside antibiotic phosphotransferase

phenotype	genotype
(1) KM, NM, RM, PM, LM	*aph*(3′)-I
(2) KM, NM, RM, PM, BT	*aph*(3′)-II
(3) RM	*aph*(5″)
(4) SM	*aph*(3″)

2. Aminoglycoside antibiotic acetyltransferase

phenotype	genotype
(1) KM-A, KM-B	*aac*(6′)-I
(2) (1), GM-C_{1a}-, -C_2	*aac*(6′)-II
(3) (1), (2), DKB	*aac*(6′)-III
(4) (1), (2), DKB, AK	*aac*(6′)-IV
(5) GM	*aac*(2′)-I
(6) LV, RM, BT, GM, KM-B, -C, DKB	*aac*(2′)-II
(7) GM-C_1	*aac*(3)

3. Aminoglycoside antibiotic adenyltransferase

phenotype	genotype
(1) SM	*aad*(3″)
(2) GM, KM, DKB	*aad*(2″)

Abbreviation:

1. KM, kanamycin; NM, neomycin; RM, ribostamycin; PM, paromomycin; LV, lividomycin; BT, butirosin; SM streptomycin.

Proposal for a Rational Nomenclature for Aminoglycoside-Aminocyclitol Modifying Enzymes

Enzyme Activity	Enzyme Trivial Name	Enzyme Systematic Name	Abbrevition of Enzyme Name	Gene Symbol
1. PHOSPHORYLATION:				
1-(1) Phosphorylation of 3'' -hydroxyl of streptomycin	Aminoglycoside 3'' – phosphotransferase	ATP: Aminoglycoside 3'' -phosphotransferase	APH(3'')	*aph*(3'')
1-(2) Phosphorylation of 3' -hydroxyl of kanamycin, ribostamycin (not butirosin) and 5'' -hydroxyl of lividomycin	Aminoglycoside 3' – phosphortransferase I	ATP: Aminoglycoside 3' -phosphostransferase	APH(3')-I	*aph*(3')-I
1-(3) Phosphorylation of 3' -hydroxyl of kanamycin, ribostamycin and butirosin	Aminoglycoside 3' – phosphotransferase II	ATP: Aminoglycoside 3' -phosphotransferase	APH(3')-II	*aph*(3')-II
1-(4) Phosphorylation of 5'' -hydroxyl of ribostamycin	Aminoglycoside 5'' – phosphotransferase	ATP: Aminoglycoside 5'' -phosphotransferase	APH(5'')	*aph*(5')
2. ACETYLATION:				
2-(1) Acetylation of 6' -amino of kanamycin, gentamicin C1 and ribostamycin	Aminoglycoside' 6 – acetyltransferase	Acetyl CoA: Aminoglycoside 6' -N-acetyltransferase	AAC(6')	*aac*(6')
2-(2) Acetylation of 3-amino of gentamicin (not tobramycin or ribostamycin)	Aminoglycoside 3 – acetyltransferase	Acetyl CoA: Aminoglycoside 3-N-acetyltransferase	AAC(3)	*aac*(3)
2-(3) Acetylation of 2' -amino of gentamicin, tobramycin and lividomycin	Aminoglycoside 2' – acetyltransferase	Acetyl CoA: Aminoglycoside 2' -N-acetyltransferase	AAC(2')-I	*aac*(2')-I
2-(4) Acetylation of 2' -amino various aminoglycoside antibiotics except kanamycin A and amikacin (BB-K8)	Aminoglycoside 2' – acetyltransferase	Acetyl CoA: Aminoglycoside 2' -N-acetyltransferase	AAC(2')-II	*aac*(2')-II
3. NUCLEOTIDYLATION:				
3-(1) Adenylylation of 3'' -hydroxyl of streptomycin and adenylylation of spectinomycin	Aminoglycoside 3'' – adenylyltransferase	ATP: Aminoglycoside 3'' -adenylyltransferase	AAD(3'')	*aad*(3'')
3-(2) Nucleotidylation of 2'' -hydroxyl of gentamicin and kanamycin	Aminoglycoside 2'' – nucleotidyltransferase	Nucleoside triphosphate: Aminoglycoside 2'' – nucleotidyltransferase	ANT(2'')	*ant*(2'')

2. GM, gentamicin; DKB, 3′,4′-dideoxy-kanamycin B; AK, amikacin (BB-K8); *aph*, aminoglycoside antibiotic phosphotransferase; *aac*, aminoglycoside antibiotic acetyltransferase; *aad*, aminoglycoside antibiotic adenyltransferase.

REFERENCES

I AMINOGLYCOSIDE ANTIBIOTIC PHOSPHOTRANSFERASE

BRZEZINSKA, M. and DAVIES, J. (1973). Two enzymes which phosphorylate neomycin and kanamycin in *Escherichia coli* strains carrying R factors. *Antimicrob. Ag. Chemother.* 3, 266–269.

DOI, O., KONDO, S., TANAKA, N. and UMEZAWA, H. (1969). Purification and properties of kanamycin-phosphorylating enzyme from *Pseudomonas aeruginosa. J. Antibiotics* 22, 273–282.

DOI, O., MIYAMOTO, M., TANAKA, N. and UMEZAWA, H. (1968). Inactivation and phosphorylation of kanamycin by drug-resistant *Staphylococcus aureus. Appl. Microbiol.* 16, 1282–1284.

DOI, O., OGURA, M., TANAKA, N. and UMEZAWA, H. (1968). Inactivation of kanamycin, neomycin, and streptomycin by enzymes obtained in cells of *Pseudomonas aeruginosa. Appl. Microbiol.* 16, 1276–1281.

KAWABE, H., KOBAYASHI, F., YAMAGUCHI, M., UTAHARA, R., and MITSUHASHI, S. (1971). 3″-Phosphoryldihydrostreptomycin produced by the inactivating enzyme of *Pseudomonas aeruginosa. J. Antibiotics* 24, 651–652.

KIDA, M., IGARASHI, S., OKUTANI, T., ASAKO, T., HIRAGA, K. and MITSUHASHI, S. (1974). Selective phosphorylation of the 5″-hydroxy group of ribostamycin by a new enzyme from *Pseudomonas aeruginosa. Antimicrob. Ag. Chemother.* 5, 92–94.

KOBAYASHI, F., KOSHI, T., EDA, J., YOSHIMURA, Y. and MITSUHASHI, S. (1973). Lividomycin resistance in staphylococci by enzymatic phosphorylation. *Antimicrob. Ag. Chemother.* 4, 1–5.

KOBAYASHI, F., YAMAGUCHI, M., and MITSUHASHI, S. (1971). Inactivation of dihydrostreptomycin by *Pseudomonas aeruginosa. Japan. J. Microbiol.* 15, 381–382.

KOBAYASHI, F., YAMAGUCHI, M., and MITSUHASHI, S. (1971). Phosphorylated inactivation of aminoglycosidic antibiotics by *Pseudomonas aeruginosa. Japan. J. Microbiol.* 15, 265–272.

KOBAYASHI, F., YAMAGUCHI, M., SATO, J. and MITSUHASHI, S. (1972). Purification and properties of dihydrostreptomycin-phosphorylating enzyme from Pseudomonas aeruginosa. *Japan. J. Microbiol.* 16, 15–19.

KONDO, S., OKANISHI, M., UTAHARA, R., MAEDA, K. and UMEZAWA, H. (1968). Isolation of kanamycin and paromamine inactivated by *E. coli* carrying R factor. *J. Antibiotics* 21, 22–29.

KONDO, S., YAMAMOTO, H., NAGANAWA, H., UMEZAWA, H. and MITSUHASHI, S. (1972). Isolation and characterization of lividomycin A inactivated by *Pseudomonas aeruginosa* and *Escherichia coli* carrying R factor. *J. Antibiotics* 25, 483–484.

MAEDA, K., KONDO, S., OKANISHI, M., UTAHARA, R. and UMEZAWA, H. (1968). Isolation of paromamine inactivated by *Pseudomonas aeruginosa. J. Antibiotics* 21, 458–459.

NAGANAWA, H., KONDO, S., MAEDA, K. and UMEZAWA, H. (1971). Structure determination of enzymatically phosphorylated products of aminoglycosidic antibiotics by proton magnetic resonance. *J. Antibiotics* 24, 823–829.

OKANISHI, M., KONDO, S., UTAHARA, R. and UMEZAWA, H. (1968). Phosphorylation and inactivation of aminoglycosidic antibiotics by *E. coli* carrying R factor. *J. Antibiotics* 21, 13–21.

OZANNE, B., BENVENISTE, R., TIPPER, D. and DAVIES, J. (1969). Aminoglycoside antibiotics: Inactivation by phosphorylation in *Escherichia coli* carrying R factors. *J. Bacteriol.* 100, 1144–1146.

UMEZAWA, H., DOI, O., OGURA, M., KONDO, S. and TANAKA, N. (1968). Phosphorylation and inactivation of kanamycin by *Pseudomonas aeruginosa. J. Antibiotics* 21, 154–155.

UMEZAWA, H., MATSUHASHI, Y., YAGISAWA, M., YAMAMOTO, H., KONDO, S. and TAKEUCHI, T. (1974). Immobilization of phosphotransferases obtained from resistant bacteria. *J. Antibiotics* 27, 358–360.

UMEZAWA, H., OKANISHI, M., KONDO, S., HAMANA, K., UTAHARA, R., MAEDA, K. and MITSUHASHI, S. (1967). Phosphorylative inactivation of aminoglycosidic antibiotics by *Escherichia coli* carrying R factor. *Science* 157, 1559–1561.

UMEZAWA, H., YAMAMOTO, H., YAGISAWA, M., KONDO, S., TAKEUCHI, T. and CHABBERT, Y. A. (1973). Kanamycin phosphotransferase I: Mechanism of cross resistance between kanamycin and lividomycin. *J. Antibiotics* 26, 407–411.

WALKER, J. B., and SKORVAGA, M. (1973). Phosphorylation of streptomycin and dihydrostreptomycin by *Streptomyces*. *J. Biol. Chem.* 248, 2435–2440.
YAGISAWA, M., YAMAMOTO, H., NAGANAWA, H., KONDO, S., TAKEUCHI, T. and UMEZAWA, H. (1972). A new enzyme in *Escherichia coli* carrying R factor phosphorylating 3′-hydroxyl of butirosin A, kanamycin, neamine and ribostamycin. *J. Antibiotics* 25, 748–750.
YAMAGUCHI, M., KOSHI, T., KOBAYASHI, F., and MITSUHASHI, S. (1972). Phosphorylation of lividomycin by *Escherichia coli* carrying an R factor. *Antimicrob. Ag. Chemother.* 2, 142–146.

II AMINOGLYCOSIDE ANTIBIOTIC ACETYLTRANSFERASE

BENVENISTE, E. and DAVIES, J. (1971). Enzymatic acetylation of aminoglycoside antibiotics by *Escherichia coli* carrying an R factor. *Biochemistry* 10, 1787–1796.
BENVENISTE, R. and DAVIES, J. (1973). Aminoglycoside antibiotic-inactivating Enzymes in actinomycetes similar to those present in clinical Isolates of antibiotic-resistant bacteria. *Proc. Nat. Acad. Sci. USA* 70, 2276–2280.
BRZEZINSKA, M., BENVENISTE, R., DAVIES, J., DANIELS, P. J. L. and WEINSTEIN, J. (1972). Gentamicin resistance in strains of *Pseudomonas aeruginosa* mediated by enzymatic N-acetylation of the deoxystreptamine moiety. *Biochemistry* 11, 761–765.
CHEVEREAU, M., DANIELS, P. J. L., DAVIES, J. and LE GOFFIC, F. (1974). Aminoglycoside resistance in bacteria mediated by gentamicin acetyltransferase II, an enzyme modifying the 2′-amino group of eminoglycoside antibiotics. *Biochemistry* 13, 598–603.
KAWABE, H., and MITSUHASHI, S. (1972). Acetylation of dideoxykanamycin B by *Pseudomonas aeruginosa*. *Japan. J. Microbiol.* 16, 436–437.
KOBAYASHI, F., YAMAGUCHI, M., EDA, J., HIRAMATSU, M. and MITSUHASHI, S. (1972). Gentamicin C acetylating enzyme from resistant strains of *Pseudomonas aeruginosa* and the structure of the acetylated product. *Gunma Rept. Med. Sci.* 5, 291–301.
LE GOFFIC, F. and MOREAU, N. (1973). Purification by affinity chromatography of an enzyme involved in gentamicin inactivation. *FEBS Lett.* 29, 289–291.
LE GOFFIC, F. (1974). Resistance of moraxella to tobramycin, kanamycin and BB-K8. Second international symposium on antibiotic resistance: Drug-inactivating enzymes and other problems of resistant bacteria, June, Czechoslovakia.
MITSUHASHI, S., KOBAYASHI, F. and YAMAGUCHI, M. (1971). Enzymatic inactivation of gentamicin C components by cell-free extract from *Pseudomonas aeruginosa*. *J. Antibiotics* 24, 400–401.
MITSUHASHI, S., KOBAYASHI, F., YAMAGUCHI, M., O'HARA, K. and KONO, M. (1972). Enzymatic inactivation of aminoglycoside antibiotics by resistant strains of bacteria. *Advances in Antimicrobiol. and Antineoplasmic Chemotherapy* (Proceeding of the VIIth international congress of Chemotherapy), 561–564.
MITSUHASHI, S., KAWABE, H., (1974). Comparative studies of antibacterial activity of various aminoglycoside antibiotics. Second international symposium on antibiotic resistance: Drug-inactivating enzymes and other problems of resistant Bacteria. June, Czechoslovakia.
MOREL, C., FREYMUTH, F. and VILLEMON-LEMOSQUET, M. (1973). Nouveau mécanisme enzymatique de résistance aux antibiotiques aminoglucosidiques. *Ann. Biol. Clin.* 31, 353–357.
O'HARA, K., KONO, M., and MITSUHASHI, S. (1974). Structure of enzymatically acetylated sisomicin by *Pseudomonas aeruginosa*. *J. Antibiotics* 27, 349–351.
O'HARA, K., KONO, M. and MITSUHASHI, S. (1974). Enzymatic inactivation of a new aminoglycoside antibiotic, sisomicin, by resistant strains of *Pseudomonas aeruginosa*. *Antimicrob. Ag. Chemother.* 23, 551–554.
OKAMOTO, S. and SUZUKI, Y. (1965). Chloramphenicol-, dihydrostreptomycin- and kanamycin-inactivating enzymes from multiple drug-resistant *Escherichia coli* carrying episome 'R'. *Nature* 208, 1301–1303.
OKANISHI, M., KONDO, S., SUZUKI, Y., OKAMOTO, S. and UMEZAWA, H. (1967). Studies on inactivation of kanamycin and resistances of *E. coli*. J. *Antibiotics* 20, 132–135.
UMEZAWA, H., OKANISHI, M., UTAHARA, R., MAEDA, K. and KONDO, S. (1967). Isolation and structure of kanamycin inactivated by a cell free system of kanamycin-resistant *E. coli*. *J. Antibiotics* 20, 136–141.
UMEZAWA, H., YAGISAWA, M., MATSUHASHI, Y., NAGANAWA, H., YAMAMOTO, H., KONDO, S., TAKEUCHI, T. and CHABBERT, Y. A. (1973). Gentamicin acetyltransferase in *Escherichia coli* carrying R factor. *J. Antibiotics* 26, 612–614.
YAGISAWA, M., NAGANAWA, H., KONDO, S., TAKEUCHI, T. and UMEZAWA, H. (1972). 6′-N-Acetylation of 3′,4′-dideoxykanamycin B by an enzyme in a resistant strain of *Pseudomonas aeruginosa*. *J. Antibiotics* 25, 495–496.
YAMAMOTO, H., YAGISAWA, M., NAGANAWA, H., KONDO, S., TAKEUCHI, T. and UMEZAWA, H.

(1972.) Kanamycin 6′-acetate and ribostamycin 6′-acetate, enzymatically inactivated products by *Pseudomonas aeruginosa*. *J. Antibiotics* 25, 746–747.
YAMAGUCHI, M., MITSUHASHI, S., KOBAYASHI, F. and ZENDA, H. (1974). A 2'-N-acetylating enzyme of aminoglycosides. *J. Antibiotics* 27, in press.

III AMINOGLYCOSIDE ANTIBIOTIC ADENYLYLTRANSFERASE

BENVENISTE, R. and DAVIES, J. (1971). R-factor mediated gentamicin resistance: A new enzyme which modifies aminoglycoside antibiotics. *FEBS Lett.* 14, 293–296.
BENVENISTE, R., YAMADA, T. and DAVIES, J. (1970). Enzymatic adenylylation of streptomycin and spectinomycin by R-factor-resistant *Escherichia coli*. *Infect. Immunity* 1, 109–119.
HARWOOD, J. and SMITH, D. H. (1969). Resistance factor-mediated streptomycin resistance. *J. Bacteriol.* 97, 1262–1271.
KAWABE, H. and MITSUHASHI, S. (1971). Inactivation of dihydrostreptomycin by *Staphylococcus aureus*. *Japan. J. Microbiol.* 15, 545–548.
KAWABE, H., INOUE, M., and MITSUHASHI, S. (1974). Inactivation of dihydrostreptomycin and spectinomycin by *Staphylococcus aureus*. *Antimicro. Ag. Chemother.* in press.
KOBAYASHI, F., YAMAGUCHI, M., EDA, J., HIGASHI, F. and MITSUHASHI, S. (1971). Enzymatic inactivation of gentamicin C components by cell-free extract from *Klebsiella pneumoniae*. *J. Antibiotics* 24, 719–721.
NAGANAWA, H., YAGISAWA, M., KONDO, S., TAKEUCHI, T. and UMEZAWA, H. (1971). The structure determination of an enzymatic inactivation product of 3′, 4′-dideoxykanamycin B. *J. Antibiotics* 24, 913–914.
SMITH, D. H., JANJIGIAN, J. A., PRESCOTT, N. and ANDERSON, P. W. (1970). Resistance factor-mediated spectinomycin resistance. *Infect. Immunity* 1, 120–127.
TAKASAWA, S., UTAHARA, R., OKANISHI, M., MAEDA, K. and UMEZAWA, H. (1968). Studies on adenylylstreptomycin, a product of streptomycin inactivated by *E. coli* carrying the R factor. *J. Antibiotics* 21, 477–484.
UMEZAWA, H., TAKASAWA, S., OKANISHI, M. and UTAHARA, R. (1968). Adenylylstreptomycin, a product of streptomycin inactivated by *E. coli* carrying R factor. *J. Antibiotics* 21, 81–82.
YAGISAWA, M., NAGANAWA, H., KONDO, S., HAMADA, M., TAKEUCHI, T. and UMEZAWA, H. (1971). Adenylyldideoxykanamycin B, a product of the inactivation of dideoxykanamycin B by *Escherichia coli* carrying R factor. *J. Antibiotics* 24, 911–912.
YAMADA, T., TIPPER, D. and DAVIES, J. (1968). Enzymatic inactivation of streptomycin by R factor-resistant *Escherichia coli*. *Nature* 219, 288–291.

OPENING LECTURE

SOME ASPECTS OF ANTIBIOTIC RESISTANCE IN BACTERIA*

JULIAN DAVIES

Department of Biochemistry, University of Wisconsin, Madison, U.S.A.

It has become clear that resistance to aminoglycoside antibiotics in clinical isolates of bacteria is generally associated with the presence of enzymes which can modify hydroxy- or amino-groups on the antibiotic; the enzymes that have been characterized are listed in Table 1 (Benveniste and Davies, 1973a). Since most of the enzymes have multiple substrates we can ask if the enzymes exist in different forms (Izozymes) with different substrate ranges. For example, a variety of β-lactamases are known, which differ extensively in their ability to hydrolyse different β-lactam antibiotics, and respond differently to inhibitors (Richmond, *et al.*, 1971).

TABLE 1

Enzymes which modify the aminoglycoside aminocyclitol antibiotics

Enzyme	Substrates
3″-*0*-phosphotransferase	streptomycin
3′-*0*-phosphotransferase	neomycin, kanamycin
5″-*0*-phosphotransferase	lividomycin
3″-*0*-adenylyltransferase	streptomycin, spectinomycin
2″-*0*-nucleotidyltransferase	gentamicin, kanamycin, tobramycin
6′-*N*-acetyltransferase	gentamicin, kanamycin, tobramycin
2′-*N*-acetyltransferase	gentamicin, tobramycin
3-*N*-acetyltransferase	gentamicin

Phosphorylation of the 3′-hydroxyl of the neomycin antibiotics can be catalyzed by two enzymes (APH(3′)-1 and APH(3′)-II) wich differ in their ability to phosphorylate butirosin and lividomycin and also respond differently to the inhibitory action of 3′-deoxy-compounds (Table 2) (Brzezinska and Davies, 1973; Umezawa *et al.*, 1973a). With the

* I wish to acknowledge the necessary and excellent contributions of the following people to this work: S. Biddlecome, P. J. L. Daniels, J. E. Dowding, D. Hochkeppel, M. J. Haas, and D. I. Smith. Financial support was provided by grants from NIH and NSF.

expectation that other 3′-*O*-phosphorylating enzymes might exist, we have surveyed some fifty different kanamycin-resistant strains of *E. coli* covering the whole range of plasmid compatibility types which are currently recognized. Among these strains, the predominant enzyme was APH (3′)-1 with the alternate form APH (3′)-II being present

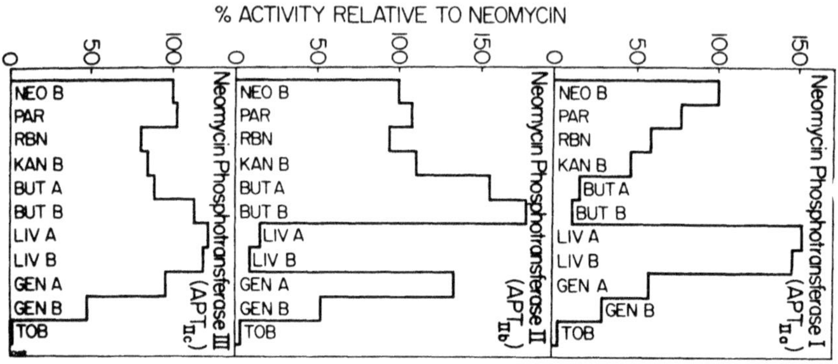

Figure 1. Substrate ranges for neomycin-kanamycin phosphorylating enzymes found in bacteria.

TABLE 2

Properties of Aminoglycoside 3′-0-Phosphotransferases

	Substrates				Inhibitors	
	Neomycin	Kanamycin	Butirosin	Lividomycin	Gentamicin	Tobramycin
APH(3′)-I	+	+	—	+	+	+
APH(3′)-II	+	+	+	—	—	—

TABLE 3

*Substrate ranges of aminoglycoside 3-N-acetyltransferases**

	AAC(3)-I	AAC(3)-II
	P. aeruginosa PS130	*P. aeruginosa* (Travers)
Kanamycin A	0	44
B	18	58
C	0	35
Dideoxykanamycin B	17	67
Gentamicin C_{1a}	107	100
Gentamicin C_1	100	69
Sisomicin	128	130
Tobramycin	14	75
Amikacin (BBK-8)	0	0
Lividomycin	0	77
Neomycin	0	123
Butirosin	0	0

* Results expressed as extent of reaction (%) in 20 minute incubations.

in only two or three isolates. A number of neomycin/kanamycin-resistant *Staphylococcus aureus* strains have also been surveyed and APH (3')-I was once more the predominant type although several strains were found to possess an activity which included substrates

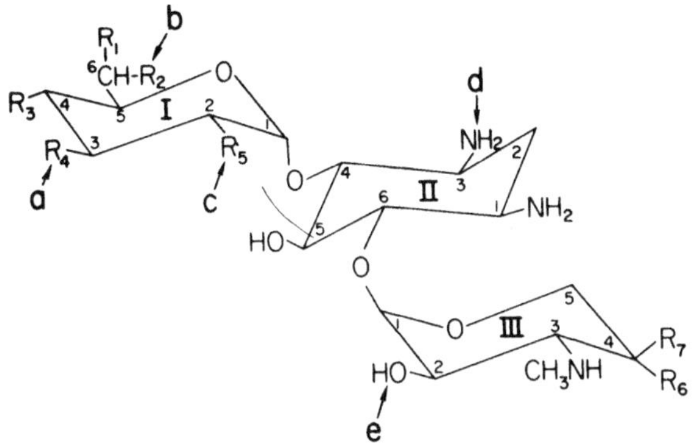

Figure 2. Enzymatic modification of gentamicin antibiotics. The arrows indicate the positions of modification a. phosphorylation (does not apply to gentamicin C complex); b, c, d, acetylation; e, adenylylation.

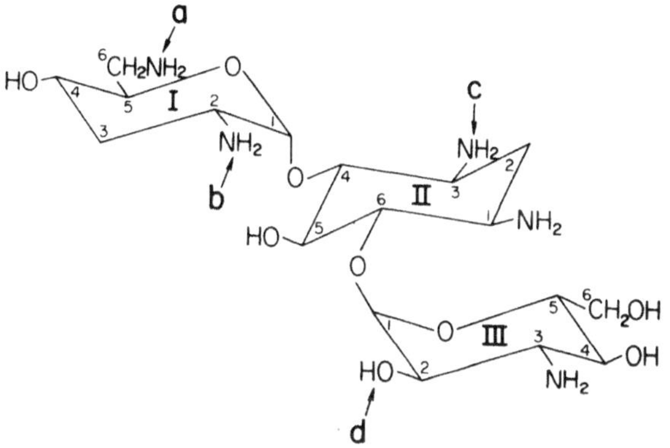

Figure 3. Enzymatic modification of tobramycin. The arrows indicate the positions of modification a, b, c, acetylation; d, adenylylation.

for APH (3')-I and APH (3')-II (Figure 1); it is not known if this is a new (hybrid) enzymatic form (APH (3')-III?) or simply the presence of both enzymes in the same bacterial isolate.

We have also screened other mechanisms for isozymic forms and have found a new 3-*N*-acetylating enzyme of broad specificity in a clinical isolate of *Pseudomonas aeruginosa*. Modification of the 3-amino group of 2-deoxystreptamine has been reported previously in *P. aeruginosa* (Brzezinska *et al.*, 1972; Kobayashi *et al.*, 1972) and also in R-factor carrying *E. coli* (Umezawa, *et al.*, 1973b; Biddlecome, 1973). *P. aeruginosa* (Travers) was found to be resistant to 3'-deoxy antibiotics of the gentamicin and kanamycin classes and thus clearly different from strains carrying AAT (3)-1. When cell-free extracts were examined, an acetylating enzyme of wide substrate specificity was found (Table 3); a sample of sisomicin was acetylated and NMR and mass spectrographic analysis indicated that modification occurred on the 3-amino group. This was confirmed by comparison with an authentic sample of 3-*N*-acetylsisomicin prepared by acetylating sisomycin with AAE (3)-I. (Biddlecome *et al.*, in preparation). We propose to name this enzyme AAT (3)-II; this brings to 4 the number of positions that can be modified on both sisomicin and tobramycin (Fig. 2, 3).

O-Adenylylating enzymes also exist in a variety of forms; an activity which catalyzes *O*-adenylylation of spectinomycin but not streptomycin has been found in several strains of *S. aureus* (Mitsuhashi, personal communication; Davies, unpublished observations). Previously, an enzyme that adenylylates both streptomycin and spectinomycin had been reported from a variety of bacterial sources. We believe that it is very likely that many of the aminoglycoside-modifying enzymes will have forms which differ in their substrate ranges; for example, the substrates activities shown in Table 4 indicate that *P. aeruginosa*

TABLE 4

Substrate range of acetyltransferase from P. aeruginosa 3796

Substrate	% Reaction
Gentamicin C_1	7
Gentamicin C_{1a}	35
Sisomicin	44
Kanamycin A	100
Kanamycin C	0
Tobramycin	24
Amikacin	2
Lividomycin	0
Neomycin	100
Butirosin	7

3796 possesses an acetylating enzyme which differs from those previously reported. We have discovered N-acetylating enzymes in certain Streptomycetes which have different substrate ranges from the normal AAC found in gram-negative organisms (Dowding, unpublished observations). Thus a large number of antibiotic-modifying enzymes exist in nature to complicate the clinical use of aminoglycosides; it is no longer possible to determine the nature of enzymatic modifications simply on the basis of resistance phenotypes or substrate ranges.

There have been many reports on the role of the host bacterium in the expression of resistance-determinants (Smith, 1969). One of the most extreme examples of this phenomenon is the behavior of R-factors in *P. mirabilis* in which enhanced expression of r-determinants occurs, due to replicative amplification of the plasmid (Rownd *et al.*, 1973). Mutants have been described in which extra copies of the R-factor DNA are made and resistance levels are concomitantly increased (Nordstrom *et al.*, 1972; Odakwa *et al.*, 1973). Transfer of an R-factor from one *E. coli* host to another or from *E. coli* to *S. typhimurium* can also give rise to altered resistance levels. We would like to cast our hat into the ring in this area by describing the isolation of mutants in which the bacterial host alters the resistance spectrum of R-factors, by mechanisms which are currently unknown.

E. coli R5/W677 (R5 determines a 6′-*N*-acetylating enzyme AAT(6′)) is normally resistant to 20–30 μg/ml kanamycin (Benveniste and Davies, 1971). We have isolated spontaneous mutants resistant to 300 μg/ml kanamycin at a frequency of approximately $10^{-7} - 10^{-8}$ by plating 10^9 cells of *E. coli* R5/W677 on rich agar containing 300 μg/ml kanamycin. Two clones JR134 and JR135 were examined for resistance phenotype; as can been seen in Table 5, the mutant strains were resistant to neomycin, tobramycin and amikacin (BBK.8) whereas R5/W677 is only weakly resistant to these drugs. The AAT (6′) in the mutant and wild-type strains appear to be identical in substrate range (Table 6) and other properties.

TABLE 5

Resistance Phenotypes of E. coli R5/W677 (6′-N-acetyltransferase) and High Level Resistant Mutants

	R5/W677	*JR134*	*HL97*
Ampicillin	res	res	res
Chloramphemicol	res	res	res
Kanamycin	res (25)	res (300)	res (300)
Streptomycin	res	res	res
Tetracycline	res	res	res
Gentamicin (Complex)	sens	weak res	weak res
Sisomicin	sens	res	res
Amikacin	weak res	res	res
Tobramycin	weak res	res	res

TABLE 6

*Substrate ranges of acetyltransferases from E. coli R5/W677 and high-level resistant mutant JR134**

	Enzyme	
Substrate	R5/W677	JR134
Kanamycin A	100	100
Kanamycin B	49	75
Kanamycin C	0	0
Gentamicin C_{1a}	66	42
Sisomycin	29	31
Tobramycin	99	107
Neomycin B	54	60
Butirosin	36	47
Amikacin	59	64

* Expressed as % relative to kanamycin A.

We have been able to show that *E. coli* JR134 and JR135 contain mutations in the host bacterium which control expression of plasmid-determined resistance. The R5 plasmids from JR134 and JR135 show normal, low levels of resistance when transferred to another host and, in addition, when *E. coli* JR134 and JR135 were "cured" of their plasmids by ethidium bromide treatment, the two recipients amplified the levels of expression of several other resistance plasmids that were transferred into them (Table 7). The location and nature of the mutations (we propose to call these *P*lasmid-*am*plification—*pam* mutants) is being further investigated.

There are a number of mechanisms which could account for increases in resistance level and phenotype that are determined by the host organism. One possibility is that the host controls the number of R-factor copies that can be made or that host components (such as RNA polymerase factors) can control the transcription of r-determinants.

Another mechanism which could control resistance levels would be to adjust the amounts of ATP or acetyl coenzyme A which are available as co-substrates.

I would like to suggest yet another possibility, that resistance levels are determined by certain host membrane components and that the basic mechanism of resistance to aminoglycoside antibiotics is *not* antibiotic inactivation, but inhibition of transport of the antibiotic into the cell. How could this come about? We know that modifying enzymes are required for resistance since R-factor mutants lacking these enzymes are drug-sensitive. Complete detoxification of the antibiotic is not essential in the case of the aminoglycoside antibiotics since we can find no change in antibiotic concentration in the medium. Although there is little evidence to support this notion at the moment, I would like to suggest that in an R-factor culture, a small amount of antibiotic enters the cell periplasmic space, is modified by an R-factor enzyme on the inner membrane and remains associated with the inner membrane to block the transport of the antibiotic through the membrane (pores?) to the ribosome target site.

TABLE 7

Resistance levels of R-factor s-a in three different hosts

	E. coli		
	W677	*K802N*	*135–100*
Chloramphenicol	sens	sens	sens
Butirosin	sens	sens	sens
Amikacin	sens	sens	res
Gentamicin C_1	sens	sens	res
Neomycin	sens	sens	weak res
Tobramycin	sens	sens	res

Finally, I would like to discuss the nature of resistance-determinants. For any given R-factor the number of genes present on its DNA is unknown. An R-factor with a molecular weight of 50×10^6 would be expected to code for 50–70 proteins, and this probably includes about 20 resistance-determinants. For most R-factors we can recognize five or six resistance-determinants but the identity of the remaining genes is unknown. Since, in *E. coli*, R-factors are stable and do not segregate unused resistance-determinants we have considered the possibility that some r-determinants may exist which are of no use to their host.

TABLE 8

"Useless" resistance determinants

Strains	Antibiotic*					
	Fus	*Cli*	*Ery*	*Pac*	*Pur*	*Nov*
CF.1	19	20	20	20	21	16
R699a/CF.1	0	0	12	18	0	0
JR200/CF.1	0	0	15	15	10	0
R806a/CF.1	11	20	30	22	24	20
R57b/CF.1	0	0	12	20	8	0
R805ak/CF.1	0	0	14	15	10	0

* Results are expressed as kill zones around a 20 μg disc. Fusidic acid-Fus; Clindamycin-Cli; Erythromycin-Ery; Pactamycin-Pac; Puromycin-Pur; Novobiocin-Nov.

Several workers have described mutants of *E. coli* that have altered outer surfaces and that are sensitive to a variety of agents (Ennis, 1971; Tamaki *et al.*, 1971). These mutants become sensitive to antibiotics such as fusidic acid and clindamycin, which are effective against gram-positive organisms such as *S. aureus* but not normally active against *E. coli*. We have previously described R-factor-determined fusidic acid resistance in *E. coli* (Datta *et al.*, 1974). A super-sensitive mutant of *E. coli* (CF/1) was obtained by R. Aline following mutagenesis with bacteriophage mμ; we introduced a variety of R-factors into this strain and tests for antibiotic resistance were carried out; the results are shown in Table 8. To our suprprise, a number of R-factors were found to determine resistance to antibiotics that are effective against gram-positive organisms but do not normally affect *E. coli*. This raises the possibility that resistance plasmids in *E. coli* and *S. aureus* may be closely related and share a common origin. As we have discussed in connection with the possible origin of antibiotic resistance genes in streptomyces (Benveniste and Davies, 1973b; Davies and Benveniste, 1974) two difficulties in considering the origin of R-factors are: (1) mechanisms of genetic transfer between unrelated species and (2) the selective forces which operate to generate stable collections of resistance-determinants on plasmids. If staphylococcal plasmids and gram-negative R-factors are related, these problems will be further complicated.

REFERENCES

BENVENISTE, R. and DAVIES, J. (1971). Acetylation of aminoglycoside antibiotics by R-factor enzymes. Biochemistry **10**, 1787.

BENVENISTE, R. and DAVIES, J. (1973). Mechanisms of antibiotic resistance in bacteria. Ann. Rev. Biochem. **42**, 471.

BENVENISTE, R. and DAVIES, J. (1973). Aminoglycoside antibiotic inactivating enzymes in Actinomycetes similar to those present in clinical isolates of antibiotic-resistant bacteria. Proc. Nat. Acad. Sci. U.S., **70**, 2276.

BIDDLECOME, S. (1973). M.S. Thesis. University of Wisconsin.

BRZEZINSKA, M., BENVENISTE, R., DAVIES, J., DANIELS, P. J. L., and WEINSTEIN, J. (1972). Gentamicin resistance in strains of *Pseudomonas aeruginosa* mediated by enzymatic *N*-acetylation of the 2-deoxystreptamine moiety. Biochemistry **11**, 761.

BRZEZINSKA, M. and DAVIES, J. (1973). Two enzymes which phosphorylate neomycin and kanamycin in *Escherichia coli* strains carrying R-factors. Antimicrob. Ag. Chemother. **3**, 266.

DAVIES, J. and BENVENISTE, R. E. (1974). Enzymes inactivating antibiotics in transit to their targets. Ann. N.Y. Acad. Sci., in press.

DATTA, N., HEDGES, R. W., BECKER, D., and DAVIES, J. (1974). Plasmid-determined fusidic acid resistance in the *Enterobacteriacae*. J. Gen. Microbiol. **83**, in press.

ENNIS, H. I. (1971). Mutants of *Escherichia coli* sensitive to antibiotics J. Bacteriol. **107**, 486.

KOBAYASHI, F., YAMAGUCHI, M., EDA, J., HIRAMATSU, M., and MITSUHASHI, S. (1972). Gentamicin C acetylating enzyme from resistant strains of *Pseudomonas aeruginosa* and the structure of the acetylated product. Gunma Rep. Med. Sci. **5**, 291.

NORDSTROM, K., INGRAM, L. C., and LUNDBÄCK, A. (1972). Mutations in R-factors of *Escherichia coli* causing an increased number of R-factor copies per chromosome. J. Bacteriol. **110**, 562.

ODAKURA, Y., TANAKA, T., HASHIMOTO, H., and MITSUHASHI, S. (1973). Mutation of R-factors capable of specifying hypersynthesis of penicillinase. Antimicrob. Ag. Chemother. **3**, 315.

RICHMOND, M. H., JACK, G. W., and SYKES, R. B. (1971). Mechanisms of drug resistance. The beta-lactamases of gram-negative bacteria including Pseudomonads. Ann. N.Y. Acad. Sci. **182**, 243.

ROWND, R., PERLMAN, D., HASHIMOTO, H., MICKEL, S., APPLEBAUM, E., and TAYLOR, D. (1973). Dissociation and reassociation of the transfer factor and resistance determinants of R-factors as a mechanism of gene amplification in bacteria. From Proceedings of the 6th Miles Symposium on Molecular Biology, ed. R. F. Beers, Jr. and R. C. Tilghman, The Johns Hopkins University Press, P. 115.

SMITH, J. T. (1969). R-factor gene expression in gram-negative bacteria. J. Gen. Microbiol. **55**, 109.

TAMAKI, S., SATO, T. and MATSUHASHI, M. (1971). Role of lipopolysaccharides in antibiotic resistance and bacteriophage adsorption in *Escherichia coli* K-12. J. Bacteriol. **105**, 968.

UMEZAWA, H., YAMAMOTO, H., YAGISAWA, M., KONDO, S., TAKEUCHI, T., and CHABBERT, Y. A. (1973). Kanamycin phosphotransferase I: mechanism of cross resistance between kanamycin and lividomycin. J. Antibiot. **26**, 407.

UMEZAWA, H., YAGISAWA, M., MATSUHASHI, Y., NAGANAWA, H., YAMAMOTO, H., KONDO, S., TAKEUCHI, T., and CHABBERT, Y. A. (1973). Gentamicin acetyltransferase in *Escherichia coli* carrying R-factor. J. Antibiot. **26**, 612.

BB-K8 (AMIKACIN), A SEMISYNTHETIC KANAMYCIN DERIVATIVE

H. KAWAGUCHI

Bristol-Banyu Research Institute, Meguro, Tokyo, Japan

Resistance to the aminoglycoside antibiotics mediated by the R-factor or plasmid has increasingly been seen in clinical strains and is probably responsible for most of the resistance to this group of antibiotics[26]. Umezawa and his associates have carried out extensive studies in characterizing the inactivation products of aminoglycoside antibiotics modified by the bacterial enzymes obtained from the resistant organisms. Ivestigations

Structure of butirosins and ribostamycin

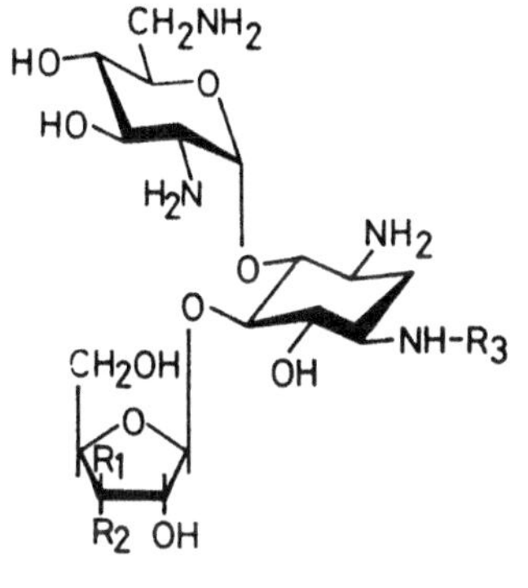

	R_1	R_2	R_3
Butirosin A	OH	H	$CO{-}\underset{OH}{CH}{-}CH_2{-}\underset{NH_2}{CH_2}$
Butirosin B	H	OH	$CO{-}\underset{OH}{CH}{-}CH_2{-}\underset{NH_2}{CH_2}$
Ribostamycin	H	OH	H

Fig. 1

in this area have also been actively pursued in other academic institutions, especially in the laboratories of Davies et al., and Mitsuhashi et al., and it has been well established that the major inactivation mechanisms of the aminoglycoside-resistant organisms are N-acetylation[2, 5, 12, 30, 41, 43], O-phosphorylation[6, 7, 9, 10, 15, 16, 19, 20, 29, 32, 34, 42] and O-adenylylation[3, 31, 40]. Based on this knowledge of the resistance mechanism, Umezawa et al. synthesized 3′-deoxykanamycin A[35] and 3′,4′-dideoxykanamycin B (DKB)[33, 36], which were in fact found to be active against kanamycin-resistant *E. coli* and *P. aeruginosa*[17].

Another type of chemical modification of the aminoglycoside antibiotics was revealed by the discovery of butirosin[8, 37, 38, 39], which was isolated from the fermentation broth of *Bacillus circulans* by Parke-Davis workers and also independently at our laboratories[28] (Fig. 1). Butirosin was the first aminoglycosidic antibiotic produced by a bacterial strain and also has an amino acid moiety in the molecule. Moreover, butirosins have significant

Structure of BB-K 8 (Amikacin*)

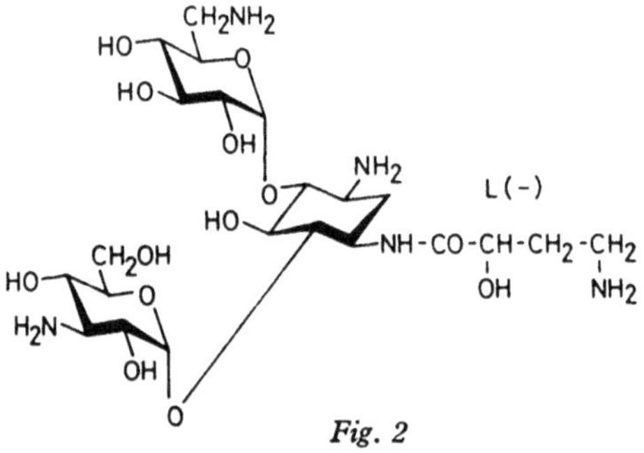

Fig. 2

Preparation of BB-K 8 (Amikacin)

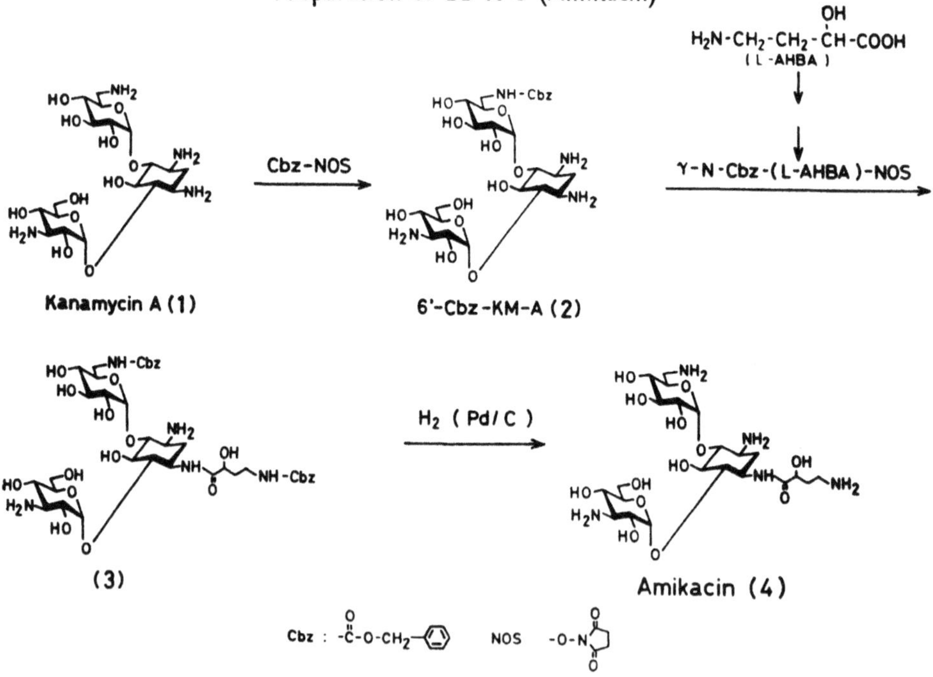

Fig. 3

anti-pseudomonal activity and inhibit some aminoglycoside-resistant organisms. The des-acyl derivative of butirosin B is ribostamycin[1, 25], a natural antibiotic produced by a *Streptomyces* strain. A comparison of the antibacterial spectra of butirosin and ribostamycin (Table 1) suggested that the selective acylation of a specific amino group of an

TABLE 1

Comparative Antibacterial Spectra of Butirosin, Ribostamycin and Kanamycin

	MIC (mcg/ml)		
	Butirosin	Ribostamycin	Kanamycin
S. aureus Smith	0.8	1.6	0.8
S. aureus A 20239	1.6	>100	100
E. coli Juhl	0.8	1.6	0.8
E. coli K 12	0.4	0.8	0.8
E. coli A 20665	0.4	>100	>100
E. coli A 20635	50	>100	100
P. aeruginosa D 15	6.3	>100	25
P. aeruginosa H 9	>100	>100	>100

TABLE 2

Physico-chemical Properties of Amikacin Base

Appearance	White needle crystals
Melting point	203—204 °C
Specific rotation	$[\alpha]_D^{23} = +99°$ (*c* 1.0, H_2O)
Formula	$C_{22}H_{43}N_5O$
Infrared spectrum (KBr)	major absorption bands at 3350, 2930, 1640, 1580, 1350 cm^{-1}
NMR spectrum (100 MHz in D_2O, δ in ppm, J in Hz):	1.1–2.3 (4H, m), 2.6–3.95 (19H, m), 4.13 (1H, d—d, J = 4.5 & 8.5), 4.97 (1H, d, J = 3.5), 5.21 (1H, d, J = 3.5).

aminoglycoside antibiotic with a specific amino acid might improve the activity and broaden the spetrum. These findings led us to investigate new types of aminoglycoside modification which yielded a new kanamycin derivative, BB-K8[13], now given the generic name "Amikacin". The structure of amikacin is shown in Fig. 2. It is a derivative of kanamycin A acylated with the L(—)-γ-amino-α-hydroxybutyryl(l-AHB) side chain at the C-1 amino group of the deoxystreptamine moiety.

Amikacin was synthesized[13] by the sequence of reactions shown in Fig. 3. The 6′-amino group of kanamycin A (*1*), which was found to be the most reactive amino function among the four amino groups of the kanamycin molecule, was protected first with a carbobenzoxy (Cbz) group by the activated ester method to give 6′-Cbz-kanamycin A (*2*). The acylating agent was prepared by protecting the γ-amino group of the amino acid (L-AHBA) with a Cbz group followed by reaction with N-hydroxysuccinimide (NOS) to form the active ester. The acylation occurred predominantly at the second most reactive amino function of kanamycin to give compound *3*. Subsequent hydrogenolysis to remove both of the protecting groups on the 6′- and the γ-amino groups yielded amikacin. The physico-chemical properties of the crystalline base of amikacin are shown in Table 2.

The site of acylation in the amikacin molecule was unequivocally determined[13] by chemical degradation studies to be the C-1 amino group of the 2-deoxystreptamine moiety as shown in Fig. 2. Consequently, three positional isomers of amikacin are possible, which are acylated by the same amino acid at any one of the three amino groups other than the C-1 position. As will be discussed in the latter part of this paper, these three isomers were actually obtained but it was rather surprising that none of these compounds except amikacin showed substantial antibiotic activity.

TABLE 3

In Vitro Activity against E. coli (17 strains)

BBRI Code	Test organism	MIC (mcg/ml)		MIC Ratio
		Amikacin	Kanamycin	
Ec-1	*Escherichia coli* NIHJ	1.6	3.1	2
Ec-3	*Escherichia coli* Juhl	3.1	6.3	2
Ec-4	*Escherichia coli* A15169	1.6	1.6	1
Ec-5	*Escherichia coli* ML-1630	1.6	>100	>64
Ec-6	*Escherichia coli* A9844	1.6	1.6	1
Ec-7	*Escherichia coli* A20365	0.2	>100	>128
Ec-8	*Escherichia coli* K 12	1.6	1.6	1
Ec-9	*Escherichia coli* NR79/W677	3.1	12.5	4
Ec-10	*Escherichia coli* JR35/C600	0.2	>100	>128
Ec-44	*Escherichia coli* A9535	1.6	3.1	2
Ec-45	*Escherichia coli* A15148	1.6	3.1	2
Ec-50	*Escherichia coli* A20109	3.1	3.1	1
Ec-52	*Escherichia coli* W677	1.6	1.6	1
Ec-53	*Escherichia coli* JR66/W677	6.3	>100	>16
Ec-58	*Escherichia coli* A9675	1.6	3.1	2
Ec-59	*Escherichia coli* A20766	3.1	>100	>32
Ec-60	*Escherichia coli* A20898	12.5	>100	>8

TABLE 4

In Vitro Activity against Klebsiella (8 strains), Enterobacter (3 strains) and Serratia (5 strains)

BBRI Code	Test organism	MIC (mcg/ml)		MIC Ratio
		Amikacin	Kanamycin	
Kp-1	*Klebsiella pneumoniae* D-11	0.4	1.6	4
Kp-2	*Klebsiella pneumoniae* A9678	3.1	3.1	1
Kp-3	*Klebsiella pneumoniae* A9977	0.2	1.6	8
Kp-4	*Klebsiella pneumoniae* A15130	3.1	3.1	1
Kp-8	*Klebsiella pneumoniae* Type 22 -3038	1.6	>100	>64
Kx-1	*Klebsiella* sp. A9661	0.8	3.1	4
Kx-2	*Klebsiella* sp. A9662	1.6	1.6	1
Kx-3	*Klebsiella* sp. A20346	1.6	1.6	1
El-1	*Enterobacter cloacae* A9656	6.3	3.1	1/2
El-2	*Enterobacter cloacae* A20364	1.6	>100	>64
El-4	*Enterobacter cloacae* A20650	6.3	6.3	1
Sm-1	*Serratia marcescens* A20019	3.1	3.1	1
Sm-2	*Serratia marcescens* A20335	3.1	>100	>32
Sm-4	*Serraria marcescens* A20442	1.6	3.1	2
Sm-5	*Serratia marcescens* A20222	0.8	3.1	4
Sm-6	*Serratia marcescens* A20460	0.8	>100	>128

Amikacin exhibited a broad spectrum of antimicrobial activity against both kanamycin-sensitive and resistant organisms[13, 21]. Its intrinsic activity against kanamycin-sensitive strains was generally equal to or somewhat higher than that of kanamycin A. In addition, amikacin showed inhibitory activity against various types of aminoglycoside-resistant organisms including *Pseudomonas* species.

The minimum inhibitory concentration (MIC) determined by a serial dilution method in Mueller-Hinton agar was compared with that of kanamycin against several groups

TABLE 5

In Vitro Activity against Proteus (16 strains)

BBRI Code	Test organism	MIC (mcg/ml)		MIC Ratio
		Amikacin	Kanamycin	
Pm-1	*Proteus mirabilis* A9554	3.1	3.1	1
Pm-2	*Proteus mirabilis* A9900	0.8	1.6	2
Pm-3	*Proteus mirabilis* A20119	12.5	3.1	1/4
Pm-4	*Proteus mirabilis* A20454	1.6	1.6	1
Pg-1	*Proteus morganii* A9553	3.1	3.1	1
Pg-2	*Proteus morganii* A20031	1.6	3.1	2
Pg-3	*Proteus morganii* A9636	3.1	3.1	1
Pg-4	*Proteus morganii* A15153	3.1	3.1	1
Pg-5	*Proteus morganii* A15166	1.6	1.6	1
Pr-1	*Proteus rettgeri* A15167	1.6	3.1	2
Pr-2	*Proteus rettgeri* A9637	0.8	1.6	2
Pv-1	*Proteus vulgaris* A9436	0.4	0.8	2
Pv-2	*Proteus vulgaris* A9526	1.6	3.1	2
Pv-3	*Proteus vulgaris* A9699	0.4	0.8	2
Pv-4	*Proteus vulgaris* ATCC 9920	0.4	0.8	2
Pv-5	*Proteus vulgaris* A9537	1.6	3.1	2

TABLE 6

In Vitro Activity against P. aeruginosa (18 strains)

BBRI Code	Test organism	MIC (mcg/ml)		MIC Ratio
		Amikacin	Kanamycin	
Pa-1	*Pseudomonas aeruginosa* D-15	6.3	50	8
Pa-3	*Pseudomonas aeruginosa* A9930	1.6	12.5	8
Pa-4	*Pseudomonas aeruginosa* H9 D-113	6.3	>100	>16
Pa-5	*Pseudomonas aeruginosa* A15150	3.1	100	32
Pa-7	*Pseudomonas aeruginosa* A9843	6.3	>100	>16
Pa-11	*Pseudomonas aeruginosa* A21616	12.5	>100	>8
Pa-12	*Pseudomonas aeruginosa* A20653	12.5	>100	>8
Pa-13	*Pseudomonas aeruginosa* A9843A	6.3	100	16
Pa-15	*Pseudomonas aeruginosa* A20717	6.3	100	16
Pa-16	*Pseudomonas aeruginosa* A20718	6.3	100	16
Pa-19	*Pseudomonas aeruginosa* A20229	0.8	12.5	16
Pa-20	*Pseudomonas aeruginosa* A20325	3.1	100	32
Pa-21	*Pseudomonas aeruginosa* A20601	12.5	100	8
Pa-23	*Pseudomonas aeruginosa* A20741	12.5	>100	>8
Pa-24	*Pseudomonas aeruginosa* A20896	12.5	>100	>8
Pa-25	*Pseudomonas aeruginosa* A20897	12.5	>100	>8
Pa-27	*Pseudomonas aeruginosa* GN-315	50	>100	>2
Pa-28	*Pseudomonas aeruginosa* CPH10527/72	6.3	>100	>16

TABLE 7

In Vitro Activity against S. aureus (15 strains)

BBRI Code	Test organism		MIC (mcg/ml)		MIC Ratio
			Amikacin	Kanamycin	
Sa-1	*Staphylococcus aureus*	209P	0.8	0.8	1
Sa-2	*Staphylococcus aureus*	Smith	0.4	0.4	1
Sa-3	*Staphylococcus aureus*	#193	0.4	0.8	2
Sa-10	*Staphylococcus aureus*	A20239	3.1	>100	>32
Sa-11	*Staphylococcus aureus*	BX-1633-2	0.4	0.4	1
Sa-33	*Staphylococcus aureus*	Terajima	0.2	0.4	2
Sa-34	*Staphylococcus aureus*	A15092	0.8	0.8	1
Sa-35	*Staphylococcus aureus*	A15094	0.1	0.8	8
Sa-36	*Staphylococcus aureus*	Russell	0.4	0.8	2
Sa-37	*Staphylococcus aureus*	A9524	0.4	0.8	2
Sa-38	*Staphylococcus aureus*	A9534	1.6	1.6	1
Sa-39	*Staphylococcus aureus*	A9578	3.1	1.6	$^1/_2$
Sa-40	*Staphylococcus aureus*	A9601	1.6	3.1	2
Sa-41	*Staphylococcus aureus*	A9602	1.6	6.3	4
Sa-49	*Staphylococcus aureus*	A20240	6.3	>100	>16

TABLE 8

In Vitro Activity against M. tubaculosis (11 strains) in Dubos Liquid Medium

BBRI Code	Test organism		MIC (mcg/ml)*		MIC Ratio
			Amikacin	Kanamycin	
Mt-1	*Mycobacterium tuberculosis*	$H_{37}Rv$	0.6	0.6	2.7
Mt-2	*Mycobacterium tuberculosis*	Schacht	0.2	0.6	3
Mt-3	*Mycobacterium tuberculosis*	Sekiya	>100	>100	—
Mt-4	*Mycobacterium tuberculosis*	Ohyabu	>100	>100	—
Mt-5	*Mycobacterium tuberculosis*	Ogawa	10.5	50	4.8
Mt-6	*Mycobacterium tuberculosis*	Arikawa	1.3	1.0	0.8
Mt-7	*Mycobacterium tuberculosis*	Fujiwara	0.3	0.7	2.3
Mt-8	*Mycobacterium tuberculosis*	Kamio	0.8	1.4	1.8
Mt-9	*Mycobacterium tuberculosis*	Nakamura	0.4	1.1	2.7
Mt-10	*Mycobacterium tuberculosis*	Ohno	1.3	5.3	4.1
Mt-11	*Mycobacterium tuberculosis*	Tarumi	1.1	5.3	4.8

* geometric mean of 4 tests.

of clinically important gram-negative bacteria and *S. aureus*. The results are shown in Tables 3–7. The last column of these tables shows the ratio of the kanamycin MIC to that of amikacin.

The MIC against tubercle bacilli was determined by a serial tube dilution method in Dubos liquid medium after 3 weeks incubation at 37°C. The MIC data shown in Table 8 are geometric means of four independent tests. Amikacin was on the average about two to four-fold more active than kanamycin A against the kanamycin-sensitive strains of *M. tuberculosis* though two of the highly kanamycin-resistant strains were also refractory to amikacin.

Table 9 compares the activity of amikacin with four reference antibiotics, kanamycin A, gentamicin C, tobramycin and butirosin A, against the aminoglycoside-resistant orga-

TABLE 9

Comparative Activity against Aminoglycoside-resistant Organisms

Organism	Enzyme	Inoculum* size (dilution)*	MIC (mcg/ml) Amikacin	Kana-mycin A	Genta-micin C	Tobra-mycin	Butirosin A
E. coli NIHJ	—	10^0	6.3	6.3	3.1	3.1	6.3
		10^2	3.1	3.1	1.6	1.6	3.1
		10^4	1.6	3.1	0.8	0.8	3.1
E. coli ML-1630[19])	NPT I	10^0	3.1	>100	1.6	3.1	3.1
		10^2	1.6	>100	0.8	1.6	3.1
		10^4	1.6	>100	0.8	1.6	3.1
E. coli A20107[23])	NPT II	10^0	3.1	>100	3.1	3.1	>100
		10^2	1.6	>100	1.6	1.6	50
		10^4	1.6	>100	0.8	1.6	50
E. coli A20732[23])	GAS	10^0	1.6	100	50	50	3.1
		10^2	0.8	50	25	25	1.6
		10^4	0.8	50	25	25	1.6
E. coli A20895[23])	GAT I	10^0	3.1	3.1	50	1.6	3.1
		10^2	1.6	3.1	25	0.8	1.6
		10^4	1.6	3.1	25	0.8	1.6
P. stuartii A20894[22])	GAT II	10^0	6.3	6.3	50	50	>100
		10^2	3.1	3.1	25	25	>100
		10^4	1.6	1.6	6.3	12.5	>100
E. coli K-12. R-5[35])	KAT	10^0	12.5	50	0.4	12.5	12.5
		10^2	3.1	25	0.2	6.3	6.3
		10^4	1.6	25	0.2	0.4	6.3
P. aeruginosa GN315[41, 43])	KAT	10^0	>100	>100	25	>100	>100
		10^2	50	>100	12.5	100	>100
		10^4	50	>100	12.5	100	>100

* multiple of dilution of standard broth culture containing ca. 5×10^8 cells/ml.

TABLE 10

In Vitro Studies of Amikacin in Clinical Institutions

Investigator	Institution	No. of strains tested	References
Karney et al. (1973)	Univ. Washington, Seattle	186	Antimicrob. Ag. Chemoth. 3: 388–342
Ries at al. (1973)	Med. Coll. Pennsylvania	200	Antimicrob. Ag. Chemoth. 3: 532–553
Yu & Washington (1973)	Mayo Clinic, Rochester	1037	Antimicrob. Ag. Chemoth. 4: 133–139
Bodey & Stewart (1973)	Univ. Texas	466	Antimicrob. Ag. Chemoth. 4: 186–192
Young & Hewitt (1973)	Univ. California, L. A.	400	Antimicrob. Ag. Chemoth. 4: 617–625
Mitsuhashi et al. (1974)	Gunma Univ. Japan	1208	Jap. J. Antibiotics 27: 189–192

nisms whose mechanisms of resistance have been published[19, 22, 30, 41, 43] or determined at Bristol Laboratories by a method utilizing radio-active ATP or acetyl coenzyme A (K. E. Price, personal communication). Mueller-Hinton agar medium and Steers' multi-inoculator apparatus[27] were used for the MIC determination. In order to assess the relative resistance of these antibiotics to the action of the plasmid-mediated enzymes, the MICs were determined at three different inocula of each strain, namely, 1-, 100- and 10,000-fold dilutions of the standard broth culture which contains approximately 5×10^8 viable cells per ml. The results shown in Table 9 clearly indicated that amikacin is a poor substrate for most of the aminoglycoside-inactivating enzymes including neomycin-kanamycin phosphotransferases I and II (NPT_I and NPT_{II}), gentamicin adenylylate synthetase (GAS), and gentamicin acetyltransferases I and II (GAT_I and GAT_{II}). With the two strains which produce kanamycin acetyltransferase (KAT), amikacin inhibited *E. coli* K12 R-5[30] but not the KAT-producing *Pseudomonas* strain GN315[41, 43]. The MIC of amikacin was influenced by the inoculum size of the resistant organisms to about the same extent as that of the other reference aminoglycoside antibiotics tested.

Comparative susceptibility of a large number of clinical bacterial isolates to amikacin and other aminoglycoside antibiotics has been published from several clinical institutions[4, 11, 17a, 24, 44, 45] (Table 10). Some of these reports indicated that amikacin was active against a substantial number of aminoglucoside-resistant strains which were resistant to either gentamicin or tobramycin, or both. Such resistant organisms included *Enterobacter* sp., *Proteus rettgeri*, *Providencia stuartii*, *Pseudomonas aeruginosa* and *Serratia marcescens*. The mechanisms of resistance or the inactivating enzymes associated with the resistance were not identified in these papers.

Price et al.[22] of Bristol Laboratories in Syracuse, N.Y., have examined 152 strains of the aminoglycoside-resistant isolates which were collected from various clinical sources in Canada, Europe, Japan and the United States. The susceptibility of these resistant strains was determined against nine selected aminoglycoside antibiotics including amikacin, and the organisms were classified into 16 types of "resistance profile" according to the aminoglycoside-inactivating enzyme(s) produced or assumed to be produced by the organisms. Some of their results are reproduced and shown in Tables 11, 12 and 13. Table 11 shows the 16 resistance profiles that would be expected for resistant organisms which produce the enzyme or combination of the enzymes listed in the last column of the table. As shown in the table, the two phosphorylation enzymes, NPT_I and NPT_{II}, can be differentiated by their different reaction to butirosin and lividomycin. In this connection, a new aminoglycoside antibiotic designated as Bu 1975[14] which has recently been isolated in our laboratories and identified as 4′-deoxybutirosin, was found to be active against both NPT_I and NPT_{II}-producing organisms, and therefore it might be useful as an additional tool for the identification of inactivating enzymes.

Table 12 shows the distribution of the 152 resistant strains composed of nine different genera among the 16 resistance profiles. The results suggested that certain resistance profiles often occur specifically in a limited group of bacterial species. For instance, while profile No. 1 (NPT_I) was widely distributed and found among five different genera, the distribution of profile No. 2 (NPT_{II}) occurred only in *E. coli* and *Enterobacter* sp. Profile No. 6 ($GAS+NPT_{II}$) was found predominantly in *K. pneumoniae*, profile No. 8 ($GAT_I+NPT_{I}{}^{+}{}_{II}$) exclusively in P. *aeruginosa*, and profile No. 9 (GAT_{II}) only in *Proteus* and *Providencia* sp.

Antibiotic resistance of the above 152 bacterial strains to amikacin was compared with that to kanamycin, gentamicin and tobramycin (Table 13). In this study, organisms were considered resistant to kanamycin and amikacin if they were not inhibited at 20 mcg/ml, and the MIC limit was taken at 8 mcg/ml for gentamicin and tobramycin. These criteria

TABLE 11

Resistance Profiles of Aminoglycoside Antibiotics

	RESISTANCE PROFILE									Enzyme(s) responsible for profile
Profile No.	Kana-mycin	Genta-micin	Tobra-mycin	Dideoxy-kana B	Amikacin	Butirosin	Livido-mycin	Paromo mycin	Neomycin	
1	R	—	—	—	—	—	R	R	R	NPT_{I}
2	R	—	—	—	—	R	—	R	R	NPT_{II}
3	R	—	—	—	—	R	R	R	R	NPT_{I+II}
4	R	R	R	R	—	—	—	—	—	GAS
5	R	R	R	R	—	—	R	R	R	$GAS+NPT_{I}$
6	R	R	R	R	—	R	—	R	R	$GAS+NPT_{II}$
7	—	R	—	—	—	—	—	—	—	GAT_{I}
8	R	R	—	—	—	R	R	R	R	$GAT_{I}+$ $+ NPT_{I+II}$
9	—	R	R	R	—	R	R	R	R	GAT_{II}
10	R	R	R	R	R	R	—	—	R	KAT
11	R	—	R	R	—	R	R	R	R	unknown
12	R	—	R	R	—	R	—	—	—	unknown
13	R	R	—	R	—	R	R	R	R	unknown
14	R	R	—	—	R	R	R	R	R	unknown
15	R	R	R	R	—	R	R	R	R	unknown
16	R	R	R	R	R	R	R	R	R	unknown

K. E. Price et al (1974)

TABLE 12

Resistance Profiles of 152 Aminoglycoside-resistant Isolates

	No. of organisms with indicated profile										
Profile No.	Esch. coli	Enterobact. sp.	Kleb. pneum.	Proteus sp.	Salmonella sp.	Serr. marces.	Provid stuart.	Pseud. aerug.	Staph. aureus		Inactivating Enzyme(s)
1	4	5	3	—	3	3	—	—	—	18	NPT_I
2	4	3	—	—	—	—	—	—	—	7	NPT_{II}
3	—	—	—	—	—	—	6	—	7	13	NPT_{I+II}
4	1	2	—	—	—	—	—	—	—	3	GAS
5	—	1	2	—	—	3	—	1	—	7	GAS + NPT_I
6	2	—	7	—	—	—	—	1	—	10	GAS + + NPT_{II}
7	1	—	—	—	—	—	1	—	—	2	GAT_I
8	—	—	—	—	—	—	—	15	—	15	GAT_I + + NPT_{I+II}
9	—	—	—	6	—	—	13	—	—	19	GAT_{II}
10	1	—	—	—	—	—	—	—	—	1	KAT
11	—	—	—	1	—	—	1	2	—	4	unknown
12	—	—	—	—	—	6	—	—	—	6	unknown
13	2	—	1	—	—	3	—	—	—	6	unknown
14	—	—	—	—	—	—	—	1	—	1	unknown
15	3	—	—	9	—	3	3	10	—	28	unknown
16	3	4	—	1	—	—	—	4	—	12	unknown
Total No.	21	15	13	17	3	18	24	34	7	152	

K. E. Price et al (1974)

TABLE 13

Incidence of Resistance to Kanamycin, Gentamicin, Tobramycin and Amikacin among 152 Aminoglycoside-resistant Isolates

Organisms	No. of Sources	No. of Strains	No. of Resistant Organisms			
			Kanamycin	Gentamicin	Tobramycin	Amikacin
Escherichia coli	13	21	20	11	12	3
Enterobacter sp.	8	15	15	7	17	4
Klebsiella pneumoniae	6	13	13	9	10	0
Proteus sp.	8	17	11	17	16	1
Providencia stuartii	6	24	10	18	16	0
Serratia marcescens	9	18	18	6	15	0
Salmonella sp.	1	3	3	0	0	0
Pseudomonas aeruginosa	11	34	34	34	16	5
Staphylococcus aureus	3	7	7	0	0	0
Total No.		152	131	102	92	13
%		100	86.2	67.1	60.5	8.6

K. E. Price et al. (1974)

were based on the expected peak serum levels produced following a standard clinical dose of these antibiotics. Thus, the results indicated that 86.2% of the isolates were resistant to 20 mcg/ml of kanamycin, 67.1% to 8 mcg/ml of gentamicin, 60.5% to 8 mcg/ml of tobramycin, and only 8.6% to 20 mcg/ml of amikacin.

Finally, the chemistry and activity of amikacin isomers[18] will be discussed. Since there are four acylable amino groups in kanamycin A, two in the 2-deoxystreptamine (DOS) and one each in the 6-amino-6-deoxy-D-glucose (6-AG) and 3-amino-3-deoxy-D-glucose (3-AG) moieties, three other positional isomers of amikacin are possible (Fig. 4). Direct acylation of intact kanamycin with the N-protected L-AHBA yielded a compound desig-

Positional Isomers of Amikacin

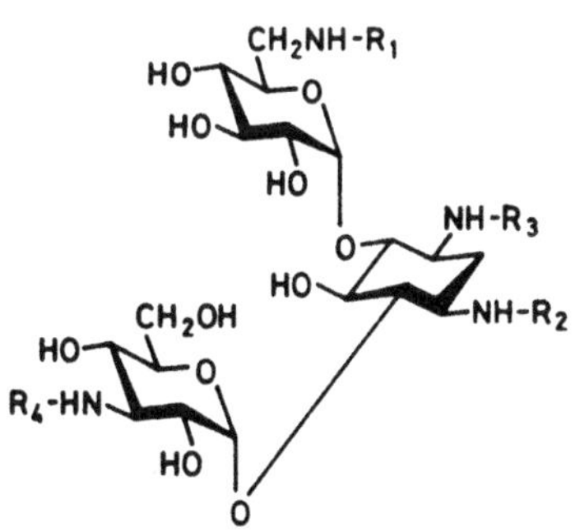

Code #	R1	R2	R3	R4	Relative Activity **
Amikacin	H	L-AHB*	H	H	1,000
BB-K 6	L-AHB	H	H	H	33
BB-K 29	H	H	L-AHB	H	15
BB-K 11	H	H	H	L-AHB	10

* L(-)-γ-amino-α-hydroxybutyryl

** agar diffusion assay (*B. subtilis* plate)

Fig. 4

nated as BB-K6, which was shown to be acylated at the C-6′ amino group of the moiety. The other two isomers, BB-K29 and BB-K11, were synthesized from the suitably N-protected kanamycins (1,6′-di-Cbz-kanamycin and 1,3,6′-tri-Cbz-kanamycin, respectively), and the site of acylation was determined to be C-3 amino group of the DOS and C-3″ amino group of the 3-AG moiety, respectively. In marked contrast to amikacin, the positional isomers are all very weakly active, with about 1–3% of the activity of amikacin by the plate difusion assay. Although the intrinsic activities of BB-K29 and BB-K11 were much lower than kanamycin, they still showed appreciable activity against some of the aminoglycoside-resistant organisms.

Configurational Isomers of Amikacin

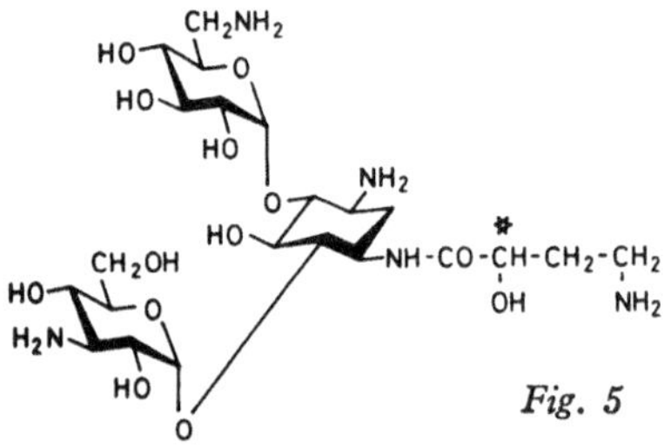

Code #	α-Configuration
Amikacin	L(-)
BB-K 19	DL
BB-K 31	D(+)

Fig. 5

The configurational isomers of amikacin (Fig. 5), which are acylated at the C-1 amino group of kanamycin with DL-AHBA (BB-K19) or D(+)-AHBA (BB-K31) instead of L(—)-AHBA in amikacin, were also prepared. The antibacterial spectra of these configurational isomers were similar to that of amikacin. However, the intrinsic activity of BB-K19 was approximately one-half of amikacin, while the isomer with the amino acid of D-configuration was only one-fourth as active as amikacin.

REFERENCES

1. Akita, E., Tsurucka, T., Ezaki, N. and Niida, T. (1970). Studies on antibiotic SF-733, a new antibiotic. II. Chemical structure of antibiotic SF-733. J. Antibiotic **23**, 173–183.
2. Benveniste, R. and Davies, J. (1971). Enzymatic acetylation of aminoglycoside antibiotics by *Escherichia coli* carrying an R factor. Biochemistry **10**, 1787–1796.
3. Benveniste, R. and Davies, J. (1971). R-factor mediated gentamicin resistance: A new enzyme which modifies aminoglycoside antibiotics. FEBS Letters **14**, 293–296.
4. Bodey, G. P. and Stewart, D. (1973). *In vitro* studies of BB-K8, a new aminoglycoside antibiotic. Antimicrob. Ag. Chemother. **4**, 186–192.
5. Brzezinska, M., Benveniste, R., Davies, J., Daniels, P. J. L. and Weistein, J. (1972). Gentamicin resistance in strains of *Pseudomonas aeruginosa* mediated by enzymatic N-acetylation of the deoxystreptamine moiety. Biochemistry **11**, 761–765.
6. Brzezinska, M. and Davies, J. (1973). Two enzymes which phosphorylate neomycin and kanamycin in *Escherichia coli* strains carrying R factors. Antimicrob. Ag. Chemother., **3**, 266–269.
7. Davies, J., Brzezinska, M. and Benveniste, R. (J971). R factors: Biochemical mechanisms of resistance to aminoglycoside antibiotics. Ann. N. Y. Acad. Sci. **182**, 226–233.
8. Dion, H. W., Woo, P. W. K., Willmer, N. E., Kern, D. L., Onaga, J. and Fusari, S. A. (1972). Butirosin, a new aminoglycoside antibiotic complex: Isolation and characterization. Antimicrob. Ag. Chemother., **2**, 84–88.

9. DOI, O., OGURA, M., TANAKA, N. and UMEZAWA H. (1968). Inactivation of kanamycin, neomycin and streptomycin by enzymes obtained in cells of *Pseudomonas aeruginosa.* Appl. Microbiol., **16**, 1276–1281.
10. DOI, O., KONDO, S., TANAKA, N. and UMEZAWA, H. (1979). Purification and properties of kanamycin-phosphorylating enzyme from *Pseudomonas aeruginosa.* J. Antibiotics **22**, 273–282.
11. KARNEY, W., HOLMES, K. K. and TURCK, M. (1973). Comparison of five aminocyclitol antibiotics *in vitro* against *Enterobacteriaceae* and *Pseudomonas.* Antimicrob. Ag. Chemother. **3**, 338–342.
12. KAWABE, H. and MITSUHASHI, S. (1972). Acetylation of dideoxykanamycin B by *Pseudomonas aeruginosa.* Japan. J. Microbiol. **16**, 436–437.
13. KAWAGUCHI, H., NAITO, T., NAKAGAWA, S. and FUJISAWA, K. (1972). BB-K8, a new semisynthetic aminoglycoside antibiotic. J. Antibiotics **25**, 695–708.
14. KAWAGUCHI, H., TOMITA, K., HOSHIYA, T., MIYAKI, T., FUJISAWA, K., KIMEDA, M., NUMATA, K., KONISHI, M., TSUKIURA, H., HATORI, M. and KOSHIYAMA, H. (1974). Aminoglycoside antibiotics. V. The 4′-deoxybutirosins (Bu-1975 C_1 and C_2), new aminoglycoside antibiotics of bacterial origin. J. Antibiotics, *in press.*
15. KIDA, M., IGARASHI, S., OKUTANI, T., ASAKO, T., HIRAGA, K. and MITSUHASHI, S. (1974). Selective phosphorylation of the 5″-hydroxy group of ribostamycin by a new enzyme from *Pseudomonas aeruginosa.* Antimicrob. Ag. Chemother. **5**, 92–94.
16. KONDO, S., YAMAMOTO, H., NAGANAWA, H., UMEZAWA, H. and MITSUHASHI, S. (1972). Isolation and characterization of lividomycin A inactivated by *Pseudomonas aeruginosa* and *Escherichia coli* carrying R factor. J. Antibiotics **25**, 483–484.
17. MITSUHASHI, S., KOBAYASHI, F., YAMAGUCHI, M., O'HARA, K. and KONO, M. (1972). Enzymatic inactivation of aminoglycoside antibiotics by resistant strains of bacteria. Adv. in Antimicrob. and Antineophastic Chemother., Prague, 1971. pp. 561–564.
17a. MITSUHASHI, S., KAWABE, H., IYOBE, S., TANAKA, T. and INOUE, M. (1974). Antibacterial activity of a new semisynthetic aminoglycoside antibiotic amikacin (BB-K8). Japan J. Antibiotics **7**, 189–192.
18. NAITO, T., NAKAGAWA, S., ABE, Y., TODA, S., FUJISAWA, K., MIYAKI, T., KOSHIYAMA, H., OHKUMA, H. and KAWAGUCHI, H. (1973). Aminoglycoside antibiotics. II. Configurational and positional isomers of BB-K8. J. Antibiotics **26**, 297–301.
19. OKANISHI, M., KONDO, S., UTAHARA, R. and UMEZAWA, H. (1968). Phosphorylation and inactivation of aminoglycosidic antibiotics by *E. coli* carrying R factors. J. Antibiotics **21**, 13–21.
20. OZANNE, B., BENVENISTE, R., TIPPER, D. and DAVIES, J. (1969). Aminoglycoside antibiotics: Inactivation by phosphorylation in *Escherichia coli* carrying R factors. J. Bacteriol. **100**, 1144–1146.
21. PRICE, K. E., CHISHOLM, D. R., MISIEK, M., LEITNER, F. and TSAI, Y. H. (1972). Microbiological evaluation of BB-K8, a new semisynthetic aminoglycoside. J. Antibiotics **25**, 709–731.
22. PRICE, K. E., PURSIANO, T. A., DEFURIA, M. D. and WRIGHT, G. E. (1974). Activity of BB-K8 (Amikacin) against clinical isolates resistant to one or more aminoglycoside antibiotics. Antimicrob. Ag. Chemother. **5**, 143–152.
23. PRICE, K. E. Personal communication.
24. RIES, K., LEVISON, M. E. and KAYE, D. (1973). *In vitro* evaluation of a new aminoglycoside derivative, BB-K8: a comparison with tobramycin and gentamicin.Antimicrob. Ag. Chemother. **3**, 532–533.
25. SHOMURA, T., EZAKI, N., TSURUOKA, T., NIWA, T., AKITA, E. and NIIDA, T. (1970). Studies on antibiotic SF-733, a new antibiotic. I. Taxonomy, isolation and characterization. J. Antibiotics, **23**, 155–161.
26. SMITH, D. H. (1969). R factors for aminoglycoside antibiotics. J. Infect. Dis. 119, 378–380.
27. STEERS, E., FÖLTZ, E. L. and GRAVES, B. S. (1959). An inocula replicating apparatus for routine testing of bacterial susceptibility of antibiotics. Antibiot. Chemoth. **9**, 307–311.
28. TSUKIURA, H., FUJISAWA, K., KONISHI, M., SAITO, K., NUMATA, K., ISHIKAWA, H., MIYAKI, T., TOMITA, K. and KAWAGUCHI, H. (1973). Aminoglycoside antibiotics. III. Bio-active degradation products from butirosins and semi-synthesis of butirosin analogs. J. Antibiotics **26**, 351–357.
29. UMEZAWA, H., OKANISHI, M., KONDO, S., HAMANA, K., UTAHARA, R., MAEDA, K. and MITSUHASHI, S. (1967). Phosphorylative inactivation of aminoglycosidic antibiotics by *Escherichia coli* carrying R factor. Science **157**, 1559–1561.
30. UMEZAWA, H., OKANISHI, M., UTAHARA, R., MAEDA, K. and KONDO, S. (1967). Isolation and

structure of kanamycin inactivated by a cell free system of kanamycin-resistant *E. coli*. J. Antibiotics Ser. A **20**, 136–141.

31. UMEZAWA, H., TAKASAWA, S., OKANISHI, M. and UTAHARA, R. (1968). Adenylylstreptomycin, a product of streptomycin inactivated by *E. coli* carrying R factor. J. Antibiotics **21**, 81–82.
32. UMEZAWA, H., DOI, O., OGURA, M., KONDO, S. and TANAKA, N. (1968). Phosphorylation and inactivation of kanamycin by *Pseudomonas aeruginosa*. J. Antibiotics **21**, 154–155.
33. UMEZAWA, H., UMEZAWA, S., TSUCHIYA, T. and OKAZAKI, Y. (1971). 3′,4′-dideoxykanamycin B active against kanamycin-resistant *Escherichia coli* and *Pseudomonas aeruginosa*. J. antibiotics **24**, 485–487.
34. UMEZAWA, H., YAMAMOTO, H., YAGISAWA, M., KONDO, S., TAKEUCHI, T. and CHABBERT, Y. A. (1973). Kanamycin phosphotransferase I: mechanism of cross resistance between kanamycin and lividomycin. J. Antibiotics **26**, 315–319.
35. UMEZAWA, S., TSUCHIYA, T., MUTO, R., NISHIMURA, Y. and UMEZAWA, H. (1971). Synthesis of 3′-deoxykanamycin effective against kanamycin-resistant *Escherichia coli* and *Pseudomonas aeruginosa*. J. Antibiotics **24**, 274–275.
36. UMEZAWA, S., UMEZAWA, H., OKAZAKI, Y. and TSUCHIYA, T. (1972). Studies on aminosugars XXXII. Synthesis of 3′,4′-dideoxykanamycin B. Bull. Chem. Soc. Jap. **45**, 3624–3628.
37. WOO, P. W. K., DION, H. W. and BARTZ, Q. R. (1971). Butirosins A and B, aminoglycoside antibiotics. I. Structure units. Tetrahedron Letters No. 28: 2617-2620.
38. WOO, P. W. K. (1971). Butirosins A and B, aminoglycoside antibiotics. II. Mass spectrometric study. Tetrahedron Letters No. 28: 2621–2624.
39. WOO, P. W. K., DION, H. W. and BARTZ, Q. R. (1971). Butirosins A and B, aminoglycoside antibiotics. III. Structures. Tetrahedron Letters No. 28: 2625–2628.
40. YAGISAWA, M., NAGANAWA, H., KONDO, S., HAMADA, M., TAKEUCHI, T. and UMEZAWA, H. (1971). Adenylyldideoxykanamycin B, a product of the inactivation of dideoxykanamycin B by *Escherichia coli* carrying R factor. J. Antibiotics **24**, 911–912.
41. YAGISAWA, M., NAGANAWA, H., KONDO, S., TAKEUCHI, T. and UMEZAWA, H. (1972). 6′-N-Acetylation of 3′,4′-dideoxykanamycin B by an enzyme in a resistant strain of *Pseudomonas aeruginosa*. J. Antibiotics **25**, 495–496.
42. YAGISAWA, M., YAMAMOTO, H., NAGANAWA, H., KONDO, S., TAKEUCHI, T. and UMEZAWA, H. (1972). A new enzyme in *Escherichia coli* carrying R-factor phosphorylating 3′-hydroxyl of butirosin A, kanamycin, neamine and ribostamycin. J. Antibiotics **25**, 748–750.
43. YAMAMOTO, H., YAGISAWA, M., NAGANAWA, H., KONDO, S., TAKEUCHI, T. and UMEZAWA, H. (1972). Kanamycin 6′-acetate and ribostamycin 6′-acetate, enzymatically inactivated products by *Pseudomonas aeruginosa*. J. Antibiotics 25, 746–747.
44. YOUNG, L. S. and HEWITT, W. L. (1973). Activity of five aminoglycoside antibiotics *in vitro* against gram-negative bacilli and *Staphylococcus aureus*. Antimicrob. Ag. Chemother. **3**, 617–625.
45. YU, P. K. W. and WASHINGTON II, J. A. (1973). Comparative *in vitro* activity of three aminoglycoside antibiotics: BB-K8, kanamycin and gentamicin. Antimicrob. Ag. Chemother. **4**, 133–139.

MICROBIOLOGICAL ASPECTS OF AMINOGLYCOSIDE RESISTANCE

J. A. WAITZ
Schering Corp., Bloomfield, N.J. U.S.A.

A number of enzymes have been isolated and described from a variety of gram-negative bacteria which can inactivate aminoglycoside antibiotics. Studies with isolated enzymes and extracts of bacteria have been useful for the elucidation of the spectrum and mechanism of inactivation. However, the bacterial host which harbors the enzymes can have marked effects on the degree of inactivation.

TABLE 1
AMINOGLYCOSIDE INACTIVATION ENZYMES

ENZYME	SOURCE	SITE OF ACTION
$NPT_{I\&II}$	R^+ E. COLI P.. AERUGINOSA S. AUREUS	3-HYDROXYL GROUP OF AN AMINO HEXOSE IS PHOSPHORYLATED
SPT	R^+ E. COLI S. AUREUS P. AERUGINOSA	3-HYDROXYL GROUP OF N-METHYL L-GLUCOSAMINE IS PHOSPHORYLATED
L_VPT	P. AERUGINOSA R^+ E. COLI	5-HYDROXYL GROUP OF D-RIBOSE IS PHOSPHORYLATED
SAdT	R^+ E. COLI	HYDROXYL GROUP OF A D-THREO METHYLAMINO ALCOHOL MOIETY IS ADENYLATED
GAS	R^+ E. COLI R^+ K. PNEUMONIAE	2-HYDROXYL GROUP OF AN AMINO-HEXOSE IS ADENYLATED
GAT_I	P. AERUGINOSA R^+ K. PNEUMONIAE R^+ E. COLI	3-AMINO GROUP OF 2-DEOXY STREPTAMINE IS ACETYLATED
GAT_{II}	PROVIDENCIA	2-AMINO GROUP OF AN AMINO HEXOSE IS ACETYLATED
GAT_{III}	P. AERUGINOSA	3-AMINO GROUP OF 2-DEOXY STREPTAMINE IS ACETYLATED
KAT	$^+$R E. COLI P. AERUGINOSA	6-AMINO GROUP OF AN AMINO HEXOSE IS ACETYLATED

From Benveniste & Davis, Ann. Rev. Biochem. *42*, 471–506 (1973).

The purpose of this presentation is to present results of some typical microbiological studies examining the effects of a number of aminoglycosides against a group of gram-negative bacteria containing several known and unknown resistance mechanisms. Our results therefore reflect what a clinical microbiology laboratory may find using conventional bacteriological methods in their routine tests of resistant gram-negative bacteria. The antibiotics studied include gentamicin, kanamycin, tobramycin, neomycin, BBK-8 or amikacin, and verdamicin. Verdamicin which was described last year is the double bonded sisomicin type analogue of gentamicin C_2.

Table 1 shows the aminoglycoside inactivation enzymes described to date from a number of bacterial sources. This is modified from a similar table of Benveniste and Davies. Neomycin phosphotransferase has been found in a number of gram-negative bacteria and *Staphylococcus* and phosphorylates a 3-hydroxyl group of an amino hexos. Streptomycin phosphotransferase, SPT, phosphorylates the 3-hydroxy group of the N-methyl L-glucosamine ring of streptomycin. Lividomycin phosphotransferase phosphorylates a 5-hydroxyl group of D-ribose. Streptomycin adenyltransferase adenylates a hydroxy group of a D-threo methylamino alcohol moiety. Gentamicin adenylsynthetase adenylates a 2-hydroxyl group of an amino hexose. Gentamicin acetyltransferase 1 and gentamicin acetyltransferase 3 both acetylate the 3-amino group of the 2-deoxy streptamine moiety of gentamicin and other aminoglycoside antibiotics. Gentamicin acetyltransferase 2 found in *Providencia* acetylases a 2-amino group of an amino hexose while kanamycin acetyltransferase acetylates the 6-amino group of an amino hexose.

TABLE 2

AMINOGLYCOSIDE RESISTANCE PATTERNS (152 STRAINS)

PROFILE	No.	%
KNOWN		
$NPT_{I\,\&\,II}$ALONE	38	25
$NPT_{I\,\&\,II}$ + OTHER(s)	70	46
GAS ALONE	3	2
GAS + OTHER(s)	20	13
GAT_{I}ALONE	2	1.3
GAT_{I} + OTHER(s)	15	9.8
GAT_{II} ALONE	19	12.5
KAT	1	0.7
UNKNOWN		
6 PATTERNS	57	37.5

From Price, K. E. et al., Antimicrob. Ag. Chemother. *5*, 143–152 (1974).

Little information is available on the relative incidence of these aminoglycoside inactivating enzymes. A number of surveys on sensitivity of gram-negative bacteria to a variety of aminoglycosides show that, although more and more inactivation mechanisms are described each year, the relative percent of various clinical isolates resistant to kanamycin, gentamicin, and tobramycin has not markedly changed over the past years. Many of these resistance mechanisms have been described from isolates outbreaks of resistance.

Table 2 shows the results of one study on the relative prevalence of inactivating enzymes in a group of mostly gram-negative bacteria. This is from an article by Price published this year. Price evaluated 152 strains of aminoglycoside resistant bacteria provided from a number of sources and amongst these several could be identified with inactivation mechanisms. Of the 152 strains, approximately 46% contained neomycin phosphotransferase 1 and 2 alone or with other inactivation enzymes, representing 70

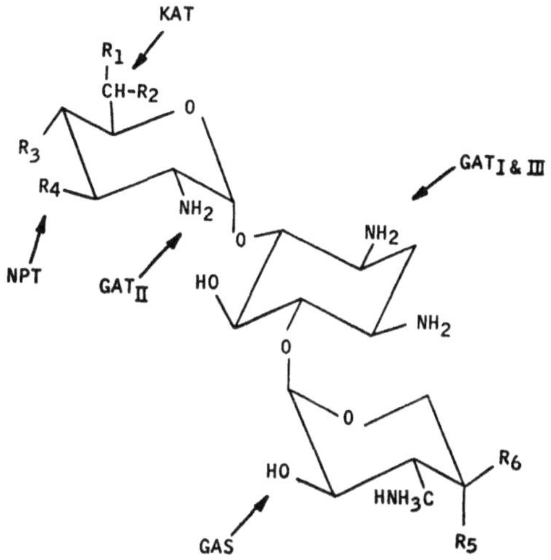

Fig. 1

of the 152 strains. Three of the 152 strains or 2% contain gentamicin adenylsynthetase alone and when these were considered with those that contain other enzymes, as well the total is 13%. Gentamicin acetyltransferase either alone or with other enzymes accounted for approximately 10% of the strains or 15 of the 152, while GAT 2 was found in 12.5 of the strains. Kanamycin acetyltransferase was found in 1 of the strains. Six different patterns of resistance were seen which did not fit any of the known categories and may represent further combination of the above. These accounted for 57 of the 152 or 37.5% of the strains.

Figure 1 shows a generalized aminoglycoside structure indicating the sites of inactivation of the several known mechanisms which will be addressed further. These are the kanamycin acetylating enzyme (KAT) which acetylates a 6-amino group, neomycin phosphorylating enzyme (NPT) which phosphorylates a 3-hydroxyl group, gentamicin acetyltransferase 2 (GAT II) which acetylates a 2-amino group, gentamicin acetyltransferase 1 and 3 (GAT I and III) which acetylate the 3-amino group of 2-deoxy streptamine and gentamicin adenylsynthetase (GAS) which adenylates a hydroxyl group. Bacteria containing these enzymes as well as some unknown mechanisms are the subject of this paper.

Table 3 shows the methods used for agar dilution MIC's. Mueller-Hinton agar with a pH of 7.2–7.4 was poured into plastic petri dishes and left to dry overnight. The inoculum used was an overnight Mueller-Hinton broth culture diluted in water to produce a final inoculum of 10^3–10^4 colony forming units/spot, and the inoculum was placed on the plates using a Steers-Foltz replicator. These were incubated at 37°C overnight and the MIC endpoint was the lowest concentration which had less than 6 isolated discrete colonies growing. Concentrations used in conventional testing are shown on the bottom.

Table 4 shows the procedure used for broth dilution MIC's. Mueller-Hinton broth was used with a volume of 4 mls in glass test tubes. The inoculum was an overnight Mueller-Hinton broth culture diluted in water to produce a final inoculum of 5×10^4–1×10^5 colony forming units/tube. These were incubated at 37°C overnight and the MIC endpoint is the lowest concentration for which no visual turbidity is seen.

TABLE 3

AGAR DILUTION

MUELLER HINTON AGAR pH 7.2–7.4
  30 ml IN 100 mm SQUARE INTEGRID PLASTIC PETRI DISHES DRY OVERNIGHT
INOCULUM — OVERNIGHT MUELLER HINTON BROTH
  CULTURE DILUTE IN H_2O
  FINAL - 10^3 - 10^4 CFU/SPOT
  STEERS-FOLTZ REPLICATOR
  INCUBATE 37°C, 18–24 HOURS
MIC END POINT - <6 ISOLATED, DISCRETE COLONIES
CONCENTRATIONS (mcg/ml)
  64, 32, 16, 8, 4, 2, 1, 0.5, 0.12, 0.06, 0

TABLE 4

BROTH DILUTION

MUELLER HINTON BROTH pH 7.2–7.4
  4 ml IN 150×16 mm GLASS TUBES
INOCULUM — OVERNIGHT MUELLER-HINTON BROTH
CULTURE
  DILUTE IN H_2O
  FINAL - 5×10^4 - 1×10^5 CFU/TUBE
  INCUBATE 37°C, 18–24 HOURS
MIC END POINT — NO VISUAL TURBIDITY
CONCENTRATIONS (mcg/ml)
  64, 32, 16, 8, 4, 2, 1, 0.5, 0.25, 0.12, 0.06, 0

Table 5 shows the method used for disc testing which is the method given in the U.S. Code of Federal Regulations and is the modified method of Bauer *et al.* Mueller-Hinton agar is used in a 150 mm diameter plastic petri dish, the media depth is 4 mm. Inoculum is standardized by comparison with a turbidity standard streaked on with a swab, dried for 3–5 minutes and the disc applied. The disc for kanamycin and neomycin contained 30 mcg, and for the other aminoglycoside antibiotics, 10 mcg discs were used. The endpoint is based on the zone diameter achieved under these test conditions.

Table 6 gives the results of agar dilution MIC determinations for a variety of organisms containing the neomycin phosphorylating enzyme. The pattern on all the subsequent slides is the same, the different bacterial strains are shown vertically, while horizontally across the top are the results with neomycin, kanamycin, gentamicin, tobramycin, verdamicin and amikacin or BBK-8. The *E. coli* MIC values for a sensitive strain ATCC

10536 is shown in contrast to clinical isolates with NPT. MIC's for neomycin and kanamycin are markedly higher for these strains than for normal sensitive *E. coli*. A similar pattern is shown for *Pseudomonas* with typical results of the sensitive strain as opposed to strain DZ which can phosphorylate neomycin and kanamycin. *Klebsiella* strains are

TABLE 5
*DISC TEST**

MUELLER HINTON AGAR pH 7.2–7.4
  150 mm DIAMETER PLASTIC PETRI DISH
  50 ml MEDIA (DEPTH 4 mm)
INOCULUM — STANDARDIZED BY TURBIDITY
STREAK WITH SWAB
DRY 3–5 MINUTES
DISC STRENGTH — 7 mm DISC
NEOMYCIN & KANAMYCIN - 30 mcg
ALL OTHERS - 10 mcg
END POINT - ZONE DIAMETER

*Federal Register modified method of Bauer et al.

TABLE 6
NPT

STRAIN	NM	KM	GM	TM	VM	AMK
AGAR DILUTION - MIC (mcg/ml)						
E. COLI						
SENS 10536	1	4	0.5	0.5	0.5	2
ST. M 589	>32	>32	1	2	1	2
BAK 2	>32	>32	2	2	1	4
PSEUDOMONAS						
SENS St. 20	0.5	16	1	0.25	1	0.5
D-Z	32	>32	4	4	1	4
KLEBSIELLA						
SENS B-63	1	4	0.5	0.5	0.5	1
AD 17	>32	>32	1	1	0.25	2
AD 18	>32	>32	0.5	1	0.5	4
SERRATIA						
SENS 78	1	4	1	4	0.5	2
128	>32	>32	1	4	0.5	2
DALT	>32	>32	1	8	0.5	4
ENTEROBACTER						
GAS B-9	>32	>32	4	8	1	16
JACKS	>32	>32	1	4	0.5	4

also shown with a sensitive strain and typical MIC values in contrast to resistant isolates. Similar results are shown for *Serratia* and *Enterobacter*. It is interesting that *Enterobacter* GAS B-9 has rather high MIC values for amikacin and tobramycin using this method.

TABLE 7
NPT

STRAIN	NM	KM	GM	TM	VM	AMK
	BROTH DILUTION - MIC (mcg/ml)					
E. COLI						
SENS 10536	1	4	0.5	1	0.5	2
ST. M 589	>32	>32	2	4	2	4
BAK 2	>32	>32	2	4	1	4
PSEUDOMONAS						
SENS St. 20	1	8	1	0.5	0.5	0.5
D-Z	16	>32	2	0.5	2	2
KLEBSIELLA						
SENS B-63	0.5	1	0.5	0.5	0.25	1
AD 17	>32	>32	0.5	0.5	0.25	2
AD 18	>32	>32	0.5	1	0.25	4
SERRATIA						
SENS 78	0.5	1	1	4	0.5	2
128	16	>32	1	2	0.12	2
DALT	>32	>32	2	4	0.5	4
ENTEROBACTER						
CAS B-9	>32	>32	4	8	2	8
JACKS	>32	>32	1	4	0.5	2

Table 7 shows the same organisms evaluated by broth dilution procedures. Generally the results are quite similar to those found in agar dilution procedures. Organisms with neomycin phosphorylating activity are highly resistant to neomycin, kanamycin, and quite sensitive to the other aminoglycoside antibiotics. Exceptions were seen in the increased sensitivity of *Serratia* 128 and *Pseudomonas* DZ to neomycin with broth dilution procedures relative to agar dilution procedures. Again a slight decrease in sensitivity of *Enterobacter* GAS B-9 to tobramycin and amikacin is apparent.

Table 8 shows the results of disc sensitivity tests with the same organisms. These results are shown in terms of zone diameters in mm. Again the neomycin phosphorylating organisms can be clearly spotted as resistant to neomycin and kanamycin, while they retain sensitivity to the other aminoglycosides. Interestingly, the increase in MIC values in both agar dilution and broth dilution tests for the *Enterobacter* GAS B-9 for tobramycin and amikacin are not detected using disc sensitivity tests.

Table 9 gives the results of agar dilution MIC determinations for organisms having the gentamicin adenylsynthetase or gentamicin adenylating enzyme. These again are contrasted with typical sensitive strains. The *E. coli*'s having this resistance pattern are typically resistant to kanamycin, gentamicin, tobramycin, and verdamicin and retain

TABLE 8
NPT

STRAIN	NM	KM	GM	TM	VM	AMK
		ZONE DIAMETER DISC TEST (mm)				
E. COLI						
SENS 10536	25	31	30	30	31	31
ST. M 589	10	0	26	25	27	26
BAK 2	0	0	27	27	28	26
PSEUDOMONAS						
SENS St. 20	26	23	36	36	34	36
DZ	12	0	20	28	23	23
KLEBSIELLA						
SENS B-63	19	23	21	24	25	22
AD 17	0	0	25	27	27	26
AD 18	0	0	25	25	27	24
SERRATIA						
SENS 78	22	23	24	21	22	23
128	12	0	23	21	29	23
DALT	0	0	24	22	27	25
ENTEROBACTER						
CAS B-9	0	0	21	21	25	21
JACKS	0	0	20	20	22	20

TABLE 9
GAS

STRAIN	NM	KM	GM	TM	VM	AMK
		AGAR DILUTION – MIC (mcg/ml)				
E. COLI						
SENS 10536	1	4	0.5	0.5	0.5	2
677/R55	2	>32	>32	>32	>32	2
JR 66	>32	>32	>32	>32	32	8
LA 290 R55	1	>32	>32	>32	32	2
KLEBSIELLA						
SENS B-63	1	4	0.5	0.5	0.5	1
G 3694	32	>32	>32	32	16	2
G 3020	32	>32	32	32	8	1
OK 6	>32	>32	>32	>32	32	2

sensitivity to amikacin and neomycin. Strain JR 66 contains an R factor that has not only gentamicin adenylating activity but also the neomycin phosphorylating activity. This then is also resistant to neomycin and has an increase in MIC values for amikacin. This R factor is the same as that found in *Klebsiella* G3694 and G3020. Interestingly, when this R factor is in a *Klebsiella*, the MIC's are reduced for verdamicin. The increased MIC seen with amikacin and *E. coli* was not detected in *Klebsiella*. OK 6 is a very good adenylator and may also have phosphorylating activity on the basis of high resistance to neomycin. This is sensitive to only amikacin.

Table 10 shows the same strains tested by broth dilution procedures. The pattern is similar to that seen with agar dilution procedures. Interestingly, the strain JR 66 which is neomycin resistant and contains not only the adenylating enzyme but the phosphoryla-

TABLE 10
GAS

STRAIN	NM	KM	GM	TM	VM	AMK
	BROTH DILUTION – MIC (mcg/ml)					
E. COLI						
SENS 10536	1	4	0.5	1	0.5	2
677/R55	2	>32	>32	>32	>32	2
JR 66	>32	>32	>32	>32	>32	8
LA 290 R55	1	>32	>32	>32	32	2
KLEBSIELLA						
SENS B-63	0.5	1	0.5	0.5	0.25	1
G 3694	32	>32	>32	32	32	2
G 3020	32	>32	32	32	8	1
OK 6	>32	>32	>32	>32	32	1

TABLE 11
GAS

STRAIN	NM	KM	GM	TM	VM	AMK
	ZONE DIAMETER DISC TEST (mm)					
E. COLI						
SENS 10536	25	31	30	30	31	31
677/R55	21	10	10	11	14	25
JR 66	0	0	0	0	13	24
LA 290 R55	23	11	11	11	17	28
KLEBSIELLA						
SENS B-63	19	23	21	24	25	22
G 3694	0	0	11	11	15	22
G 3020	13	0	12	12	15	27
OK 6	0	0	0	9	12	22

ting enzyme shows the increased MIC's with amikacin, while the same R factor in *Klebsiella* G2694 and G3030 does not show this. In addition, 2 strains of *Klebsiella* from the same source and carrying the same R factor showed different sensitivity to verdamicin.

Table 11 shows the results of disc sensitivity tests with the same organisms. These tests clearly show resistance to kanamycin, gentamicin, tobramycin, and verdamicin and sensitivity to neomycin and amikacin. The general pattern seen is the same as that seen with MIC's. However, the disc tests did not detect the difference between amikacin's activity against JR 66 and the *Klebsiella*. It is also interesting that verdamicin shows rather large zone diameters against several strains which are resistant to it. This suggests problems with the disc test methods.

TABLE 12

STRAIN	NM	KM	GM	TM	VM	AMK
AGAR DILUTION - MIC (mcg/ml)						
GAT_I						
E. COLI						
SENS 10536	1	4	0.5	0.5	0.5	2
JR 88	4	8	>32	2	>32	4
JR 90	4	8	>32	2	>32	4
PSEUDOMONAS						
SENS ST. 20	0.5	16	1	0.25	1	0.5
ST. 130	16	32	>32	0.5	>32	2
ST. 138	8	32	>32	0.5	>32	4
CAPE 18	16	32	>32	1	>32	2
GAT_{II}						
PROVIDENCIA						
163	>32	8	>32	>32	8	8
GAT_{III}						
PSEUDOMONAS						
TRAVERS	>32	>32	>32	>32	>32	4
SHREVE	32	>32	>32	>32	>32	8

Table 12 gives the results of agar dilution tests in a number of organisms containing acetyltransferase activity. The first two sets are *E. coli* and *Pseudomonas* containing gentamicin acetyltransferase 1. This gives resistance to gentamicin and verdamicin but does not inactivate neomycin, kanamycin, tobramycin, and amikacin. The *Pseudomonas* strains show a similar pattern of resistance to gentamicin and verdamicin but not tobramycin and amikacin. The increase in MIC's for *E. coli* is rather modest. The *Providencia* strain contains GAT II which acetylates the 2-amino group and is characterized by sensitivity to kanamycin, resistance to gentamicin, tobramycin, and neomycin; amikacin and verdamicin show fairly high MIC's relative to their activity against normal susceptible organisms. GAT III acetylates aminoglycosides at the same site as GAT I does. The interesting difference here is the high resistance to tobramycin in contrast to GAT I. These *Pseudomonas* organisms are resistant to all of the aminoglycosides except amikacin.

Table 13 shows the same organisms tested by broth dilution procedures. The results are not strikingly different from those observed in agar dilution tests. GAT I gives resistance to gentamicin and verdamicin in the *E. coli*, and higher MIC's with neomycin and kanamycin against *Pseudomonas*. This suggests a difference between *Pseudomonas* and *E. coli*. GAT II is resistant to neomycin, gentamicin, and tobramycin and shows reduced activity with verdamicin and kanamycin. GAT III conveys resistance to all of the aminoglycosides except amikacin.

TABLE 13

STRAIN	NM	KM	GM	TM	VM	AMK
BROTH DILUTION - MIC (mcg/ml)						
GAT I						
E. COLI						
SENS 10536	1	4	0.5	1	0.5	2
JR 88	1	4	>32	2	>32	2
JR 90	1	8	>32	1	>32	4
PSEUDOMONAS						
SENS ST. 20	1	8	1	0.5	0.5	0.5
ST. 130	8	>32	>32	1	>32	1
ST. 138	4	32	>32	1	>32	2
CAPE 18	8	>32	>32	0.25	>32	1
GAT II						
PROVIDENCIA						
164	>32	16	>32	>32	8	4
GAT III						
PSEUDOMONAS						
TRAVERS	>32	>32	>32	>32	>32	2
SHREVE	16	>32	>32	>32	>32	4

Table 14 shows the results of disc sensitivity tests with the same organisms. These results confirm that gentamicin and verdamicin are not active against GAT I organisms, containing either *E. coli* or *Pseudomonas*. Again they show the difference in kanamycin activity against GAT I in *E. coli* and GAT I in *Pseudomonas*, although this may in part be due to the relatively low incidence of *Pseudomonas* strains that are sensitive to kanamycin. The *Providencia* strain is shown very clearly on results of disc tests to be resistant to neomycin, gentamicin, tobramycin; sensitive to kanamycin and amikacin. The suggestion of increased activity of verdamicin against this strain is shown with disc tests as well. GAT III is shown in two different *Pseudomonas*. These organisms are resistant to all of the aminoglycosides except amikacin.

TABLE 14

STRAIN	NM	KM	GM	TM	VM	AMK
	ZONE DIAMETER DISC TEST (mm)					
	GAT I					
E. COLI						
SENS 10536	25	31	30	30	31	31
JR 88	22	28	0	27	0	25
JR 90	22	27	9	30	0	27
PSEUDOMONAS						
SENS ST. 20	26	23	36	36	34	36
ST. 130	14	0	0	30	0	25
ST. 138	19	12	0	31	0	25
CAPE 18	19	12	12	35	11	31
	GAT II					
PROVIDENCIA						
164	0	20	0	12	21	24
	GAT III					
PSEUDOMONAS						
TRAVERS	0	0	0	0	0	24
SHREVE	0	0	0	0	0	23

TABLE 15
KAT

STRAIN	NM	KM	GM	TM	VM	AMK
	AGAR DILUTION – MIC (mcg/ml)					
E. COLI						
SENS 10536	1	4	0.5	0.5	0.5	2
R5/677	1	>32	0.5	>32	0.25	8
HL 97/677	>32	>32	4	>32	2	>32
PSEUDOMONAS						
SENS St. 20	0.5	16	1	0.25	1	0.5
GN. 315	16	>32	2	32	2	32

Table 15 gives the results of agar dilution tests with organisms containing kanamycin acetyltransferase which acetylates the 6-amino group of kanamycin, tobramycin, and amikacin. Two *E. coli* R factor strains are shown, with clear resistance to neomycin, kanamycin, tobramycin, and amikacin. It is interesting that R 5/677 is sensitive to neomycin and much more sensitive to amikacin, although this effect was not seen with tobramycin. Verdamicin and gentamicin are active against these organisms based on the fact that 2 of the 3 gentamicin components, C_1 and C_2 are not acetylated by this organism while gentamicin C_{1a} is. Results for *Pseudomonas* strain Gn 315 again show a high level of resistance to neomycin, kanamycin, tobramycin, and amikacin.

Table 16 gives the results of broth dilution tests. KAT containing strains are resistant to kanamycin, tobramycin, and amikacin, but sensitive to gentamicin and verdamicin.

Table 17 shows the results of disc tests against the same organisms. As suggested by dilution procedures, strains containing 6 acetylating enzymes are resistant to kanamycin, tobramycin, and amikacin. The disc tests do not show as great a decrease in activity for tobramycin and amikacin as do dilution tests. This is particularly true for the *E. coli* strains.

Table 18 shows the results of agar dilution tests on a selection of organisms with unknown resistance mechanisms. *E. coli* Swindinsky is resistant to neomycin, kanamycin, tobramycin, and amikacin. It is sensitive to verdamicin but has reduced sensitivity to gentamicin. This is an R factor and has been found to be resistant to gentamicin C_1 but sensitive to C_{1a} and C_2. *Pseudomonas* Shriners are resistant to all of the aminoglycosides thus far evaluated, tobramycin shows the greatest activity against these strains which is considered to be a reflection of its greater activity against *Pseudomonas* in general.

TABLE 16
KAT

STRAIN	NM	KM	GM	TM	VM	AMK
	BROTH DILUTION – MIC (mcg/ml)					
E. COLI						
SENS 10536	1	4	0.5	1	0.5	2
R5/677	2	>32	1	32	1	32
HL 97/677	16	>32	4	>32	4	>32
PSEUDOMONAS						
SENS ST. 20	1	8	1	0.5	0.5	0.5
GN. 315	16	>32	2	32	1	16

TABLE 17
KAT

STRAIN	NM	KM	GM	TM	VM	AMK
	ZONE DIAMETER DISC TEST (mm)					
E. COLI						
SENS 10536	25	31	30	30	31	31
R5/677	23	15	28	20	31	22
HL 97/677	20	0	26	14	30	13
PSEUDOMONAS						
SENS ST. 20	26	23	36	36	34	36
GN. 315	13	0	21	12	24	15

Proteus Hard is fairly sensitive to the aminoglycosides using agar dilution procedures. *Proteus* OK-1 and Rand are highly resistant to all of the aminoglycosides except amikacin, while the Woosely strain and R membel show reduced sensitivity to amikacin but increased sensitivity to verdamicin.

TABLE 18
UNKNOWN

STRAIN	NM	KM	GM	TM	VM	AMK
AGAR DILUTION – MIC (mcg/ml)						
E. COLI						
SENS 10536	1	4	0.5	0.5	0.5	2
SWIDINSKY	>32	>32	8	>32	4	16
PSEUDOMONAS						
SENS ST. 20	0.5	16	1	0.25	1	0.5
SHRINERS 10099	>32	>32	>32	8	32	>32
SHRINERS 10006	>32	>32	>32	8	32	>32
PROTEUS						
HARD	4	4	2	2	0.5	4
OK-1	>32	>32	>32	>32	>32	1
V WOOSELY	1	4	0.5	2	0.5	8
R MEMB.	4	4	8	16	0.5	8
R AND.	>32	>32	>32	>32	32	4

TABLE 19
UNKNOWN

STRAIN	NM	KM	GM	TM	VM	AMK
BROTH DILUTION – MIC (mcg/ml)						
E. COLI						
SENS 10536	1	4	0.5	1	0.5	2
SWIDINSKY	>32	>32	8	>32	4	32
PSEUDOMONAS						
SENS ST. 20	1	8	1	0.5	0.5	0.5
SHRINERS 10099	>32	>32	>32	8	8	32
SHRINERS 10006	>32	>32	>32	8	8	32
PROTEUS						
HARD	4	16	8	8	2	16
OK-1	>32	>32	>32	32	32	1
V WOOSELY	8	4	4	8	1	16
R MEMB.	2	1	32	16	0.5	16
R AND.	>32	>32	>32	>32	>32	8

Table 19 shows the results of broth dilution tests. The Swidinsky strain is resistant to neomycin, kanamycin, tobramycin, and amikacin; has reduced sensitivity to gentamicin and is quite sensitive to verdamicin. The Shriners strain with broth dilution procedures show greater sensitivity to tobramycin and verdamicin, although relative to normally sensitive strains, these can be considered resistant. *Proteus* strain Hard has an unusual pattern of resistance to kanamycin and amikacin, reduced sensitivity to gentamicin and tobramycin, and is sensitive to verdamicin and neomycin. Woosely shows much greater resistance using broth dilution procedures than agar dilution procedures.

TABLE 20
UNKNOWN

STRAIN	NM	KM	GM	TM	VM	AMK
ZONE DIAMETER DISC TEST (mm)						
E. COLI						
SENS 10536	25	31	30	30	31	31
SWIDINSKY	0	0	20	12	23	18
PSEUDOMONAS						
SENS ST. 20	26	23	36	36	34	36
SHRINERS 10099	0	0	0	20	12	0
SHRINERS 10006	0	0	0	19	13	0
PROTEUS						
HARD	24	22	23	24	24	22
OK-1	0	0	9	10	14	23
V WOOSELY	19	24	23	24	25	23
R MEMB.	19	25	21	24	28	24
R AND.	0	0	0	0	12	22

Table 20 shows the results of disc tests against the strains with unknown resistance mechanisms. *E. coli* Swindinsky results generally confirm those of the dilution tests indicating resistance to neomycin, kanamycin, tobramycin, and amikacin, and reduced activity of gentamicin. The Shriners *Pseudomonas* strains appear on the basis of disc tests to be sensitive to tobramycin, although this is not supported by the broth dilution results. The *Proteus* strains suggest general sensitivity of Hard, Woosely, and Memb, but this is not supported by the results of dilution tests. This may be a reflection of the unusually large zone sizes produced against *Proteus* by aminoglycoside antibiotics in general.

In summary, we have evaluated a group of gram-negative bacteria with a number of known and unknown resistance mechanisms using conventional broth dilution, agar dilution, and disc test methods as might be done in a clinically oriented microbiology laboratory. Among the antibiotics evaluated, have been gentamicin, kanamycin, tobramycin, BBK-8 or amikacin, verdamicin, and neomycin. Our results suggest the presence of bacterial host effects particularly with regard to degree of inactivation, and substantial method effects on the MIC values obtained. While studies of isolated enzymes may be useful in understanding inactivation mechanisms, conventional antibacterial tests may be more meaningful from the viewpoint of clinical utility, although these may produce quite different results depending upon the method used.

COMPARATIVE STUDIES OF ANTIBACTERIAL ACTIVITY OF VARIOUS AMINOGLYCOSIDE ANTIBIOTICS

S. MITSUHASHI and H. KAWABE
Department of Microbiology, School of Medicine, Gunma University
Maebashi, Japan

Since the discovery of streptomycin (SM), many aminoglycoside antibiotics have been introduced into clinical medicine and have been found to be one of the most effective antibiotics against infections with various pathogenic bacteria. Epidemiologic studies have disclosed, however, that the isolation frequency of aminoglycoside-resistant bacteria is now increasing in response to the universal widespread use of the drugs. According to increase in the incidence of infections caused by drug-resistant bacteria, especially by R factor-resistant ones, many studies have been focused on the biochemical mechanisms and genetics of drug-resistance.

Recent studies have disclosed the following: (1) transmissible (R) or nontransmissible (*r*) plasmids are widely distributed among many species of resistant bacteria (Mitsuhashi, 1972; Mitsuhashi et al., 1973), and (2) biochemical mechanisms of resistance in clinical isolates are mostly different from those of *in-vitro* developed resistant mutants and are generally common in many bacterial strains isolated in different places all over the world.

TABLE 1
Enzymatic inactivation of aminoglycoside antibiotics

Mechanism	Positions modified	Inactivated Product [a]
Phosphorylation	3-OH of N-methyl-L-glucosamine	A-3″-phosphate
	3-OH of 6-aminoglucose	A-3′-phosphate
	3-OH of 2,6-diaminoglucose	A-3′-phosphate
	3-OH of glucosamine	A-3′-phosphate
	5-OH of ribose	A-5″-phosphate
Acetylation	6-NH_2 of 6-aminoglucose	A-6′-acetate
	6-NH_2 of 2,6-diaminoglucose	A-6′-acetate
	6-NH_2 of 3,4-dideoxy-2,6-diaminoglucose	A-6′-acetate
	6-NH_2 of purpurosamine	A-6′-acetate
	6-NH_2 of 4,5-didehydropurpurosamine	A-6′-acetate
	3-NH_2 of 2-deoxystreptamine	A-3-acetate
Adenylylation	3-OH of N-methyl-L-glucosamine	A-3″-adenylylate
	2-OH of 3-aminoglucose	A-2″-adenylylate
	2-OH of garosamine	A-2″-adenylylate
	2-OH of 2,6-diaminoglucose	A-2′-adenylylate

[a]) A stands for aminoglycoside.

As shown in Table 1, enzymatic inactivation of aminoglycoside antibiotics is considered to be the main mechanism of resistance to the drugs (Umezawa, 1974; Benveniste and Davies, 1973; Mitsuhashi et al., 1974). Based on these findings, a new scientific approach has been introduced into chemotherapeutic science and the known aminoglycoside antibiotics have been tailored to render them once more effective against resistant bacteria. 3′,4′-Dideoxykanamycin B (DKB) was synthesized from KM-B and was found to be

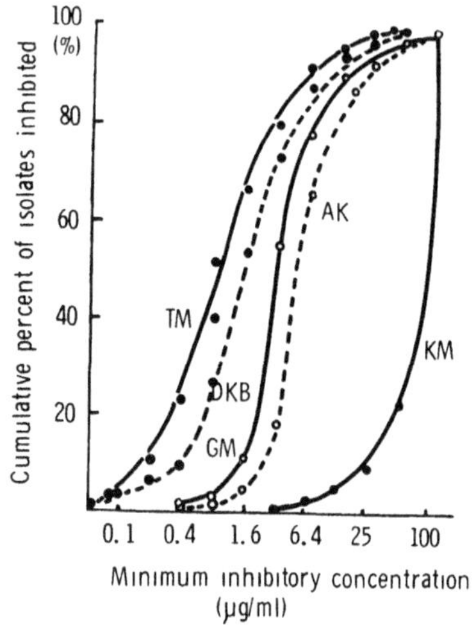

Fig. 1. Comparative susceptibility of *P. aeruginosa* to various aminoglycoside antibiotics. Results based on surveys of 650 strains of *P. aeruginosa*.

effective against *P. aeruginosa* as well as KM-resistant strains of bacteria; this being partly due to resistance to aminoglycoside 3′-O-phosphotransferase (Umezawa et al., 1971).

Amikacin (AK), a semisynthetic aminoglycoside antibiotic was produced by acylation of the 1-amino group of the 2-deoxystreptamine moiety of KM-A with L(—)-γ-amino-α-hydroxy butyric acid (Kawaguchi et al., 1972). Thus, amikacin (BB-K8) has been found to be effective even against *Pseudomonas aeruginosa* in addition to kanamycin-resistant strains of bacteria. Effectiveness of gentamicin (GM)-C complex (Weinstein et al., 1963) and tobramycin (Thompson and Prsti, 1968) toward bacterial strains resistant to known aminoglycoside antibiotics can be also explained by knowledge of the biochemical mechanisms of aminoglycoside resistance.

As shown in Fig. 1, the newly introduced aminoglycoside antibiotics are found to be quite effective against *Pseudomonas aeruginosa* strains of clinical origin, which are known to be highly resistant to various chemotherapeutic agents. It should be noted however that some of *P. aeruginosa* strains are already resistant to DKB, tobramycin (TB) or amikacin (AK) before the introduction of these drugs into practical use and the isolation frequency of GM-resistant strains is now increasing in many countries owing to widespread use of the drug.

We collected GM-resistant strains of *P. aeruginosa* and their susceptibility to DKB was examinated (Table 2). It was found that most of the GM-resistant strains were sensitive to DKB and AK. Next, we selected *P. aeruginosa* strains resistant to these new aminoglycoside antibiotics from our stock cultures and their comparative susceptibility to various aminoglycoside antibiotics was examined. As shown in Table 3, the KM-, (KM.GM)-, (KM.GM.DKB)- and (KM.GM.DKB.AK)-resistant strains of *P. aeruginosa* can be demonstrated, indicating a tendency of the multiplicity of resistance.

TABLE 2

Antibacterial activity of new aminoglycoside antibiotics against GM-resistant strains of P. aeruginosa

Microorganism		Susceptibility to				
isolated in	No.	SM	KM	GM	DKB	AK
Europe	8	r	r	r	s	s
	1	r	r	r	r	s
USA	10	r	r	r	s	s

r, resistant; s, sensitive.

TABLE 3

Susceptibility of P. aeruginosa strains to various aminoglycoside antibiotics
r, resistant; s, sensitive

Bacterial strain	KM-A	GM-C	DKB	AK
GN3433	r	s	s	s
GN4043	r	s	s	s
GN4044	r	s	s	s
GN1567	r	r	s	s
GN4468	r	r	s	s
GN4471	r	r	s	s
GN3049	r	r	r	s
GN3059	r	r	r	s
GN4494	r	r	r	s
GN 573	r	r	r	r
GN3056	r	r	r	r

r, resistant s, sensitive.

According to the biochemical studies of aminoglycoside resistance in *P. aeruginosa*, it was found that one of the inactivation mechanisms of the drugs was due to acetylation of the 6′-NH_2 group of aminoglycosides.

Then, we selected GM-C_1, 6′-N-methyl DKB and amikacin (AK) (Fig. 2) and their antibacterial activities were examined using *P. aeruginosa* strains resistant to KM, DKB and GM-C_{1a}. As shown in Table 4, we could demonstrate two types of resistant strains, i.e., (6′-N-Me DKB)^r AK^s and (6′-N-Me DKB)^s AK^r (r, resistant; s, sensitive; Me, methyl). To know the biochemical mechanism of resistance, cell-free extracts were prepared from GN4925 and GN315, and purified by affinity chromatography.

TABLE 4

Susceptibility of P. aeruginosa strains to various aminoglycoside

Bacterial strain	KM-A	DKB	GM-C_{1a}	GM-C_1	6′-N-Me DKB	AK
GN4925	r	r	r	r	r	s
GN5462	r	r	r	s	r	s
GN 315	r	r	r	s	s	r
GN 269	r	r	r	s	s	r

r, resistant; s, sensitive; Me, methyl.

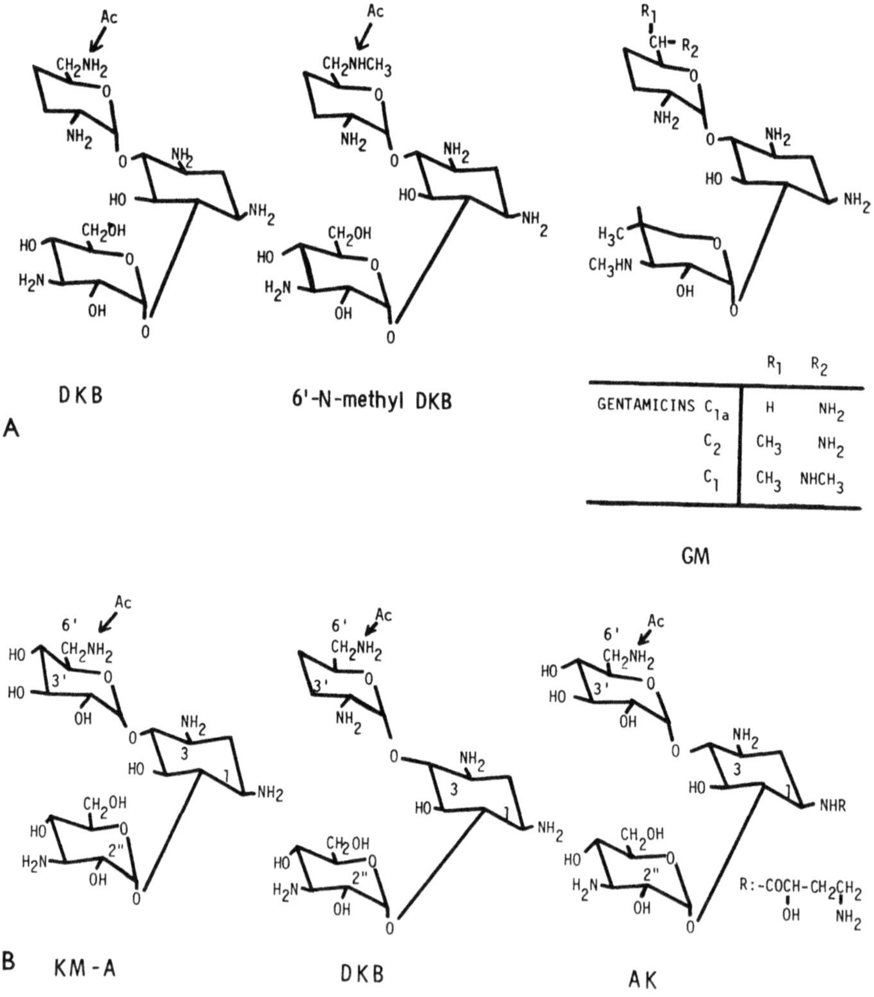

Fig. 2. Chemical structure of aminoglycoside antibiotics. Arrow indicates the acetylated position.

As shown in Fig. 3, a purified sample of aminoglycoside acetyltransferase (AACase) from GN4925 could inactivate both DKB and 6′-N-Me DKB, but did not both GM-C_1 and AK. But we do not know at present why GN4925 is resistant to GM-C_1 but can not

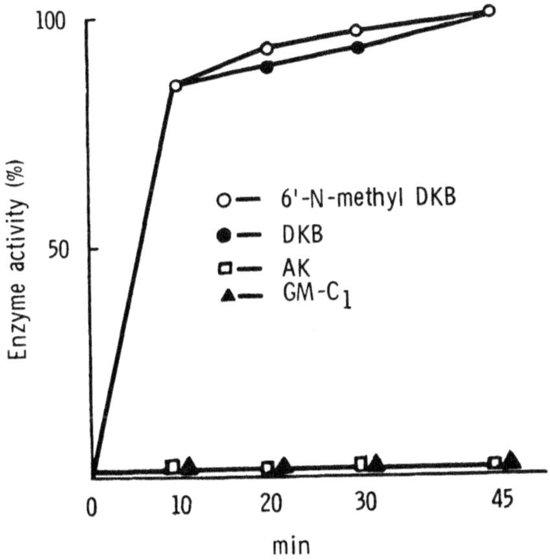

Fig. 3. Acetylation of aminoglycoside antibiotics by a purified sample of aminoglycoside acetyltransferase (AACase) from GN4925. GN4925 AACase was purified by affinity chromatography. The reaction mixture contained the following materials in 0.5 ml of final volume; 0.2 ml of 0.2M acetate buffer (pH 6.0), 0.05 ml of 0.5mM substrate, 0.05 ml of 0.5mM acetyl-CoA, 0.1 ml of purified enzyme (protein concentration, 150 mg/ml), and 0.1 ml of distilled water.

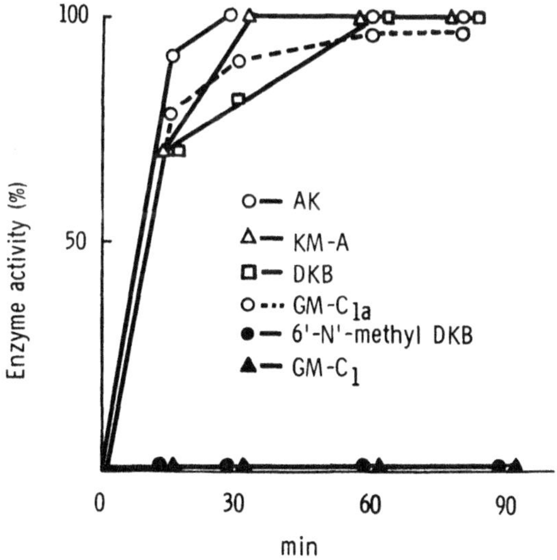

Fig. 4. Acetylation of aminoglycoside antibiotics by a purified sample of aminoglycoside acetyltransferase (AACase) from GN315. GN315 AACase was purified by affinity chromatography. Reaction mixture, see the footnote of Fig. 3.

inactivate the drug. Resistance mechanism of GN4925 to GM-C_1 is predicted to be due to impermeability of the drug.

By contrast, a purified GN315 AACase inactivated AK, KM, DKB and GM-C_{1a}, but did not inactivate both 6′-N-Me DKB and GM-C_1 (Fig. 4). These results coincide with their MIC levels toward GN4925 and GN315.

The purified GN315 AACase and GN4925 AACase showed a single peak of activity

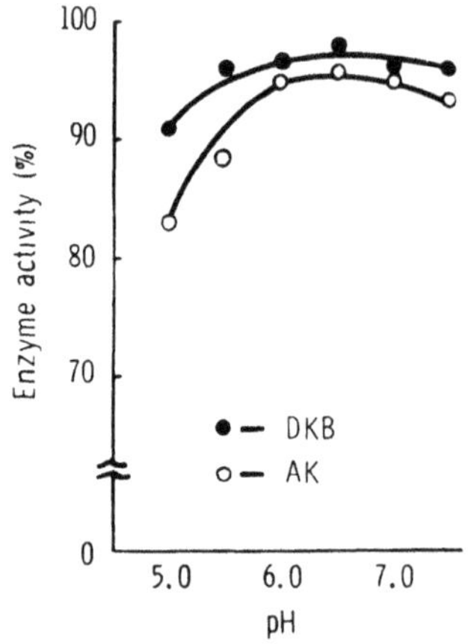

Fig. 5. Effect of pH on the activities of aminoglycoside acetyltransferase (AACase) from GN315. A purified sample of AACase by affinity chromatography was used.

curves toward (KM, DKB and AK) and (KM, DKB and 6′-N-Me DKB), respectively, suggesting that these enzymes were purified as an almost homogeneous state. Effect of pH values on the activities of the purified GN315 AACase and GN4925 AACase was studied; the activities curves against each drug being almost the same at wide range of pH value. One of the representative results is shown in Fig. 5. These results indicated that the aminoglycoside resistance of GN315 and GN4925 was due to acetylation of the drugs and they produced their own type of aminoglycoside acetyltransferase (AACase).

The inactivated products of DKB, 6′-N-Me DKB and AK were purified through column chromatography. Chemical analysis of the inactivated products indicated to be the monoacetylated products of DKB, 6′-N-Me DKB and AK. Paper chromatography, thin layer chromatography, mass spectra and nuclear magnetic resonance (nmr) spectra of the inactivated products indicated that the structure of the inactivated products was

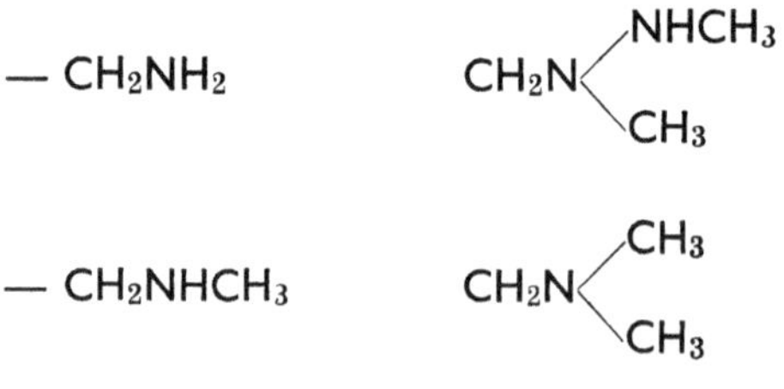

Fig. 6. Chemical modification of the 6′-amino group in the aminosugar linked to 2-deoxystreptamine moiety

6′-N-acetyl aminoglycosides. Therefore, it is concluded that there are two types of 6′-N-acetyltransferase; (1) one enzyme acetylates both 6′-NH_2 and 6′-N-Me, and (2) another one acetylates only 6′-NH_2 but does not 6′-N-Me (Fig. 6).

ACKNOWLEDGEMENT

The authors wish to express deepest appreciation to H. Umezawa, H. Kawaguchi and M. J. Weinstein for supply of DKB, 6′-N-Me DKB, AK and each component of gentamicin-C complex, respectively. We wish to thank H. Umezawa, S. Kondo and H. Kawaguchi for their helpful advice in carrying out these experiments.

REFERENCES

BENVENISTE, R. and DAVIES, J. (1973). Mechanism of antibiotic resistance in bacteria. Ann. Rev. Biochem. in press.

KAWAGUCHI, H., NAITO, T., NAKAGAWA, S. and FUJISAWA, K. (1972). BB-K8, a new semisynthetic aminoglycoside antibiotic. J. Antibiotics, 25, 695.

MITSUHASHI, S. (1971). Transferable drug resistance factor R. University of Tokyo Press, Tokyo; University Park Press, Baltimore and London.

MITSUHASHI, S., INOUE, M., KAWABE, H., OSHIMA, H. and OKUBO, T. (1973). Genetic and biochemical studies of drug resistance in staphylococci. Microbiology and Immunology, Vol. 1. Staphylococci and Staphylococcal infections (Karger, Basel 1973), pp. 144–165.

MITSUHASHI, S., HASHIMOTO, H., TANAKA, T. and IYOBE, S. (1974). Genetics and Biochemistry of R factors. Adv. Antimicrob. and Antineoplastic Chemoth.-Athens, 1973, in press.

THOMPSON, R. Q. and PRSTI, E. A. (1968). Nebramycin, a new broad-spectrum antibiotic complex. III. Isolation and chemical properties. Antimicrob. Ag. Chemoth. – 1967, pp. 332–340.

UMEZAWA, H. (1974). Biochemical mechanisms of resistance to aminoglycoside antibiotics. Drug action and drug resistance in bacteria. II. Aminoglycoside antibiotics. University of Tokyo Press. in press.

UMEZAWA, H., UMEZAWA, S., TSHUCHIYA, T. and OKAZAKI, Y. (1971). 3′, 4′-Dideoxykanamycin B active against Kanamycin-resistant *Escherichia coli* and *Pseudomonas aeruginosa*. J. Antibiotics 24, 485.

WEINSTEIN, M. J., LUEDEMANN, G. M., ODEN, E. M. and WAGMAN, G. H. (1963). Gentamicin, a new broad spectrum antibiotic complex. Antimicrob. Ag. Chemoth. – 1963. pp. 1–13.

RESISTANCE OF MORAXELLA TO TOBRAMYCIN, KANAMYCIN AND BBK 8 (AMIKACIN)

F. LE GOFFIC and ANNIE MARTEL
Laboratoire de Chimie, Ecole Normale Supérieure, Paris, France

There are now three mechanisms by which aminosidic antibiotics are inactivated by resistant bacteria isolated in clinical situations (Benveniste et *al.*, 1973).

The present communication deals with the isolation and the characterization of a new enzyme harboured by Moraxella strains isolated in two different hospitals in France. This enzyme confers resistance to: kanamycin A (32 γ), BBK 8 (32 γ), tobramycin (32 γ), whereas gentamicin, sisomicin, neomycin remain active on these strains.

The Moraxella are grown in the usual way and collected by centrifugation. They are washed with an appropriate buffer then disrupted by the combination of lysozyme digestion and freezing thawing technique. The supernatant thus obtained after centrifugation at 20,000 rpm is incubated at 37°C in the usual way by using kanamycin A as substrate. When C^{14} Acetyl CoA is the cofactor, the antibiotic incorporates radioactivity whereas neither phosphorylating nor adenylating activity is detected. Thus the enzyme responsible of this resistance pattern is an acetyltransferase.

With a more accurate characterization of this acetylating activity in view the enzyme is purified by affinity chromatography (Le Goffic et *al.*, 1973). The resin used is In-

TABLE 1

Each compound listed in incubated with the enzyme preparation coming from affinity chromatography and C^{14} CoASAc at the pH listed (Tris maleate buffer). Aliquots are withdrawn at 2, 5, 15, 30 and 45 minutes to ensure that a linear assay is always occurring. The values represent the average of at least three different incubation mixtures.

Substrates	% Acetylated compound			
	pH 5.3	pH 6.5	pH 7.5	pH 8.4
Tobramycin	100	10	0	0
Kanamycin A	42	85	61	16
Kanamycin B	65	77	38	4
BBK 8	0	30	19	0
Gentamicin C_{1a}	79	20	7	0
Sisomicin	13	52	16	9
Neomycin	51	25	10	0
Neamine	12	55	30	11
Gentamicin A	0	0	0	0
Gentamicin C_1	0	0	0	0
Paromomycin	0	0	0	0

dubiose A 4 — gentamicin complex. The chromatography is performed in the cold and is followed by continual determination of the U.V. absorption at 280 nm. The enzymatic activity may be monitored in the usual way (Benveniste et *al.*, 1973). The eluent used is a gradient of NH_4Cl containing 10 mM Tris, 10 mM $MgCl_2$ (pH 7.4).

As shown in Figure 1 when the chromatography is conducted with untreated indubiose as stationary phase there is a complete overlap of protein nucleic acids and the enzymatic activity. When, however, the chromatography is done on indubiose activated by gentamicin, the elution pattern is quite different. First, those substances which do not bind to the resin are eluted, then on raising the concentration of ammonium chloride to 0.5 M the enzymatic activity appears. This activity is in advance on a peak which is due to the U.V. absorption of the ribosome.

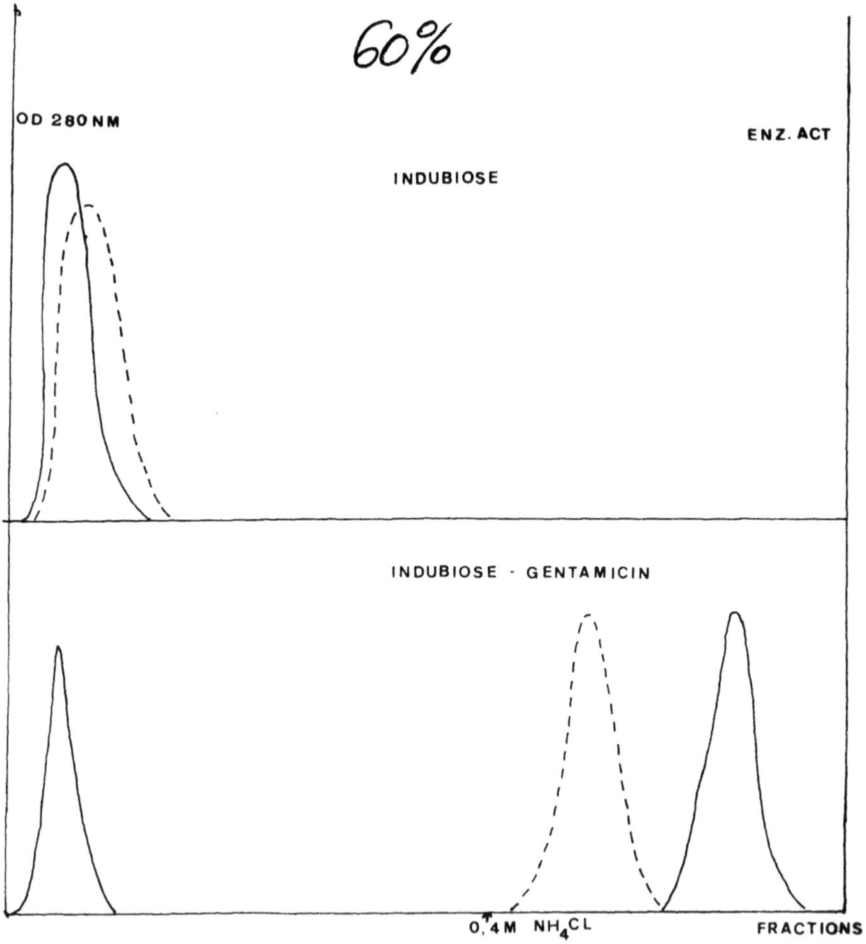

FIGURE 1

Figure 1. Elution patterns of the acetylase of Moraxella on indubiose (a), indubiose-gentamicin (b): absorbance at 280 nm (———), enzymatic activity (- - - - -).

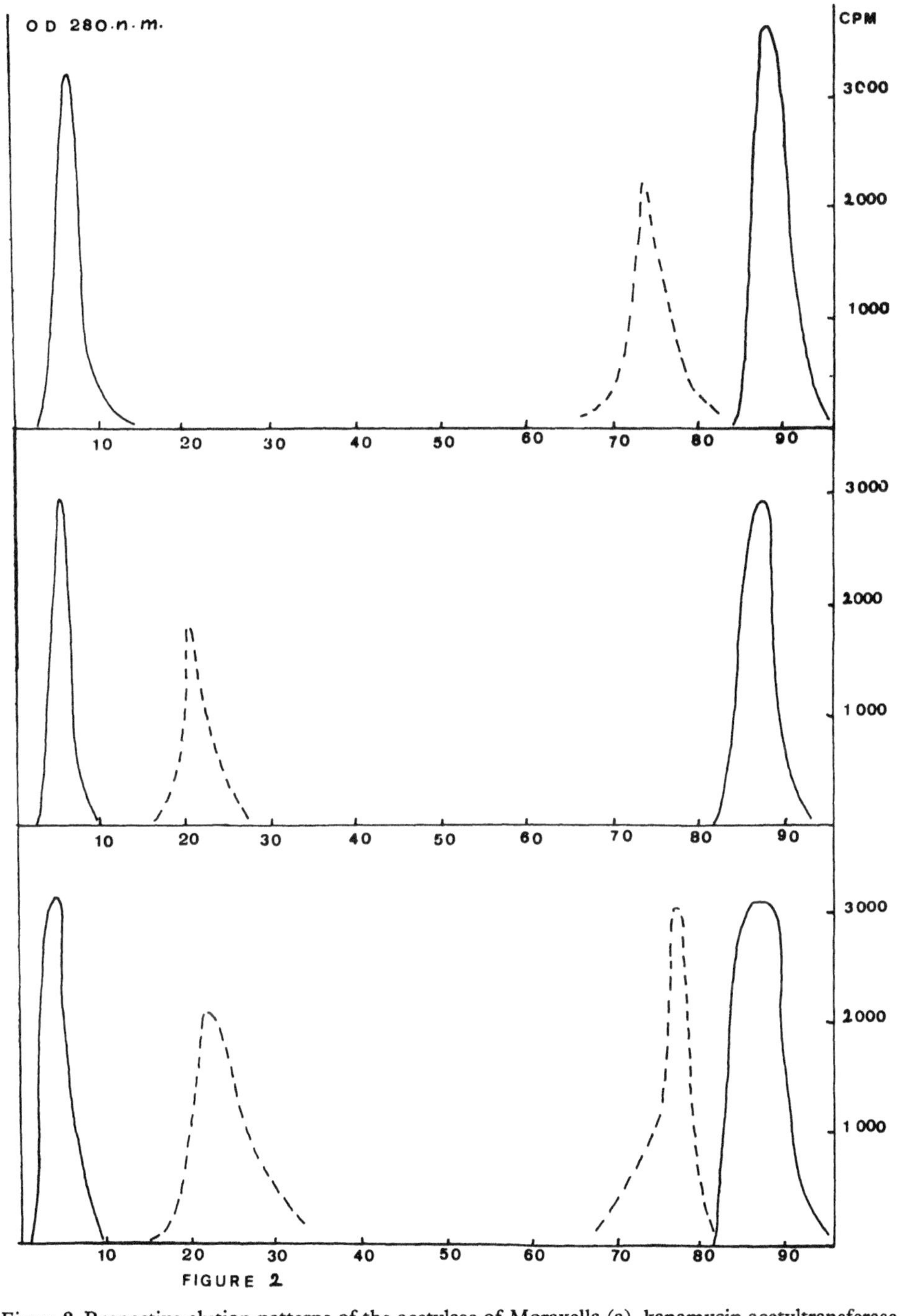

Figure 2. Respective elution patterns of the acetylase of Moraxella (a), kanamycin acetyltransferase (b) and the mixture of these enzymes (c) on indubiose gentamicin column: absorbance at 280 nm (———), enzymatic activity (- - - - - -).

The specifity of this acetylating activity and the influence of the pH are summarized in Table 1. The substitution of the 6′ amino group and its neighbourhood appear to be very important, for steric hindrance abolishes completely enzymatic activity. Gentamicin C_1 is not acetylated, whereas gentamicin C_{1a} is well recognized by the enzyme at pH 5.3.

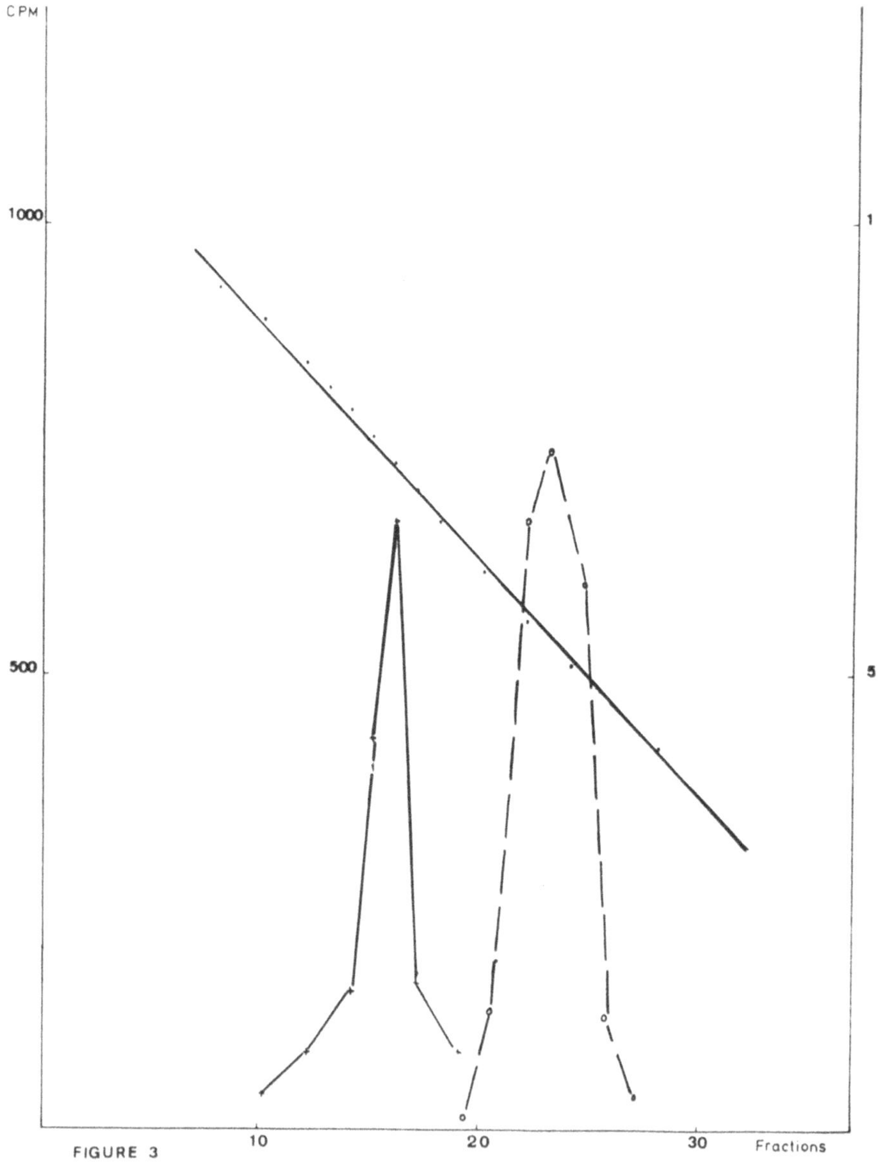

Figure 3. Electrofocusing comparison of acetylase of Moraxella and kanamycin acetyltransferase: pH gradient (———), enzymatic activity of kanamycin acetyltransferase I (- - - - -), enzymatic activity of kanamycin acetyltransferase II (—X—).

This, of course, suggests that this acetyltransferase acetylates the 6′ amino group of the aminohexose ring of the aminosidic antibiotics.

This point of view is confirmed by preparing 10 mg of acetyl kanamycin A which is identical to a sample of 6′ acetyl kanamycin A prepared by using the well known kanamycin acetyltransferase as catalyst.

This now raises the question of the comparison of these two enzymes.

They have been compared from three points of view.

1) *Affinity chromatography behaviour*:

The chromatography of both enzymatic extracts is performed under the same conditions and as shown in Figure 2, kanamycin acetyltransferase is eluted with 0.25 M NH_4Cl whereas 0.5 M NH_4Cl is necessary to remove acetyltransferase of Moraxella from the column.

2) *Electrofocusing*:

The respective isoelectric points of these enzymes have been determined in the usual way. The values found, as shown in Figure 3, are respectively 7.6 for acetyltransferase of Moraxella and 5.4 for kanamycin acetyltransferase.

3) *Michaelis Menten constants*:

These constants have been determined in the usual way. The results are summarized in Table 2.

TABLE 2

Each compound listed is incubated with the enzyme preparations purified by affinity chromatography. The affinity constant Km is determined in the classical way by the Lineweaver Burk technique.

Substrates	Km en M.	
	Kana Act.	Moraxella Act.
Tobramycin	22	4
Kanamycin	4	8
BBK 8	57	10

In conclusion it is clear from these results that the acetyltransferase found in Moraxella which confers resistance to kanamycin A, BBK 8 and tobramycin is a new enzyme quite different from kanamycin acetyltransferase previously described. We suggest to call this new enzyme kanamycin acetyltransferase II.

REFERENCES

Benveniste, R. and Davies, J. (1973). Ann. Rev. Biochem., **42**, 471–506.
Le Goffic F. and Moreau, N. (1973). FEBS Letters, **29**, 289–291.
Le Goffic, F., Moreau, N. and Chevereau, M. (1973). Biochimie, **55**, 1183–1185.

MECHANISM OF INCREASED RATE OF APPEARANCE OF HIGHLY STREPTOMYCIN-RESISTANT AND CHLORAMPHENICOL-RESISTANT VARIANTS IN *E. COLI* R+

J. HUBÁČEK, H. BRANÁ and K. ČEJKA
Department of Bacterial Genetics, Institute of Microbiology, Czechoslovak Academy of Sciences, Prague, Czechoslovakia

R factors acquired by the bacterial cell are known to influence the response of the cell to different antibiotics. They increase the sensitivity of the cell to rifampicin, actinomycin and flavomycin (Romero et al., 1971, Watanabe et al., 1972, Braná et al., 1973) but, on the other hand, they confer resistance to a broad range of antibiotics (Richmond, 1973). In bacteria with an R factor which determines resistance to low concentrations of streptomycin (str) or chloramphenicol (cam), an increased rate of appearance of highly streptomycin-resistant (str-r) or chloramphenicol-resistant (cam-r) variants can be observed. Ginoza and Painter (1964) found that the mutagenic effect of R factor was limited to drugs to which the R factor genes determined some degree of resistance. They conclude that a new chromosomal mutation to low-level resistance to streptomycin appears in highly resistant bacteria as a result of genetic recombination between the chromosome and the R factor. The co-operation of the two genes, chromosomal and plasmid, produces an organism which is highly resistant.

Analyzing this phenomenon, Pearce and Meynell (1968a) explained the results in that the R factor acts simply by providing an initial degree of resistance and, at a sufficient cell concentration, a chromosomal mutation to low degree resistance can occur. As the result of a synergistic action of the resistance determinant with the second resistance gene, either on the R factor or on the chromosome, a highly resistant organism can be obtained.

In our experiments we analyzed the mechanisms of appearance of highly resistant variants to streptomycin and chloramphenicol in *E. coli* R^+ strains.

MATERIALS AND METHODS

Media: As a defined medium, the M9 medium was used, appropriately supplemented with aminoacids and thiamine (20 μg/ml) and sugars as carbon source (0.25% w/v). Either the L-broth or Penassay broth were employed. The agar media were solidified with Oxoid agar No. 3 (1.5% w/v).

Bacterial strains: Using *E. coli K-12* strains AB 1157 and AB 2463 (Howard-Flanders and Theriot, 1966), streptomycin-sensitive (str-s) derivatives were constructed: ED 2421 donor strain (kindly supplied by Dr. P. Broda) with the direction of chromosome transfer argG, strA, xyl... was crossed with AB 1157 or AB 2463 recipients and among xyl^+ recombinants selected str-s colonies were detected. The other *E. coli* strains used were JC 5455 and JC 5466 recA56 (kindly supplied by Dr. N. Willetts), C600 (Appleyard,

1954), X8011 (Hubáček and Glover, 1970) and HK 26R1 (Braná et al., 1973). The R factor used was R1 drd-19 which confers resistance to cam, amp, str and sul (Meynell and Datta, 1967).

The procedure used for conjugal transfer of the R factor was the same as described by Pearce and Meynell (1968b); isolation of R^- segregants or cells with the resistance transfer factor (RTF) only was carried out according to Pearce and Meynell (1968a). The degree of str-resistance was determined as a minimal concentration of antibiotics which completely inhibits growth of bacteria in the Penassay broth (MIC).

Biochemical methods: For the determination of chloramphenicol acetyl transferase (CAT) activity the cell-free extracts were prepared. The bacteria were harvested, washed and disintegrated in a MSE 20kc disintegrator (7×20 sec); after centrifugation for 30min at $60{,}000 \times g$ the enzyme activity in the supernatant was assayed by a colorimetric method (Shaw and Brodsky, 1968). Streptomycin adenylate transferase (SAT) was determined by the method described by Benveniste et al. (1970). Proteins were estimated by the method of Lowry et al. (1951).

Chemicals: Adenosine triphosphate (ATP), 5,5′-dithiobis-(2-nitrobenzoic acid) (DTNB) and S-acetyl coenzyme A (ac-CoA) were obtained from the Sigma Chemical Company; ^{14}C-ATP was from the Institute for Research, Production and Uses of Radio-isotopes, Prague; chloramphenicol, streptomycin and cephaloridine were commercially purchased.

RESULTS

a) *High-level streptomycin resistance in R^+ cells*

The frequency of spontaneous str-r mutants in *E. coli* strains used is about 1.0×10^{-9}. After infection with the R1 plasmid the increased rate of appearance of highly str-r variants was observed when the bacteria were plated on LA with higher than minimal inhibitory concentration of streptomycin. The frequency of such variants on the plates containing 250–500 μg/ml of the antibiotic was about thousand-times higher in the plasmid-infected strain than in the uninfected one. The rate of appearance of the high-level resistant variants increased only in cells infected with the whole R1 plasmid but not with drug-sensitive segregant carrying RTF (Table 1).

TABLE 1

Frequency of highly str-r mutants in E. coli $R1^+$ and R^- strains

Strain	Rec function	Plasmid R1	Frequency of str-r mutants
		—	4.1×10^{-8}
JC 5455	+	+	5.5×10^{-5}
		RTF only	9.4×10^{-8}
AB 1157	+	—	7.3×10^{-8}
		+	7.2×10^{-5}
JC 5466	recA56	—	3.1×10^{-9}
		+	3.7×10^{-5}
AB 2463	recA13	—	6.6×10^{-9}
		+	3.0×10^{-6}

An attempt was made to shed some light on the mechanism of this phenomenon. In R+ cells the genetic change can occur on the plasmid itself, on the chromosome or simultaneously on both the chromosome and the plasmid. When R+ strain and the high-level str-r variant of R+ strain were conjugated with *E. coli* C600 recipient, the level of resistance was the same in both cases (Table 2). This indicates that the hereditary change

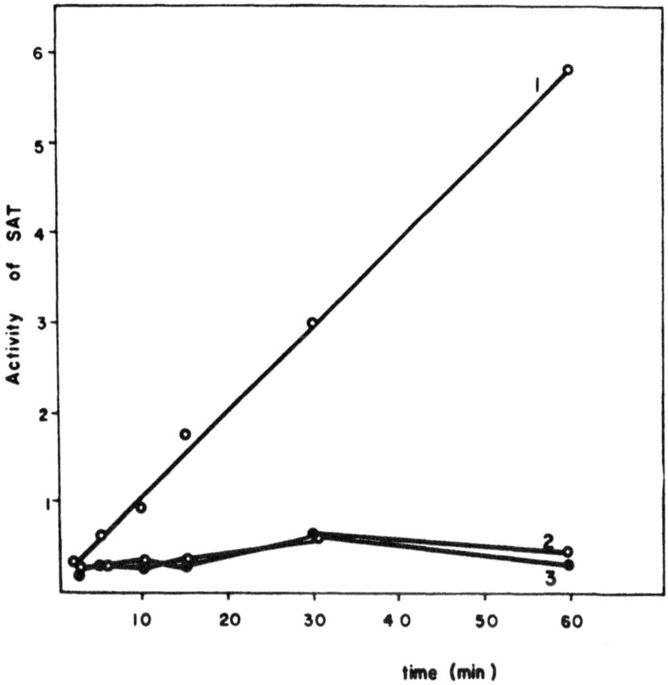

Fig. 1. The activity of SAT in cell-free extracts of *E. coli*. The activity of the enzyme was assayed by the method of Benveniste et al. (1970) and expressed in μmoles of ^{14}C-adenylylated streptomycin/mg proteins. Curve 1 – JC 5455 R1+ strain, curve 2 – JC 5455 R− strain, and curve 3 – JC 5455 high-level resistant R1− segregant.

responsible for the appearance of high-level str-resistance is not located on the plasmid. On the other hand, R− segregants isolated from the R+ str-r variant were found to be more resistant to streptomycin (MIC 22,5 μg/ml) than the original uninfected strain (MIC 10 μg/ml); after reinfection of this strain with the R1 plasmid a highly str-r variant was again obtained (Table 2). We can therefore assume, that a genetically stable low-level resistance alteration has occurred on the chromosome.

The alteration would be the result of a direct chromosomal mutation to a low-level str-resistance or may have its origin on the plasmid, and the chromosomal location is acquired during the process of genetic recombination between chromosome and R factor. To distinguish between the two possibilities we have established the frequency of high-level resistant variants in isogenic rec+ and recA− strains. As can be seen in Table 1 the rate of appearance of highly str-r variants is not affected by recA mutation in the cell.

The biochemical mechanism of a new chromosomal low-level str-resistance mutation was studied. The kinetics of the adenylylation of streptomycin as a function of time in the

TABLE 2

The level of str-resistance in different E. coli R1+ and R− strains

Strain	Obtained from	Obtained by	MIC μg/ml str	Further procedure	Resulting derivative	MIC μg/ml str
JC 5455	—	—	10	—	—	—
JC 5455 R1	JC 5455	infection with R1	200	selection on streptomycin	JC 5455 R1-r	2.500
JC 5455-r	JC5455R1-r	curing of R1	22.5	reinfection with R1	JC 5455-r R1	2.500
				infection with R1 from JC 5455 R1	C600 R1	500
C600	—	—	25	infection with R1 from JC 5455 R1-r	C600 R1	500

reaction mixture with ^{14}C-ATP and cell-free extract from osmotically shocked cell culture is shown in Fig. 1. In comparison with a linear increase in the activity of SAT in a cell-free extract from the R$^+$ strain (curve 1), there are no measurable activities of the enzyme in the uninfected *E. coli* strain (curve 2) or in the R$^-$ segregant from the high-level resistant strain (curve 3). The results support the idea that a new low-level resistance information, which occurs in R$^+$ cells in the presence of higher concentration of streptomycin, is not identical with the str locus on the plasmid but is rather the direct result of chromosomal mutation.

b) High-level chloramphenicol resistance in R$^+$ cells

In a *E. coli* R$^+$ culture, the mutability to high-level chloramphenicol resistance was obvious. Under the conditions of drug-resistance selection, when the bacteria were plated on synthetic agar containing graded levels of cam (2 μg/ml up to 1,000 μg/ml), resistant variants were obtained. The number of resistant colonies gradually decreased with higher concentrations of the antibiotic used, and at 1,000 μg/ml only about 3 per cent of colony-forming units survived (Fig. 2). The CAT was isolated from cells grown on different concentrations of cam and its activity was assayed. Resistance to a higher level of cam was found to be accompanied by an increase in the cell content of the enzyme. The same

TABLE 3

Stability of high-level cam-resistance in E. coli R1+

Subcultures in drug-free M9 medium	No. of generations	No. of resistant colonies on M9 plates: without cam	No. of resistant colonies on M9 plates: with cam 500 μg/ml
0	0	5.5×10^7	5.6×10^7
1	5.6	5.6×10^7	3.0×10^7
2	6.2	8.9×10^7	1.6×10^7
3	5.1	6.2×10^7	1.0×10^7

phenomenon was observed when the bacteria were grown on a synthetic medium with glycerol as a carbon source where the regulation of enzyme synthesis by catabolite repression was eliminated (Fig. 2). If the bacteria selected on plates with cam (500 μg/ml) were subcultivated in drug-free medium and plated again on 500 μg/ml of cam, the ability to grow was lost; after approximately 12 generations in drug-free medium only about 17 per cent of bacteria in the R$^+$ population retained the high-level resistance (Table 3).

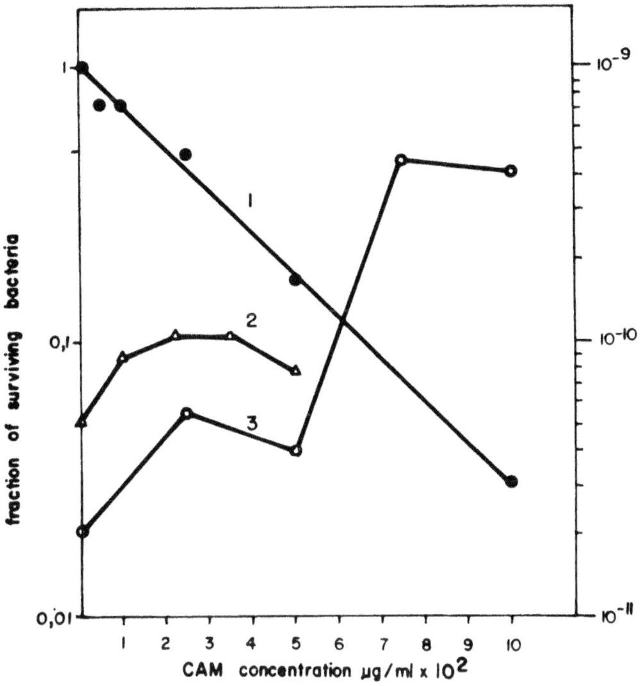

Fig. 2. The survival of *E. coli* X 8011 R1$^+$ cells and the activity of CAT. The enzyme activity, assayed as described in Methods, was expressed as unit/cell, the unit being calculated as the amount of the enzyme converting 1 μmol of DTNB/min. Curve 1 – the fraction of surviving bacteria at different concentrations of cam, curve 2 – enzyme activity in bacteria grown on M9 plates with glycerol, curve 3 – enzyme activity in bacteria grown on M9 plates with glucose.

DISCUSSION

The str- and cam-resistance determinants on an R factor are the genes that specify the synthesis of enzymes that confer relatively low-level resistance to antibiotics. Ginoza and Painter (1964) and Pearce and Meynell (1968a) described the increased rate of appearance of highly str-r and cam-r variants in E. coli R$^+$ strains. The results obtained in our studies of this phenomenon indicate the different mechanisms of str- and cam-high-level resistance. The str-resistance would be the result of a synergistic action of two low-level resistance determinants; one presented in the R plasmid determines the synthesis of str-adenylylating enzyme, the other is a new chromosomal mutation which occurs in R$^+$ cells in the presence of higher concentration of streptomycin. The mutation is not acquired by direct genetic recombination between plasmid and chromosome, as postulated by Ginoza and Painter (1964), because a recA mutation has no effect on the rate of appearance of highly str-r variants (Table 1); recombinations between the chromosome and R plasmids depend on a fully functioning bacterial rec-system (Tschäpe and Böhme, 1972). The biochemical mechanism of this chromosomal resistance is not identical with plasmid resistance (Fig. 1) and remains unknown.

In the case of cam resistance our data support the idea that the change in the regulation of synthesis of CAT plays a certain role. It was assumed that under protein synthesis inhibition the synthesis of an inhibitor of plasmid replication would be reduced in *Proteus mirabilis* (Punch and Kopecko, 1972). Disproportionate replication of the R-determinants and the increase in the number of resistance genes per cell leads to an increase in the amount of antibiotic-inactivating enzyme (Rownd et al., 1971). Some derepression of plasmid replication in a fraction of bacterial population would operate also in *E. coli* at high cam concentrations. This regulatory mechanism might play a role in the selection of highly cam-resistant and genetically unstable variants in an *E. coli* population.

In practical terms, in spite of the different mechanisms of appearance of highly resistant bacteria, this phenomenon shows how bacterial R^+ cells may respond to a changing environment. Under the selection pressure exerted by the higher concentrations of the antibiotic on plasmid-infected bacterial populations, more resistant variants would emerge and complicate even further the rather involved epidemiological situation in bacterial resistance to antibiotics.

REFERENCES

Appleyard, R. K. (1954). Segregation of new lysogenic types during growth of doubly lysogenic strain derived from *Escherichia coli* K12. Genetics 39, 440.

Benveniste, R., Yamada, T. and Davies, J. (1970). Enzymatic adenylylation of streptomycin and spectinomycin by R-factor-resistant *Escherichia coli*. Infection and Immunity 1, 109.

Braná, H., Hubáček, J. and König, J. (1973). The effect of actinomycin D and flavomycin on *Escherichia coli* R^+ strains. Folia microbiol. 18, 257.

Ginoza, H. S. and Painter, R. B. (1964). Genetic recombination between the resistance transfer factor and the chromosome of *Escherichia coli*. J. Bacteriol. 87, 1339.

Howard-Flanders, P. and Theriot, L. (1966). Mutations of *Escherichia coli* K 12 defective in DNA repair and genetic recombination. Genetics 53, 1137.

Hubáček, J. and Glover, S. W. (1970). Complementation analysis of temperature-sensitive host specificity mutations in *Escherichia coli*. J. Mol. Biol. 50, 111.

Lowry, H. O., Rosebrough, N. J., Farr, A. L. and Randall, R. (1951). Protein measurement with the folin phenol reagent. J. biol. Chem. 193, 265.

Meynell, E. and Datta, N. (1967). Mutant drug resistance factors of high transmissibility. Nature, Lond. 214, 885.

Pearce, L. E. and Meynell, E. (1968a). Mutation to high-level streptomycin-resistance in R^+ bacteria. J. gen. Microbiol. 50, 173.

Pearce, L. E. and Meynell, E. (1968b). Specific chromosomal affinity of a resistance factor. J. gen. Microbiol. 50, 159.

Punch, J. D. and Kopecko, D. J. (1972). Positive and negative control of R-factor replication in *Proteus mirabilis*. J. Bacteriol. 109, 336.

Richmond, M. H. (1973). Resistance factors and their ecological importance to bacteria and to man, p. 191. In Davidson, J. N. (ed.), Progress in Nucleic Acid Research and Molecular Biology, vol. 13. Academic Press Inc., New York.

Romero, E., Riva, S., Fietta, A. M. and Silvestri, L. G. (1971). Effect of R factors on rifampicin resistance in *Escherichia coli*. Nature New Biol. 234, 56.

Rownd, R., Kasamatsu, H. and Mickel, S. (1971). The molecular nature and replication of drug resistance factors of the enterobacteriaceae. Ann. N.Y. Acad. Sci. 182, 188.

SHAW, W. W. and BRODSKY, R. F. (1968). Characterization of chloramphenicol acetyl-transferase from chloramphenicol-resistant *Staphylococcus aureus*. J. Bacteriol. 95, 28.

TSCHÄPE, H. and BÖHME, H. (1972). Recombinations between different R(f) and R(i) plasmids and between R-plasmids and the chromosome in *Escherichia coli* and *Proteus mirabilis*, p. 217. In Krčméry, V., Rosival, L., Watanabe, T. (ed.)., Bacterial plasmids and antibiotic resistance. Avicenum – Czechoslovak Medical Press, Prague and Springer-Verlag, Berlin, Heidelberg, New York.

WATANABE, T., OGATA, Y., SUGAWARA, K. and ODA, K. (1972). Increase of flavomycin sensitivity of bacteria by R factors, p. 105. In Krčméry, V., Rosival, L., Watanabe, T. (ed.), Bacterial plasmids and antibiotic resistance. Avicenum – Springer – 1972.

R FACTORS TO AMINOGLYCOSIDE ANTIBIOTICS

E. J. PEREA, ROSA M. DAZA

Service of Microbiology. Clinica Puerta de Hierro. Medical School.
Autonomous University. Madrid, Spain

This research was supported by a grant, No. 12-085-72, of Comisión de Descuento Complementario. Instituto Nacional de Previsión. Spain.

The existence of R factors and their transmission through conjugation was studied in 50 strains of Gram-negative bacilli isolated at our hospital. The recipient strain was *E. coli* K12 E711 and the selecting characteristic was resistance to kanamycin (25 mcg/ml). The conjugation times were 20 minutes and 18 hours.

We describe the frequency of transfer of the R factors as a function of the time of conjugation, and we determined the degree of resistance to aminoglycosides transferred (testing spectinomycin (SP), streptomycin (SM), paromomycin (PM), neomycin (NM), kanamycin (KM), gentamicin A (GM-A), gentamicin C (GM-C), tobramycin (TB) and BB-K8).

Resistance factors to gentamicin C were transferred from six strains. On two occasions, this resistance did not appear in the donor strain. These two factors conferred a moderate resistance to tobramycin and BB-K8.

We mapped the genetic factors R 19 and R 38 by transduction with phage P1Kc as well as by deletions occurring from spontaneous loss or treatment with acridine orange or rifampicin. We found that R 19 is "fi^{+}" and R 38 "fi^{-}".

Finally, we try to interpret these transferred resistance in terms of inducible enzymes.

R FACTORS TO AMINOGLYCOSIDE ANTIBIOTICS

The mechanisms of antibiotic resistance can be divided into two large groups: those due to changes in cellular components that stop the antibiotic from acting on its target, and those that involve chemical changes and inactivation of the antibiotic (Benveniste and Davies, 1973). This second type of resistance is usually found in the strains isolated clinically. It is very common in Gram-negative bacteria and is associated with an extrachromosomic element or plasmid.

We used 50 strains of bacteria to study the possible existence of R factors and their rate of transmission as a function of the time of conjugation. Special attention was paid to the levels of resistance to aminoglycoside antibiotics. Using transduction and deletion, we made genetic maps of factors R 19 and R 38. We attempted to cure these factors with acridine orange (Hirota, 1960) and rifampicin (Johnston and Richmond, 1970).

MATERIAL AND METHODS

Bacteria. — The possible existence of R factors to aminoglycoside antibiotics was studied in 25 strains of *E. coli*, 11 of *Klebsiella pneumoniae* and 14 strains of *Pseudomonas aeruginosa* isolated in our hospital and selected by their resistance to 512 mcg/ml of kanamycin.

E. coli K 12 strains: E 711 Lac⁻ his, pro, trp, nalidixic acid resistant (Moreno et al., 1970); UTH 1038, Lac^+, F^- (Goldschmidt et al., 1970); a rifampicin-resistant mutant of HfrC met (Clowes and Hayes, 1968) isolated by us.

Phages. — P1Kc, kindly supplied by Dr. Martin del Rio and Dr. Fernandez, C.S.I.C., Spain; MS2 (Davis, Strauss and Sinsheimer, 1961), kindly supplied by Dr. N. Datta, London.

Media. — The nutrient broth was Oxoid #2 CM 67; MacConkey agar was Difco CM0075-01; Mueller Hinton was Oxoid CM 337.

Transfer of factors by conjugation. — The methods were as described by Clowes and Hayes (1968), Meynell and Meynell (1970) and Datta and Hedges (1972). The recipient strain was K 12 E 711 and the mixtures were incubated for 20 minutes and 18 hours.

Transduction. — The methods were as described by Lennox (1955). The donor strains were resistant recombinants of K 12 E 711 and the recipient was UTH 1038. The transducing phage was P1Kc.

Determination of antibiotic resistance. — The antibiotic sensitivity of all the strains was determined by the Bauer-Kirby method, with discs of the following concentrations: kanamycin 30 mcg, streptomycin 5 mcg, chloramphenicol 10 mcg, tetracycline 10 mcg, gentamicin 5 mcg, and sulfamethoxypyridazine 150 mcg. We determined the M.I.C. to kanamycin, streptomycin, spectinomycin, gentamicin A and C, tobramycin and BB-K8 (amikacin), using the agar dilution method (Ericsson and Sherris, 1971) with a range of 0.5 to 512 mcg/ml and an inoculum of about 1×10^5 bacteria.

fi character. — R factors were transferred to an HfrC, rifampicin-resistant mutant. An undiluted lysate of phage MS2 (titre approximately 10^{11} plaque-forming units/ml) was spotted on the R^+ culture. Visible lysis indicated that the plasmid was fi^- (Coetzee, Datta and Hedges, 1972).

Spontaneous segregants. — Two recombinants, R 19 and R 38, of the K 12 E 711 that had acquired the resistance, were studied after 20 generations in antibiotic-free culture media to check for spontaneous segregants. Some 1000 colonies were examined for loss of resistance markers.

Elimination of antibiotic resistance. — The action of acridine orange and rifampicin in "curing" the antibiotic resistances was determined. The methods were as described by Bouanchaud, Scavizzi and Chabbert (1969).

RESULTS

The results obtained are shown in Table I. As can be seen, none of the strains of *Ps. aeruginosa* transferred, and the highest percentage of transfer was offered by *E. coli*. With regard to the duration of contact, more strains transferred after 18 hours as could be expected.

The rate of transfer at 20 minutes ranged from 3×10^{-4} to 1×10^{-7}; at 18 hours the range was 2×10^{-3} to 2×10^{-7}. Except for three cases in which it was similar, the rate was always higher after longer conjugation.

In the resistance factors presented by us in Table 2, we considered the kanamycin marker to be representative of the resistance to paramomycin and neomycin as well. The factors transferred run from resistance to kanamycin and sulfamides only up to resistances that include kanamycin, sulfamides, chloramphenicol, tetracycline and gentamicin. The most consistently transferred block of resistances was the one that included kanamycin and sulfamides.

TABLE I - TRANSFER OF R FACTORS AS A FUNCTION OF THE CONJUGATION TIME.

	Strain nº	Recombinant strain	
		20 minutes	18 hours
E coli	25	13 (52%)	21 (84%)
Klebsiella pneumoniae	11	3 (27%)	6 (55%)
Pseudomonas aeruginosa	14	-	-
TOTAL	50	16 (32%)	27 (54%)

TABLE II.- R FACTORS TRANSFERRED

	Nº
KM - SU	2
KM - SU - CB	1
KM - SU - APC	1
KM - SU - CM	1
KM - SU - SM	1
KM - SU - SM - APC	1
KM - SU - SM - APC - CB - CM	8
KM - SU - SM - APC - CB - CM - TC	9
KM - SU - SM - CM - TC - GM-C	6

KM Kanamycin, SU Sulphamethoxypyridacine, CB Carbenicillin, APC Ampicillin, CM. Chloramphenicol, SM: Streptomycin, TC Tetracycline, GM-C: Gentamicin

In cases of transfer at 20 minutes and 18 hours, the R factor transferred was the same, except for the three cases shown in Table 3 in which the factor was larger when the contact was more prolonged. Chloramphenicol was one of the most constant factors to transfer late.

The six factors, R 6, R 19, R 38, R 39, R 40, and R 42, that yielded resistances to KM, SU, SM, CM, TC and GM-C were the ones we studied in greatest detail (Table 4).

TABLE III.- DIFFERENCES IN THE R FACTORS TRANSFERRED FROM THE SAME STRAIN AS A FUNCTION OF THE CONJUGATION TIME

Strain	R FACTOR TRANSFERRED Recombinant 20 minutes	Recombinant 18 hours	Difference
E. coli	KM - SU - APC	KM-SU-APC-SM-CB-CM	SM-CB-CM
E. coli	KM - SU - APC - SM	KM-SU-APC-SM-CB-CM-TC	CB-CM-TC
E. coli	KM - SU	KM-SU-CM	CM

TABLE IV.- M.I.C. OF THE DONOR STRAINS AND THEIR RECOMBINANTS TO AMINOGLYCOSIDE ANTIBIOTICS

STRAIN	KM		NM		PM		SM		SP		GM-C		GM-A		TB		BB-K8	
	D	R	D	R	D	R	D	R	D	R	D	R	D	R	D	R	D	R
E. coli nº 6	512	512	128	32	512	512	256	256	64	64	1	64	512	512	2	2	4	4
E. coli nº 19	512	512	128	32	512	512	256	128	64	64	1	32	512	512	2	2	2	2
E. coli nº 38	512	512	64	8	512	512	128	32	64	8	128	8	512	512	1	0,5	1	1
Kl. pneumoniae nº 39	512	256	64	8	512	512	128	32	128	32	128	16	512	512	1	0,5	0,5	0,5
Kl. pneumoniae nº 40	512	64	64	8	512	128	128	16	128	8	128	16	512	512	1	0,5	0,5	0,5
Kl. pneumoniae nº 42	64	64	16	4	256	128	8	8	64	8	128	8	512	512	1	0,5	0,5	0,5

M I C en mcg/ml , KM Kanamycin , NM Neomycin , PM Paronomycin ;
SM Streptomycin , SP Spectinomycin , GM Gentamicin , TB Tobramycin
D - Donor strain
R - Recombinant strain

With kanamycin, paramomycin and gentamicin A, the resistance levels acquired by the recipient strain at 20 minutes and 18 hours were the same as those of the donor. This was not the case with streptomycin, neomycin and gentamicin C, where the resistance levels in the recipients were always lower than those of the donor. Some of these recombinants presented a lower degree of resistance after 20 minutes of conjugation than after 18 hours. However, in strain 6 and 19, the recombinants showed a resistance to gentamicin C that was not found in the donor strain.

Two strains of K 12 E 711 that received factors R 19 and R 38 were selected for transduction by phage P1Kc, using K 12 UTH 1038 as the recipient strain. The rate of transduction was always about 4×10^{-7}.

TABLE V.- DELETIONS DUE TO SPONTANEOUS LOSS, ACRIDINE ORANGE AND RIFAMPICIN IN R19 AND R38.

Antibiotype KM-NM-SM-GMC-SU-CM-TC	E coli K12 E711 R19 (1000 colonies)			E coli K12 E711 R38 (1000 colonies)		
	Spontaneous loss	acridine orange	rifampicin	Spontaneous loss	acridine orange	rifampicin
KM-NM GMC-SU-CM-TC	46,6%	96%	91,7%	20,8%	8,3%	98,4%
NM GMC-SU-CM-TC	41,7%					
GMC-SU-CM-TC	8,3%					1,6%
KM-NM-SM SU-CM-TC					91,7%	
RATE OF LOSS	96,6%	96%	91,7%	20,8%	100%	100%

In the first factor, we achieved the combined transduction of all the resistances except GM-C, thus obtaining the transductants SM KM, SM KM NM, SM KM NM CM SU and SM KM NM CM SU TC. From this, we constructed the genetic map of R 19 (Fig. 1).

We transduced factor R 38 in bloc and also obtained the following transductants: TC CM SM, SM NM KM, SM NM KM TC CM and SM NM KM TC CM SU. The genetic map of this factor is shown in Fig. 2.

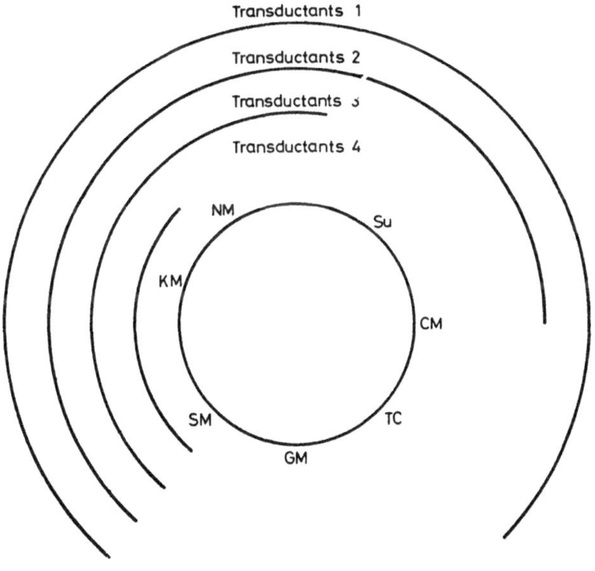

Fig. 1. R 19 factor. Deletion during transduction. The inner circle is the parent. The next four circles are transductants obtained with phage P1Kc. Transductants 1 were selected for CM and NM, transductants 2 with selection for CM and 3 and 4 transductants were selected for SM. No region appears essential for replication. The CM SU position can be inverted. Distances between genetic markers are arbitrary.

We also derived the genetic maps of these two R factors from deletions occurring from spontaneous loss or after treatment with acridine orange and rifampicin (Table 5). In a total of 1000 colonies from each of the recombinants, K 12 E 711 R 19 and K 12 E 711 R 38, simultaneous loss of all the markers never occurred.

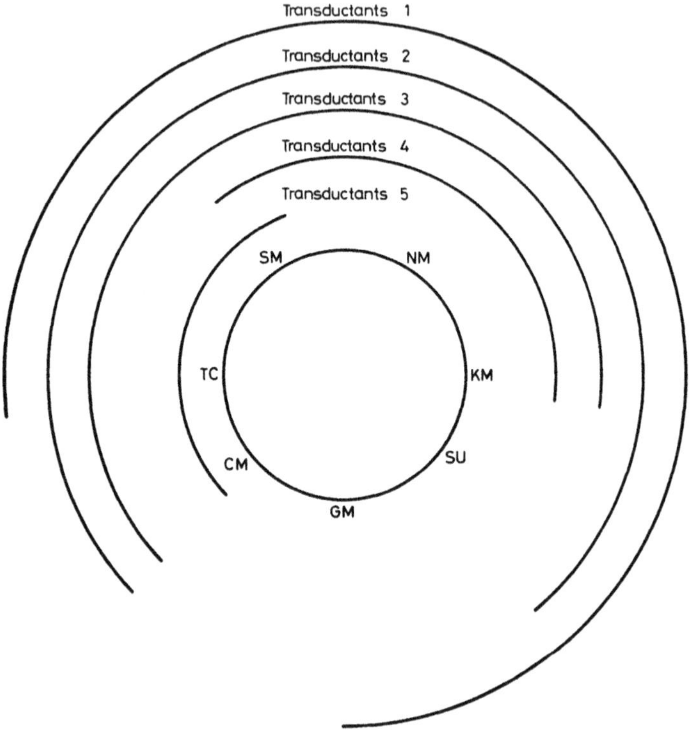

Fig. 2. R 38 factor. Deletion during transduction. The inner circle is the parent. The next circles are transductants obtained with phage P1Kc; transductants 1 were selected for NM, transductants 2 were selected for TC and CM, transductants 3 and 5 were selected for CM, transductants 4 were selected for SM. Distances between genetic markers are arbitrary.

The most frequently lost factor for R 19 was SM (46.6 per 100 colonies by spontaneous loss, 96% with acridine orange and 91.7% with rifampicin). The KM and SM markers were spontaneously lost together in 41.7% and with NM in 8.3%.

Factor R 38 lost the SM marker spontaneously in 20.8%, with acridine orange in 8.3% and with rifampicin in 98.4%. The KM NM SM markers were lost together with rifampicin in only 1.6% while the GM-C marker was lost with acridine orange in 91.7%.

The total rate of loss in R 19 was practically the same from spontaneous loss as from treatment with acridine orange or rifampicin. Factor 38, however, had a 100% loss with acridine orange and rifampicin and only 20.8% spontaneously.

The *E. coli* K 12 HfrC R 38 was found to be sensitive to the male specific phage MS2, so that the R 38 factor does not repress the synthesis of F pili and can be considered as a "fi⁻" factor. On the other hand, transfer of factor R 19 to the HfrC does not cause its lysis by phage MS2; R 19 is thus a "fi⁺" factor.

DISCUSSION

The block of sulfamides, ampicillin and chloramphenicol appeared very frequently (17 of the 30 factors transferred). According to other authors (Witchitz and Chabbert 1971b) this group of resistances is rather rare, appearing in only 2% of the factors isolated.

The fact that more resistances were transferred after 18 hours than after 20 minutes (Table 3) can be explained in two ways: We may have performed an interrupted mating of the R factor and the recipient strain carries an autonomous replicon which rescues individual transferred genes by recombination (Yoshikawi and Sakai, 1970). Alternatively, the R factor may have a number of replicons (Meynell, 1972). The second hypothesis seems more probable since we found that no part of the factor was indispensable for replication of the plasmid.

We found resistance factors to gentamicin C in three strains of *Klebsiella pneumoniae* and three of *E. coli*. The factors with these resistances isolated by Witchitz and Chabbert also carried resistances to three or four other antibiotics; ours were larger.

The resistances to aminoglycosides transferred by *E. coli* numbers 6 and 19 were practically the same (Table 4) since they gave *E. coli* K 12 E 711 high resistance to kanamycin, paramomycin, streptomycin, spectinomycin and gentamicin A; the resistance to neomycin was lower. Low levels of resistance to tobramycin and BB-K8 were also transferred.

The most interesting point is the appearance in the recombinants of a resistance to gentamicin C that was not shown by the donors. This paradoxical result could be interpreted assuming that the host bacteria influences the degree of resistance conferred by an R factor. Along these lines, Watanabe (1971) found that the same R factor produced a resistance to 100 mcg/ml of tetracycline when transferred to a *Salmonella*, but only to 10 mcg/ml in a *Shigella*. The resistance to gentamicin C was in our cases better expressed by the *E. coli* K 12 E 711 than by the donor. Witchitz and Chabbert (1971a) described a strain of *Providencia*, with 16 mcg/ml M.I.C. to gentamicin C, that transferred a resistance level of 64 mcg/ml to K 12 LA 290. The opposite result (higher levels in the donor than in the recipient *E. coli* K 12 E 711) is a common finding and occurs frequently with neomycin, streptomycin, spectinomycin, and gentamicin C. There was no significant difference between the level of resistance acquired after 20 minutes and that transferred after 18 hours of conjugation.

The factors transferred from strains number 38, 39, 40 and 42 (Table 4) yielded the same level of resistance to kanamycin, paramomycin and gentamicin A in the recipient as in the donor. However, the resistance levels to the aminoglycosides, neomycin, streptomycin, spectinomycin and particularly gentamicin C were lower in the recombinant.

These donor strains (38, 39, 40, 42), while resistant to gentamicin C, were sensitive to tobramycin and BB-K8; strains 6 and 19 were sensitive to gentamicin C and were slightly resistant to tobramycin and BB-K8. These resistance levels were transferred intact to the *E. coli* K 12 E 711.

With regard to the genetic maps of factors R 19 and R 38, we found a correlation between the co-transduced markers and the deletions obtained spontaneously and with curing agents. We feel that these two mapping methods are complementary. While we obtained more data with transduction, the low number of deletions found from spontaneous loss and curing agents may be due to the small number of colonies checked.

In *E. coli*, "fi^+" factors are in the majority (Meynell et al., 1968). Factor R 19 is "fi^+" while R 38 is "fi^-", in spite of the fact that they both have the same antibiotic resistance markers, both came from strains of *E. coli* and both were isolated at our hospital at the same time. The R factors to gentamicin C that have thus far been isolated were all "fi^-".

although they generally came from *Klebsiella pneumoniae* (Withchitz and Chabbert, 1971a; Martin et al., 1971).

Resistance to an antibiotic does not necessarily define a gene; an enzyme may inactivate more than one antibiotic if they all have similar structures. There are also a number of different enzymes that can act on the same antibiotic; there are four different enzymes that can inactivate gentamicin (Benveniste and Davies, 1973).

All the aminoglycosides presently in use are inactivated by acetylation, phosphorylation or adenylation produced by enzymes from the resistant bacteria.

TABLE VI - INTERPRETATION OF THE ENZYMATIC INACTIVATION OF AMINOGLYCOSIDE ANTIBIOTICS BY THE R FACTORS TRANSFERRED.

	KM	NM	PM	SM	SP	GM-C	GM-A	TB	BB-K8	ENZYMES
E.coli nº 6	+*	(+)*	+°	+■	+■	+*	+°	(+)*	+*	KAcT* NPT I° SAdT■
E. coli nº 19	+	(+)	+	+	+	+	+	(+)	(+)	
E. coli nº 38	+°	(+)°	+°	+■	+■	+**	+°	-	-	
K. pneumoniae nº 39	+	(+)	+	+	+	+	+	-	-	NPT I°
K. pneumoniae nº 40	+	(+)	+	+	+	+	+	-	-	GAcT II**
K pneumoniae nº 42	+	(+)	+	+	+	+	+	-	-	SAdT■

(-) antibiotic only partially inactivated
The abbreviations for the antibiotics are explained in Table IV
KAcT Kanamycin acetyltransferase, NPT I Neomycin-Kanamycin phosphotransferase; SAdT Streptomycin--spectinomycin adenyltransferase; GAcT Gentamicin acetyltransferase

Of the nine enzymes described by Benveniste and Davies (1973), due to substrate specificity, those that inactivate streptomycin and spectinomycin do not inactivate the other aminoglycosides and vice versa. This is undoubtedly due to structural differences: the first group has streptidine in its molecule and the second has deoxystreptamine.

Only one of these enzymes has been found in any one R^+ strain examined, although there is no reason why they could not all be present on the same R factor. In view of what we have just discussed, and Benveniste's and Davies' (1973) outline of the enzymatic inactivation of aminoglycosides, it is difficult to interpret extrachromosomal resistance to these antibiotics in enzymatic terms.

In Table 6 we summarize the spetrum of each of the transferred factors. A cross indicates resistance to the antibiotic and a cross in parenthesis indicates that the resistance transferred was low. If we assume that the transferred resistances represent the production of inactivating enzymes, the first two factors transferred would represent the production of kanamycin acetyl-transferase (KAcT), neomycin-kanamycin phosphotransferase I (NPT I) and streptomycin-spectinomycin adenyltransferase (SAdT).

In the four remaining strains, resistance could be due to the following enzymes:

neomycin-kanamycin phosphotransferase I (NPT I), gentamicin acetyl-transferase II (GAcT II) and streptomycin-spectinomycin adenyl transferase (SAdT).

In conclusion, we found three types of resistance factors to aminoglycosides:

a) Those that confer resistance only to the neomycin, kanamycin and gentamicin A group. This was the least common factor.
b) Those with the aforementioned resistances and resistance to streptomycin. When "group A" resistances are present in the recipient to the same degree as in the donor, the streptomycin resistance level is lower.
c) R factors that include the aforementioned groups and gentamicin C. In some cases the resistance is expressed better in the donor than in the recipient, but the opposite is true in some other cases; this fact was expressed in the appearance of a resistance in the recombinant that was not seen in the donor. This raises the problem that the expression of a plasmid may depend on the host that houses it. These last factors also yielded a low level of transferable resistance to tobramycin and BB-K8. Factors R 19 and R 38, which we studied most extensively, are included in this group and are fi^+ and fi^- respectively.

REFERENCES

ARBER, W. (1969) in Bacterial Episomes and Plasmides. Ciba Foundation Symposium. J. A. Churchill Ltd. London.

BAUER, A. W., KIRBY, W. M., SHERRIS, J. C., and TURCK, M. (1966). Antibiotic susceptibility testing by a standardized single disk method. Am. J. Clin. Path. 45, 493.

BENVENISTE, R. and DAVIES, J. (1973). Mechanisms of Antibiotic Resistance in Bacteria. Annual Review of Biochemistry. 42, 471–506.

BOUANCHAUD, D. H., SCAVIZZI, M. R. and CHABBERT, Y. A. (1969). Elimination by Ethidium Bromide of Antibiotic Resistance in Enterobacteria and Staphylococci. Jour. Gen. Microbiol. 54, 417–425.

BRZEZINSKA, M., BENVENISTE, R., DAVIES, J., DANIELS, J. L. and WEINSTEIN, J. (1972). Gentamicin Resistance in Strain of *Pseudomonas Aerugirosa* Mediated by Enzymatic N-Acetylation of Deoxyestreptamine Moiety. J. Biochemistry, 11, 5, 761–765.

CLOWES, R. C. and HAYES, W. (1968). Experiments in Microbial Genetics. Backwell Sci. Pub. Oxford.

COETZZE, J. N., DATTA, N. and HEDGES, R. W. (1972). R factors from *Proteus rettgeri*. Jour. Gen. Microbiol. 72, 543–552.

DATTA, N. and HEDGES, R. W. (1972). Host ranges of R factors. Jour. Gen. Microbiol. 70, 453–460.

DAVIES, J. E., STRAUSS, J. H. and SINSHEIMER, R. L. (1961). Bacteriphage MS2: another RNA phage. Science, New York, 134–1427.

ERICSSON, H. M. and SHERRIS, J. C. (1971). Antibiotic Sensitivity Testing. Acta. Path. Microbiol. Scand. Supp. 217.

GOLDSCHMIDT, E. P., CARTER, M. S., MATNEY, T. S., BUTTLER, M. A., and GREEN, A. (1970). Genetic analysis of the histidine operon in *E. coli* K 12. Genetics 66, 219–229.

HIROTA, Y. (1960). The effect of acridine dyes on mating type factors in *Escherichia coli*. Proc. Nat. Acad. Sci. U.S.A. 46, 57.

JOHNSTON, J. H. and RICHMOND, M. H. (1970). The increase rate of loss penicillinase plasmids from *Staphylococcus aureus* in the presence of rifampicin. Jour. Gen. Microbiol. 60, 137–139.

LENNOX, E. S. (1955). Transduction of linked genetic characters of the host by bacteriophage T1. Virology 1, 190–206.

MARTIN, C. M., IKARI, N. S., ZIMMERMAN, J. and WAITZ, A. J. (1971). A virulent nosocomial *Klebsiela* with a transferable R factor for Gentamicin: emergence and suppression. Jour. of Infect. Diseases 124, Suppl. 24–29.

MEYNELL, G. G. (1972). Bacterial plasmids. Ed. MacMillan. London.

MEYNELL, G. G. and MEYNELL, E. (1970). Theory and Practice in Experimental Bacteriology. 2nd Edition. Cambridge University Press. London.

MEYNELL, E., MEYNELL, G. G., and DATTA, N. (1968). Phylogenetic relationships of drug resistance factors and other transmissible bacterial plasmids. Bacteriological Reviews 32, 35–83.
MITSUHASHI, S. (1971). Transferable drug resistance factor R. University Park Press. Tokyo.
MORENO, M., DAMASO, D., PEREA, E. J., MARCO, M. L., MANCHADO, P. (1971). Transmision de factores de resistencia a los antibioticos entre bacterias: conjugacion y transduccion. Antib. y Quimiot. I, 5, 211–220.
PEREA, E. J., DAZA, R. M. (1974). Factores de Resistencia a Antibioticos Aminosidos. Jornada Internacional sobre antibioticos aminosidos. Ed. Farmitalia. Barcelona.
WATANABE, S. (1971). Infectious Drug Resistance in Bacteria. Current Topics in Microbiology and Immunology. New York.
WITCHITZ, J. L. and CHABBERT, Y. A. (1971a). Resistance transferible a la Gentamicine I. Ann. Ins. Pasteur. 121, 733–742.
WITCHITZ, J. L. and CHABBERT, Y. A. (1971b). High level transferable resistance to gentamicin. The Jour. Antibiotics (Tokyo). 24, 137–139.
WITCHITZ, J. L. and CHABBERT, Y. A. (1972). Resistance transferible a la gentamicine II. Ann. Ins. Pasteur. 122, 367–378.
YOSHIKAWA, M. and SAKAI, K. (1970). Facilitation of the transfer of R factor by the resident R factor. J. Bact. 102, 285–287.

DEMONSTRATION OF A NEW TYPE OF R PLASMID IN E. COLI CONFERRING GENTAMICIN RESISTANCE

*V. KRČMÉRY, K. HASUDA, S. IYOBE and S. MITSUHASHI
Department of Microbiology, Gunma University, Maebashi 371, Japan

Knothe et al. (1973) isolated from the urine a strain of *Escherichia coli* 027 showing high-level kanamycin and low-level gentamicin resistance but a complete susceptibility to streptomycin. Kanamycin (KM) and Gentamicin (GM) resistances were transferred always together to *E. coli* K12. As the resistance pattern and its potential enzymatic background seemed to be interesting, we undertook conjugal and transductional experiments with this strain.

MATERIALS AND METHODS

Bacterial strains

A clinical isolate in Frankfurt by H. Knothe, *E. coli* F-027, resistant to tetracycline (TC), chloramphenicol (CM), sulfanilamide (SA), kanamycin (KM), 3′,4′-dideoxy kanamycin B (DKB), lividomycin (LV), gentamicin (GM) and ampicillin (APC) was used for the donor strain of R factor.

For the recipient strains *E. coli* K12 ML1410 *met*$^-$ *nal-r* (nalidixic acid resistant) and *E. coli* K12 W3630 *mal*$^-$ were used.

Media

Brain heart infusion (BHI, difco) broth was used routinely. For the determination of drug-resistance, heart infusion (HI) agar (Eiken, Tokyo) and peptone water were used. Peptone water consisted of 1,000 ml of distilled water, 5 g of NaCl and 10 g of peptone. Semi-synthetic medium was used for the assay of SA resistance.

Eosin methylene blue (EMB)-maltose agar (Lederberg, 1950) containing both 50 μg/ml of nalidixic acid (NA) and one of drugs to select drug-resistance marker on R factor was used for the selection of R transfer from F-027 R$^+$ to ML1410. L-broth and L-agar (3) were used for phage propagation and transduction experiments.

Conjugal transfer of R factor

Donor and recipient cultures were mixed at early stationary phase of growth. After 30 min, the bacterial conjugation in the mixture was interrupted by blender treatment. Then the appropriate dilutions of the mixture were spread on the selective plates.

Transduction experiment

Transduction was carried out by the methods of Lennox (1955). For the phenotypic expression of drug resistance, the recipient was kept at 37 C for 30 min in fresh broth

* Present adress: Research Institute of Hygiene, Bratislava, Czechoslovakia.

before spreading on the selective plate. The number of input phage and recipient cells was 1×10^9 p.f.u./ml and 5×10^9 cells/ml, respectively. Transduction frequency was expressed by the number of transductants per that of input phages.

RESULTS

1. Conjugation of R_{ms151} into E. coli K12 ML1410 nal^r or W3630

From the conjugational experiments, summarised in Tables 1 and 2, it may be seen that the plasmid called R_{ms151} is usually transferred en block. Only few TC-mono-resistant or CM.KM.LV.SA.GM.DKB (without TC) exconjugants could be scored when selected on individual antibiotics concerned.

TABLE 1

Transfer of drug-resistance from E. coli F-027

Selection (μg/ml)	Transfer frequency	Exconjugants [a)]: Number of exconjugants obtained	Exconjugants [a)]: Drug-resistance pattern
TC (25)	9.3×10^{-5}	45	TC.CM.KM.LV.SA.GM.DKB.APC
		5	TC
CM (12.5)	9.6×10^{-5}	50	TC.CM.KM.LV.SA.GM.DKB.APC
KM (50)	9.9×10^{-5}	50	TC.CM.KM.LV.SA.GM.DKB.APC
SA (200)	5.6×10^{-5}	50	TC.CM.KM.LV.SA.GM.DKB.APC
GM (3.2)	5.6×10^{-5}	49	TC.CM.KM.LV.SA.GM.DKB.APC
		1	CM.KM.LV.SA.GM.DKB.APC

Donor, *E. coli* F-027; Recipient, *E. coli* K12 ML1410 nal^r
Conjugation experiment, see Materials and Methods.

a) Fifty exconjugants were picked at random from selective plate and purified by successive inoculation on the same selective plate.

TABLE 2

Resistance of E. coli strains with or without R- $_{151}$ factor

Host	R_{ms151}	TC	CM	KM	LV	SA	GM_{C1}	DKB	APC	SM	BB-K8
027	+ a)	≥50	≥25	≥800	≥100	≥800	12.5	25	≥50	<1.56	0.8
	— b)	<1.56	1.56	<6.25	6.25	25	<0.8	<1.56	3.12	<1.56	0.8
ML1410	+ c)	≥50	≥25	≥400	≥100	≥200	12.5	12.5	≥50	≥1.56	0.8
	—	<1.56	1.56	<6.25	1.56	<12.5	<0.4	<3.12	<1.56	<1.56	0.8

Transfer of R factor, see Materials and Methods.

a) Original donor of R factor.
b) 027 strain which had lost R factor during storage.
c) Exconjugants which acquired R factor from *E. coli* F-027 R_{ms151}^+.

2. Transduction of R_{ms151} into E. coli K12 W3630

From the Table 3 it may be seen that, as expected, it was possible to separate genes for KM resistance from those for GM resistance. Transductional experiments led to a construction of a tentative genetic map of this plasmid as suggested on Fig. 1.

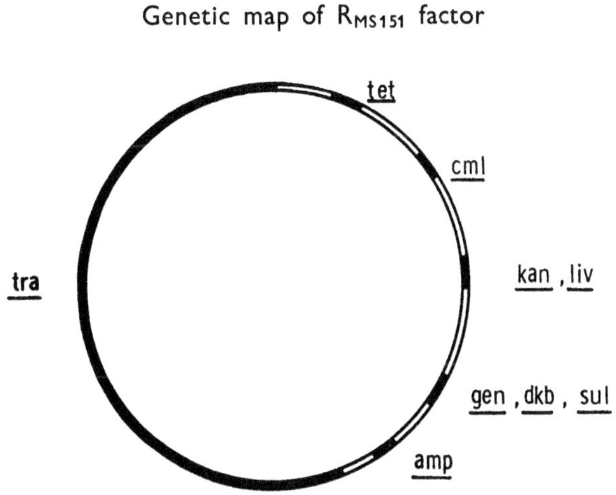

Fig. 1. The genetic map of R_{ms151} factor from *E. coli* F-027.
The distances between each marker are drawn arbitrarily. Genetic symbols of each marker; *tet*, resistance to TC; *cml*, resistance to CM; *kan*, resistance to KM; *liv*, resistance to LV; *gen*, resistance to GM; *dkb*, resistance to DKB; *sul*, resistance to SA; *amp*, resistance to APC; *tra*, conjugal transferability.

TABLE 3

Segregation of R_{ms151} factor after transduction with bacteriophage P1.

Selection (μg/ml)	Genetic properties of transductants	(Isolation freq. %)
	TC *tra*$^+$	(94.9)
TC(25)	TC.CM.APC *tra*$^+$	(3.8)
	TC.CM.KM.LV.SA.GM.DKB.APC *tra*$^+$	(1.3)
	TC.CM.APC *tra*$^+$	(44.4)
	CM.KM.LV.SA.GM.DKB.APC *tra*$^+$	(28.9)
CM (12.5)	TC.CM.KM.LV.SA.GM.DKB.APC *tra*$^+$	(12.2)
	CM.KM.LV.SA.GM.DKB.APC *tra*$^-$	(6.7)
	TC.CM.KM.LV *tra*$^+$	(6.7)
	TC.CM.KM.LV.APC *tra*$^+$	(1.1)
	CM.KM.LV.SA.GM.DKB.APC *tra*$^-$	(32.5)
	CM.KM.LV.SA.GM.DKB.APC *tra*$^+$	(27.5)
KM (50)	TC.CM.KM.LV *tra*$^+$	(27.5)
	TC.CM.KM.LV.SA.GM.DKB.APC *tra*$^+$	(7.5)
	KM.LV.SA.GM.DKB.APC *tra*$^-$	(5.0)

Donor of resistance, ML1410 R_{ms151}^+ which was obtained in the experiment shown in Table 1.
Transduction, see Materials and Methods.
Transduction frequency: ca. 10^{-7}.

3. *Enzymatic background of aminoglycoside and ampicillin resistances*

Enzymatic studies in transductants with separated resistance to drugs involved indicate that ampicillin is decomposed by a transferable high-level lactamase. Kanamycin and lividomycin are inactivated by phosphorylation while GM and DKB are inactivated by an adenylation. The results of these studies will be reported in detail elsewhere.

4. *DNA studies of R_{ms151}*

Studies of the satellite DNA obtained from the original strain as well as respective exconjugants, which will be reported elsewhere, showed that R_{ms151} forms covalently closed circles.

DISCUSSION

Plasmid R_{ms151} is a complex multiple-resistance plasmid occurring in a strain of *E. coli* from the urine of patients from the same clinic where *P. aeruginosa* strains with R plasmids conferring resistance to GM, KM and SM also occur. Enzymatic pattern and the properties of R_{ms151} however are significantly different from those of *P. aeruginosa* in this area which will be reported elsewhere. R_{ms151} does not code for streptomycin-adenyltransferase or streptomycin-phosphotranspherase enzymes attacking streptomycin, but can inactivate KM or LV by a high-level phosphorylation and GM or DKB by a low-level adenylation. BB-K8, however, is not a substrate for any of these enzymes since the R-plasmid bearing parent strain 027 is highly susceptible to this drug.

REFERENCES

SIETZEN, W., KNOTHE, H., KRČM'RY, V., 1973. Simultaneous transfer of gentamicin and kanamycin resistance from an urine strain of *E. coli* Z. allg. Mikrobiol. 13(3): 277–278.

LEDERBERG, J. 1950. Isolation and characterization of biochemical mutants of bacteria. Methods. Med. Res. 3: 5–22.

LENNOX, E. S. 1955. Transduction of linked genetic characters of the host by bacteriophage Pl. Virology 1: 190–206.

III.

RESISTANCE TO OTHER DRUGS

CHAIRMEN: B. WEISBLUM (U.S.A.) and S. M. NAVASHIN (USSR)

OPENING LECTURE

ALTERED METHYLATION OF RIBOSOMAL RNA IN ERYTHROMYCIN-RESISTANT STAPHYLOCOCCUS AUREUS

B. WEISBLUM

Pharmacology Department, University of Wisconsin Medical School, Madison, U.S.A.

This work was supported by a Research Grant from the National Science Foundation (GB-17108) and from Funds provided by the Upjohn Co. and from Eli Lilly and Co.

"This is not an antagonism in the ordinary sense, but a kind of interaction of which, so far as I am aware, no previous example has been seen."

L. P. Garrod, Brit. Med. J. *2*: 57 (1957)

Methylation of 23S ribosomal RNA, a structural component of the 50S ribosome subunit, the receptor for macrolide and lincosamide and streptogramin-B type ("MLS") antibiotics has been identified as the chemical change responsible for resistance to these antibiotics in clinical isolates of *Staphylococcus aureus* and *Streptococcus pyogenes.* In this work, I would like to review microbiological, biochemical, and genetic studies of resistance to these antibiotics. This review is organized under four main headings:

(A) Conclusions regarding rRNA methylation and erythromycin resistance.
(B) A summary of works of previous investigators interpreted in terms of the methylation reaction.
(C) A summary of work on which the conclusions regarding rRNA methylation are based.
(D) Epidemiological and clinical implications of these studies.

A. MY CONCLUSIONS ARE THE FOLLOWING:

(1) Certain strains of *S. aureus* have the capacity to N^6-dimethylate specific adenine residues in 23S rRNA.
(2) Ribosomes in which adenine is specifically methylated bind MLS antibiotics with reduced affinity.
(3) Cells in which the dimethylation reaction occurs are co-resistant to all MLS antibiotics.

(4) Methylation can be expressed in an inducible or a generalized constitutive mode. In either case, induced cells are co-resistant to *all* MLS antibiotics; when erythromycin is removed, inducible cells cease to methylate adenine.
(5) In the natural resistant isolate, erythromycin is the most effective inducer compared to other commonly available MLS antibiotics.

The features of this model are summarized in the Figure 1 below.

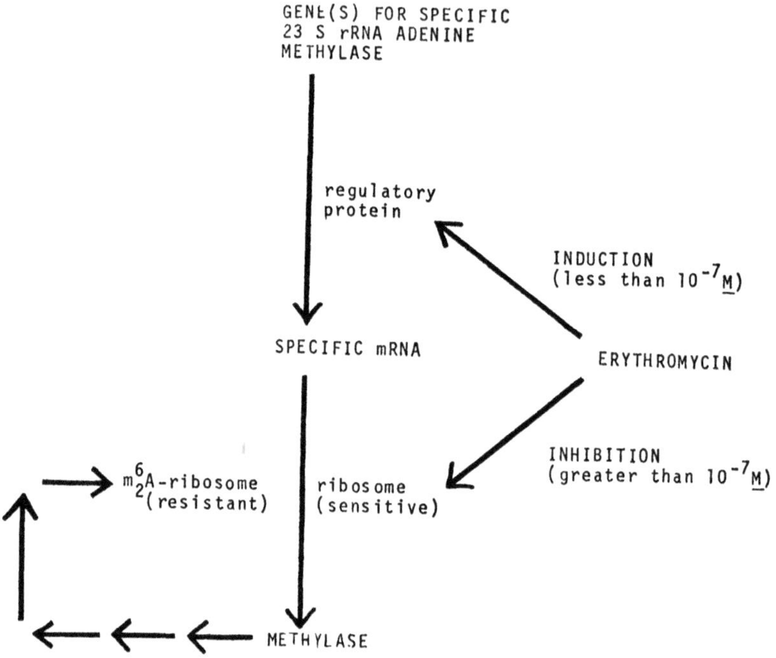

Fig. 1

B. PREVIOUS WORK

In a study of antibiotic resistance in *S. aureus*, Chabbert (1956) noted that certain clinical isolates resistant to erythromycin were sensitive to spiramycin. He also observed that low concentrations of erythromycin apparently antagonized the action of spiramycin because of the reduced size of the inhibition zone produced by a spiramycin-impregnated filter paper strip in proximity to an erythromycin-impregnated strip on solid medium innoculated with the test organism. Garrod (1957) extended these observations to another macrolide pair, erythromycin-oleandomycin. By means of a slightly different assay, it was shown that erythromycin-resistant staphylococci could grow on oleandomycin-containing solid medium only in a circular zone surrounding the erythromycin disc. Elsewhere, on the plate oleandomycin inhibited growth. The strains resistant to erythromycin but sensitive to either oleandomycin or spiramycin were said by Garrod to have "dissociated" resistance to erythromycin.

These observations can be interpreted as follows: In the dissociated strains, erythromycin serves as an inducer *in situ*, i.e., the low initial levels of erythromycin initially attained as erythromycin diffuses from the test disc into the agar, induce cells such that an otherwise inhibitory concentration (>1 µgm/ml) once attained is no longer effective. Surrounding the erythromycin test disc is a circular zone of cells in which adenine methylation has been induced. These cells, susceptible to erythromycin prior to testing, actually become resistant as a consequence of the testing procedure. Spiramycin is ineffective as an inducer at subinhibitory concentrations and so produces a zone of inhibition. Genetic information which specifies or regulates the methylation reaction is at least partially encoded by a plasmid in strains such as RN 453 [Rush *et al.* (1969)]; such information cannot be obtained by the process of random mutation plus selection in the laboratory, but can only be obtained in nature from another bacterium which acts as donor either directly by conjugation or indirectly by phage-mediated transduction. The phylogenetic origins of this type of genetic information remain obscure. Erythromycin-resistant mutant of *S. aureus* can be selected in the laboratory; however, from more definitive studies of *E. coli* ribosomes [Otaka *et al.* (1970)] such mutants are presumed to have altered ribosomal proteins as a consequence of which macrolide antibiotics are bound with reduced affinity. In this way, co-resistance to erythromycin and spiramycin seen in mutant strains can arise by the mutational alteration of a preexisting gene template for a ribosomal protein with a resultant amino acid substitution.

In order to explain the behavior of the inducible strains, it is postulated that a circular zone about 2 cm in diameter surrounding the erythromycin disc consists of cells induced *in situ* by the erythromycin which diffuses from the sensitivity disc. Erythromycin thus provides a protective "umbrella" of methylation. When this zone intersects the zone of potential inhibition by spiramycin, induced cells proximal to the erythromycin disc which are covered by this umbrella grow despite the presence of spiramycin because of a continuous inductive stimulus provided by erythromycin. Other geometric arrangements of testing materials are possible in studies of antibiotic antagonisms. Chabbert (1957), for example, placed two antibiotic impregnated paper strips perpendicular to each other on a lawn of uninduced cells. Discs may be preferable because their use allows testing several different antibiotics simultaneously per plate.

Barber and Waterworth (1964) further extended the studies of Garrod (1957) to include erythromycin-induced resistance to lincomycin and pristinamycin (presumably "component I"), while Bourse and Monier (1967) reported that *only* pristinamycin I (the Streptogramin B component) and not pristinamycin II (the Streptogramin A component) was involved. Griffith *et al.*, (1965) similarly noted antagonism (i.e., erythromycin-induced resistance) if a lincomycin disc was placed in the proximity of an erythromycin disc.

For antagonism by erythromycin to occur, it is necessary (but not sufficient) that the test antibiotic act on the 50S ribosome subunit. In the works cited above, the involvement of two new classes of 50S subunit inhibitors, distinct from each other as well as from the macrolides, were discovered. The pristinamycins are members of a group of antibiotics known as the streptogramins; a review of the streptogramins has been prepared by Vazquez (1967). The streptogramins have been subdivided into two classes — one class consists of large heterocyclic lactones, while the other class consists of cyclic polypeptides. These have been referred to as the streptogramin A and streptogramin B groups, respectively. Erythromycin antagonizes the action of streptogramin B-group antibiotics but not streptogramin A-group antibiotics. Both the A and B-group antibiotics inhibit 50S subunit function. Pristinamycin I belongs to streptogramin B-group while pristinamycin II belongs to the streptogramin A-group. To make matters more confusing, another pair of streptogramin antibiotics, the staphylomycins, named M-type and S-type, have been described; they correspond structurally and functionally to the streptogramin A and B-groups, respectively. A possible source of confusion is the fact that staphylomycin M was so designated because the heterocyclic lactone was found to consist of a ring-structure about 20 atoms in length. Such structures are referred to as "macrolides" by chemists; however, in the context of this discussion, it is important to recognize that the staphylomycin M antibiotics are macrolides in name only and that from a functional point of view it is the staphylomycin S-group along with other streptogramin B-type antibiotics as well as lincosamides and macrolides (sensu strictu) to which erythromycin induces resistance. Other streptogramins are designated as A-type or B-type in direct parallel to the name compounds. These include respectively, mikamycin A and B, PA 114 A and B, ostreogrycin A and B, and many more.

The antagonism described by Chabbert (1956) when assayed by the test disc method provides a simple way to determine whether erythromycin induces resistance to a given antibiotic. We first undertook a survey using the disc method of many different antibiotics in order to determine the range of antibiotics to which erythromycin induced resistance [Weisblum and Demohn (1969)]. The conclusion that emerged from these studies was that erythromycin induced resistance to macrolides, lincosamides, and streptogramin B-type antibiotics but not to streptogramin A-type antibiotics, nor to chloramphenicol, amicetin, or several other classes of 50S subunit inhibitors. These findings pointed to the fact that only antibiotics which act on the 50S subunit were involved and that these belonged to three out of at least eight classes of antibiotics which act on the 50S subunit.

If erythromycin-induced resistance was related to some type of ribosome modification, the modification could either involve ribosomal RNA or ribosomal protein or both. In preparation for such analytical studies, we undertook to determine the requirements for induction, namely, those optimal conditions of cultivation which one would employ in order to obtain induced cells efficiently [Weisblum *et al.* (1971)]. Moreover, it was also important to isolate constitutively resistant mutants of an inducible parent strain. Biochemical studies of such mutants grown in the absence of erythromycin would have to reveal the same chemical modifications as those induced by erythromycin in the inducible parent strain.

From these studies, we concluded that: (a) growth in the presence of 0.1 γ/ml (10^{-7} M) erythromycin for 1 hour a culture of *S. aureus* could be induced in such a way that every cell in the population was capable of forming a colony when plated on solid medium containing 50 γ/ml erythromycin; (b) the optimum for induction was in the range 0.01 to 0.1 γ/ml erythromycin while higher or lower concentrations of erythromycin used for induction produced a sharp decline in the number of colony forming units on 50 γ/ml containing solid medium without a corresponding effect on cell viability; (c) inhibition of RNA and protein synthesis by the antibiotics streptovaricin and chloramphenicol respectively inhibited induction (measured by colony-forming ability on 50 γ/ml erythromycin medium) without proportional reduction in cell viability; on the other hand, inhibition of DNA synthesis by novobiocin inhibited induction only insofar as it reduced cell viability. These observations suggested that RNA and protein-synthesis but not DNA synthesis were required for induction.

In a search for the biochemical alteration responsible for erythromycin inducible resistance, Mitsuhashi and his coworkers reported that induced and constitutively resistant cells showed a reduced uptake of erythromycin and furthermore, that ribosomes from these strains had reduced affinity for macrolides. Their results have been summarized by Saito *et al.* (1972). Studies of Nakajima *et al.* (1968) were unable to demonstrate erythromycin metabolites if labelled erythromycin was incubated with erythromycin-resistant cells or cell-free extracts derived from them which was consistent with the idea that a structural modification of the ribosome rather than an "erythromycinase" might be responsible for the resistant phenotype. Ribosomal proteins extracted from sensitive and resistant cells failed to show any differences when fractionated on CM cellulose [Saito *et al.* (1972)] or by electrophoresis on acrylamide gels [Lai (1972)]. Neither of these techniques completely resolved all the ribosomal proteins (approximately 35), and even if they could, mutations involving uncharged amino acids might result in undetectable electrophoretic differences in the ribosomal proteins.

C. STUDIES OF rRNA METHYLATION

In parallel with these studies of ribosomal proteins, we also compared the extent of rRNA methylation with ^{3}H methyl-methionine as methyl donor in uninduced inducible, induced inducible, and constitutively resistant cells [Lai and Weisblum (1971)]. In preliminary studies a 10–20% consistently higher degree of agregate methylation (methyl cpm/OD_{260} unit) of 23S rRNA from induced and constitutive by resistant cells compared to 23S rRNA from uninduced cells was noted, while 16S rRNA showed only 0–5% increase. Although the observed changes were reproducible, we inferred that a more unequivocal type of change might be seen if methyl-labelled RNASs were degraded and differences in the extents of methylation of individual bases were compared. For this purpose, a double-label experiment was performed in which rRNA's from induced or constitutive cells grown in the presence of [Me-^{3}H]-methionine were compared with rRNA's from uninduced cells grown in the presence of [Me-^{14}C]-methionine. In these studies, we discovered a dramatic change involving N^6,N^6-dimethyl adenine (m_2^6A) which went from <0.1 residue per 23S to approximately 1–2 residues per 23S rRNA. No detectible changes in methylation of guanine or of the pyrimidines was seen, nor were any changes detectible in the 16S rRNA, particularly m_2^6A which is a "normal" constituent of 16S rRNA in induced, uninduced, and constitutively resistant cells.

In an attempt to identify other possible changes in the pattern of 23S rRNA methylation, [Me-^{14}C]-labelled 23S rRNA from sensitive and resistant cells was prepared and degraded to oligonucleotides with T_1 ribonuclease. The digests were fractionated by two dimensional fingerprinting techniques and methyl-labelled fragments located by autoradiography [Lai *et al.* (1973a)]. Seven methyl-labelled fragments were found to be common to rRNA from uninduced, induced, and constitutive cells, while a single additional fragment was found only in the digests from the induced and constitutively resistant cells. Analysis of the 8 fragments revealed that the additional fragment was derived by dimethylation of a single adenine residue in the sequence AAAG, that this fragment was the only one which contained m_2^6A, and that m_2^6A was the only methyl-labelled component present in this fragment. Further structural studies in which the [Me-^{14}C]-labelled RNA was degraded with pancreatic RNAase permitted identification of a heptamer sequence Pu_6Py (tentatively identified as GGAAAGC or GAAAGGC) as the site of adenine methylation in 23S rRNA (Tanaka and Weisblum, unpublished data). In order to determine whether altered methylation of adenine was the cause or the result of erythromycin resistance we attempted to reconstitute 50S ribosome subunits using proteins and 5S rRNA from sensitive cells and 23S rRNA from sensitive or resistant cells. From the work of Nomura and his colleagues, it had recently become technically feasible to reconstitute functionally active 50S subunits from constituent proteins and RNA's in *Bacillus stearothermophilus*. Attempts to reconstitute 50S subunits from *S. aureus* were unsuccessful so that heterologous reconstitution was attempted in which proteins and 5S rRNA were purified from an erythromycin-sensitive strain of *B. stearothermophilus* while 23S rRNA was purified from erythromycin-sensitive or resistant *S. aureus* [Lai *et al.* (1973)]. This type of reconstruction was found to possess a phenotype which corresponded to that of the *S. aureus* preparation from which the 23S rRNA was obtained. These studies pointed to the fact that the 23S rRNA plays an important role in specifying the state of sensitivity or resistance of the ribosome toward erythromycin.

Constitutive methylators which also display the MLS-resistant phenotype can be isolated in the laboratory from inducible strains by placing the latter on solid medium containing 10 μgm/ml lincomycin, carbomycin, streptogramin B, or any non-inducing MLS antibiotic. Strains isolated in this manner fall into three classes:

(1) Generalized constitutives which show co-resistance to all MLS antibiotics tested.

(2) Partial constitutives which show co-resistance to several but not all MLS antibiotics tested. In these strains, resistance is inducible by erythromycin to those MLS antibiotics to which the organism remains sensitive.

(3) Mutants with reversed specificity of induction. In one such example, a colony which grew up on carbomycin medium (10 μgm/ml) was found to be inducible by carbomycin and lincomycin, and by erythromycin to a much lesser degree, if at all [Tanaka and Weisblum (1974)].

So far, constitutive methylation has only been examined in a strain which shows generalized constitutive resistance. In the strain studied, it was found that m_2^6A as well as the tetranucleotide sequence (A, A, m_2^6A) G were present without a requirement for induction by erythromycin [Lai *et al.* (1973a)]. From a biochemical point of view, constitutive mutants are useful because they permit us to study the interaction of erythromycin with a system not previously perturbed by erythromycin.

D. EPIDEMIOLOGICAL AND CLINICAL IMPLICATIONS OF THESE STUDIES

1. *Erythromycin resistance in other bacteria.* The MLS phenotype has been found in *Streptococcus pyogenes* group A [Lai (1972), Clewell and Franke (1974)] and group D [Courvalin *et al.* (1972)]. The studies of Clewell and Franke and of Courvalin *et al.* demonstrated the association of MLS resistance with the presence of a plasmid. Upon elimination of the plasmid, cells displayed MLS sensitivity. In the studies of Lai (1972) m_2^6A was shown to be present in 23S rRNA of an MLS-resistant strain but absent from the 23S rRNA of a (non-isogenic) sensitive strain. By extrapolation from our experience with *S. aureus*, these strains are all presumed to be constitutive methylators. Apparently an erythromycin-inducible strain of *S. pyogenes* group A has been found by Hyder and Streitfeld (1973); unfortunately, they only tested erythromycin and lincomycin but did not test streptogramins.

By means of a simple sensitivity disc assay, a presumptive identification of methylating strains can be made. The ease of identification of this unique property in common organisms pathogenic to man is of obvious epidemiological importance. The exact relationship between the chemical structures of MLS antibiotics and resistance of ribosomes which contain methylated 23S rRNA is not yet clear; however, in view of the ease with which the determination of MLS-resistance can be made, it is unfortunate to note the number of independent reports calling attention to the pattern of co-resistance to erythromycin and lincomycin which omit an additional test of streptogramins. It might be desirable to call a moratorium on further partial descriptions of this sort.

A study of the organization of genetic elements which specify inducible resistance has not been possible because of: (a) the inherent difficulties involved in genetic studies of *S. aureus* and (b) the failure to isolate an extrachromosomal element involved in this process. The only hard evidence available concerns a form of erythromycin resistance conferred by the plasmid pI 258 in *S. aureus*. The resistance specified by this plasmid appears to be of the constitutive type both with respect to MLS resistance (disc method) and 23S rRNA methylation, and by the process of deletion mapping, Novick (1967) has ordered a set of genes which specify resistance to erythromycin, penicillin, and metal ions. This plasmid has a molecular weight of approximately 18×10^6 daltons.

In attempting to ascertain whether erythromycin induces the synthesis of specific messenger RNA, unlabelled RNA from uninduced, induced and constitutively resistant cells has been used in competition hybridization assays involving pI 258 and its ^{3}H-labelled transcript (deHaseth and Weisblum, unpublished data). Preliminary results show RNA from resistant cells to be more effective in the competitive reaction.

2. *Recognition of the MLS phenotype.* In view of the prevalence of Staphylococci, it is of interest to have epidemiological information concerning the type of erythromycin resistance present when resistant organisms are obtained from a patient. Clinical isolates of erythromycin-resistant *S. aureus* appear to fall into at least two classes. These are: (a) Inducible methylators, and (b) Constitutive methylators. Having defined these phenotypes, it would be of interest to determine whether additional mechanisms of clinical resistance exist. The ease with which disc testing can be done suggests a means by which methylating resistant strains can be identified.

In the early studies of Garrod (1957) and of Barber and Waterworth (1964) on erythromycin resistance in clinical isolates of *S. aureus* strains were described as having either "dissociated" resistance on the one hand or "double" resistance on the other. The basis for this designation suggested by Garrod (1957) was respectively sensitivity or resistance to spiramycin or oleandomycin in addition to erythromycin resistance. Unfortunately the term "double" resistance was applied both to *natural* isolates co-resistant to both erythromycin and spiramycin as well as to mutant strains selected in the laboratory from initially sensitive (but erythromycin-inducible) parent strains. In view of our present understanding of the role of rRNA methylation, it is likely that the double resistant strains obtained as such from natural isolates were constitutive methylators while the double resistant strains selected in the laboratory, acquired the resistant phenotype by mutation in a ribosomal protein structural gene with consequent amino acid replacement. In any case, both the dissociated and double resistant natural isolates appear to be methylators, differing only in the regulation of the same biochemical reactions.

Presumptive identification of an inducible methylator can easily be made with sensitivity discs appropriately placed on a test plate, as described above. From making inquiries at a local hospital diagnostic bacteriology laboratory, it was learned that technicians place erythromycin and lincomycin test discs on diametrically opposite sides of a test plate because of distortions occasionally seen in the inhibitory zones which preclude measurement of the inhibitory zone diameter. In point of fact, even the most extreme examples of "antagonism" should not preclude making such a measurement. Much useful epidemiological information pertaining to erythromycin resistance is thus deliberately discarded. The possible clinical significance of knowing whether a particular isolate of *S. aureus* is an inducible methylator will be discussed below.

3. *An exception to the Kirby-Bauer method of antibiotic sensitivity determination.* The "Kirby-Bauer method" [Bauer *et al.* (1966)] is a widely used technique for ascertaining clinical sensitivity of a bacterial strain to a particular antibiotic. This technique is based on the assumption that under defined conditions (medium, innoculum size, disc potency) the diameter of an inhibitory zone produced by a specific antibiotic can be correlated with the minimum inhibitory concentration (m.i.c.) of that antibiotic determined in liquid culture. It is assumed that this latter figure, considered in the light of attainable serum levels of antibiotic, affords some idea whether a specific organism can be treated with a specific antibiotic. The case of inducible methylators provides an instructive *exception* to the Kirby-Bauer method.

For most (if not all) inducible methylators, the diameter of the inhibitory zone surrounding a lincomycin disc is large enough that the strain is reported by the clinical bacteriology laboratory to be lincomycin-sensitive. If a determination of m.i.c. in broth is attempted, there is a high probability that a mutant constitutive methylator is present which would then be selected by the presence of lincomycin. Such mutants are capable of growing at lincomycin concentrations of at least 50 γ/ml and overgrow the culture, but if present in the innoculum used for sensitivity disc testing only appear as single colonies arising within the inhibitory zone and do not interfere with measurement of the inhibitory zone diameter. In the treatment of infections caused by erythromycin-resistant lincomycin-sensitive staphylococci, it would be useful to know whether the causative agent is a methylator and to anticipate the possible replacement of the original population by constitutive methylators if lincomycin is used.

4. *Macrolides in the environment.* Although not as widely used in agriculture as the tetracyclines, the macrolide antibiotic tylosin has found wide application in the control of mycoplasma infections of livestock. Resistance to macrolides has not yet assumed the proportions of resistance to tetracyclines found in gram-negative fecal flora of livestock, however, the widespread use of this antibiotic would be expected to introduce some selective pressure for the emergence of methylating strains.

CRITIQUE

The role of erythromycin as inducer of rRNA methylation has been conceived in terms of model similar to the extensively studied system of induced beta galactosidase synthesis. To date, there is no direct evidence for the existence of a putative repressor nor for a methylase. Only preliminary evidence is available for the existence of RNA whose synthesis is induced by erythromycin.

REFERENCES

BARBER, M. and WATERWORTH, P. M. (1964). Antibacterial activity of lincomycin and pristinamycin: A comparison with erythromycin. Brit. Med. J. **2**: 603–606.

BAUER, A. W., KIRBY, W. M. M., SHERRIS, J. C. and TURCK, M. (1966). Antibiotic susceptibility testing by a standardized single disc method. Amer. J. Clin. Pathol. **45**: 493–496.

BOURSE, R. and MONIER, J. (1967). Effet de l'erythromycine sur la croissance de *S. aureus* "resistant dissocié" en bacteriostase par un autre macrolide ou un antibiotique apparenté. Ann. Inst. Pasteur **110**: 67–79.

CHABBERT, Y. (1956). Antagonisme *in vitro* entre l'erythromycine et la spiramycine. Ann. Inst. Pasteur **90**: 787–790.

CLEWELL, D. B. and FRANKE, A. E. (1974). Characterization of a plasmid determining resistance to erythromycin, lincomycin, and vernamycin Bα in a strain of *Streptococcus pyogenes*. Antimicrobial Ag. Chemother., in press.

COURVALIN, P. M., CARLIER, C. and CHABBERT, Y. A. (1972). Plasmid-linked tetracycin and erythromycin resistance in group D "Streptococcus". Ann. Inst. Pasteur **123**: 755–759.

GARROD, L. P. (1957). The erythromycin group of antibiotics. Brit. Med. J. **2**: 57–63.

GRIFFITH, L. J., OSTRANDER, W. E., MULLINS, C. G. and BESWICK, D. E. (1965). Drug antagonism between lincomycin and erythromycin. Science **147**, 746–747.

HYDER, S. L. and STREITFELD, M. M. (1973). Inducible and constitutive resistance to macrolide antibiotics and lincomycin in clinically isolated strains of *Streptococcus pyrogenes*. Antimicrobial Ag. Chemother. **4**: 327–331.

LAI, C. J. (1972). Thesis. University of Wisconsin.

LAI, C. J., DAHLBERG, J. and WEISBLUM, B. (1973a). Structure of an inducibly methylatable nucleotide sequence in 23S ribosomal ribonucleic acid from erythromycin-resistant *Staphylococcus aureus*. Biochemistry **12**: 457–460.

LAI, C. J., WEISBLUM, B., FAHNESTOCK, S. and NOMURA, M. (1973b). Alteration of 23S ribosomal RNA and erythromycin-induced resistance to lincomycin and spiramycin in *Staphylococcus aureus*. J. Mol. Biol. **74**: 67–72.

LAI, C. J. and WEISBLUM, B. (1971). Altered methylation of ribosomal RNA in an erythromycin-resistant strain of *Staphylococcus aureus*. Proc. Nat. Acad. Sci. U.S.A. **68**: 856–886.

NAKAJIMA, Y., INOUE, M., OKA, Y. and YAMGISHI, S. (1968). A mode of resistance to macrolide antibiotics in *Staphylococcus aureus*. Japan J. Microbiol. **12**: 248–250.

NOVICK, R. P. (1967). Penicillinase plasmids of *Staphylococcus aureus*. Fed. Proc. **26**: 29–38.

OTAKA, E., TERAOKA, H., TAMAKI, M., TANAKA, K. and OSAWA, S. (1970). Ribosomes from erythromycin-resistant mutant of *Escherichia coli*. J. Mol. Biol. **48**: 499–510.

RUSH, M. G., GORDON, C. N., NOVICK, R. P. and WARNER, R. C. (1969). Penicillinase plasmid DNA from *Staphylococcus aureus*. Proc. Nat. Acad. Sci. U.S.A. **63**: 1304–1310.

SAITO, T., SHIMIZU, M. and MITSUHASHI, S. (1972). Macrolide resistance in Staphylococci. Ann. N.Y. Acad. Sci. **182**: 267–278.

TANAKA, T. and WEISBLUM, B. (1974). Mutant of *Staphylococcus aureus* with lincomycin- and carbomycin-inducible resistance to erythromycin. Antimicrobial Ag. Chemother. **5**: in press.

VAZQUEZ, D. (1967). The Streptogramin family of antibiotics. in Antibiotics I, Mechanisms of Action (D. Gottlieb and P. D. Shaw, eds.) Springer Verlag, N. Y.

WEISBLUM, B. and DEMOHN, V. (1969). Erythromycin-inducible resistance in *Staphylococcus aureus*: Survey of antibiotic classes involved. J. Bacteriol. **98**: 447–452.

WEISBLUM, B., SIDDHIKOL, C., LAI, C. J. and DEMOHN, V. (1971). Erythromycin-inducible resistance in *Staphylococcus aureus*: Requirements for induction. J. Bacteriol. **106**: 835–847.

OPENING LECTURE

SOME PROPERTIES OF PHENOTYPE IN TETRACYCLINE RESISTANT ORGANISMS

S. N. NAVASHIN, YU. O. SAZYKIN, I. P. FOMINA
National Institute of Antibiotics, Moscow, USSR

Tetracycline antibiotics are still of great importance in modern medicine. At the same time, resistant strains of many pathogens mainly bearing R-factors are widespread in various regions of the world. The decreased efficiency of tetracycline therapy with respect to diseases caused by tetracycline resistant strains resulted in a limitation of their use. Studies on the mechanism of action, the nature of resistance, the epidemiology and ecology of resistant organisms are the fundamental basis for the synthesis of new tetracycline derivatives.

Though the present communication is concerned with the problem of tetracycline resistance, it is not yet possible to give positive answers to many questions, and certain aspects still seem to be obscure. We decided to discuss here data on the epidemiology of tetracycline resistance of some pathogens isolated during a long period of observations, as well as the preliminary results of studies on some mechanisms of resistance to this group of antibiotics.

The dynamics of antibiotic sensitivity of the two main patogens causing purulent infections, that is *Staph. aureus* and *E. coli*, has been studied for about 10 years in two groups of patients.

The first group was concerned with strains isolated from patients in divisions of general surgery treated with antibiotics for a short time during the post-operative periods.

The second group was concerned with strains isolated from patients with chronic inflammatory processes of the respiratory tracts or with severe septicopyemia complicated with acute kidney insufficiency.

Tetracycline resistance was noticed in more than 70 per cent of the strains of *Staph. aureus* and registered in practically all multiple resistance strains (Tables 1, 2).

The correlation between the frequency of tetracycline resistant staphylococci and the progress character of the pathological process was shown. Tetracycline resistance was found in 83 per cent of the staphylococci isolated from the patients with severe septic processes as compared to 75 per cent of such strains isolated from patients in the divisions of general surgery. The methicillin resistant strains of *Staph. aureus* were resistant to tetracycline in 97 per cent of the cases at resistance levels of 100 to 250 mcg/ml. Semi-synthetic tetracyclines and their 6-desoxyderivatives: minocycline and doxicycline had marked advantages over the natural antibiotics. The minocycline resistant strains amounted to 12.5 and 40 per cent respectively. Minocycline was greatly superior to other tetracyclines in the comparative frequency of highly sensitive strains. 6–7.5 per cent of

TABLE 1

Sensitivity to antibiotics of clinical strains Staph. aureus isolated in general departments and from patients with sepsis

Antibiotics	Per cent of resistant strains isolated from patients	
	in surgical departments	from patients with sepsis
Benzylpenicillin	72.7	90.6
Methicillin	17	30.3
Ampicillin	32.6	—
Erythromycin	39.7	57.6
Oleandomycin	29,6	49.4
Fusidine	0	0
Streptomycin	67.4	79
Chloramphenicol	49.7	66.4
Tetracycline	75.0	83.0
Gentamycin	0	0

TABLE 2

Antibiotic resistance of clinical strains of Staph. aureus

Number of determinants	Determinants of antibiotic resistance	Per cent of strains
1	TC; EM; CM	5,4
2	TC; CM; PC; TC; PC; SM; PC; EM	17,8
3	PC; CM; SM; PC; TC; CM; PC; TC; PC; SM; EM	15,5
4	PC; TC; SM; CM; EM; CM; TC; SM; PC; TC; EM; CM	30,1
5	PC; EM; TC; SM; CM; PC; TC; SM; CM; KM; EM; TC; CM; SM; KM	23.0
6	PC; EM; CM; TC; SM; KM	8.2

TABLE 3

Degree of sensitivity to tetracyclines of 130 clinical E. coli strains

Antibiotics	MIC . mcg/ml										Per cent of resistant strains
	0.05–1	2	4	8	16	31	62	125	250	500	
Oxytetracycline	8	25	19	7	—	2	16	17	16	20	54.7
Chlortetracycline	3	5	6	12	14	7	11	21	37	15	81.1
Methacyclin	6	18	5	23	1	4	11	9	24	29	60.3
Doxycycline	10	24	21	6	20	14	9	7	12	7	53.4
Minocycline	79	19	14	9	6	2	1	—	—	—	6.5

the strains were sensitive to 0.05–2 mcg/ml of minocycline, while the same concentrations of doxycycline and tetracycline inhibited the growth of 21.2 and 27.5 per cent of the strains respectively (Table 3).

Multiple resistant and tetracycline resistant staphylococci were mainly sensitive to minocycline (Table 4). An analogous tendency was also observed with respect to the strains of *E. coli* isolated from the two groups of patients.

TABLE 4

Action of minocycline on polyresistant strains of Staph. cureus

Number of resistance determinants	Determinants of antibiotic resistance	Number of strains	MIC minocycline mcg/ml 0.05-2	3.I-8
2	Tc.Cm PC.Tc	7	9	—
3	Pc.Tc.Sm Pc.Tc.Sm	I2	9	3
4	Pc.Tc.Sm.Cm PC.TC.M.SM. Em.PC.Tc.Sm. Pc.TC.Em.Sm.	51	38	13
5	Em.Tc.Cm.Sm.Pc. Pc.Tc.Sm.Cm.Km Em.Tc.Cm.Sm.Km	4	—	4
6	Pc.Tc.Cm.Sm.Em.Km Em.TC.Sm.Cm.Km.M. Pc.Tc.Sm.Cm.Em.m.	6	—	6

TABLE 5

Sensitivity of E.coli strains isolated from patients in surgical departments and from patients with sepsis

Antibiotics	Per cent of resistant strains isolated from patients	
	in surgical departments	from patients with sepsis
Tetracycline	54,7	86,9
Chloramphenicol	38,6	73,6
Streptomycin	53,3	88,9
Kanamycin	28,9	48,1
Gentamicin	15,9	35,8
Ampicillin	41,8	72,9
Carbenicillin	43,1	72,3

The frequency of the tetracycline resistant strains isolated from patients in the divisions of general surgery was 54.7 per cent against 86.9 per cent of the strains isolated from septic cases or cases with urinary infections. The frequency of the tetracycline resistant *E. coli* was lower than in the case of staphylococci. However, strains resistant to tetracycline were more frequent than those resistant to aminoglycosides and semisynthetic penicillins (Table 5).

The same as against staphylococci, minocycline was the most active against *E. coli*. While the resistance of *E. coli* strains to natural tetracyclines ranged within 54.7 to 81.0 per cent, only 6.9 per cent of the strains were resistant to minocycline. The values of the

MIC of minocycline for the resistant strains of *E. coli* did not generally exceed 16–64 mcg/ml. while those of tetracycline were 32 to 500 mcg/ml. The multiple resistant strains carrying the determinants of resistance to streptomycin, tetracycline, chloramphenicol were as a rule sensitive to minocycline. A comparative analysis of the sensitivity of clinical strains of *Staph. aureus* and *E. coli* therefore showed a correlation between the frequency of the tetracycline resistant strains and the duration of the antibiotic therapy, as well as an absence of cross resistance and difference in the activity levels of tetracycline and its structural analogues.

The data are indicative of the following:

(a) The need of a detailed genetic analysis of the nature of the tetracycline resistant strains isolated from various sources, since they are characterised by differences in the structure and antibiotic resistance levels.

(b) Multiplicity of the resistance mechanisms, confirmed by the absence of complete cross resistance to minocycline (in cases with a common mechanism of action) in the tetracycline resistant strains, exists in two species, that is the Coli and staphylococci.

It does not seem expedient to discuss in this communication the data on the molecular mechanism of tetracycline action. It should be only mentioned that the effect of tetracyclines on the ribosome-matrix systems of the protein synthesis in the microbial cell has not yet been elucidated, the same as that of aminoglycosides, chloramphenicol or puromycin. Up to the present, the studies have concentrated on 30S subunits. However, just as before, there is no agreement on the number of tetracycline molecules being bound by the ribosome, the path of the binding, etc. The concept of the formation of magnesium bridges during the final process is the most common, but other fixation mechanisms are not excluded either.

The assertion that tetracyclines reduce the interaction of the RNA messenger codone with the RNA anticodone for which 30S subunits are responsible is probably of value (Hogenauer G., Turnowsky F., 1972).

It was shown (Hirashima A. et al., 1973) that tetracycline was more active in inhibiting the protein synthesis by the microbial cell membranes than proteins of the cytoplasm. It is not clear whether this was associated with the ribosome differentiation, with peculiar features of the intracellular distribution of ribosomes or some other factors. However, it is important that tetracyclines had a specific effect on the synthesis of the membrane protein components responsible for the antibiotic transport into the cells and defining tetracycline resistance after changes in their properties.

It is at present difficult to correlate such data with the mechanisms of tetracycline resistance, whose main aspects to be investigated are (1) characteristics of the cell wall permeability and (2) changes in the system of the protein synthesis.

Beginning from the first half of the Sixties, the mechanism of active transport of tetracyclines into the microbial cell, the effect of various factors on such transport and finally the causes of lower permeability of tetracyclines into the resistant cell as compared to the sensitive one are being intensively studied in a number of laboratories (Izaki K., Arima K., 1963; Franklin T. T., Higginson B., 1969).

The kinetics of the antibiotic absorption was studied using both their labeled substances (Reynard A. et al., 1971, Lapatik M. D., Plakunov V. K., 1969) and luminiscence microscopy (Chanina M. F., et al., 1970; Korn M. Y., Chanina M. F., 1970; Kersten 1971). Studies in the same directions are being continued (Dockter, 1973).

Though attempts are made to create three-dimension models reflecting tetracycline binding with ribosomes (Smythiers et al., 1972), there are still no three-dimension models reflecting various stages of active tetracycline transfer through the membranes.

At the same time, the systems of active transport of tetracyclines are of interest from the point of view of their genetic control (Sompolinsky D. et al., 1970) and the mechanism of tetracycline resistance induction (Inoue M., Hashimoto H., Mitsuhashi S., 1970).

One of the major problems is now identification of the membrane proteins, the production of which is coded by R-factors providing tetracycline impermeability of the cell membrane and consequently the cell resistance.

The cytoplasmic membrane of the bacterial cell fulfils the function of the main barrier of permeability. At the same time it is the area of localization of many enzymes and their organization into complex systems. According to the data published recently by Spenger and Guest (1974) 56 protein fractions were found in the membrane of *E. coli*, while the number of the individual proteins is certainly much higher. There are rather clear indications that the membrane contains not less than 100 functionally active proteins including the permease complexes, electron-transport systems, DNA-reducing enzymes, etc. High lability of the protein set of the cytoplasmic membrane as a subcellular structure is one of its peculiar properties; the composition of the nutrient medium markedly affects, both quantitatively and qualitatively, the protein membranes (Spenger M. E., Guest T. R., 1974). As for the facultative anaerobes, they are in addition markedly affected by the aeration level. Therefore, the presence of a new protein in the membrane, the formation of which is coded by R-factor and induced by tetracycline, is of little significance for the cell from the physiological point of view, since the molecular organization of the membrane is subjected to constant variation.

It is particularly difficult to find the membrane component causing the tetracycline resistance of the cell, since the system of its functional activity control *in vitro* has not yet been elaborated. The multistage process of isolation of the above individual substance cannot be at present controlled by the method used for example in the control of isolation and identification of antibiotic inactivating enzymes.

For finding the R-factor coding membrane component under modern experimental conditions one should try to limit the synthesis of the products of the chromosome gene activity. When such conditions are provided, the cell is incorporated into a limited number of substances, each of which can be identified more easily than in the case of cell fractionation accompanied by violent synthetic processes.

Attempts were recently described at finding specific membrane components in tetracycline resistant organisms. One of these studies (Franklin T. J., Rownd R., 1973) deals with *Proteus mirabilis* carrying R-factor.

In order to create conditions for better detecting protein presumably coded by the tetracycline resistance determinant, R^+ cells of *Pr. mirabilis* were used in the stationary growth phase when protein synthesis dependent on the chromosome genes was weakened. Under such conditions, induction of tetracycline resistance in combination with the radioisotope label provided detection of the protein fraction present only in the induced R^+ cells by using electrophoresis in polyacrylamide gel.

However, the minicells of *E. coli* described for the first time by Adler et al. in 1967 proved to be more suitable for studies on the protein product coded by the tetracycline resistant determinant than the cells of the stationary growth phase. An anomalous cell division observed in definite strains of *E. coli*, Bac. subtilis (Mendelson et al., 1974) and other microorganisms resulted in the formation, along with ordinary cells, of small spheres surrounded by membrane and cell walls containing normal ratios of RNA and protein and no noticeable amounts of chromosome DNA. Such facts of cardinal importance indicate that the above peculiar natural cell models are extremely suitable for studying extrachromosomal hereditary elements and their function, provided that it is possible to introduce into them — by conjugation or some other similar means — certain plasmids prior to or after their formation. The absence of synthesis of the proteins coded by the chromosomal genes resulted specifically in the possibility of successfully revealing protein products of the resistance determinants in R^+ minicells by ordinary biochemical methods. Although the minicells have been used until now as testing material only in studies on the mechanism of tetracycline resistance (Embden van J., Cohen S. N., 1973; Levy, Mc. Murry, 1974, 1974), it may be assumed than their use will in future enable significant advances in the differentiation of proteins coded by the determinants of resistance in relation to other antimicrobial agents.

We consider the data published in 1974 by Levy and Mc Murry to be of utmost interest; further important developments in this field have been presented here. By using a system without synthesis of the proteins coded by the chromosome genes, the authors succeeded in describing several specific proteins coded by R-factor DNA and bound with it, as well as several proteins associating with membranes and being also coded by R-factor DNA.

By using electrophoresis in polyacrylamide gel, a new protein with a molecular weight of about 50,000 was found at rather significant amounts in minicells carrying an R-factor (R-222) and selected in the presence of tetracycline. Insignificant amounts of the protein were also found in minicells carrying the same R-factor but not subjected to induction, that is which were in no contact with tetracycline. Up to 70 per cent of the protein was localized in the membrane fraction.

The authors proved that the regularity they have described, i. c. a marked increase in the amounts of the specific protein after tetracycline induction, was not a particular fact, and this is of significant interest. Experiments with ten R-factors of various origin which possessed the determinant of tetracycline resistance showed that the presence of the new protein in the minicells was observed in six of them.

The authors have undoubtedly significantly advanced in the identification of the cell membrane responsible for tetracycline resistance of the microbial cell at the level of the phenotype.

It should be noted that other studies on the R-factor carrying minicells (Van Embden J., Cohen S. N., 1974) have also shown that the synthesis of four polypeptides out of five whose production was coded by the given R-factor are liable to the tetracycline inducing effect. The study did not, however, reveal such a pronounced formation of a new protein component on tetracycline induction as the previous one.

The data confirm that the ketoenol groups of DCB rings, the most important elements of the structure for protein synthesis inhibition (A. I. Gurevich et al., 1974) are indentical in minocycline and natural tetracyclines.

In spite of the importance of the data on the new protein components in the membranes of the tetracycline resistant R^+ cells of *E. coli*, the changes in the membranes after tetracycline resistance expression may be more complex since, for example, increased contents of lipids were noted in the membranes of the tetracycline resistant cocci (Norrington F. E., James, 1970).

The analysis of the latest data on the mechanism of tetracycline resistance is also indicative of further developments in the concept of the mechanism of tetracycline penetration through the microbial cell membrane (V. K. Plakunov). According to this author, tetracycline penetration into the cell is achieved by a stereospecific system of transfer of the normal metabolite of the component of the tricarboxylic acid oxalacetic acid cycle. The stereospecific mechanism of the antibiotic absorption by means of the above system was also preserved with respect to the tetracycline resistant strains of *E. coli* and *Staphylococcus*, including the clinical ones. However, the resistant strains — unlike the sensitive ones — are capable of maintaining the intracellular concentration of tetracycline for a prolonged period of time at lower levels than those in the medium. This is due to the particularly intensive functioning of the system of active elimination of tetracycline from the resistant cells.

This system is energy dependent and its functioning is suppressed in the presence of cyanide and dinitrophenol. The above mechanism of tetracycline resistance is according to this author characteristic of various microbial species and resistance coded by both the chromosome and extrachromosome determinants. These concepts are of definite interest although there is in some cases only indirect evidence that they are correct.

In connection with the problem discussed it is necessary to show the same with semisynthetic tetracyclines as compared to natural tetracyclines. According to some authors (for instance Speciale A., 1971 and others) and our own observations presented above, minocycline is active against clinical tetracycline resistant strains of *E. coli* and *Staphylococci*.

The cause of the higher minocycline activity in comparison to natural tetracyclines is not yet clear, although certain data indicate that this semisynthetic antibiotic is a less active inducer of tetracycline resistance than tetracycline (Robertson, Reeve, 1972). However, the same authors have shown that *E. coli* cells carrying various R-factors previously induced by tetracycline to the maximum resistance level preserved a relatively high sensitivity level to minocycline: its concentrations inhibiting by 50 per cent the protein synthesis dependent on the R-factor were 10–20 times lower than those of tetracycline. Non-induced R^+ cells of *E. coli* were also more sensitive to minocycline than to tetracycline. It follows that the difference in the inducing capacity does not — or at least not completely — explain the higher activity of minocycline. In our experiments with a cell-free system of polypeptide synthesis prepared with sensitive and resistant strains of *E. coli*, minocycline was not sufficiently more efficient than oxytetracycline, a representative of natural tetracyclines.

Both tetracycline drugs inhibited in particular the synthesis of polyphenylalanine on the ribosomes of the clinical strains of *E. coli* practically to the same degree, in spite of the fact that there was some difference in the MIC of oxytetracycline. This suggested that minocycline penetrated into the microbial cells in greater amounts than the natural antibiotic and could explain some differences in the values of the MIC of the two antibiotics.

The minocycline capacity for competition with C^{14} oxytetracycline after their joint binding by the cells of *E. coli* was in the experiments with the tetracycline-resistant strains indeed greatly lower than in the experiments with the sensitive strains. In other words, the systems of minocycline and oxytetracycline transfer into the resistant cells are rather far from being identical.

It is probably because alternative paths are available for penetration of minocycline into the cell through its membrane that high levels of the antibiotic accumulation are observed as compared to the natural tetracyclines.

At the same time, it is of interest that minocycline was not more active than tetracycline against pneumococci resistant to the latter (Gopalakrishna, Lerner, 1973). This may be associated with an important role played in the development of resistance by capsules or some other formations specific for these microbes.

It is indubitable that changed permeability of the cell membrane in R^+ bacteria and consequently decreased absorption of C^{14}-oxytetracycline is the most common mechanism of tetracycline resistance. But some strains resistant to oxytetracycline at concentrations of 250 mcg/ml absorbed the antibiotic at many times higher amounts than other strains resistant to the same concentration of the antibiotic.

It should be noted that the existence of analogous strains was shown by Reynard et al. (1971).

We have carried out some experiments with one of the tetracycline resistant "highly absorbing" strains of *E. coli*, trying to elucidate the cause of its resistance to the antibiotic penetrated into the cell. The study on the distribution of C^{14}-oxytetracycline among various cell fractions of such strains showed no peculiarities in the distribution of the label. In addition, it was not possible to demonstrate a degradation of the C^{14}-oxytetracycline molecule or its modification followed by activity loss by using either microbiological titration, or radiochromatography. We had no evidence proving the existence of an oxytetracycline inactivating enzymatic system.

In this connection, a number of experiments have been performed using cell free ribosome-matrix systems. We tested various variants of the standard system of polypeptide synthesis, that is combinations of ribosomes and the supernatant of the sensitive and resistant strains with absorption capacities for oxytetracycline. Furthermore, a cell free system of protein synthesis with an endogenic matrix was also tested. When the ribosome of the resistant strain was combined with the supernatant of the sensitive one, the system of the polypeptide synthesis was just as sensitive to oxytetracycline as the system containing the ribosomes and the supernatant of the sensitive strain alone. Similar results were also observed with the cell free system not of polypeptide but of protein. In other words, the origination of the ribosomes from the resistant strain did not decrease the sensitivity of polyphenylalanine and protein synthesis to tetracycline. The presence in the system of the supernatant of the resistant strain cells resulted in a definite decrease in the oxytetracycline effect, so that suppression of the polypeptide synthesis by the antibiotic decreased. Interpretation of the above data is at present difficult since one could suspect rather changes in the properties of the ribosomes to be the cause of tetracycline resistance in the strain of E. coli tested by analogy with other antibiotic inhibitors of protein synthesis.

The set of the enzymes participating in the synthesis of protein is rather numerous in such cell fractions. The problem requires further investigation.

At the end of the discussion on some of the modern aspects of tetracycline resistance we should like to underline once more the fact that tetracycline antibiotics are rather difficult to investigate. This applies also to the molecular mechanism of their action. This concerns such a problem as special localization of the tetracycline resistance determinant in R-factors in immediate proximity to RTF or even in the very transfer unit. Finally it applies to the mechanisms of tetracycline resistance mainly associated with the peculiar properties of the molecular organization of the biological membranes, etc.

REFERENCES

ADLER, H. I., FISHER, W. D., COHEN, A., HARDIGREE, A. A., (1967) Miniature. *E. coli* cells deficient in DNA. Proc. Nat. Acad. Sci. USA, 57, 321–326.

CHANINA, M., POTAPOV, V., TERECHOV, A. (1970). Appearance of multiple resistance caused by repeated treatment of bacteria by tetracycline. Antibiotiki, 15, 1, 34.

DOCKTER, M. E., MAGNUSON, T. A. (1973). Kinetics of chlortetracycline uptake in *Staphylococcus aureus* by a fluorescence technique. Biochem. Biophys. Res. Communs, 54, No. 2, 790–795.

EMBDEN van J., COHEN, S. N. (1973). Molecular and genetic studies of an R-factor or system consisting of independent transfer and drug resistance plasmids. J. Bact., 116, No. 2, 699–709.

FRANKLIN, T. J., GODFREY, A. (1965). Resistance of *Escherichia coli* to tetracyclines. Biochem J., 94, No. 1, 54–60.

FRANKLIN, T. J. (1967). Biochemical mechanisms of bacterial resistance to the tetracyclines. Biochem. J., 102, No. 1, 2P.

Franklin, T. J., Higginson, B. The "active" accumulation of tetracycline by *E. coli*. Biochem. J. 1969, 112, No. 1, 12P.

Franklin, T. J., Higginson, B. (1970). The active accumulation of tetracycline by *E. coli*. Biochem. J., 116, No. 2, 287–297.

Franklin, T. J., Rownd, R. (1973). R-Factor mediated resistance to tetracycline in *Proteus mirabilis*. J. Bact., 115, I, 235–242.

Gurevitch, A., Kiseleva, O., Kolosov, M., Schemjakin, M. (1974). The influence of tetracycline analogues on protein and nucleic acids byosynthesis in *E. coli*. Biochimia, 39, 2, 421.

Hirachima, A., Childs, G., Inoue, M. (1973). Differential inhibitory effects of antibiotics on the biosynthesis of envelope proteins of *Escherichia coli*. J. Mol. Biol., 79, No. 2, 373–389.

Högenauer, G., Turnowsky, E. (1972). The effect of streptomycin and tetracycline on codonanticodon interactions. FEBS Letters, 26, No. 1, 185–188.

Hutchings, B. L. (1969). Tetracycline transport is *Staphylococcus aureus* H. Biochim. et biophys. acta, 174, No. 2, 734–748.

Inoue, M., Hashimoto, H., Mitsuhashi, S. (1970). Mechanism of tetracycline resistance in *St. aureus*. I. Inducible resistance to tetracycline. J. Antibiot., 23, No. 2, 68–74.

Gopalakrishna, K. V., Lerner, Ph. I. (1973). Tetracycline-resistant pneumococci. Increasing incidence and cross resistance to newer tetracyclines. Amer. Rev. Respirat. Disease, 108, No. 4, 1107–1010.

Izaki, K., Arima, K. (1963). Disappearance of oxytetracycline accumulation in the cells of multiple drug – resistant *Escherichia coli*. Nature, 200, No. 49k4, 384–385.

Korn, M., Chanina, M. F. (1970). Luminiscent microscopy study of tetracycline binding by sensitive and resistant cells. Microbiologia, 39, No. 3, 517.

Levy, S. B., Mc Murry, L. (1974). Detection of an inducible membrane protein associated with R-factor – mediated tetracycline resistance. Biochem. Biophys. Res. Communs, 56, No. 4, 1060–1068.

Lopatik, M., Plakunov, V. (1973). A rapid evacuation of antibiotic from tetracycline resistant strain of E. coli K 12. Microbiologia, 42, 316.

Mendelson, N. H., Reeve, T. N., Cole, R. M. (1974). Physiological studies of *Bacillus subtilis* minicells. J. Bact., 117, No. 3, 1312–1319.

Norrington, F. E., James, A. M. (1970). The cell-wall lipids of cells of tetracycline-sensitive and tetracycline-resistant strains of *Streptococcis pyogenes*. Biochim. Biophys. Acta, 218, No. 2, 269–277.

Plakunov, V., Lopatik, M., Milnikova, S. (1973). The nature of tetracycline transfer system in the microbial cells. Microbiologia, 42, 70.

Robertson, T. M., Reeve, E. C. R. (1972). Analysis of the resistance mediated by several R-factors to tetracycline and minocycline. Genet. Res., 20, No. 2, 239–252.

Sompolinsky, D., Zaidenzaig, V., Ziegler-Scholomowitz, R., Abramova, N. (1970). Mechanism of tetracycline resistance in *Staphylococcus aureus*. "J. Gen. Microbiol.", 62, No. 3, 351–362.

Smythies, T. R., Benington, F., Morin, R. D. (1972). On the molecular mechanism of action of the tetracyclines. Experientia, 28, No. 10, 1253-1254.

Spenger, M. E., Guest, T. R. (1974). Proteins of the inner membrane of *E. coli*: identification of succinate dehydrogenase by polyacrylamide gel electrophoresis with amber mutants. J. Bact., 117, No. 3, 947–953.

THE RELATION OF A TETRACYCLINE-INDUCED R FACTOR MEMBRANE PROTEIN TO TETRACYCLINE RESISTANCE*

S. B. LEVY

Department of Molecular Biology and Microbiology and the Department of Medicine, Tufts University School of Medicine and the Tufts-New England Medical Center, Boston, U.S.A.

The tetracyclines are widely used antibiotics whose antimicrobial action arises from an inhibition of protein synthesis by interference with the binding of aminoacyl transfer RNA to the 30S ribosome (Suzuki et al., 1966; Weisblum and Davies, rev., 1968). This effect requires the entry of tetracycline into the cell (Laskin and Chen, 1964). Although some tetracycline may bind to the cell surface (DeZeeuw, 1968) most, if not all, of it enters the cell by an active transport mechanism (Arima and Izaki, 1963; DeZeeuw, 1968; Franklin and Godfrey, 1965). Uptake of the drug is inhibited by conditions and agents which block metabolic activity (Arima et al., 1963).

Resistance to the tetracyclines is widespread in bacteria (Davies and Rownd, 1972; Franklin and Godfrey, 1965; Izaki et al., 1966; Sompolinsky et al., 1970). The principal genetic basis of this resistance lies in extrachromosomal DNA elements called plasmids or episomes. In the gram-negative organism this resistance resides on resistance transfer factors (R factors) and can generally be transmitted by conjugation from one bacterium to another, even between bacterial species (Davies and Rownd, 1972; Watanabe, 1963). In the organism containing a tetracycline-resistant R factor, degradation of the antibiotic is not found, but there is a decrease in tetracycline uptake into the cell (Franklin, 1967; Izaki et al., 1966). The level of resistance to tetracycline can be increased by incubating the organism in sub-inhibitory amounts of tetracycline (Franklin, 1967; Izaki et al., 1967; Unowsky and Rachmeler, 1966). Such experiments have illustrated the inductive nature of this drug resistance.

Detection of a Tetracycline-Inducible Membrane Protein in E. Coli Minicells Containing Tetracycline-Resistant R Factors

We have been studying R factor-specific proteins in an *E. coli* minicell system where the only significant DNA present is the R factor (Levy, 1970; 1971a; 1971b; 1974; Levy and McMurry, 1974; Levy et al., 1974). Minicells are tiny cells formed during abnormal cell division of an *E. coli* mutant isolated and described by Adler (Adler et al., 1967). They can be separated from whole cells by differential centrifugation (Adler et al., 1967; Levy and Norman, 1970; Roozen et al., 1971) and/or a penicillin treatment (Levy, 1970).

* Supported in part by USPHS grant AI 10648, a grant from the University Cancer Committee (American Cancer Society, grant IN 23-0) and a Research Career Development Award.

Minicells generally contain no DNA, but when R factors are introduced by conjugation into the minicell-producing strain every minicell produced appears to contain an R factor (Levy and Norman, 1970; Levy, 1971a). In the minicell R factors code for the synthesis of R factor DNA, RNA and protein (Inselburg, 1971; Levy, 1971b; Roozen et al., 1971). We have developed a system whereby a large amount of radioactivity (^{35}S-methionine) is incorporated into R factor protein in R^{+} minicells. Using this R factor-in-minicell system, we can detect certain membrane-associated, and DNA-binding proteins of R factor origin (Levy, 1974; Levy et al., 1974). Many of these proteins appear to be specific for particular R factor compatibility groups (Chabbert et al., 1972; Datta and Hedges, 1971) (see Table 1).

TABLE 1

E. coli Minicell Strains[a]	R Factor	Known R Factor Resistance[b]	Compatibility Type[c]	Presence of tet Protein
x 984	—			
D1-7	222	Cm, Tc, Su, Sm	FII	+
D1-8	222-R^{3}	Cm, Su, Sm	FII	—
D1-26	R1	Cm, Amp, Su, Sm	FII	—
D1-39	pSL-222 tet-2	Cm, Tc^{ts}, Su, Sm	FII	+
D1-41	R124	Tc	FIV	+
D1-42	R386	Tc	FI	+
D2-14	R64	Sm, Tc	I	+
D3-2	RM98	Amp	N	—
D3-9	N-3	Tc	N	+
D4-2	AM20	Sm, Spc, Tc, Cm, Su Gen, Km/Neo	?	+
D6-2	CF-2	Tc	?	+
D7-2	RAI	Tc	A	NE[d]
D8-2	R725	Tc	H	NE
D8-4	R27	Tc	H	NE
D9-2	R446b	Tc	M	+

[a] All R^{+} strains are derivatives of R^{-} minicell producing strain x 984 received from R. Curtiss III.
[b] Amp = ampicillin; Cm = chloramphenicol; Gen = gentamycin; Km/Neo = Kanamycin/Neomycin; Su = sulfonamides; Sm = streptomycin; Spc = spectinomycin; Tc = tetracycline.
[c] Datta and Hedges, 1971; Chabbert et al, 1972.
[d] Ne = not yet examined.

In studies attempting to correlate the R factor gene products synthesized in the minicell with a known function, we examined R factors bearing tetracycline resistance. Since resistance is related to cell permeability, initial studies examined the membranes of minicells harboring tetracycline-sensitive and tetracycline-resistant R factors. In experiments using radioactive glycerol or ^{32}P to label lipids synthesized in these R^{+} minicells, no difference in the major phospholipids was noted. Studies of the membrane proteins also illustrated no detectable differences until these same minicells were examined following pre-incubation in the presence of small amounts of tetracycline. Two samples of purified minicells harboring an R factor bearing resistance to tetracycline were labeled by incubating with ^{35}S-methionine in the presence or absence of tetracycline (10 μg/ml). Analysis of the radioactive protein on sodium dodecyl sulfate (SDS) polyacrylamide gels demonstrated a new protein of rather major proportion visible in only the extracts from minicells labeled in the presence of tetracycline (Figure 1). This protein (designated *tet*) had an estimated molecular weight of 50,000 in SDS-polyacrylamide gels and remained associated with the minicell membrane upon lysis of the minicells (Levy and McMurry,

1974). Further studies demonstrated that minicells containing R factors of many different origins and compatibility types could be induced by tetracycline to synthesize this protein if the R factor carried resistance to tetracycline (Figure 1, Table 1). R factors which lost this resistance, lost the ability to code for this protein (Levy and McMurry, 1974). Although easily seen in the minicells, this protein was not visualized in cells similarly

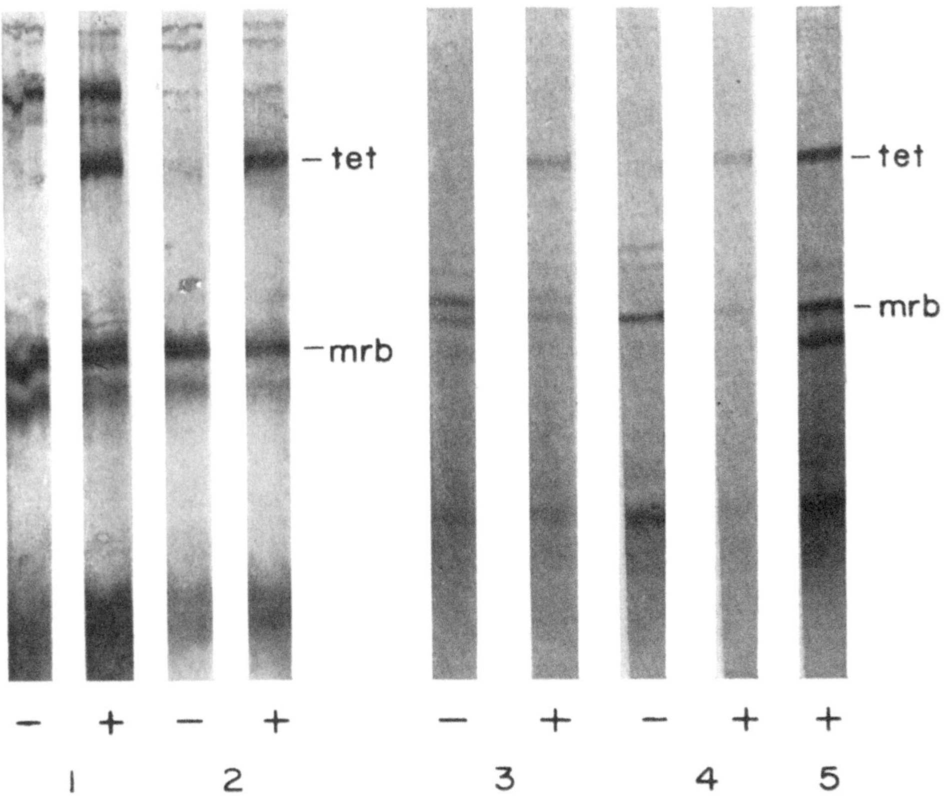

Figure 1. SDS-polyacrylamide gel analysis of proteins synthesized in different R^+ minicells in the presence (+) or absence (—) of tetracycline. Minicells (4–6 × 10^9/ml) containing different R factors were labeled by ^{35}S-methionine in the presence or absence of tetracycline (10 μg/ml). The protein synthesized was electrophoresed in SDS-polyacrylamide gels (Levy and McMurry, 1974) which were subsequently autoradiographed. *tet* = tetracycline — induced membrane protein; *mrb* = = major R factor membrane protein band of molecular weight 28–30,000 seen in minicells containing R factors of many different compatibility types (Levy et al, 1974). 1 = D1-41; 2 = = D1-42; 3 = D4-2; 4 = D6-2; and 5 = D1 = 7.

labeled in ^{35}S-methionine in the presence of tetracycline. Such results demonstrated the relatively small amount of this R factor protein in proportion to the total proteins synthesized by the cell. In examining another R factor membrane protein, *mrb* (Figure 1) which is our designation for a major R factor membrane protein seen on SDS-polyacrylamide gels for a large number of R factors (Levy and McMurry, 1974; Levy et al., 1974), a similar situation was found. Although the *mrb* represented as much as 35% of the radioactivity incorporated into membrane protein by an R^+ minicell it was not detected in whole cells until purified extracts of the membranes were made (Levy et al., 1974). Presumably we shall be able to detect this *tet* protein when suitable cell membrane extracts are made. The tetracycline-induced membrane protein appears to be located in the inner membrane since it is extracted with sodium lauryl sarcosinate, an agent which has been shown to selectively solubilize the cytoplasmic membrane (Filip et al., 1973).

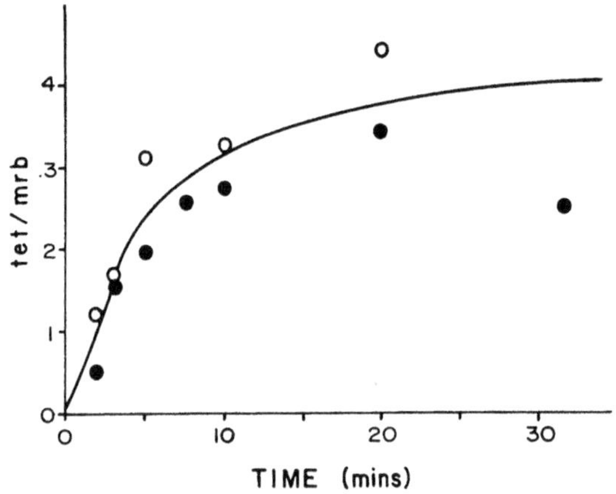

Figure 2. Kinetics of induction of the *tet* protein by tetracycline (10 μg/ml). D1-7 minicells were suspended in supplemented methionine assay media (Levy and McMurry, 1974). At different times after addition of tetracycline (10 μg/ml) and ^{35}S-methionine (100 μCi/ml), samples were removed, prepared, and analyzed by SDS-polyacrylamide gel electrophoresis. Gels were dried under vacuum before autoradiography. The kinetics of induced synthesis were analyzed by comparing the amounts of the *tet* protein with another major R factor protein being synthesized constitutively in the minicell, the *mrb* (see Figure 1). Relative amounts of each were determined by weighing appropriate areas from densitometer tracings of the gels. Open and closed circles refer to calculations from different gels.

Its synthesis is inhibited by rifamycin and chloramphenicol indicating *de novo* synthesis from DNA. It is not extracted into organic solvents, nor is its mobility in gel electrophoresis altered by butanol extraction of the membranes. In order to determine the time required for induction of synthesis of this protein, R factor-containing minicells were induced with tetracycline in the presence of high specific activity radioactive ^{35}S-methionine. Synthesis of the tetracycline-induced protein band was detected as early as two minutes after contact with tetracycline (10 μg/ml). The kinetics of induction indicated that within 10 minutes its synthesis reached equilibrium with the rest of minicell protein synthesis (Figure 2) using *mrb* as an indicator. The rapidity of induction of this protein reflects the time of induction of tetracycline resistance in whole cells (Levy, unpublished data; Franklin, 1967). The ratio of this protein relative to *mrb* for different R factors and under different labeling conditions ranges from 30–80%.

To determine whether induction of synthesis of this protein resulted in a stable change in transcription, minicells harvested from an R^+ minicell-producing culture (D1–7) grown up in the presence of tetracycline were labeled in the absence of tetracycline. Despite multiple washings and a purification of minicells through sucrose gradients which should have removed tetracycline, the minicells retained an ability to synthesize the *tet* protein though in reduced amounts (Figure 3). Minicells similarly prepared from cultures which never saw tetracycline, synthesized little if any *tet* protein (Figure 3). From other studies it appeared that synthesis decreased and stopped after 30 minutes of incubation

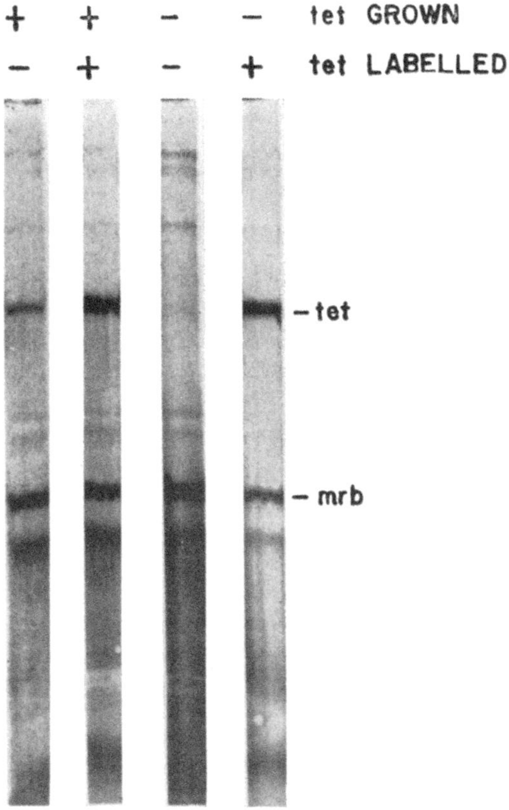

Figure 3. Synthesis of the *tet* protein in minicells harvested from cultures grown in the presence or absence of tetracycline.

D1-7 cells were grown up in L broth in the presence (+, induced) or absence (—, uninduced) of tetracycline. The minicells were purified and labeled for 120 minutes in ^{35}S-methionine in the presence (+) or absence (—) of tetracycline (10 μg/ml). Samples were prepared and analyzed by SDS-polyacrylamide gel electrophoresis. +, — refer to presence or absence of tetracycline: 1) during growth of the cell culture from which the minicells were harvested; 2) during labeling of the purified minicells. *tet, mrb* (see Figure 1). The same amount of radioactivity was placed on each gel.

suggesting among possibilities the loss of inducer or slow degradation of the induced mRNA. These studies demonstrated that the induction process had resulted in continued synthesis of this protein for a limited time in the absence of tetracycline although it is possible that significant amounts of tetracycline (greater than 0.10 μg/ml, see below) were retained in the minicell.

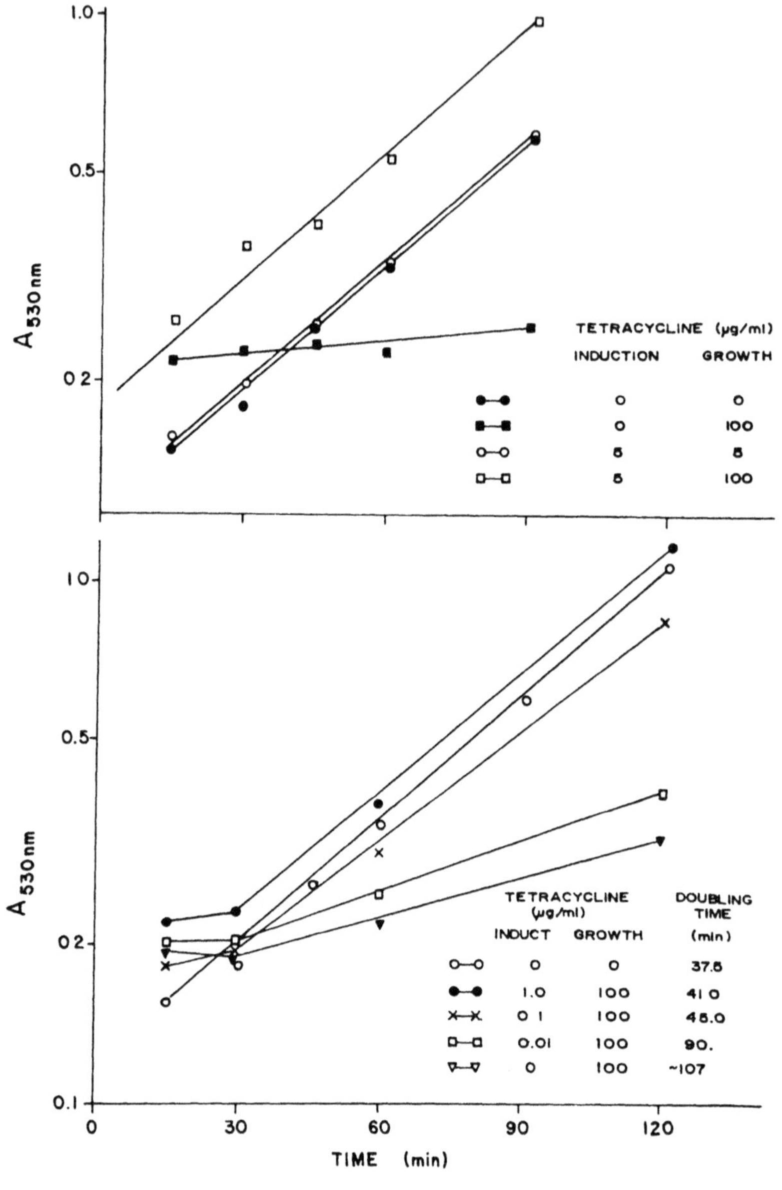

Figure 4. Induction of tetracycline resistance by tetracycline: 4A Induction of resistance to high levels of tetracycline by incubation in subinhibitory levels of tetracycline (5 µg/ml). Duplicate cultures of D1-7 cells were pre-incubated for 30 minutes in the presence or absence of tetracycline (5 µg/ml). One culture of each was then challenged by increasing the tetracycline concentration to 100 µg/ml. Growth was recorded by optical density (A_{530} nm, Coleman 101 spectrophotometer). 4B Minimal amount of tetracycline capable of inducing increased resistance to tetracycline. The experimental protocol was as outlined in 4A. Pre-incubation lasted 60 minutes in the presence of designated amounts of tetracycline. Doubling time refers to doubling of optical density, A_{530} nm

Tetracycline Resistance Evaluated in Whole Cells and Minicells

R factor 222 imparts a relatively high resistance to tetracycline in *E. coli* cells. In order to correlate resistance in these cells to induction of synthesis of this protein, the phenotypic expression of resistance to tetracycline in these cells and minicells was examined. *E. coli* minicell strain D1–7 containing R factor 222 was grown up in L broth and suspended at an optical density (A_{530} nm) of 0.1 in half-strength L broth (Levy, 1974). Following growth of the cells for 60 minutes in the presence or absence of tetracycline (5 μg/ml), a challenging dose of tetracycline (100 μg/ml) was introduced into the culture medium. Cells which had been incubated in sub-inhibitory concentrations of tetracycline were relatively unaffected by this large amount of tetracycline (Figure 4). Growth of the same bacteria, not pre-incubated with tetracycline, was severely depressed by the tetracycline despite the presence in these bacteria of an R factor bearing tetracycline-resistance. At least 0.1 μg of tetracycline protected cells against this challenge with high levels of tetracycline (Figure 4B). In a companion experiment, minicells containing R factor 222 were labeled with ^{35}S-methionine in the presence of different concentrations of tetracycline. Analysis of the proteins synthesized showed a similar requirement for at least 0.1 μg of tetracycline in order to induce synthesis of the *tet* protein. Incubation with 0.01 μg or less tetracycline was insufficient both for induction of synthesis of the *tet* protein in minicells and for induction of tetracycline resistance in cells. These studies demonstrated another correlation between phenotypic expression of tetracycline resistance in cells and the induction of synthesis of the *tet* protein in minicells.

Since the minicell is not able to grow, radioactive labeling studies of protein synthesis were performed in order to examine its resistance to tetracycline. Minicells from tetracycline-resistant R^+ strains showed significant resistance to tetracycline (Table 2).

TABLE 2

SENSITIVITY OF CELLS AND MINICELLS TO TETRACYCLINE AND MINOCYCLINE

	% Inhibition[a]	
R^+ Cells	(10 μg/ml) Tetracycline	(10 μg/ml) Minocycline
D1-7	0	25
D8–4	11	38
D9–2	48	69
R^+ Minicells		
D–7	0	62
D8–4	0	6
D9–2	46	73

[a] ^{35}S-methionine incorporation into protein in 60 minutes compared with the same cells or minicells without the tetracycline.

Induction to high level resistance in minicells, however, differed from the results with cells. Protein synthesis in cells incubated in sub-inhibitory amounts of tetracycline was resistant to large amounts of tetracycline. Similar studies with minicells were not striking and induction to resistance was reduced and not always seen.

Using another means to evaluate resistance to tetracycline, we examined the uptake of ^{3}H-tetracycline (New England Nuclear) in R$^-$ cells and R$^+$ cells induced to resistance by tetracycline. Binding of tetracycline to both R$^+$ and R$^-$ cells was detected at 0°, but upon raising the temperature to 37°, label was released from the R$^+$ cells, but was taken up by the R$^-$ cells (Figure 5). Similar results were found with minicells harvested from induced cultures (Figure 5). Although there was significantly less uptake in uninduced R$^+$

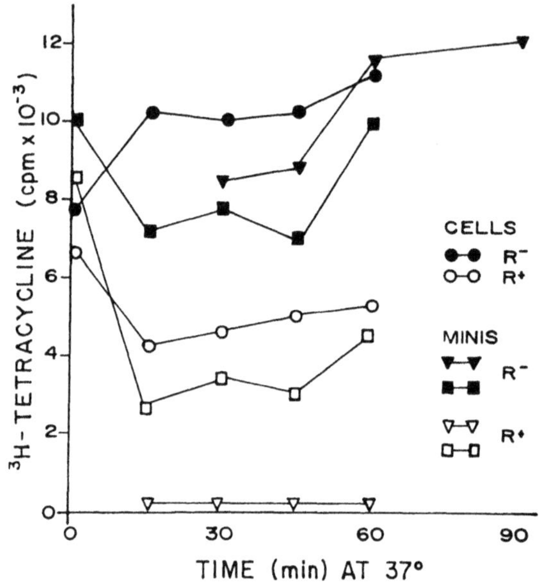

Figure 5. Uptake of ^{3}H-tetracycline by sensitive and resistant cells and minicells. Cells and minicells from a D1-7 cell culture grown in the presence of tetracycline (10 μg/ml) and its R$^-$ isogenic strain, *x* 984, were separated and the uptake of ^{3}H-tetracycline (1 μCi/2 mg tetracycline) studied. Cells and minicells were incubated at 0° for 30 minutes in the presence of ^{3}H-tetracycline (0 point) to evaluate nonspecific binding. The temperature was then raised to 37° C and subsequent samples were taken, filtered on Millipore 0.45 mμ filters and counted in a liquid scintillation counter. In a separate experiment, minicells were incubated with ^{3}H-tetracycline at 37° C without a 0° preincubation period.

minicells as compared with R$^-$ minicells, a 60-minute induction incubation of uninduced R$^+$ minicells with tetracycline only slightly decreased the uptake and the amount of ^{3}H-tetracycline taken up was still more than found for R$^+$ minicells harvested from cell cultures grown in the presence of tetracycline (i.e., induced minicells).

A reduced inductive effect by tetracycline in minicells has been recently reported by Franklin and Foster (1974). Their results as well as ours suggest a requirement for a cell product or greater cell wall synthesis in minicells in order to fully express the induction to high level tetracycline resistance despite the induced synthesis of the *tet* protein. We have some indication that the age of the minicell, reflecting its metabolic state and/or cell wall growth potential, may be involved. It is possible that although the *tet* protein is synthesized, its most effective placement in the cell membrane requires a certain stage of cell membrane growth.

Resistance of Cells and Minicells to Other Tetracycline Derivatives

Several derivatives of tetracycline have been found with increased anti-bacterial activity, e.g., doxycycline and minocycline. Recently, reports have indicated that R factors can be distinguished by their relative resistance to the different tetracyclines, especially between tetracycline and minocycline (Robertson and Reeve, 1972; Scavizzi, 1972). Scavizzi describes two types of R factors differing in their relative level of resistance to the two drugs. Robertson and Reeve showed that in some instances minocycline failed to induce the cells to higher level resistance, and although a cell was induced to high level resistance by tetracycline, it remained sensitive to minocycline. We have begun looking at various tetracycline-resistant R factors in order to relate tetracycline resistance to induced synthesis of the tetracycline-membrane protein. Minicells containing R factor 222 showed equal resistance to both tetracycline and doxycycline at 10 μg/ml, but were quite sensitive to minocycline (10 μg/ml). Other R factors tested showed differing sensitivities to the drugs. SDS-polyacrylamide gel electrophoresis of the proteins synthesized by R factor 222 in the minicell in the presence of these other tetracyclines demonstrated the presence of the 50,000 molecular weight protein with doxycycline, tetracycline and minocycline even though minicells were relatively sensitive to the latter drug. The amount of this *tet* protein synthesized was reduced in minicells in contact with minocycline, but whether decreased resistance was related to decreased production of this membrane protein or to the entry of the drug into the cell leading to reduced total protein synthesis is not yet known.

Recently, we have begun to analyze in a similar fashion R factors which we have isolated with temperature-sensitive expression of tetracycline resistance. At both 30° and 42°, R factor 222 is highly resistant to tetracycline and produces in relatively large amounts the *tet* protein. In contrast, two mutant 222 R factors, pSL 222-tet 1, pSL 222-tet 2 show resistance only at the lower temperatures. When we examined one of these R factors, pSL 222-tet 2, we found the *tet* protein produced at both 30° and 42°, but its synthesis at 42° relative to *mrb* was about half that seen for the 222 wild type. Since the *tet* protein was synthesized, but in relatively less amounts in minicells containing the mutant R factor as compared with wild type 222, the temperature sensitivity may reflect a mutation in a regulatory protein not yet identified. The fact that the protein was detected is also consistent with an amino acid substitution in the *tet* protein which would not change its molecular weight size in the SDS-gels, but could affect its function. Whatever the mechanism, the temperature sensitivity mutation is located on the R factor since it is transmissible to other bacteria. Additional studies using other gel systems and analysis of peptide digests are in progress to enable us to compare the *tet* protein produced by the temperature-sensitive and wild type R factors.

SUMMARY

We have used the R factor-in-minicell system to detect and describe a new protein associated with R factor-mediated tetracycline resistance. This protein is located in the inner membrane of the *E. coli* minicell and presumably of the cell, although its absolute amounts are too small to be yet detected in the whole *E. coli* cell. Its production has been found in minicells containing R factors of many different compatibility groups. As with phenotypic expression of resistance to tetracycline this protein is induced in the presence of small amounts of tetracycline at levels mirroring those required to get phenotypic expression of resistance. Loss of tetracycline resistance results in loss of synthesis of this protein. The protein is synthesized in the presence of analogues of tetracycline, including minocycline despite relative sensitivity to this drug. ^{3}H-tetracycline uptake studies with cells and minicells from induced cell cultures show decreased uptake of tetracycline. Cell cultures pre-incubated with subinhibitory amounts of tetracycline show decreased uptake and increased resistance to tetracycline as compared with uninduced cultures; in contrast, this induction is less effective in minicells.

These results imply a role for this tetracycline-induced membrane protein in tetracycline resistance, although the mechanism is not yet clear.

ACKNOWLEDGEMENT

I greatly appreciate the excellent assistance of L. McMurry and S. Rosbroj in these studies. I thank the E. Lilly and C. Pfizer Laboratories for supporting my travel to this symposium. Minocycline was kindly provided by Lederle Laboratories.

REFERENCES

ADLER, H. I., FISHER, W. D., COHEN, A. and HARDIGREE, A. A. (1967). Miniature *E. coli* cells deficient in DNA. Proc. Nat. Acad. Sci. USA 57: 321–326.

ARIMA, K. and IZAKI, K., 1963. Accumulation of oxytetracycline relevant to its bactericidal action in the cells of *Escherichia coli*. Nature (London), 200: 192–193.

CHABBERT, Y. A., SCAVIZZI, M. R., WITCHITZ, J. L., GERBAUD, G. R. and BOUANCHAUD, D. H., 1972. Incompatibility groups and the classification of fi$^-$ resistance factors. J. Bacteriol. 112: 666–675.

DATTA, N. and HEDGES, R. W., 1971. Compatibility types among fi$^-$ R factors. Nature (London) 234: 222–223.

DAVIES, J. E. and ROWND, R., 1972. Transmissible multiple drug resistance in Enterobacteriaceae. Science 176: 758–768.

DEZEEUW, J. R., 1968. Accumulation of tetracyclines by *Escherichia coli*, 1968. J. Bacteriol. 95: 498–506.

FILIP, C., FLETCHER, G., WULFF, J. L. and EARHART, C. F., 1973. Solubilization of the cytoplasmic membrane of *Escherichia coli* by the ionic detergent, sodium-lauryl sarcosinate. J. Bacteriol. 115: 717–722.

FRANKLIN, T. J. and GODFREY, A., 1965. Resistance of *Escherichia coli* to tetracyclines. Biochem. J. 94: 54–60.

FRANKLIN, T. J., 1967. Resistance of *Escherichia coli* to tetracyclines. Changes in permeability to tetracyclines in *Escherichia coli* bearing transferable resistance factors. Biochem. J. 105: 371–378.

FRANKLIN, T. J. and FOSTER, S. J., 1974. Expression of R factor-mediated resistance to tetracycline in *Escherichia coli* minicells. Antimicrob. Agents and Chem. 5: 194–195.

INSELBURG, J., 1971. R factor deoxyribonucleic acid in chromosomeless progeny of *E. coli*. J. Bacteriol. 105: 620–628.

IZAKI, K., KIUCHI, K. and ARIMA, K., 1966. Specificity and mechanism of tetracycline resistance in a multiple drug resistant strain of *Escherichia coli*. J. Bacteriol. 91: 628–633.

LASKIN, A. I. and CHAN, V. M., 1964. Inhibition by tetracyclines of polyuridylic acid directed phenylalanine incorporation in *Escherichia coli* cell-free systems. Biochem. Biophys. Res. Commun. 14: 137–142.

LEVY, S. B., 1970. Resistance of minicells to penicillin lysis: a method of obtaining large quantities of purified minicells. J. Bacteriol. 103: 836–839.

LEVY, S. B. and NORMAN, P., 1970. Segregation of transferable R factors into *Escherichia coli* minicells. Nature (London), 227: 606–607.

LEVY, S. B. 1971a. Studies on R factors segregated into *E. coli* minicells. Ann. N.Y. Acad. Sci. 182: 217–225.

LEVY, S. B., 1971b. Physical and functional characteristics of R factor deoxyribonucleic acid segregated into *Escherichia coli* minicells. J. Bacteriol. 108: 300–308.

LEVY, S. B. and MCMURRY, L., 1974. Detection of an inducible membrane protein associated with R factor-mediated tetracycline resistance. Biochem. Biophys. Res. Commun. 56: 1060–1068.

LEVY, S. B., 1974. R factor proteins synthesized in *E. coli* minicells. I. Incorporation studies with different R factors and detection of DNA-binding proteins. J. Bacteriol. (in press).
LEVY, S. B., MCMURRY, L. and PALMER, E., 1974. R factor proteins synthesized in *E. coli* minicells. II. Membrane-associated R factor proteins. J. Bacteriol. (in press).
ROBERTSON, J. M., and REEVE, E. C. R., 1972. Analysis of the resistance mediated by several R factors to tetracycline and minocycline. Genet. Res. Camb. 20: 239–252.
ROOZEN, K. J., FENWICK, R. G., Jr. and CURTISS, R. III, 1971. Synthesis of ribonucleic acid and proteins in plasmid-containing minicells of *Escherichia coli* K12. J. Bacteriol. 107: 21–33.
SCAVIZZI, M. R., 1972. Existence de deux caractères transferables de résistance aux tetracyclines. Ann. Inst. Pasteur 122: 3–17.
SOMPOLINSKY, D., ZAIDENZAIG, Y., ZIEGLER-SCHLOMOWITZ, R. and ABRAMOVA, N., 1970. J. Gen. Microb. 62: 351–362.
SUZUKI, I., KAJI, H. and KAJI, A., 1966. Binding of specific sRNA to 30s ribosomal subunits: effects of 50s ribosomal subunits. Proc. Nat. Acad. Sci. USA 55: 1483–1490.
UNOWSKY, J. and RACHMELER, M., 1966. Mechanisms of antibiotic resistance determined by resistance-transfer factors. J. Bacteriol. 92: 358–365.
WATANABE, T., 1963. Infective hereditary of multiple drug resistance in bacteria. Bact. Rev. 27: 87–115.
WEISBLUM, B. and DAVIES, J., 1968. Antibiotic inhibitors of the bacterial ribosome. Bacteriol. Rev. 32: 493–528.

TETRACYCLINE RESISTANCE UNASSOCIATED WITH A CHANGE OF CELL WALL PERMEABILITY IN E. COLI

S. M. NAVASHIN, I. V. BELIAVSKAYA, Y. O. SAZYKIN,
N. S. GRYAZNOVA
National Research Institute of Antibiotics, Moscow, USSR

The biochemical mechanisms of bacterial resistance to antibiotics inhibiting protein synthesis may be divided into 3 fundamentally different types: enzymatic inactivation of the drug, an alteration of ribosomes which makes them incapable to bind the antibiotic, and a selective decrease in the permeability of the cell membrane.

Investigators working with tetracyclines encountered, as a rule, the third type of bacterial resistance, regardless of the chromosomal or extrachromosomal localization of the respective determinants. In studies of multi-resistant *E. coli* strains isolated in a surgical department from cases of postoperative complications, we observed, in conformity with the data published in literature, a diminished capacity to absorb ^{14}C-oxytetracycline (Gryaznova et al., 1974). However, some highly resistant strains took up a significant amount of the antibiotic. Fig. 1 demonstrates the absorption capacities of a susceptible strain, a "typical" resistant strain with a low uptake of ^{14}C-oxytetracycline and a strain characterized by the same degree of resistance but intensively absorbing the

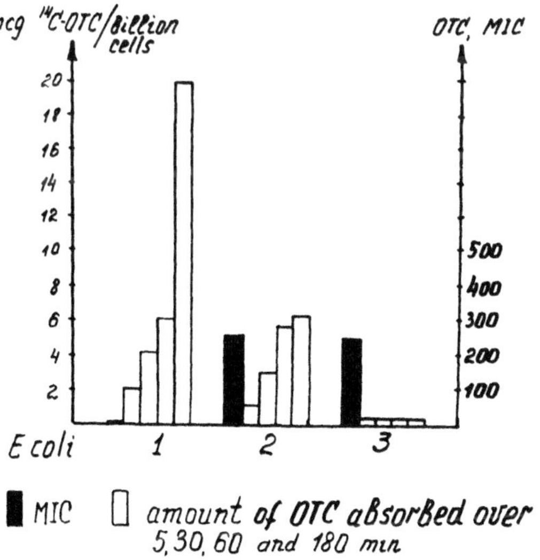

Fig. 1. Capacity of E. coli strains absorb ^{14}C-OTC as compared to MIC of the antibiotic

antibiotic. These data suggested that the resistance of the latter strain was due either to intracellular inactivation of oxytetracycline or to the ability of the ribosomes to function in its presence. There was also a third possibility, i.e. a different intracellular distribution of absorbed oxytetracycline, or rather its fixation by some undefined substances existing only in resistant cells, so that it could not react with the ribosomes. We explored these possibilities, and the following are some of our results.

Cells treated with ^{14}C-oxytetracycline were fractionated according to the well-known method described by Park and Hancock, which separates a pool of low molecular weight substances, ethanol soluble proteins, nucleic acids, trypsin soluble proteins and cell wall peptidoglycan.

No differences as to the distribution of the label were observed between susceptible and resistant cells.

TABLE 1

Effect of oxytetracycline on polyphenylalanine synthesis in a system containing ribosomes of a susceptible or a resistant strain and the supernatant of the susceptible strain

OTC concentration	Inhibition of polyphenylalanine synthesis (%)	
	Ribosomes of susceptible strain	Ribosomes of resistant strain
10^{-4}M	93	95
5×10^{-5}M	87	85
10^{-5}M	62	56
5×10^{-6}M	44	45
10^{-}M	12	9

TABLE 2

Effect of oxytetracycline on protein synthesis in a system containing ribosomes of a susceptible or a resistant strain and the supernatant fraction of the susceptible strain

OTC concentration	Inhibition of protein synthesis (%)	
	Ribosomes of susceptible strain	Ribosomes of resistant strain
10^{-3} M	35	50
10^{-4} M	30	47
10^{-5} M	20	40
10^{-6} M	12	30
10^{-7} M	8	14

In view of this, we subsequently used cell-free systems of polyphenylalanine synthesis directed by polyuridylic acid or of protein synthesis with endogenous mRNA. Table 1 compares the effect of oxytetracycline (OTC) on polyphenylalanine synthesis in systems containing ribosomes of a tetracycline susceptible strain (No. 1) or those of a tetracycline resistant intensively absorbing strain (No. 2). The systems included ribosomes, the supernatant fraction (only from the susceptible strain), ^{14}C-phenylalanine, tRNA, polyuridylate, GTP, ATP, phosphoenolpyruvate, phosphoenolpyruvate kinase and Tris buffer.

As shown in the Table, the susceptible and resistant strains had ribosomes with an equally high sensitivity to oxytetracycline.

Table 2 illustrates oxytetracycline activity in a system with endogenous mRNA. The system included ribosomes, the supernatant fraction (only from the susceptible strain), a mixture of ^{14}C-amino acids, GTP, ATP, phosphoenolpyruvate, phosphoenolpyruvate kinase and Tris buffer.

Thus, similar results were obtained in the system with endogenous mRNA: the protein synthesis was equally sensitive to oxytetracycline if effected by ribosomes of a resistant intensively absorbing strain or by those of a susceptible strain.

TABLE 3

Effect of oxytetracycline on polypeptide synthesis in a system containing ribosomes of a susceptible strain and the supernatant of a susceptible or a resistant strain

OTC concentration	Inhibition of polypeptide synthesis (%)	
	Supernatant of susceptible strain	Supernatant of resistant strain
10^{-4} M	62	32
10^{-5} M	68	25
10^{-6} M	55	15
10^{-7} M	35	6

Then we attempted to ascertain whether the supernatant fraction of a resistant strain could affect the oxytetracycline sensitivity of ribosomes.

The modulating influence of this fraction on the inhibition of polypeptide synthesis by oxytetracycline is demonstrated in Table 3; the system contained ribosomes of a susceptible strain.

Table 3 indicates that the supernatant from the cells of a resistant and intensively absorbing *E. coli* strain significantly modified the inhibition of polypeptide synthesis by oxytetracycline. In percentages, the inhibition fell 2 to 5 times in its presence.

TABLE 4

Effect of oxytetracycline on polypeptide synthesis in a system containing ribosomes of a resistant strain and the supernatant of a resistant or a susceptible strain

OTC concentration	Inhibition of polypeptide synthesis (%)	
	Supernatant of susceptible strain	Supernatant of resistant strain
10^{-3} M	50	27
10^{-4} M	42	25
10^{-5} M	38	23
10^{-6} M	27	22

In the next experiment we added to the system of polypeptide synthesis ribosomes extracted from the cells of a resistant strain and the supernatant from a resistant or a susceptible strain (Table 4).

With this modification of the cell-free system of polypeptide synthesis, the supernatant from a resistant strain also reduced the inhibitory effect of oxytetracycline.

Since the inhibition of polypeptide synthesis was reduced by the supernatant from an *E. coli* strain which was resistant to the antibiotic while absorbing it at considerable amounts, we searched for signs of a possible inactivation of oxytetracycline. However, microbiologic assays did not reveal any decline in oxytetracycline activity after its exposure to the supernatant with or without ribosomes.

No products of inactivation were found when ^{14}C-oxytetracycline was subjected to paper radiochromatography following incubation in these systems.

The mode of action of tetracycline antibiotics on the molecular level is as yet considerably less known than that of certain other inhibitors of protein synthesis, such as chloramphenicol or puromycin. In some cases of tetracycline resistance, along with the ribosomes per se, studies should be made of the enzyme systems and co-factors required for binding aminoacyl-tRNA to the ribosome (the step inhibited by tetracyclines).

REFERENCE

GRYAZNOVA, N. S., BELIAVSKAYA, I. V., PETYUSHENKO, R. M., SAZYKIN, Y. O., NAVASHIN, S. M Antibiotiki, 1974, No. 4, 332–336.

UPTAKE OF OXYTETRACYCLINE AND MINOCYCLINE BY E. COLI CELLS AND THEIR EFFECT ON PROTEIN SYNTHESIS

I. V. BELIAVSKAYA, S. M. NAVASHIN, N. S. GRYAZNOVA,
Y. O. SAZYKIN
National Institute of Antibiotics, Moscow, USSR

In recent years various laboratories have been engaged in a search for semi-synthetic analogues of natural tetracycline antibiotics. This trend in modern research has already produced promising results, as several new drugs have been successfully introduced into clinical practice. One of them is minocycline (7-dimethylamine-6-deoxy-6-demethyl-tetracycline) (Martell et al., 1967).

As shown by microbiological studies, the antimicrobial efficacy of minocycline is 5 to 10 times greater than that of natural tetracyclines (Renzini et al., 1967).

Due to the wide prevalence of tetracycline resistant bacterial strains, the potential activity of semi-synthetic analogues against such organisms (Fedorko et al., 1968) deserves special attention.

Proceeding on the basis of the current concept that tetracycline antibiotics are bound to ribosomes and act as inhibitors of protein synthesis, we compared the effect of oxy-tetracycline and minocycline on the production of polyphenylalanine in a cell-free system of polypeptide synthesis.

The system consisted of ribosomes (isolated in a centrifuge at 105,000 g), soluble enzyme fraction, ^{14}C-phenylalanine-tRNA, polyuridylate, GTP, ATP and the necessary salt components.

TABLE 1

Effect of oxytetracycline and minocycline on polyphenylalanine synthesis in a cell-free system containing ribosomes of a tetracycline sensitive E. coli strain (OTC MIC 2 mcg/ml, MC MIC 0.5 mcg/ml). Incubation with antibiotics 30 min.

Concentration of OTC or MC	Inhibition of polyphenylalanine synthesis (%)	
	OTC	MC
5.10^{-6} M	67	91
10^{-6} M	63	85
5.10^{-7} M	60	72
10^{-7} M	55	44

Table 1 demonstrates the inhibition of polyphenylalanine synthesis by oxytetracycline (OTC) and minocycline (MC) in a system containing ribosomes of a tetracycline sensitive strain.

Thus, in an experiment with ribosomes of a tetracycline sensitive strain the activities of oxytetracycline and minocycline were quite similar.

These antibiotics were also compared in a system containing ribosomes of a tetracycline resistant strain (Table 2).

TABLE 2

Effect of oxytetracycline and minocycline on polyphenylalanine synthesis in a cell-free system containing ribosomes of a tetracycline resistant E. coli strain (OTC MIC 250 mcg/ml, MC MIC 31 mcg/ml). Incubation with antibiotics 30 min.

Concentration of OTC or MC	Inhibition of polyphenylalanine synthesis (%)	
	OTC	MC
10^{-4} M	90	93
5.10^{-5} M	62	68
5.10^{-6} M	56	45
10^{-6} M	23	32

These data show that oxytetracycline and minocycline were equally active as inhibitors of polypeptide synthesis by ribosomes of a tetracycline resistant strain.

Taken as a whole, the results of the experiments with cell-free systems of polypeptide synthesis do not permit to explain the higher antibacterial efficacy of minocycline, as compared to oxytetracycline, by a more powerful action on the ribosomes.

Alternatively it may be supposed that minocycline penetrates into the bacterial cell more readily than natural tetracyclines.

Tetracycline antibiotics are known to be transferred through the bacterial membrane by a stereospecific system of active transport.

TABLE 3

Effect of minocycline on the binding of ^{14}C-oxytetracycline to intact cells of E. coli and to the fraction of high molecular weight substances. Preincubation with 500 mcg/ml of minocycline, incubation with 50 mcg/ml of ^{14}C-oxytetracycline

E. coli strain No.	Preincubation with MC, min.	Incubation with ^{14}C-OTC, min.	Amount of ^{14}C-OTC (% of control) bound to	
			intact cells	high molecular weight substances
1	5	5	26.8	—
	5	60	4.5	1.7
	60	5	22	10.4
	60	60	10.4	6.0
2	5	5	—	16.2
	5	60	23.5	16.2
	60	5	42.5	31.5
	60	60	24.7	9.1
3	5	5	71	62
	5	60	58	44
	60	5	60	40
	60	60	65	25

Taking into account the closely similar structures of natural tetracyclines and minocycline, it seems probable that all these substances are transferred into the cell by a common transport system. On the other hand, it is tempting to speculate that minocycline, a more effective antibiotic, may have greater opportunities for cell penetration: for instance, along with the path shared with all tetracyclines, there may exist a specific system transporting only minocycline.

Using ^{14}C-oxytetracycline, we attempted, firstly, to reveal a competitive effect of minocycline on its uptake by the bacterial cell and, secondly, to compare this effect in tetracycline susceptible and resistant cells of E. coli.

Along with the amount of the label in intact cells, as influenced by minocycline, we measured also the radioactivity in the total fraction of high molecular weight substances. This fraction was isolated from *E. coli* cells after exposure to ^{14}C-oxytetracycline with or without minocycline.

The results are presented in Table 3.

Table 3 shows that minocycline markedly decreased the uptake of ^{14}C-oxytetracycline. It should be noted, in particular, that the high molecular weight fraction isolated from the cells had also less radioactivity than control preparations. Consequently, competition between minocycline and ^{14}C-oxytetracycline affected both penetration through the cell membrane and the binding to intracellular acceptors. Comparative experiments with a susceptible strain (No. 1) and two resistant strains (Nos. 2 and 3) showed that at the step of ^{14}C-oxytetracycline penetration the competitive effect of minocycline was more pronounced in the first case.

Thus minocycline competes with oxytetracycline more effectively if the bacterial strain is tetracycline susceptible.

This suggests that in tetracycline resistant strains the uptake of minocycline depends more on its specific transport system than on the system shared with all natural tetracyclines.

This supposition should be obviously tested by different methods, and primarily by means of ^{14}C-minocycline to compare its uptake by tetracycline susceptible and resistant strains.

REFERENCES

Martell, M. J., Boothe, J. H., J. Med. Chem., 10, 44–46, 1967.
Renzini, G., Bevilacqua, R. L., Boemi, G., Ravagnan, L., Antibiotica, 5, 241, 1967.
Fedorko, J., Katz, S., Allnoch, H., Am. J. Med. Sci., 255, No. 4, 1968.

MODE OF ACTION OF CURING COMPOUNDS

F. E. HAHN and JENNIE CIAK
Department of Molecular Biology, Walter Reed Army Institute of Research
Washington, D.C., U.S.A.

Following the discovery of R-factors, several Japanese investigators reported in 1961 (Watanabe and Fukasawa, 1961; Mitsuhashi, Harada and Kameda, 1961) that growing R-factor-containing bacteria in the presence of certain aminoacridines resulted in a loss of drug resistance properties in small percentages of such bacterial populations. The effect became known as "curing". These experiments were patterned after the work of Hirota (1960) who had shown that growth of *Escherichia coli* in the presence of proflavine or acridine orange eliminated the F-episome from these bacteria. At the time of these early studies, the distinction between episomes and plasmids was not yet sharply drawn.

During the following decade, numerous additional studies of curing of R-factors were reported from different laboratories, and it is now possible to group the agents which eliminate resistance determinants from R-factors into two major categories. The first one includes substances which act upon the bacterial envelope or on the assembly of the cell-wall polymer, such as sodium dodecyl sulfate (Tomoeda, Inuzuka, Kubo and Nakamura, 1968) or penicillin (Kawakami and Landman, 1965). To this one might add instances in which plasmids render bacteria more sensitive to the actions of inhibitors of cell-wall polymer formation such as macarbomycin (Mitsuhashi, Hashimoto and Umezawa, 1970) or flavomycin (Watanabe, Ogata, Sugawara and Oda, 1972).

The second category includes agents which interfere with the replication of R-factor DNA. These are ultraviolet radiation (Rownd, Nakaya and Nakamura, 1966), mitomycin C (Kawakami and Landman, 1965), nalidixic acid (Chabbert, Baudens and Bouanchaud, 1969), phenethyl alcohol (Unowsky and Bontempo, 1967), trimethoprim (Pinney and Smith, 1972) which converts prototrophic bacteria into phenocopies of thymine auxotrophs, or actual thymine starvation of thymine auxotrophs (Pinney and Smith, 1971). Most important, DNA-complexing compounds, especially those which intercalate into DNA (Hahn and Ciak, 1971), possess the group property of curing. Our own interest has been in agents of this last type (Ciak, Wormley and Hahn, 1969; Hahn and Ciak, 1971; Hahn and Ciak, 1972; Hahn and Ciak, 1973). This paper is concerned with the curing action of DNA-binding compounds.

EXPERIMENTAL

Our test organism was *Salmonella typhimurium* strain LT-2 into which the R-factor from *Escherichia coli* RS-2 (Silver and Falkow, 1970) had been transferred (Hahn and Ciak, 1973). Spontaneous segregation of resistance determinants in *S. typhimurium* LT-2 (Watanabe and Lyang, 1962) and in other *Salmonella* strains (Lebek, 1963; Anderson, 1968) is a frequent occurrence and chemical curing of resistance determinants in *Salmo-*

nellae is also attained with higher frequencies than, for example, in *E. coli* (Bouanchaud, Scavizzi and Chabbert, 1969). A correlation has been recognized (Lebek, 1969) between the ability of bacteria spontaneously to segregate R-factor resistance determinants and their susceptibility to chemical curing. The original R-factor in *E. coli* RS-2 carries determinants of resistance to kanamycin, chloramphenicol, ampicillin, streptomycin and sulfonamide. Since our strain of *S. typhimurium* is naturally insensitive to sulfadiazine, our curing experiments were directed against the determinants of resistance to the four

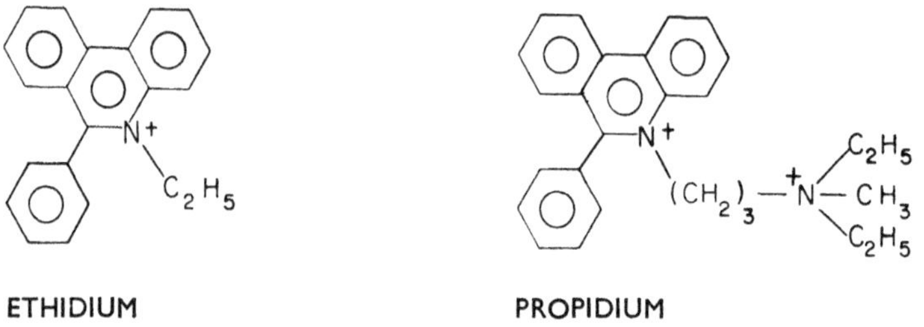

Fig. 1. Structures of ethidium and propidium.

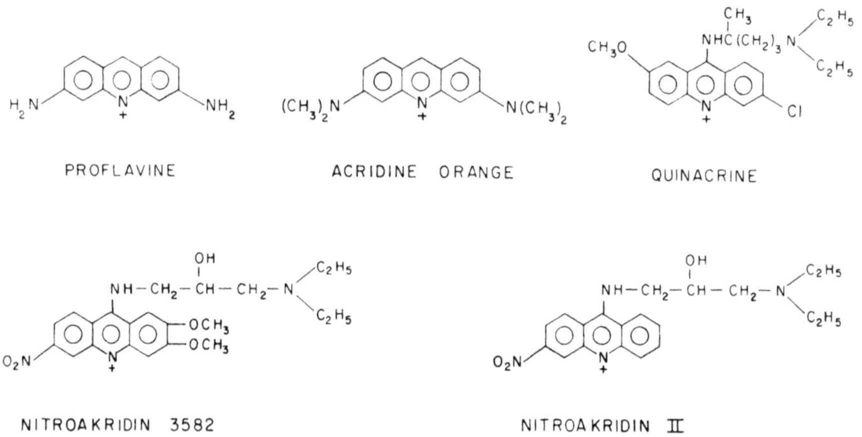

Fig. 2. Structures of substituted aminoacridines.

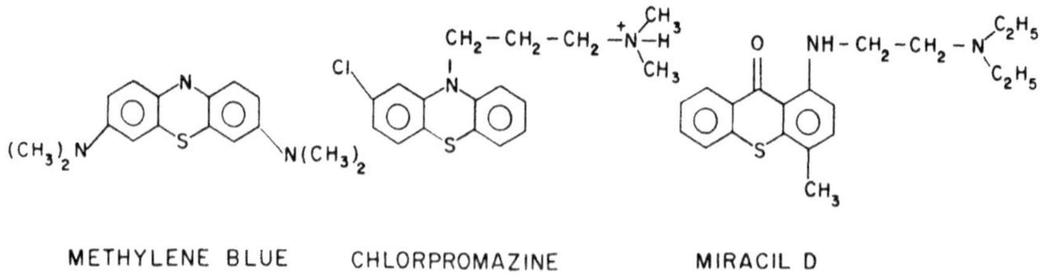

Fig. 3. Structures of heterocyclic sulfur-containing compounds.

antibiotics cited. Curing experiments were carried out by a modification of the technique of Hirota (1960) which we have described previously (Hahn and Ciak, 1971 and 1972).

The compounds which were studied belong into several sub-sets. Fig. 1 shows the phenanthridine derivatives ethidium and propidium. The second group (Fig. 2) was comprised of the acridines, proflavine, acridine orange, quinacrine, Nitroakridin 3582

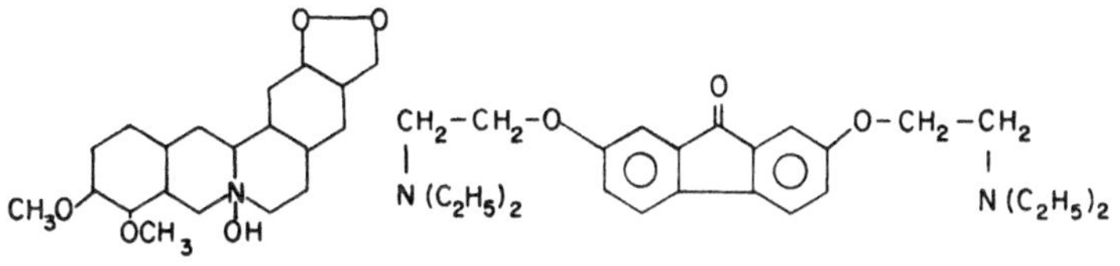

BERBERINE TILORONE

Fig. 4. Structures of berberine and tilorone.

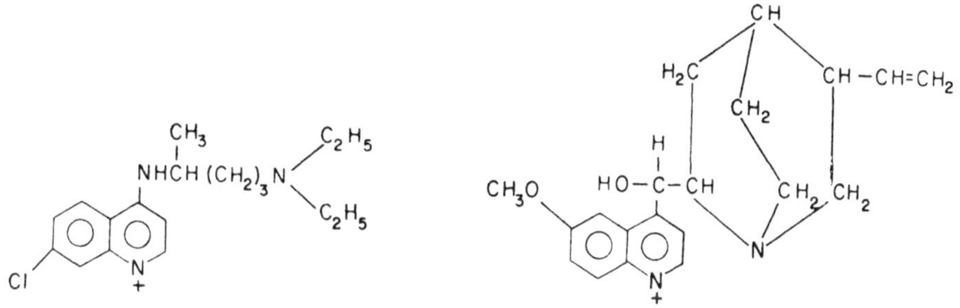

CHLOROQUINE QUININE

Fig. 5. Structures of 4-aminoquinolies.

$$H_2\overset{+}{N}-(CH_2)_3\ \overset{+}{N}H-(CH_2)_4-\overset{+}{N}H(CH_2)_3-\overset{+}{N}H_2$$

SPERMINE

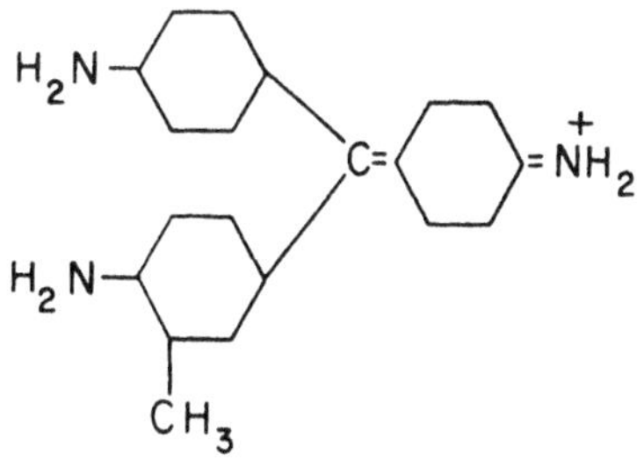

p-ROSANILINE

Fig. 6. Structures of spermine and p-rosaniline.

and 5-γ-diethylamino-β-hydroxypropylamino)-2-nitroacridine; the last two were gifts from Farbwerke Hoechst. We also studied three-membered ring systems with a sulphur heteroatom (Fig. 3) such as the phenothiazines, methylene blue and chlorpromazine and the thioxanthone derivative miracil D. Additional compounds with three or more condensed ring systems (Fig. 4) were the alkaloid, berberine, and the synthetic fluorenone derivative, tilorone. Among derivatives of 4-aminoquinoline (Fig. 5) were quinine and chloroquine, and, finally, (Fig. 6) we included in the study the DNA-complexing compounds spermine and p-rosaniline which do not intercalate.

All substances were supplied to cultures of *S. typhimurium* in subinhibitory concentrations, in most instances to a final concentration of 10^{-4} M. A few compounds were strong growth inhibitors and could not be employed at that standard concentration, the curing results with these substances are shown in Table 1. The results with all compounds which were studied at 10^{-4} M are listed in Table 2.

TABLE 1

Per Cents Elimination of Resistance Determinants in S. typhimurium.

Compounds	Kana	Chloramph	Strepto	Ampicill
6.25×10^{-6} M				
Nalidixic Acid	70	56	60	64
2.5×10^{-5}M				
Nitroacridine II	99	96	96	97
Daunomycin	86	80	77	81
Proflavine	86	79	64	68
5×10^{-5}M				
Nitroakridin 3582	82	69	70	69
8×10^{-4}M				
Coumadine	70	56	60	64

TABLE 2

Per Cents Elimination of Resistance Determinants in S. typhimurium

Compunds at 10^{-4} M	Kana	Chloramph	Strepto	Ampicill
Ethidium	96	82	84	82
Miracil D	91	82	81	82
Quinacrine	92	79	81	71
Propidium	89	78	71	71
Tilorone	85	80	74	78*
p-Rosaniline	87	71	73	66
Acridine Orange	85	71	69	68
Berberine	83	68	70	62
Quinine	81	60	66	57
Spermine	81	56*	71	58
Chlorpromazine	78	63	67	72*
Hoechst 33258	72	64	64	67*
Chloroquine	80*	58	52	53
Methylene Blue	68	50	48	41

* Significant inversions in activity sequence.

The first fact which emerges from an examination of the Tables is that the test compounds produced segregation of individual resistance determinants. Curing of the kanamycin determinant occurred invariably at higher frequencies than that of the other three determinants. In our previous work with *E. coli* RS-2 (Hahn and Ciak, 1971), we have

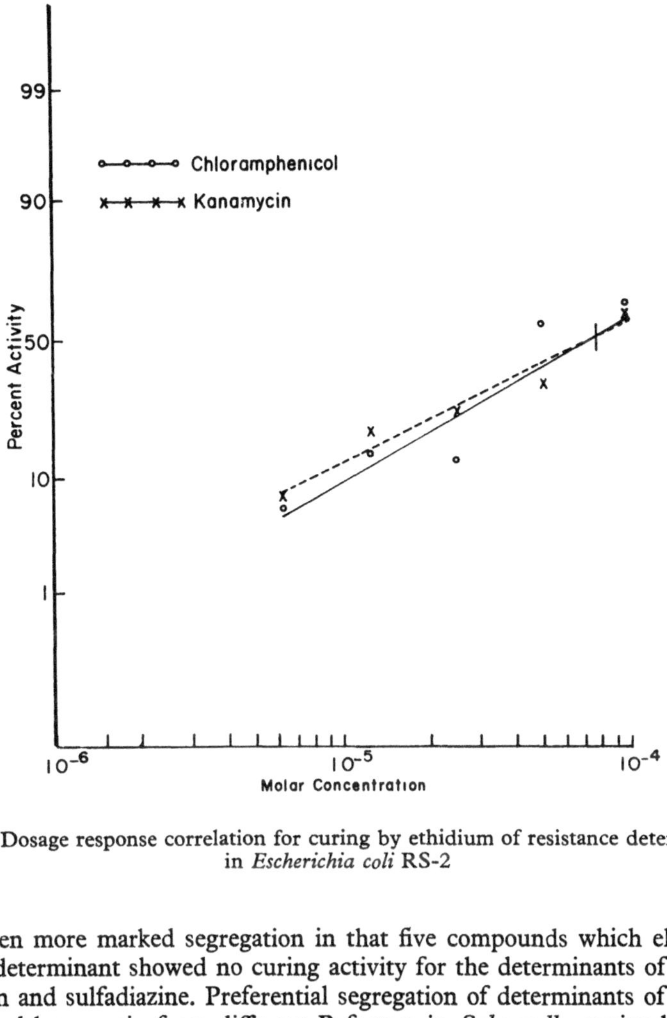

Fig. 7. Dosage response correlation for curing by ethidium of resistance determinants in *Escherichia coli* RS-2

observed even more marked segregation in that five compounds which eliminated the kanamycin determinant showed no curing activity for the determinants of resistance to streptomycin and sulfadiazine. Preferential segregation of determinants of resistance to ampicillin and kanamycin from different R-factors in *Salmonella* strains has also been noted in curing studies of other authors (Bouanchaud, Scavizzi and Chabbert, 1969). Anderson (1968) observed spontaneous segregation of resistance determinants in a number of *Salmonella* strains and interpreted these findings by assuming the existence of asynchronies in the biosynthesis of individual DNA replicons relative to the replication cycle of the host bacteria. Observations of enhanced segregation of resistance determinants by curing agents suggest that seemingly unitary R-factors, even when capable of isolation as populations of uniform DNA species or transferred *en bloc* as units of genetic transmission, are, in reality, recombinational assemblies of individually biosynthesized replicons

(Falkow, Tompkins, Silver, Guerry and LeBlanc, 1971) which differ in their susceptibility to DNA-complexing chemical and to the ensuing inhibitions of DNA replication.

The second fact which emerges from a review of Table 2 is that the curing potencies of the studied compounds fell into a systematic hierarchic sequence. The compounds

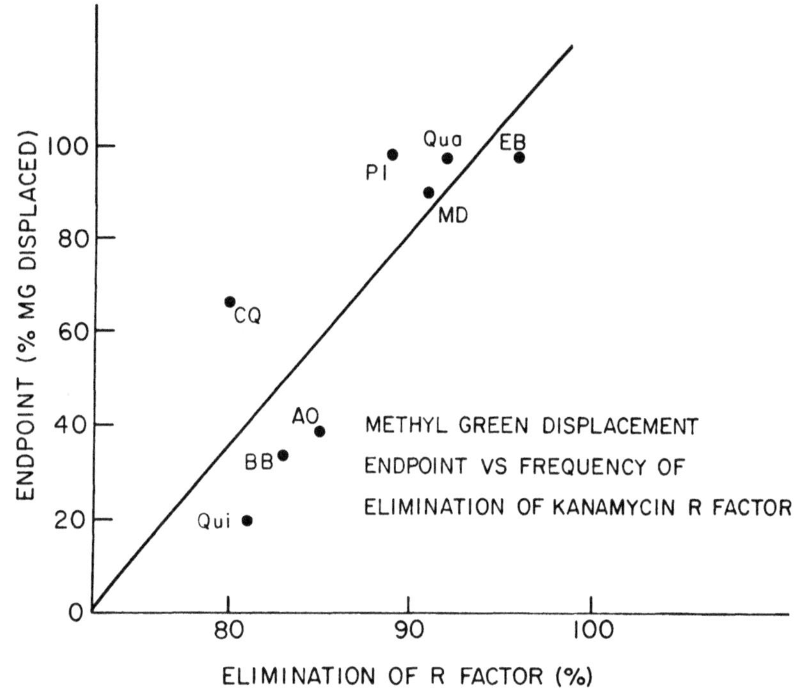

Fig. 8. Correlation between curing frequencies for the kanamycin determinant in *S. typhimurium* and the ability of compounds (5×10^{-5}M) to displace 1.9×10^{-5}M methyl green from its complex with 2.4×10^{-4}M [nucleotides] DNA. EB: ethidium bromide; Qua: quinacrine; PI: propidium iodide; MD: miracil D; CQ: chloroquine; AO: acridine orange; BB: berberine; Qui: quinine.

have been listed according to decreasing curing potencies. This sequence shows only a few inversions which are based upon potency differences which may not all be statistically significant. We assume that these systematic activity differences reflect the different extents to which curing compounds bind to R-factor DNA. A definitive test of this hypothesis requires biophysical studies on the interaction of these curing compounds with isolated R-factor DNA. Such studies have just been initiated in our laboratory. However, there exists circumstantial evidence in favor of the assumption that different curing activities are the results of differences in the binding of curing compounds to R-factor DNA.

The first line of evidence is intrinsic to our selection of compounds. For each substance selected (with the exception of spermine, coumadine, nalidixic acid and p-rosaniline), there existed biophysical evidence of intercalation binding into double helical DNA. In fact, the ability of forming complexes with DNA in this manner was a key property according to which substances were selected for study.

The second line of evidence comes from our previous work and is reproduced in Fig. 7. This diagram (Hahn and Ciak, 1971) shows the probit transformations of the dosage-response correlations which governed the elimination of the kanamycin and chloramphenicol determinants from the R-factor in *E. coli* RS-2 by ethidium. The very existence

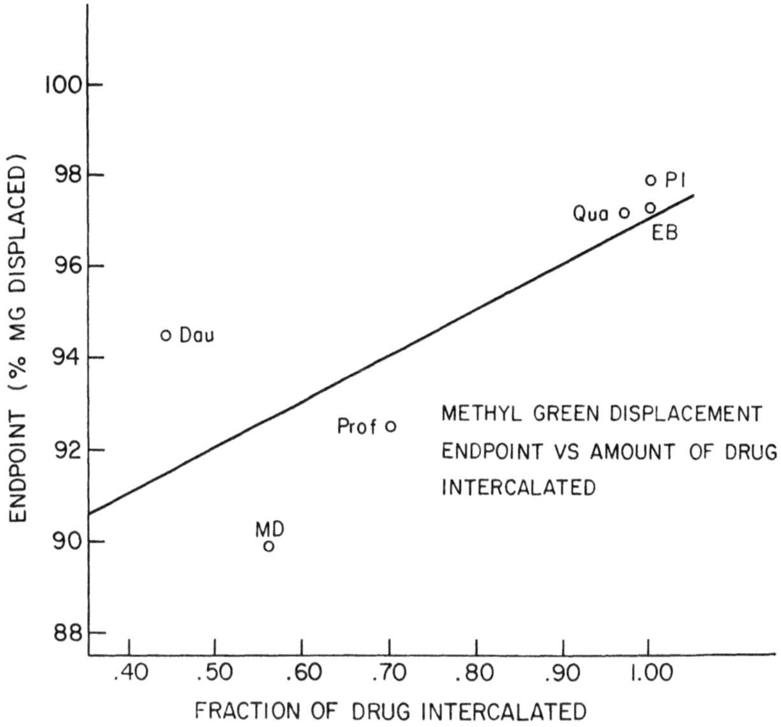

Fig. 9. Correlation between endpoints of methyl green displacement and unwinding stoichiometry for supercolied DNA (Waring, 1970). Abbreviations as in Fig. 8. Additionally, Dau: daunomycin.

of such dosage-response correlations indicates that the effect is due to the reversible occupancy of ethidium receptor sites which obeys the mass action law. These sites are believed to be on DNA.

Finally, we have taken the endpoints of the displacement of methyl green from its complex with DNA (Krey and Hahn, 1971) by our test compounds and have plotted elimination frequencies of the kanamycin determinant in *S. typhimurium* as a function of these endpoints (Fig. 8). We interpret this diagram as one indication that the extent of DNA binding of curing compounds is correlated with the extent of curing which they produce. This inference might be debated on grounds that the calf-thymus DNA-methyl green complex consists of pieces of linear DNA while all R-factor DNAs that have been isolated have been shown to be circular supercoiled in nature. However, when the equivalence points of unwinding of supercoiled DNA which have been determined by Waring (1970) were plotted as a function of the ability of the same compounds to displace methyl green, the diagram, Fig. 9, was obtained which shows that unwinding of DNA and displacement of methyl green are correlated parameters.

For these reasons, we are entertaining the working hypothesis which we have previously postulated (Hahn and Ciak, 1971) that the ability of intercalating compounds to eliminate resistance determinants from R-factors is a function of the binding of such substances to R-factor DNA. This immediately raises another question: How can curing by such a mechanism occur in bacteria at concentrations of curing compounds which do not inhibit bacterial growth and, hence, chromosomal DNA replication?

First of all, the binding affinity of closed circular DNA, for example, for ethidium, is greater than that of the affinities of nicked or unconstrained DNA molecules so that at low drug concentrations a preferential binding to circular supercoiled DNA occurs (Bauer and Vinograd, 1971). Secondly, a preferential effect of ethidium has been demonstrated *in vivo* on the functions of mitochondrial DNA in cultures of mammalian cells (Zylber, Vesco and Penman, 1969; Zylber and Penman, 1969; Knight, 1969) and of kinetoplastic DNA in trypanosomes (Riou and Delain, 1969). Finally, the ability of ethidium to act as a DNA template poison has been investigated *in vitro* with the result that the drug is about one order of magnitude more active when the template is circular supercoiled in structure as compared to unconstrained DNA (Richardson and Parker, 1973). It is, therefore, reasonable to assume that the selective toxicity of intercalating compounds for R-factors at concentrations which do not inhibit bacterial growth is the result of selective binding of these substances to circular supercoiled R-factor DNA (Hahn and Ciak, 1971).

The structural rules which determine that a compound will intercalate into DNA are established. An intercalating substance must have a flat planar aromatic ring system of a minimal area of ~ 30 Å^2 for an effective interaction with the planes of the DNA base pairs which occupy roughly 40 Å^2. Additionally, it must possess one or two protonated groups which can bind electrostatically to phosphoric acid molecules in the DNA strand backbones. Frequently, such as in the examples of the nitroacridines, quinacrine and chloroquine, the presence of a strongly electronegative substituent on the ring system, enhances affinity for DNA and biological activity which result from intercalation binding, although the reasons for this are not as yet apparent.

The curing results which we have obtained may, therefore, be considered one biological expression of structural requirements for intercalation binding as well as the beginning of the recognition of structure-activity rules for compounds which eliminate resistance determinants from R-factors.

Japanese workers have attempted earlier to establish such structure-activity rules for compounds which inhibited the *transmission* of R-factors between *Shigella flexneri* and *E. coli* (Hayashi, Kodaira, Ogawa, Baba, Kikuchi, Tada and Iwasaki, 1965). Among 17 dyes which were studied, 12 were derivatives of acridine. Unsubstituted acridine and acridine N-oxide were inactive. But substitutions of $-NH_2$ resulting in several amino-acridines produced active compounds whose activity was maintained when the amino groups were methylated or acetylated. However, activity was lost when larger substituents such as propionyl or benzoyl were introduced into these amino groups. Also, 9-amino-acridines with various alkyl substituents elsewhere on the ring system were of low activity. An exception to this is quinacrine which has been shown independently by other authors (Levy and Watanabe, 1966) to inhibit the transfer of R-factors. Without engaging in detailed theoretical-chemical discussions of these observations, it can be stated that these structure-activity rules reflect structural and electrochemical requirements for strong intercalation binding of acridines into DNA, a situation which the authors of that work did not mention or discuss.

If one considers the pertinent fact that transmission of R-factor DNA requires active conjugal DNA biosythesis (Fenwick and Curtiss, 1973a) and that an inhibition of conjugal DNA synthesis, for example, by nalidixic acid, results in failure of R-factor transmission (Fenwick and Curtiss, 1973b), it is readily apparent that the structure activity relationships which were recognized by Hayashi and his associates in 1965, from their transfer experiments, reflect, in actuality, structure-activity rules for the inhibition of R-factor DNA synthesis. It is reasonable to think that the Japanese workers would have obtained the same results if they had studies actual curing. Key evidence that curing involves inhibition of the formation of extrachromosomal genetic determinants comes from kinetic studies on the elimination of the lac marker from the F'lac episome by acridine orange (Hohn and Korn, 1969). The kinetics of loss of lac are, in fact, equal to the growth kinetics of the lac^- progeny; hence, the lac^+ parent culture was progressively diluted with new lac^- cells, owing to the inhibition of episome replication by acridine orange. Finally, Rownd and his associates (1966) selected various bacteria which had lost all resistance determinants through curing of their R-factors either by ultraviolet irradiation or by acriflavine. When the global DNA of such cured bacteria was analyzed by density gradient centrifugation, the R-factor DNAs were absent. Curing had, therefore, physically eliminated the R-factors from the test organisms.

SUMMARY

Our curing results as well as the train of reasoning presented in this paper encourage us to make the following summary concerning the mode of action of curing agents:

1. Substances which form complexes with DNA, especially by intercalation, share the group property of being capable of eliminating resistance determinants from bacterial R-factors. Selective toxicity for R-factors appears to be based on preferential association of intercalating compounds with closed circular supercoiled DNA. Isolated R-factor DNAs possess this special conformation.
2. Structural requirements for intercalation, hence, represent structure-activity rules for R-factor elimination. The same rules also apply to substituted acridines which inhibit the transfer of R-factor among enterobacteria.
3. Ability of compounds to displace methyl green from its complex with DNA correlates with their curing potency; it also correlates with their ability to cause unwinding of circular supercoiled DNA. Dosage-response correlations of curing by ethidium indicate that this effect is due to reversible occupancy of receptor sites according to the mass action law.
4. Template toxicity of intercalating compounds for R-factor DNA causes the failure of this DNA to be replicated. This is proposed to be the mode of action of curing. It also appears to underlie the inhibition of R-factor transmission which is known to require *de novo* conjugal R-factor DNA biosynthesis.

REFERENCES

ANDERSON, E. S. (1968). The ecology of the transferable drug resistance in the enterobacteria. Ann. Rev. Microbiol. 22, 131.

BAUER, W. and VINOGRAD, J. (1971). The use of intercalative dyes in the study of closed circular DNA. Progr. Mol. Subcell. Biol. 2, 181.

BOUANCHAUD, D. H., SCAVIZZI, M. R. and CHABBERT, Y. A. (1969). Elimination by ethidium bromide of antibiotic resistance in *Enterobacteria* and *Staphylococci*. J. Gen. Microbiol. 54, 417.

CHABBERT, Y. A., BAUDENS, J. G. and BOUANCHAUD, D. H. (1969). Medical aspects of transferable drug resistance. Bacterial Episomes and Plasmids, CIBA Foundation Symp. Churchill, London, 227.

CIAK, J., WORMLEY, S. and HAHN, F. E. (1969). Elimination of the F lac episome from *Escherichia coli* 3876 by substances which intercalate into DNA. Bact. Proc. 67.

FALKOW, S., TOMPKINS, L. S., SILVER, R. P., QUERRY, P. and LEBLANC, D. J. (1971). The replication of R-factor DNA in *Escherichia coli* K-12 following conjugation. Ann. New York Acad. Sci. 182, 153.

FENWICK, R. G. and CURTISS, R. (1973a). Conjugal deoxyribonucleic acid replication by *Escherichia coli* K-12: Stimulation in dnaB(ts) donors by minicells. J. Bacteriol. 116, 1212.

FENWICK, R. G. and CURTISS, R. (1973b). Conjugal deoxyribonucleic acid replication by *Escherichia coli* K-12: Effect of nalidixic acid. J. Bacteriol. 116, 1236.

HAHN, F. E. and CIAK, J. (1971). Elimination of bacterial episomes by DNA-complexing compounds. Ann. New York Acad. Sci. 182, 295.

HAHN, F. E. and CIAK, J. (1972). Elimination of bacterial episomes by DNA-complexing compounds. Bacterial Plasmids and Antibiotic Resistance, Avicenum, Prague and Springer, Berlin-Heidelberg-New York 297.

HAHN, F. E. and CIAK, J. (1973). Restoration of drug-resistant bacteria to sensitivity to chemotherapeutic drugs. Proc. 1972 Army Sci. Conf. 2, 1.

HAYASHI, K., KODAIRA, T., OGAWA, Y., BABA, K. and KIKUCHI, K. (1965). Studies of substances which act on episomic infective transfer systems. The correlation between the chemical structure of acridine derivatives and the inhibition of the transmission of multiple drug resistance. Jap. J. Bacteriol. 20, 498.

HIROTA, A. (1960). The effect of acridine dyes on mating type factors in *Escherichia coli*. Proc. Natl. Acad. Sci. U.S. 46, 57.

HOHN, B. and KORN, D. (1969). Cosegregation of a sex factor with the *Escherichia coli* chromosome during curing by acridine orange. J. Mol. Biol. 45, 385.

KAWAKAMI, M. and LANDMAN, O. E. (1965). Experiments concerning the curing and intracellular site of episomes. Biochem. Biophys. Res. Comm. 18, 716.

KNIGHT, E. (1969). Mitochondria-associated RNA of the HeLa cell. Effect of ethidium bromide on the synthesis of ribosomal and 4S RNA. Biochemistry 8, 5089.

KREY, A. K. and HAHN, F. E. (1971). Methyl green-DNA complex: Displacement of dye by DNA-binding substances. Proc. First Europ. Bioph. Cong. 1, 223.

LEBEK, G. (1963). Der spontane Verlust episomal-Übertragener Mehrfachresistenz gegen Antibiotica bei Populationen gramnegativer Darmkeime in Kulturen mit und ohne Antibiotica. Zschr. Hyg. 149, 255.

LEBEK, G. (1969). Die infektiöse bakterielle Antibiotikaresistenz. Hans Huber, Bern and Stuttgart.

LEVY, S. B. and WATANABE, T. (1966). Mepacrine and transfer of R factors. Lancet 2, 1138.

MITSUHASHI, S., HARADA, K. and KAMEDA, M. (1961). Elimination of transmissible drug-resistance by treatment with acriflavin. Nature 189, 947.

MITSUHASHI, S., IYOBE, S. and HASHIMOTO, H. (1970). Preferential inhibition of the growth of *Escherichia coli* strains carrying episomes. J. Antibiotics 23, 319.

PINNEY, R. J. and SMITH, J. T. (1971). R-factor elimination by thymine starvation. Genet. Res. Camb. 18, 173.

PINNEY, R. J. and SMITH, J. T. (1972). Elimination of an R-factor by trimethoprim. J. Pharm. Pharmacol. 24 Suppl, 126 p.

RICHARDSON, J. P. and PARKER, S. R. (1973). Effect of ethidium bromide on transcription of linear and circular DNA templates. J. Mol. Biol. 78, 715.

RIOU, G. and DELAIN, E. (1969). Abnormal circular DNA molecules induced by ethidium bromide in the kinetoplast of trypanosoma cruzi. Proc. Natl. Acad. Sci. U.S. 64, 618.

ROWND, R., NAKAYA, R. and NAKAMURA, A. (1966). Molecular nature of the drug-resistance factors of the enterobacteriacea. J. Mol. Biol. 17, 376.

SILVER, R. P. and FALKOW, S. (1970). Specific labeling and physical characterization of R-factor deoxyribonucleic acid in *Escherichia coli*. J. Bacteriol. 104, 331.

TOMOEDA, M., INUZUKA, M., KUBO, N. and NAKAMURA, S. (1968). Effective elimination of drug resistance and sex factors in *Escherichia coli* by sodium dodecyl sulfate. J. Bacteriol. 95, 1078.
UNOWSKY, J. and BONTEMPO, J. (1967). The elimination of R-factor-mediated drug resistance by phenethyl alcohol and acriflavin. Abstr. Seventh Intersci. Conf. Antimicr. Agents & Chemoth. 57.
WARING, M. (1970). Variation of the supercoils in closed circular DNA by binding of antibiotics and drugs. Evidence for molecular models involving intercalation. J. Mol. Biol. 54, 247.
WATANABE, T. and FUKASAWA, T. (1961). Episome-mediated transfer of drug resistance in *Enterobacteriaceae*. II. Elimination of resistance factors with acridines. J. Bacteriol. 81, 679.
WATANABE, T. and LYANG, K. W. (1962). Episome-mediated transfer of drug resistance in *Enterobacteriacea*. V. Spontaneous segregation and recombination of resistance factors in *Salmonella typhimurium*. J. Bacteriol. 84, 422.
WATANABE, T., OGATA, Y., SUGAWARA, K. and ODA, K. (1972). Increase of flavomycin sensitivity of bacteria by R-factors. Bacterial Plasmids and Antibiotic Resistance. Avicenum, Prague and Springer, Berlin-Heidelberg-New York.
ZYLBER, E. and PENMAN, S. (1969). Mitochondrial-associated 4S RNA synthesis inhibition by ethidium bromide. J. Mol. Biol. 46, 201.
ZYLBER, E., VESCO, C. and PENMAN, S. (1969). Selective inhibition of the synthesis of mitochondria-associated RNA by ethidium bromide. J. Mol. Biol. 44, 195.

Acknowledgements. Figs. 8 and 9 have been assembled by Anne K. Krey for inclusion in an article on methyl green displacement from DNA. The careful technical assistance of Cynthia Henry is gratefully acknowledged by the authors.

EFFECTS OF SOME BIOLOGICALLY ACTIVE COMPOUNDS ON ANTIBIOTIC RESISTANCE IN MICROORGANISMS

I. I. BELOUSOVA, E. B. LISHNEVSKAYA, I. M. TERESHIN
Research Institute of Antibiotics, Leningrad, U.S.S.R.

The problem of overcoming the development of resistance that arose at the dawn of chemotherapy has not yet been solved. Moreover, it has become quite clear that there is no integral solution to this problem, since the obstacle to it is a remarkable adaptation of microorgamisms to the very extreme conditions. The situation is still more aggravated by the widespread transmissiveness of the resistance factors.

Defensive mechanisms of the cell have been shown to result from the wide scope of genetic and biochemical processes, i.e. those ranging from the mutational DNA variations to the decrease of the permeability of the cell wall and membrane to the drugs, the production of inactivating enzymes, the intensification or induction of alternative metabolic paths, and the variations in the ability of structure-targets to interact with antibiotics.

Accordingly, two main methods to overcome the antibiotic resistance can be considered. One of them is the blocking process at genotype level that includes the inhibition of replication of the episome or certain sites of the chromosome incorporating the resistance loci, as well as the inhibition of resistance transfer to the sensitive cultures and the decrease of the development of the resistant mutants.

The other method comprises the resistance inhibition at the phenotype level by the combined use of antibiotics and compounds increasing their penetration into the cell-inhibiting the inactivation process, enhancing the affinity of antibiotics to the structure-targets, or those interfering the functioning or induction of the by-passing metabolic routes.

The search for compounds which would make antibiotic therapy more efficient proceeds in both directions. These directions also define, to a large extent, the specific characteristics of the investigated substances which comprise, firstly, those capable to interact with the genetic apparatus of the cell, and secondly those enhancing the cell permeability. Such a subdivision is rather relative since many of the substances belonging to one or the other group possess a polyfunctional effect.

The most investigated group among the DNA-tropic agents that are used to overcome bacterial resistance to antibiotics is that of acridines, i.e. compounds known for their effects on bacterial mutagenesis (Ephrussi et al., 1949; Sevag, 1964). The efficiency of acridine orange in eliminating the R-factors of bacterial cells was shown as far back as 1957 (Hirota et al., 1957). Since then numerous papers have been published attesting to the specific effects of acridines on the antibiotic resistance in various bacterial species and strains (Sevag et al., 1970; Mitsuhashi et al., 1963; Brodinova et al., 1969; Bouanchaud et al., 1971; Vartanyan, 1973).

The maximum result is usually attained in the case of simultaneous effects of acridines and antibacterial compounds on the bacterial culture. In the majority of cases the extent of elimination is not high and varies greatly (from 0 to 100%) depending on the substance or the bacterial strain.

The elimination mechanism has been studied in detail (Hahn et al., 1971). It has been shown to be based on intercalation of the acridine ring between the base pairs in the double chain of the episomal DNA, thus impeding its normal function as a matrix.

Acridines have been demonstrated to prevent the development of resistance if present in the antibiotic containing media on which sensitive organisms were grown by the passage procedure (Warren et al., 1967; Soloviev et al., 1968; Zueva et al., 1969; Podbornov et al., 1969; Moroz et al., 1971; Podbornov, 1972). These substances have been also noted to have an effect on the transfer of R-factors in cases of conjugation (Levy et al., 1966; Astapov, 1970), transformation (Ayad, 1969) and transduction (Zueva et al., 1969).

The presented data show that acridine intercalation into the structure of bacterial DNA resulted in appreciable changes in the functioning of the genetic material of bacterial cells.

There is some evidence that it is possible to select antibiotic sensitive cultures in the presence of acridine (Yoshikawa et al., 1967). Yoshikava and Sevag found that cells of some *E. coli* strains which were R-factor carriers proved to be more sensitive to the bactericidal effect of atabrin, but according to other authors this phenomenon was not observed by far in all cases.

The experiments carried out on mice by Bouanchaud and Chabbert (Bouanchaud et al., 1969) revealed a high efficiency of acriflavine *in vivo* (up to 96% of the bacteria were shown to lose their resistance to Sm-Km-Cm-Tc-Su). The elimination mechanism *in vitro* was apparently similar to that *in vivo*. Sevag et al. (1970) succeeded in obtaining good clinical results when using atabrin in combination with antibiotics for the therapy of chronic infections of the excretory system in children.

Numerous experiments were performed to study the effect of ethidium bromide on antibiotic resistance. This compound, which also belongs to the intercalating agents, was shown to be capable of interacting with nucleic acids and of inhibiting the DNA- and RNA-polymerase reaction (Waring, 1966).

According to some authors, ethidium bromide was a fairly active eliminative agent (Bouanchaud et al., 1969; Moroz et al., 1972) and could inhibit the transfer of resistance in the process of conjugation or transduction at concentrations not preventing cell propagation of the donor or recipient (Glatman et al., 1973; Antsyferova et al., 1973). However, just as in the case of acridines, its efficiency varied greatly, i.e. from 0 to 100%, according to the bacterial species or strain, as well as to the marker resistance. The cause of this phenomenon has not been definitely established. The experiments showed ethidium bromide to be bound to a similar degree to the DNAs isolated of various bacterial species, but to interact differently with the linear and circular DNAs (Hadson et al., 1967). This last-mentioned phenomenon could apparently explain the selective activity of intercalatory agents in the case of episomes.

Polyamines arose a definite interest in research workers due to their effects on the development of resistance in microorganisms. These compounds do not belong to the intercalatory agents and are known to form DNA complexes as a result of electrostatic interaction (Suwalsky et al., 1969) but produce no conformational changes in overtwisted DNA (Waring, 1970). Being a part of the ribosomes, polyamines apparently play an important role in the cell functioning (Spirin et al., 1971).

Sevag et al. (1970) showed that spermine and spermidine drastically inhibited the development of resistance if present in the media on which sensitive *Escherichia coli* or staphylococci were grown by the passage method. Pretreatment with spermine had no effect on the development of resistance in the population.

In the experiments performed by Sevag, polyamines revealed no ability to increase the antibiotic efficiency in the case of obviously resistant cultures. However some other authors (Hahn et al., 1971) found a partial elimination of the R-factors in the presence of spermine. We succeeded in increasing 1.5–4 times the effects of a number of antibiotics (Cm, Sm) on *E. coli* and *St. aureus* in case of simultaneous introduction of polyamines into the medium (Tereshin et al., in press). Apparently, the effects of such compounds on antibiotic resistance are beyond their antimutagenic effects noted by Sevag.

Of compounds capable to change the cell permeability, ethylene diamine tetraacetate (EDTA) appears to be the most extensively investigated one. By means of binding bivalent cations contained in the bacterial cell walls and membranes, EDTA was found to increase the permeability of these structures to various compounds including antibiotics. In such cases, essentially the passive transport was affected while systems of the active one remained unchanged (Levie, 1968). The effect of EDTA was shown to be reversible in 2–3 generations in absence of the preparation.

EDTA introduction into the medium was found to result in an appreciable reduction of the minimum bacteriostatic concentrations (MBC) of a number of antibiotics for *St. aureus*, both laboratory strains and those isolated from pathogenic material (Rawal, 1971), *Salmonella typhi* (Muschel, 1968), *E. coli* (Weiser et al., 1968). The EDTA combination with antibiotics exhibited a synergistic effect towards *Pseudomonas aeruginosa* (Weiser et al., 1969).

In our experiments, sub-bacteriostatic concentrations of disodium EDTA in combination with chloramphenicol, tetracycline, erythromycin increased 4–14 times the inhibition of resistant *E. coli* and *St. aureus* (Levchenko et al., in press). Less toxic calcium salts of EDTA had no effect on the resistance. A very strong inhibition (3–4 times higher) of the transfer of resistance to chloramphenicol was observed in *E. coli*.

Introduction of EDTA sub-bacteriostatic concentrations to the bacterial culture was shown to have no effect on chloramphenicolase activity in *E. coli* and *St. aureus* (Solovieva et al., 1974), while penicillinase induction in the staphylococcus was inhibited by 30–60% (Gross et al., 1970). It was suggested that this effect might result from EDTA binding of magnesium which was necessary for the enzyme induction. However, Hamilton-Miller (1965) noted an enhancement of penicillinase activity in the presence of EDTA as a result of higher cell permeability to the antibiotic-inductor.

The data obtained from the experiments with EDTA are rather contradictory. In particular, New et al. (1970) regarded the synergistic effect of EDTA and antibacterial antibiotics on the resistant strains of enterobacilli as being questionable. They found that EDTA effect was observed only at its high levels and proved to be of additive character resulting in no decrease of the resistance. It could not be ruled out that such discrepancies were associated with the diversity of EDTA effects according to the species or strain specificity and to the resistance type.

Many research workers used surface-active agents to reduce the resistance of bacterial cultures to antibiotics. Thus Elinov et al. (1966, 1968, 1969) showed that such anionic surfactants as sulphanole and alkyl-sulphate increased penicillin activity in relation to resistant staphylococci 1,000 times both in case of pretreatment of the cells and of simultaneous use with the antibiotic. They also demonstrated a considerable irreversible inhibition of the production of extracellular penicillinase by surface-active agents.

Sonstein et al. (1972) also noted the predominant efficiency of anionic surfactants, in particular of dodecyl-sulphate, in the plasmid elimination of staphylococci. In some cases the elimination reached 100% and proved to be maximal with the compounds having the longest carbon chain.

A group of Japanese workers (Tomoeda et al., 1968, Inozuka et al., 1969) found a complete or partial loss of the resistance in *E. coli* in the presence of dodecyl-sulphate, as well as of the ability for R-factor transfer in conjugation.

We observed a comparatively small (twofold) increase in the sensitivity of *E. coli* to chloramphenicol, tetracycline or streptomycin in the presence of sub-bacteriostatic levels of dodecyl-sulphate, though the same amounts were noted to increase 9–15 times the inhibition of transfer of the resistance to chloramphenicol (Tereshin et al., in press).

Tomoeda et al. (1968) assumed the resistance transfer to be inhibited due to the lysis of the pili, i.e. the most vulnerable structure of the cell wall.

The selective effect of dodecyl-sulphate noted in the case of cells containing R-factors (Inozuka et al., 1969) could be explained by the possible association of the resistance locus with the DNA site ensuring sensitivity to the detergent. However, it could not be ruled out that the interaction of R-factors with the membrane might itself result in conformational changes of some membranous sites, thus making the cell susceptible to the detergent effect. The disturbance of the R-factor interaction with the cell membrane due to the detergent effect might in such a case result both in an elimination of the R-factor and in the acquirement of resistance to dodecyl-sulphate by the membrane. But this phenomenon was not universal, since Sonstein et al. (1972) did not observe any selective effect of dodecyl-sulphate on the resistant staphylococci.

A pronounced decrease in resistance was found in the presence of another anionic detergent, i.e. sodium desoxycholeate (Tereshin et al., in press; Glatman et al., 1969), which exhibited a 12–20 times higher inhibition of the resistance transfer in conjugation.

Dimethyl-sulfoxide, a fairly low-toxic compound with bactericidal effect (Basch et al., 1968), was noted to increase the sensitivity of resistant Salmonella, Staphylococci and Streptococci by 14–36% and to inhibit 1.5–4 times stronger the transfer of R-factors (Tereshin et al., in press; Szydlovska, 1972). In its effect, dimethyl-sulfoxide is similar to surface-active agents and can substantially change the permeability of the cell walls (Chajar et al., 1968).

Present data (Jacob et al., 1965) make it possible to suggest that the effects of the above mentioned compounds are connected not only with their influence on the penetration of the antibiotic into the cell, but also with their ability to injure the replicative complex of bacteria and the interaction of genetic material with the membranous cell structures, which results in elimination of the resistance.

Since the resistance could be displayed at the level of nearly all the main cell functions, it seemed advisable, for the purpose of its control, to use such polyfunctional compounds as basic nuclear proteins, i.e. protamines and histones. These substances are characterized by rather low molecular weight (5,000–20,000) and a high content of amino acids having basic side-chains. At a neutral pH-value both histones and protamines were found to have a high positive charge which ensured a strong electrostatic interaction with negatively charged nucleic acids (Shin et al., 1970).

Basic proteins are known to be able to interact with RNA-polymerase (Spelsberg, 1969) and polysomes (Jachymczuk et al., 1972). Histone F proved to be a part of ribosome (Gurley, 1970).

Certain data prove the effect of basis proteins on the permeability of the cells (Ryser et al., 1965). Quinton et al. (1973) showed that this effect results from the interaction of polyactions, such as protamine-sulphate, arginine histone or poly-L-lysine, with negatively charged sites of the plasmatic membrane, which brings about appreciable conformational changes of the membranous structure.

Basic proteins were shown to possess an antibacterial effect (Ermolieva, 1968) associated — as demonstrated by Ashmarin et al. (1972) — mainly with changes in the permeability of cells in microorganisms.

Some protamines were found to have an antitumour efficiency and to inhibit the malignant growth by 30–70% (Pasternack et al., 1971).

There is some evidence of synergism of basic proteins and antibiotics at the level of macroorganism.

We studied the effects of basic proteins in combination with antibiotics with broad spectrum of action (Cm, Tc, Sm, Er) on the sensitive and resistant cultures of *E. coli* and *St. aureus* (Elgart, 1972; Levchenko et al., in press). Protamine hydrochloride (kindly supplied by "Spofa"), which does not exhibit any effect on the growth of bacterial cultures at the amounts of 25 to 50 mcg/ml, was found to reduce 2.5–10.0 times the minimum bacteriostatic concentration of chloramphenicol for the strains having a high (i.e. up to 1,000 mcg/ml) level of episomic resistance, and to inhibit completely the resistance in the *E. coli* strain with chromosomic resistance. The minimum bacteriostatic concentrations of tetracycline and erythromycin for *E. coli* were noted to decrease 2–4 times under the effect of protamine hydrochloride, and the resistance of styphylococci to chloramphenicol and streptomycin was inhibited 4–10 times.

Ecmolin, i.e. triprotamine containing histidine besides arginine and lysine, was shown to possess a similar effect, whereas protamine-sulphate and fractions of the thymus histones proved to be less effective. It was observed that histones appreciably (i.e. 100–200 times) inhibited the transfer of resistance in conjugation, the F_3 fraction being the most active.

Protamine hydrochloride at sub-bacteriostatic concentrations decreased the production of resistant recombinants 2–5 times. Higher amounts inhibited the process 50 times more intensively.

We did not observe any selective effect of basic proteins on the R-factor carrying strains. Under the conditions used, protamines had no influence on the resistance rate of the pretreated bacterial cultures or on the origination of the process of resistance. However, Bass (1973) showed ecmolin to inhibit 4–12 times the development of resistance to erythromycin, penicillin and tetracycline in staphylococci. In the case of streptomycin no effect was observed.

The effects of basic proteins on the resistance, as well as their bactericidal effects (Ashmarin et al., 1972) proved to be rather variable in dependence on the experimental conditions.

In order to investigate the character of the effects of basic proteins on the resistance of bacteria we examined the effects of these compounds on certain mechanisms of the cell resistance to antibiotics.

The first series of the experiments (Belousova et al., 1973) was carried out on *E. coli* strains sensitive or resistant to tetracycline and differing in their antibiotic sorption (by 4–8 times). Protamine hydrochloride added to the bacterial culture at 50 mcg/ml and at higher amounts was found to increase the antibiotic binding by the resistant strains by 60–80% and to have no effect on the sorption in sensitive organisms.

Preliminary growing of the resistant cultures in the presence of protamine resulted in enhancing the subsequent sorption in the protamine-free medium by only 15–20%.

No protamine effect was noted on the marked decrease of the antibiotic binding induced by low tetracycline concentrations, both in resistant and sensitive cultures of *E. coli*.

The second series of experiments was performed using *E. coli* and *St. aureus* resistant to chloramphenicol and producing chloramphenicolase (Solovieva et al., 1974). In the presence of protamine hydrochloride or ecmolin, the antibiotic inactivation was inhibited by 30–50% both in the induced and constitutionally produced enzymes. Fractions of the thymus histones proved to be little effective.

The third series of experiments was carried out with the ribosomes of *E. coli* sensitive or resistant to chloramphenicol (Belousova et al., 1973a, 1973b, 1974). No differences were found in the ribosomes of various strains as regards their ability for antibiotic binding, irrespective of the resistance rate or its nature.

The average rate of antibiotic binding in a cell-free system was 0.75 moleM of C^{14}-Cm per 0.1 mole of the ribosomes. In the presence of protamine hydrochloride at an amount of 25–50 mcg per 1.0 mg of ribosomes, antibiotic binding increased up to 0.99 mole per 1.0 mole of the ribosomes thus becoming nearly equimolar. Such an effect was observed with ribosomes of both the sensitive and resistant E. coli strains.

Equimolar binding was also found in the absence of protamine in case of a sufficiently high chloramphenicol concentration in the system (20 moles of Cm per 1.0 mole of the ribosomes), whereas in the presence of protamine such binding was observed at 4.5 moles of the antibiotic in the incubation mixture. Calculations showed that protamine was able to enhance the tightness of chloramphenicol binding to ribosomes (the binding constant was doubled — from $1.5 \,.\, 10^6 M^{-1}$ to $3.2 \,.\, 10^6 M^{-1}$).

This result was supported by the effect of protamine on competitive relations between chloramphenicol and erythromycin in their interactions with ribosomes. Under usual conditions, the presence of erythromycin in the system resulted in a drastic decrease of chloramphenicol sorption, while in the absence of protamine it was only a little smaller as compared with the control.

The effect of protamine on specific chloramphenicol binding ensuring its antibacterial action was also confirmed by the results obtained with a cell-free system for polyphenyl alanine synthesis by the native and protamine hydrochloride-treated ribosomes of *E. coli*. Chloramphenicol at the concentration of 10^{-4}M was shown to inhibit the *in vitro* synthesis by 22.0% in the control, in case of the protamine-treated ribosomes the inhibition amounted on the average to 72%.

Chloramphenicol resistance in the majority of *E. coli* strains that associated either with the decrease of cell permeability to the antibiotic, or with its inactivation in the culture, resulted in an appreciable reduction of its intracellular content. Accordingly, the ability of protamine-hydrochloride to increase the effect of low concentrations of chlor amphenicol on the ribosomes might play an essential part in the decrease of *E. coli* resistance.

The effect of protamine was noted to be specific in respect of 70s ribosomes, and no influence was observed on chloramphenicol interaction with 80s ribosomes of *Candila albicans*.

The effect of protamine-hydrochloride was apparently determined by its interaction with the ribosomal proteins but not with ribosomal RNA, since at concentrations sufficient for its maximum effect on the binding, protamine did not cause any changes in the melting temperature of ribosomal RNA or in their sensitivity to RNAse. At the same time, protamine was found to completely restore the ability of ribosomes for chloramphenicol binding that had been reduced as a result of trypsin treatment.

The diversity of the effects of basic proteins on cell resistance appears to make studies on these compounds rather promising.

Numerous experiments showed that, as a rule, substances effective in respect of particular species, strains or resistance markers proved to be little effective for other cultures or markers, even very similar in their origin, and present in the same episome.

This might depend on a number of factors such as different cell permeability of various bacterial species and strains to the investigated compounds, the localization of the resistance, the number of episome replicas in the cell, interaction of the genetic cell elements with the membrane. The indicated diversity of effects could be also ensured *in vivo* by the localization of microorganisms in the organs and tissues of the body, as well as by their accessibility to the investigated substance.

All the foregoing together with the possible toxicity of some compounds makes it difficult to control the resistance with the use of biologically active agents. Nevertheless, it is hardly advisable to discontinue further attempts to search for effective resistance inhibitors or attempts to introduce them into practice, since any progress in this direction might be of essential importance for antibiotic therapy.

REFERENCES

ANTSYFEROVA, N. G., BRODINOVA, N. S., MOROZ, A. F. (1973). Ethidium bromide inhibition of the transductional transfer of *E. coli* R-factor by the P1Kc phage. Antibiotiki, 10, 898.

ASHMARIN, I. P. et al. (1972). Antibacterial and antiviral functions of the cellular basic proteins and the prospects of their practical use. Izvestiya Akademii Nauk SSSR, 4, 502.

ASTAPOV, A. A. (1970). Acridine inhibition of the transfer of RTF factor. Antibiotiki, 1, 46.

AYAD, S. R. (1969). Inhibition of *B. subtilis* transforming system by acriflavine. FEBS Letters, 2, 5, 348.

BASCH, H., GADEBUSCH, H. H. (1968). *In vitro* antimicrobial activity of dimethylsulfoxide. Appl. Microbiol. 16, 12, 1963.

BASS, T. M. (1973). Ecmolin effect on the development of staphylococcal resistance to some antibiotics. Antibiotiki, 4, 350.

BELOUSOVA, I. I., ELGART, R. E., TERESHIN, I. M. (1973). The decrease of *E. coli* resistance to tetracycline under protamine effect *in vitro*. Antibiotiki, 1, 68.

BELOUSOVA, I. I., LISHNEVSKAYA, E. B., TERESHIN, I. M. (1973). The interaction of *E. coli* ribosomes with chloramphenicol. The effect of protamine hydrochloride on the tightness of association of *E. coli* ribosomes with chloramphenicol and on the antibiotic inhibition of polypeptide synthesis in a cell-free system. Antibiotiki, 5, 411.

BELOUSOVA, I. I., LISHNEVSKAYA, E. B., TERESHIN, I. M. (1973). The effects of erythromycin and its combination with protamine hydrochloride on the binding of C^{14}-chloramphenicol with *E. coli* ribosomes. Antibiotiki, 2, 129.

BELOUSOVA, I. I., LISHNEVSKAYA, E. B., LEVCHENKO, A. B., TERESHIN, I. M. (1974). The interaction of *E. coli* ribosomes with chloramphenicol. The effects of basic proteins on some properties of ribosomes and their ability for chloramphenicol binding. Antibiotiki, 3, 216.

BOUANCHAUD, D. H., CHABBERT, V. A. (1971). Practical effectiveness of agents curing R factors and plasmids. Annals N.Y. Acad. Sci., 182, 305.

BONAUCHAUD, D. H., SCAVIZZI, M. R., CHABBERT, V. A. (1969). Elimination by ethidium bromide of antibiotic resistance in *Enterobacteriaceae* and *Staphylococcus*. J. Gen. Microbiol., 54, 417.

BRODINOVA, N. S., MOROZ, A. F. (1969). Elimination of R factor in *E. coli* under the effects of acridinic dyes and ultraviolet rays. Antibiotiki, 6, 535.

CHAJAR, B. M., HARMON, S. A. (1968). The effect of dimethylsulfoxide on permeability of *Staph. aureus*. Bioch. Bioph. Res. Comm., 32, 6, 940.

ELINOV, N. P., STROYEV, S. S. (1966). Studies on the effects of penicillinase and surface-active agents. I. The inhibition of endo- and exopenicillinase by surface-active agents. Antibiotiki, 4, 358.

ELINOV, N. P., STROYEV, S. S. (1968). The effects of surface active agents on penicillin resistance in *Staphylococcus in vitro* and *in vivo*. Antibiotiki, 7, 637.

ELINOV, N. P., AFINOGENOV, G. E. (1969). The effects of sulphanole and alkyl sulphate on penicillin resistant staphylococci from bone wound. Trudy Leningradskogo Chimico-pharmacevticheskogo instituta, Issue 27, 64.

ELGART, R. E. (1972). The effects of various surface active agents and protamine on the antibiotic resistance in bacteria. Materialy V. Konferenzii molodych uchenych Leningradskogo NII antibiotikov (Trudy LNIIA, Issue 9), 127.

EPHRUSSI, B., HOTTINGER, H., CHIMENES, A. M. (1949). Action de l'acriflavine sur les levures. I. La mutation "petite colonie". Ann. Inst. Pasteur, 76, 351.

ERMOLIEVA, Z. V. (1968). Antibiotics, interferon, bacterial polysaccharides. Moscow, "Medizina", 181.

GLATMAN, L. I., MOROZ, A. F., WAINTRAUB, E. A. (1969). The inhibition of the transfer of multiple resistance factor in *E. coli* by various chemical compounds. In "Lekarstvennaya ustoychivost mikroorganismov", Erevan, 45.

GLATMAN, L. I., MOROZ, A. F., GROMOVA, T. G. (1973). The effect of ethidium bromide on R-factor transfer in *E. coli* at conjugation. Antibiotiki, I, 52.
GROSS, R., HOLZMAN, R. G. (1970). Effect of EDTA on penicillinase inductibility in *Staph. aureus*. Med. J. Austral., I, 10, 482.
GURLEY, I. R. (1970). Isolation and characterization of histone F′ in ribosomes. Bioch. Bioph. Res. Comm., 40, 2, 428.
HAHN, F. E., CHIAK, I. (1971). Elimination of bacterial episomes by DNA-complexing compounds. Annals N.Y. Acad. Sci., 185, 295.
HAMILTON-MILLER, I. M. T. (1965). Effect of EDTA upon bacterial permeability to benzylpenicillin. Bioch. Bioph. Res. Comm. 20, 6, 688.
HIROTA, V., IIJIMA, T. (1957). Acriflavine as an effetive agent for the elimination of F-factor in *E. coli*. Nature, 180, 655.
HADSON, B., VINOGRAD, I. (1967). Catenated circular DNA molecules in HeLa cell mitochondria. Nature, 216, 647.
INOZUKA, N. et al. (1969). Specific action of sodium dodecyl sulfate on the sex factor of *E. coli* K-12 Hfr strains. J. Bact., 100, 2, 827.
JACHYMCZUK, W. I., SICLIWANOWICZ, B., ZALESKI, I. (1972). The influence of basic proteins on polysomes. The interaction of histones with rat liver polysomes. Acta Biochem. Polon., 19, 2, 117.
JACOB, F., RYTER, A., CUSIN, F. (1965). On the association between DNA and membrane in bacteria. Proc. Roy. Soc. (London), Ser. B, 164, 267.
YOSHIKAWA, M., SEVAG, M. G. (1967). Sensitivity of *E. coli* to atabrine conferred by R-factor and its potential clinical significance. J. Bact., 93, 245.
LEVIE, L. (1968). Studies on the permeability change produced in coliform bacteria by EDTA. J. Biol. Chem., 243, 9, 2373.
LEVCHENKO, A. B. et al. (in press). Effects of methyl uracil, pentoxyl, lysozyme and EDTA salts on the resistance of *E. coli* and *St. aureus* to chloramphenicol, streptomycin, tetracycline and erythromycin. Antibiotiki.
LEVCHENKO, A. B. et al. (in press). Effects of protamine, ecmolin and histones on the resistance of *E. coli* and *St. aureus* to chloramphenicol, tetracycline, streptomycin and erythromycin. Antibiotiki.
LEVY, S. B., WATANABE, T. (1966). Mepacrine and transfer of R-factors. Lancet, 2, 1138.
MILLER, B. F. R. (1942). Antibacterial properties of protamine and histone. Science, 96, 428.
MITSUHASHI, S. et al. (1963). Elimination of drug resistance of *Staph. aureus* by treatment with acriflavine. J. Bact., 86, 162.
MOROZ, A. F., BRODINOVA, I. S., PODBORNOV, V. M. (1971). Prevention of the development of drug resistant forms in staphylococci by using antibiotic combinations with DNA-tropic substances. Antibiotiki, 5, 405.
MOROZ, A. F., GROMOVA, T. G., GLATMAN, L. I. (1972). Elimination of drug resistance in *E. coli* and *Shigella flexneri* by ethidium bromide. Antibiotiki, 12, 1085.
MUSCHEL, L. H., GUSTAFSON, L. (1968). Antibiotic, detergent and complement sensitivity of *Salmonella typhi* after EDTA acid treatment. J. Bact., 95, 6, 2010.
NEW, H. C., WINSHELL, E. B. (1970). Lack of synergy of EDTA with antimicrobials in resistant *Enterobacteriaceae*. Nature, 225, 5234, 763.
PASTERNACK, N. A. et al. (1971). Antitumoral activity and chemical structure of protamines of various origin. Antibiotiki, 9, 816.
PODBORNOV, V. M. et al. (1969). The effect of benzylpenicillin combination with acrichin on the development of drug resistant forms in *Staphylococcus* and on penicillinase production. In "Lekarstvennaya ustoychivost mikroorganismov", Erevan, 105.
PODBORNOV, V. M. (1972). Prevention of the development of drug resistant forms in *E. coli* by the combinations of antibiotics with DNA-tropic agents. Antibiotiki, 6, 539.
QUINTON, M., PHILPOTT, C. W. (1973). A role for anionic sites in epithelial architecture. Effect of cation polymers on cell membrane structure. J. Cell Biol., 56, 3, 787.
RAWAL, B. D. (1971). EDTA-induced variations in *St. aureus* suppression of growth and antibiotic resistance. Canad. J. Pharm. Sci., 6, 1, 22.
RYSER, H. I. P., HANCOCK, P. (1965). Histones and basic polyaminoacids stimulate the uptake of albumin by tumor cells in culture. Science, 150, 3695.
SEVAG, M. G. (1964). Prevention of the emergence of antibiotic-resistant strains of bacteria by atabrine. Arch. Biochem. Biophys., 108, 85.
SEVAG, M. G., J. DE CURSI (1970). Biochemical process providing the basis of drug resistance in microorganisms, and biochemical ways for its prevention. In "Funktsyonalnaya biochimiya kletochnych struktur", Publishing House "Nauka", 369.

SHIN, T. J., BONNER, J. (1970). Template properties of DNA-polypeptide complexes. J. Molec. Biol., 50, 2, 333.
SOLOVIEV, V. N., GALANOV, R. Ya. (1968). Acridine effect on the development of antibiotic resistance. Antibiotiki, 3, 257.
SOLOVIEV, V. N. (1969). The All-Union Symposium on Clinical Application of Antibiotics. Antibiotiki, 6, 571.
SOLOVIEVA, N. N., TERESHIN, I. M., BELOUSOVA, I. I. (1974). Studies on chloramphencolase of *E. coli* and *St. aureus*. Antibiotiki, 4, 327.
SONSTEIN, S. A., BALDWIN, I. N. (1972). Nature of elimination of the penicillinase plasmid from *Staph. aureus* by surface active agents. J. Bact., 109, 1, 262.
SPELSBERG, T. C. (1969). Inhibition of RNA-polymerase acting by arginin-rich histones. Experientia, 25, 2, 129.
SPIRIN, A. S., GAVRILOVA, L. P. (1971). The ribosome. Publishing House "Nauka", 47.
SUWALSKY, M. et al. (1969). An X-ray study of the interaction of DNA with spermine. J. Molec. Biol., 42, 2, 363.
SZYDLOWSKA, T. (1972). Studies on the role of dimethyl-sulfoxide in resensibilization of antibioticoresistant bacterial strains. Arch. immunol. et ther. exps., 20, 2, 193.
TERESHIN, I. M. et al. (in press). Effects of spermine, spermidine, dimethyl-sulfoxide, dodecylsulphate and desoxycholeate on the resistance of bacteria to some antibiotics. Antibiotiki.
TOMOEDA, M. et al. (1968). Effective elimination of drug resistance and sex factors in *E. coli* by sodium dodecylsulfate. J. Bact., 95, 3, 1078.
VARTANYAN, Zh. S. (1973). Studies on the eliminative effect of rifampicin and acriflavine on the determinants of antibiotic resistance in clinical strains of Staphylococcus. Antibiotiki, II, 994.
WARREN, G. H. et al. (1957). Effect of mixtures of atabrine and antibacterial agents on the emergence of resistant strains of *Mycobacteria tuberculosis*. Nature, 215, 5100, 526.
WARING, M. (1970). Variation of the supercoils in closed circular DNA by binding of antibiotics and drugs — evidence for molecular model involving interaction. J. Molec. Biol., 54, 247.
WARING, H. I. (1966). Cross linking and interaction in nucleic acids. Symp. Soc. Gen. Microbiol., 16, 235.
WEISER, R., ASSCHER, A. W., WIMPENNY, I. (1968). *In vitro* reversal of antibiotic resistance by ADTA. Nature, 219, 5161, 1365.
WEISER, R., WIMPENNY, I., ASSCHER, A. W. (1968). Synergistic effect of EDTA (antibiotic combination on *Pseudomonas aeruginosa*. Lancet II, 7621, 619.
ZUEVA, V. S., SOLOVIEV, V. N. (1969). Effects of acridinic compounds on the development of drug resistance to antitubercular agents in mycobacteria of bird type. Antibiotiki, 4, 355.
ZUEVA, V. S., LINEVICH, Yu. G. (1969). Disturbances of the transfer of extrachromosomal factors of resistance in staphylococcal populations *in vitro* and *in vivo*. In "Lekarstvennaya ustoychivost mikroorganismov", Erevan, 56.

PLASMID-MEDIATED TETRACYCLINE AND ERYTHROMYCIN RESISTANCE IN GROUP D "STREPTOCOCCUS":

IDENTIFICATION OF TWO PLASMIDS

P. M. COURVALIN, C. CARLIER, O. CROISSANT, and D. BLANGY
Institut Pasteur, Paris, France

ABSTRACT

Two circular DNA plasmids designated *Tet* and *Ero*, distinguishable by their molecular weights (64.5 and 17.6 × 10^6 respectively) and their guanine plus cytosine (G + C) content (36.5 and 34.5% respectively) were identified in *Streptococcus faecalis* strain KR, resistant to tetracycline and erythromycin. Loss of resistance to tetracycline, to erythromycin and to both antibiotics simultaneously in independently isolated "cured" derivatives correspond to the loss of the *Tet*, *Ero*, and of both plasmids respectively. In the singly resistant strains we estimated that 3 and 6 copies of the *Tet* and *Ero* plasmids respectively are present per chromosomal genome equivalent.

ERYTHROMYCIN AND RESISTANCE TRANSFER FACTORS IN THE INTESTINAL TRACT*

J. C. GOULD
Central Microbiological Laboratories, Western General Hospital, Edinburgh, G.B.

Erythromycin was introduced more than twenty years ago (McGuire et al., 1952) and initially most used in the treatment of staphylococcal infection (Forfar et al., 1955) since it was active against *Staphylococcus pyogenes* many strains of which were resistant to penicillin, the tetracyclines and streptomycin, the other then readily available antibiotics. Several reports indicated that resistance to erythromycin could develop rapidly in staphylococci, or appear after prolonged use (Hobson, 1954, Lepper et al., 1953, Maccabe and Gould, 1956, Maccabe et al., 1961 and Forfar et al. 1966) and this resistance was probably due to the transduction of plasmids transferring resistance from an originally small proportion of resistant strains (Lacey, 1973). The use of the antibiotic resulted in selection and spread of these resistant strains until there was a general restriction in the use of erythromycin (Lancet, 1956). Thus erythromycin-resistance is now relatively rare among staphylococci and the main indication for the antibiotic is as an alternative to penicillin in the treatment of streptococcal infections.

With the availability of a much wider range of antistaphylococcal drugs there is now less need to be concerned with erythromycin resistance and consideration can be given to the more frequent use of erythromycin for infections due to sensitive bacteria. Such infections include respiratory tract and soft tissue lesions due to streptococci, pneumococci and *Haemophilus species*, and anaerobic infections due to bacteroides.

At present the tetracyclines, sulphonamides, ampicillin and the cephalosporins are widely used in the treatment of these infections but are associated with the production of resistance transfer factors (R-factors) among the Gram-negative coliform bacilli: Thus the bowel flora of treated patients will be likely to show an increase in R-factor production and contain strains of coliform bacilli resistant to multiple antibiotics. Antibiotic-resistant strains of *Escherichia coli* and other resistant Gram-negative bacilli are frequently isolated from the faeces of patients many of whom receive these antibiotics, and it is believed that infections, particularly of the urinary tract, can result from these bacteria.

As infections requiring treatment with antibacterial drugs are common both in the general community and in hospital the continued use of antibiotics encouraging R-factor production would be expected to progressively increase the incidence of antibiotic resistance among Gram-negative bacilli. However a number of antibacterial agents such as erythromycin, the lincomycins and nalidixic acid have not yet been described as being associated with R-factor production and therefore their use may be less liable to result in the selection and spread of resistant coliform bacilli.

* This work was supported by Abbot Laboratories, Queenborough, Kent, with supplies of antibiotics and generous research assistance.

METHODS

In the present study the effect of erythromycin on the faecal flora has been examined in patients who were admitted to hospital for investigation of the intestinal tract and bowel surgery. The results obtained in these patients have been compared with a similar though larger series receiving a combination of ampicillin and neomycin for pre-operative bowel preparation.

Erythromycin, as the ethyl succinate or stearate was given in doses varying from 2 to 7 days. Samples of faeces were obtained on at least one occasion before treatment and further specimens were examined 1–2 weeks after the finish of treatment. In some patients intestinal contents obtained at operation were available for bacteriological examination. The patients receiving ampicillin-neomycin mixtures were similarly examined but only 25 such cases had their flora examined for R-factor in parallel with the patients receiving

TABLE 1

Frequency of Detected R-factor Among Antibiotic Resistant Isolates

	Percentage of isolates donating resistance to					
	Erythromycin	Ampicillin	Tetracycline	Chloramphenicol	Sulphonamide	Streptomycin
Patients before treatment	0	22	24	6	21	16
Following neomycin plus ampicillin	0*	25	30	10	25	20
Following erythromycin	0	20	28	12	18	14

* from 25 patients.

TABLE 2

Nature of the Faecal Flora Isolated from Patients

	Mean of Viable Counts per gm of Faeces				
	Number of Patients	*Esch. coli* Antibiotic sensitive	*Esch. coli* Antibiotic resistant	*Klebsiella Proteus Pseudomonas*	Anaerobic flora
On first hospital admission	210	4×10^7	3×10^4	7×10^2	10^{10}
After antibiotic treatment in hospital	156	3×10^7	8×10^5	10^5	10^{10}
After treatment with erythromycin 5–7 days	34	6×10^7	7×10^5	3×10^5	10^{10}
After erythromycin 14 days	8	10^7	5×10^6	3×10^6	10^{10}

erythromycin (Table 1). A limited number of patients received erythromycin for 14 days and are detailed in Table 2. Both aerobic and anaerobic culture of the faeces or intestinal contents was carried out using a quantitative method (Gunn and Gould, 1965, Buchan and Gould, 1967). The main groups of bacteria were identified (Table 3) and sensitivity tests to various antibiotics including erythromycin, ampicillin, streptomycin, chloramphenicol and tetracycline were performed (Gould and Bowie 1952). Subcultures of coliform bacilli were identified by biochemical analysis and *Esch. coli* further by serotyping. Antibiotic-resistant isolates were examined for R-factor by their ability to induce resistance in suitable recipient organisms by the method described by Jonsson et al. (1972).

TABLE 3

Sensitivity of Intestinal Flora to Erythromycin (μg. per ml.)

Anaerobic bacteria Bacteroides & Lactobacilli	Clostridia	Streptococci	*Esch. coli*
0.1–10 μg/ml.	0.5–10	0.25–5.0	7.5–25

Sensitivity of Respiratory tract flora to erythromycin (μg. per ml.)

Streptococcus pyogenes	Pneumococcus	*Haemophilus influenzae*	*Staphylococcus pyogenes*
0.1–0.5	0.25–1.0	1.0–5.0	0.25–12.5

RESULTS

Erythromycin had an inhibitory effect on the majority of strains of *Esch. coli* isolated at concentrations varying from 7.5 to 25 μg/ml. (Table 4); only 15 per cent of the strains were more resistant but none to greater than 100 μg per ml. In contrast the strains of *Proteus*, *Klebsiella* and *Pseudomonas* isolated were resistant to more than 250 μg per ml. However, these latter organisms were less frequently isolated and their numbers were always relatively fewer in the faeces (Table 2).

TABLE 4

Effect of Antibiotic Administration of the Faecal Flora

Treatment	Number of patients	*Esch. coli*	Streptococcus	Anaerobic organisms
	logarithmic reduction in viable count per gm of faeces			
Ampicillin plus Neomycin for 5–7 days	156	6×10^6	8×10^4	8×10^7
Erythromycin 5–7 days	34	4×10^5	10^5	4×10^8

The remaining faecal organisms, *streptococci, clostridia, lactobacilli and bacteroides* were sensitive to lower concentrations (0.1 to 10 μg per ml) and it is thus not surprising that there was a considerable depressant effect on the bowel flora of patients who were receiving large doses of erythromycin (greater than 2 gm per day) and this was clearly measurable in the reduction of both the micro-aerophilic and anaerobic organisms (Table 3). Generally in the patients under review there was little evidence of colonisation or overgrowth by minority groups of Gram-negative bacilli, *staphylococci nor Candida*, although the number of such organisms did increase in some of the patients who were treated for longer than 7 days (Table 2) and this is comparable with the results when other antibiotics are used for longer periods (Gunn and Gould, 1965, Gould 1971).

R-factors were detected in a moderately high proportion (about 40%) of antibiotic-resistant isolates of *Esch. coli* but the percentages were similar both before and after antibiotic treatment with either erythromycin or neomycin plus ampicillin. The availability of R-factor may be greater following antibiotic therapy because of the higher incidence and number of antibiotic-resistant bacteria such as *Esch. coli*, *Proteus*, *Pseudomonas* and *Klebsiella*.

Erythromycin resistance was not conferred in any transfer of resistance to the sensitive recipient organisms either from antibiotic-resistant-erythromycin-sensitive or antibiotic-resistant-erythromycin-resistant strains of *Esch. coli*, nor as far as was tested from other Gram-negative bacilli which were erythromycin-resistant (Table 1).

There was fairly frequent isolation of *Esch. coli* resistant to ampicillin, streptomycin, tetracycline and sulphonamide from patients following the use of erythromycin (Tables 5 and 6) but their numbers were usually small and on many occasions such strains were serologically different from the predominant *Esch. coli* isolated before treatment.

TABLE 5

Isolation of Antibiotic-resistant Esch. Coli from 190 Patients

	Percentage of patients from whom strains resistant to				
	Ampicillin	Tetracycline	Sulphonamide	Streptomycin	Erythromycin
	were isolated				
Before treatment	8	12	11	6	12
After treatment	20	30	18	18	15

TABLE 6

Sensitivity of Esch. coli isolated from patients

	Percentage of Strains Examined Resistant to				
	25 μg/ml Ampicillin	25 μg/ml Tetracycline	10 μg/ml Chloramphenicol	10 μg/ml Streptomycin	25 μg/ml Erythromycin
Isolates obtained before treatment	8	12	0	6	15
Isolates after erythromycin treatment	20	18	5	10	25
Isolates after neomycin plus ampicillin treatment	24	34	3	18	12

DISCUSSION

A higher proportion of strains of *Esch. coli* isolated after treatment with erythromycin or ampicillin plus neomycin were antibiotic-resistant than before treatment but there was little difference between the results using the two regimes (Table 6). Similarly the frequency of detection of R-factor was little different following the two antibiotic treatments (Table 1). This would seem to indicate that the increase in resistant organisms is dependent on selection of the resistant flora following chemotherapy as well as increase in R-factor production, particularly as there was no evidence of the transfer of erythromycin resistance to sensitive recipient strains of *Esch. coli*. Further, some of the erythromycin-resistant strains were of different serotype than the original sensitive strains and it may be assumed that they were acquired from the hospital environment, or were present in an undetectable minority in the patient before treatment began. The antibiotic then selected the resistant strains.

The conclusion is that erythromycin has not in this experiment, been found associated with R-factor production which transfers erythromycin-resistance among Gram-negative bacilli. The tendency to select strains of Gram-negative bacilli resistant to other antibiotics such as ampicillin during erythromycin therapy has been found to be no greater, and possibly less than when ampicillin, neomycin, tetracycline or sulphonamide are used. A high proportion of the antibiotic-resistant strains of coliform bacilli isolated, many of them producing R-factor, seemed to be acquired from the hospital environment and selected by the antibiotic in use.

REFERENCES

BUCHAN, K. A. and GOULD, J. C. (1967). Pre-operative Bowel Sterilisation Post.-Grad med. J. Suppl. May 14–17.

FORFAR, J. O., MACCABE, A. F., BALF, C. L., WRIGHT, H. A. and GOULD, J. C. (1955). Staphylococcal infection in the new-born treated with Erythromycin. Lancet *1*: 584.

FORFAR, J. O., KEAY, A. J., MACCABE, A. F., GOULD, J. C. and BAIN, A. D. (1966). Liberal use of Antibiotics and its effect in neonatal Staphylococcal infection with particular reference to Erythromycin. Lancet: *2*, 295.

GOULD, J. C. and BOWIE, J. H. (1952). The determination of bacterial sensitivity to Antibiotics. Edinb. med. J. *59*: 178.

GOULD, J. C. (1971). Opportunists and Opportunity in Infection Bayer Symposium III, 71–76, Springer-Verlag Berlin.

GUNN, A. A. and GOULD, J. C. (1965). Control of the Intestinal flora by means of antibacterial drugs in surgery of the colon. Gut *6*: 582.

HOBSON, D. (1954). Activity of Erythromycin against *Staphylococcus pyogenes*. Brit. med. J. *1*: 236.

JONSSON, MARIANE, RUTBERG, L., TUNEVALL, G. (1972). Transferable resistance to antibiotics in Gram-negative bacteria isolated in a hospital for infectious diseases. Scand. J. Infect. Dis. *4*: 133.

LACEY, R. W. (1973). Genetic basis, Epidemiology and future significance of antibiotic resistance in *Staphylococcus aureus*. Leading Article 1956. Restriction of Antibiotics. Lancet *1*: 193.

LEPPER, M. H., MOULTON, B., DOWLING, H. F., JACKSON, G. G., KORMAN, S. (1953). Epidemiology of Erythromycin-resistant styphylococci in a hospital population. Antibiotics Annual 1953–54. New York Med. Encyclopaedia Inc. p. 308.

MACCABE, A. F. and GOULD, J. C. (1956). The Epidemiology of an Erythromycin resistant Staphylococcus.

MCGUIRE, J. M., BUNCH, R. L., ANDERSON, R. C., BOAZ, H. E., FLYNN, E. H., POWELL, N. M. and SMITH, J. W. (1952). "Ilotycin" a new antibiotic. Antibiotics and Chemotherapy *2*: 281.

RIFAMPICIN SENSITIVITY OF CLINICAL STAPHYLOCOCCAL STRAINS AND THE DEVELOPMENT OF RESISTANCE DURING RIFAMPICIN THERAPY

L. N. SAMOILOVA, L. E. BODUNKOVA, I. P. FOMINA, S. M. NAVASHIN

National Institute of Antibiotics, Moscow, U.S.S.R.

The administration of rifampicin, a new wide-spectrum semisynthetic antibiotic, to patients with staphylococcal infection is based on its high activity against staphylococci, surpassing the effect of cephalosporins, ampicillins and benzylpenicillin. Among clinical staphylococcal strains the number of those resistant to rifampicin has been found not to exceed 21%; staphylococci endowed with multiple resistance, including methicillin resistant strains, are as a rule susceptible to the new antibiotic (Bergamini, N. et al., 1970, Dans P. E. et al. 1970).

However, numerous reports describing a rapid development of resistancc by various microbial species during rifampicin therapy cast doubt on the usefulness of this antibiotic for treating infections (Atlas ann Turck, 1968).

The purpose of the present study was to evaluate the rifampicin sensitivity of clinical staphylococcal strains and to investigate the development of resistance during rifampicin therapy of chronic staphylococcal infection.

MATERIALS AND METHODS

90 clinical strains of staphylococci from surgical patients or patients with chronic inflammatory lung disease were tested for sensitivity to rifampicin, as compared to benzylpenicillin, methicillin, tetracycline, chloramphenicol erythromycin and streptomycin, by the method of serial two-fold dilutions in meat-peptone broth (MPB) or 1.5% meat-peptone agar (MPA), pH 7.2. For sensitivity tests in MPB 4 different inocula were used: 10^3, 10^5, 10^7 and 10^8 organisms/ml. The minimal inhibitory and the minimal bactericidal concentrations of rifampicin were determined after 24 hrs and 72 hrs exposures of the test cultures to the antibiotic at 37°C.

The chemotherapeutic effect of the antibiotic was studied on white mice with chronic staphylococcal septico-pyemia induced by an intravenous injection of *Staphylococcus aureus*, strain No. 91, which was sensitive to rifampicin (MIC 0.05 mcg/ml). 5×10^8 organisms were used per animal.

Measurements were made of the doses ensuring 50% and 100% survival (CD_{50} and CD_{100}), and also of the doses required for complete sterilization of the infected animals.

During rifampicin therapy the development of resistance was tested by inoculating minced kidney tissue into a selective medium which contained 1–10 mcg/ml of the antibiotic. The mutants obtained were examined in respect to virulance and the time course of changes in sensitivity.

RESULTS AND DISCUSSION

None of the clinical staphylococcal strains were resistant to rifampicin: MIC values ranged from 0.002 to 0.02 mcg/ml. On the other hand, 94.4% of the strains were resistant to benzylpenicillin, 93.3% to erythromycin, 94.4% to chloramphenicol and tetracycline, and 53.3% to methicillin (Table 1 and Fig. 1). Most staphylococci showed multiple resistance: 79% of the strains had determinants of resistance to 5–8 antibiotics, and only 21% to 3–4 drugs (Table 2).

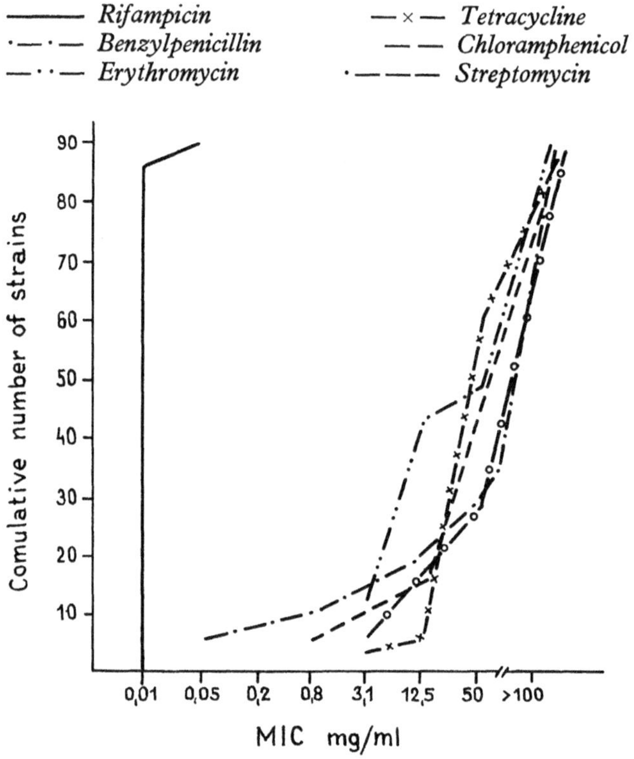

Fig. 1

TABLE 1

90 clinical staphylococcal strains distributed according to their sensitivity to various antibiotics

Antibiotic	MIC (mcg/ml)								Percentage of resistant strains
	>100–100	50–25	12.5–6.2	3.1–1.6	0.8–0.4	0.2–0.1	0.05–0.02	0.01–0.002	
Benzylpenicillin	65	10	5	5	5	—	5	—	94.4
Methicillin	40	8	30	12	—	—	—	—	53.3
Erythromycin	60	15	9	—	—	5	—	—	93.3
Chloramphenicol	30	55	2	3	—	—	—	—	94.4
Tetracycline	55	20	10	5	—	—	—	—	94.4
Streptomycin	58	14	8	6	4	—	—	—	80
Rifampicin	—	—	—	—	—	—	4	86	0

TABLE 2

Resistance structure of 90 clinical staphylococcal strains

No. of determinants	Determinants of resistance	No. of strains	Percentage of strains
3	PC TC CM; PC TC SM	4	4.4
4	PC TC EM CM; PC TC SM CM; TC SM EM CM	15	16.7
5	PC TC M EM CM; TC SM KM EM CM; PC TC SM EM CM	24	26.7
6	PC M TC SM EM CM; PC M TC SM KM EM PC TC SM KM EM CM;	43	47.8
7	PC M TC SM LM EM CM	2	2.2
8	PC M TC SM KM LM EM CM	2	2.2

Rifampicin MIC values were markedly dependent on the size of the inoculum and the period of incubation in the presence of the antibiotic. Although MIC remained practically unaffected when the inoculum was increased from 10^3 to 10^7 organisms/ml, it sharply rose (1000-fold and more) when 10^8 organisms/ml were inoculated. With a large inoculum (10^7–10^8 organisms/ml) MIC values were 4- to 10-fold higher after 72 hrs exposure of the culture to rifampicin than after 24 hrs incubation (Table 3).

TABLE 3

Rifampicin MIC values related to the size of the inoculum and the incubation period

Strain No.	Inoculum	Incubation period (hrs) 24	48	72
Staph. 209P		0.005	0.01	0.05
Staph. 58		0.005	0.01	0.01
Staph. 62	10^3	0.005	0.01	0.02
Staph. 67		0.01	0.02	0.02
Staph. 65		0.05	0.01	0.02
Staph. 209P		0.005	0.01	0.1
Staph. 58		0.01	0.01	0.02
Staph. 62	10^5	0.05	0.05	0.02
Staph. 67		0.05	0.05	0.1
Staph. 65		0.01	0.02	0.02
Staph. 209P		0.01	0.02	0.02
Staph. 58		0.01	0.02	0.02
Staph. 62	10^7	0.01	0.02	0.2
Staph. 67		0.05	0.05	0.1
Staph. 65		0.02	0.05	0.05
Staph. 209P		12.5	12.5	25
Staph. 58		12.5	12.5	12.5
Staph. 62	10^8	50	50	50
Staph. 67		6.2	12.5	12.5
Staph. 65		50	100	100

The above data suggest that staphylococcal populations are heterogenous in their susceptibility to rifampicin. This conclusion is indirectly supported by the fact that bacterial growth was observed in tubes which contained rifampicin in concentrations exceeding many times MIC for the given strain (skip tubes phenomenon). Most strains showed this phenomenon during MIC measurements, and the variants isolated were highly resistant to the antibiotic. Bacterial populations were heterogenous in respect to rifampicin sensitivity in 40 of the 90 strains, and after 18 hrs incubation in MPA with increasing rifampicin concentrations resistant mutants emerged with an average incidence of 1.7×10^8.

Non-susceptible mutants could be isolated from animal tissues as early as on the 4th or 5th day of rifampicin treatment of chronic staphylococcal infection in white mice, both with CD_{50} and CD_{100}. The rifampicin dose required for complete sterilization of the animals usually exceeded CD_{100} 6 to 8 times. For the isolated mutants of *Staphylococcus aureus* No. 91, MIC rose up to 10,000 mcg/ml, and at the same time their virulence for the animals fell 2 to 4 times.

Thus the rapid development of resistance during rifampicin therapy of staphylococcal infection seems to be caused by heterogenous sensitivity and by the frequent emergence of rifampicin resistant mutants in initially susceptible populations of clinical staphylococcal strains.

REFERENCES

BERGAMINI, N. et al. (1970). Arzneimittel Forsch., 20, 10, 1546.
DANS, P. E. et al. (1970). Am. J. Med. Sci., 259, 120.
ATLAS, E., TURCK, M. (1968). Am. J. Med. Sci., 256, 247.
McCABE, W. R., LORIAN, V. (1968). Am. J. Med. Sci., 256, 255.

IV.

ANTIBIOTIC RESISTANCE IN P. AERUGINOSA

CHAIRMAN: A. HOLLOWAY (AUSTRALIA)

OPENING LECTURE

GENETIC BASIS OF DRUG RESISTANCE IN *PSEUDOMONAS AERUGINOSA*

B. W. HOLLOWAY, P. CHANDLER, V. KRISHNAPILLAI,
B. MILLS, H. ROSSITER, V. STANISICH, J. WATSON
Department of Genetics, Monash University
Clayton, Victoria, Australia

Research in the authors' laboratory is supported by the Australian Research Grants Committee

Over the last ten years the importance of *Pseudomonas aeruginosa* in nosocomial infections has increased markedly. Two factors undoubtedly have contributed to this development. The first is the pathogenicity of *P. aeruginosa* under certain conditions particularly where natural body resistance is lowered, as in terminal carcinoma or burns. As yet it is not known if there is a genetic component to the pathogenicity of this organism. However, highly mucoid strains of *P. aeruginosa* can be readily isolated from the tracheobronchial secretions of children with cystic fibrosis, the data suggesting that such variants may have a contributing role to decline and death in such patients (Govan and Williams, 1973 and Govan, personal communication). The second is the notorious resistance to antibacterial agents shown by this bacterium. Two genetic sites for this resistance are now known — the chromosome and plasmids. The genetic analysis of how antibiotic resistance is determined in this bacterium is made possible by systems of genetic recombination including conjugation and transduction, a library of genetically marked strains and a well characterized chromosome map. This paper will examine the genetic nature of certain chromosomal borne resistance determinants, the ways in which various plasmids carrying resistant determinants can be distinguished from each other, how other chromosomal genes can affect the expression of plasmid borne characteristics, including plasmid establishment, and some speculation as to why drug resistance plasmids have only recently been detected in *P. aeruginosa*.

ANALYSIS OF GENETIC ORGANIZATION IN *P. AERUGINOSA*

Conjugation and chromosome transfer is mediated by a variety of plasmids, the most commonly used being FP2 (Holloway, 1969; Holloway, Krishnapillai and Stanisich, 1971). Close linkage can also be established using two transducing phages, F116L and G101, which transduce enough of the chromosome to cotransduce markers which are about 2 minutes apart, using the estimate of chromosomal distance as measured by interrupted conjugation procedures. Two unrelated strains of *P. aeruginosa*, denoted PAO and PAT, have been studied and most mapping work has been done in strain PAO, although where comparisons can be made, the two strains seem to have the same marker distribution (Watson and Holloway, unpublished observations). Using interrupted matings, plate matings, and transduction analysis, the following map of *P. aeruginosa* has been constructed (Fig. 1).

Genetic circularity of the chromosome has not yet been demonstrated. Two origins of chromosome transfer, one an arbitrary site, 0′, which accounts for a majority of trans-

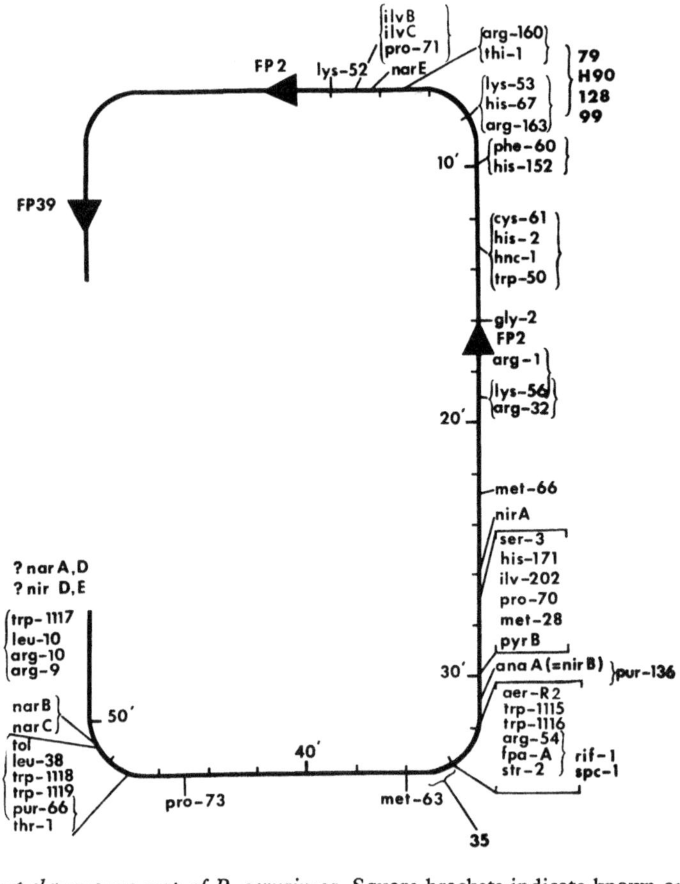

Figure 1. Current chromosome map of P. aeruginosa. Square brackets indicate known order, curved brackets indicate that the precise order remains to be determined.
Marker symbols:

aer-R2 – aeruginocinogenic determinant; *ana* – affected in anaerobic growth; *arg* – arginine biosynthesis; *cys* – cysteine biosynthesis; *fpaA* – resistance to *p*-fluorophenylalanine; *gly* – glycine biosynthesis; *his* – histidine biosynthesis; *ilv* – isoleucine and valine biosynthesis; *leu* – leucine biosynthesis; *lys* – lysine biosynthesis; *met* – methionine biosynthesis; *nar* – dissimilatory nitrate reductase activity; *nir* – dissimilatory nitrite reductase activity; *phe* – phenylalanine biosynthesis; *pro* – proline biosynthesis; *pyr* – pyrmidine biosynthesis; *rif* – rifampicin resistance; *ser* – serine biosynthesis; *str* – resistance to streptomycin; *spc* – spectinomycin resistance; *thi* – thiamine biosynthesis; *thr* – threonine biosynthesis; *trp* – tryptophan biosynthesis; *pur* – purine biosynthesis; 35, 79, H90, 99, 128 – prophage locations.

fers and another site, currently thought to be located between 10′–14′, at which transfer is less frequent. Discovery of the second origin has meant relocation of several markers, but no major revisions of the map. Times on the map are arbitrarily related to the 0′ site. The indicated origin of FP39 at about 10 minutes in advance of the FP2 0′ origin is based on times of entry of various markers and it is possible that the FP39 origin may in fact coincide with the FP2 origin but there is a 10 minute delay by FP39 in promoting chromosome transfer.

Accurate mapping by conjugation with FP2 or FP39 donors is difficult beyond 40′ and it is not known whether the gap on the left hand side of the map between markers in the 50′ region and *pro-4* near the 0′ origin site of FP2 represents mapping difficulties, or the failure to isolate mutants of those genes which are located in that region. There are no obvious similarities in marker distribution between the *P. aeruginosa* map and the maps known for *E. coli* and *S. typhimurium*. Indeed in *P. aeruginosa* there is a lack of clustering of genes determining biosynthesis of amino acids and purines in contrast to the situation found in *E. coli* and *S. typhimurium*.

Plasmids other than FP2 and FP39 have been studied for their ability to promote chromosome transfer. These include plasmids conferring antibiotic resistance and Stanisich and Holloway (1971) showed that two such plasmids, R68 and R91, could promote chromosome transfer in *P. aeruginosa*. These plasmids showed a differential ability to promote chromosome transfer in different strains of *P. aeruginosa*, and the frequencies for strain PAO were insufficient for mapping purposes.

Recently a variant of R68 has been obtained which promotes chromosome transfer in strain PAO at frequencies higher than that obtained with FP2 in that strain. Crosses between FP-derivatives of PAO and a PAO.R68 donor gave a few recombinants for chromosomal markers (frequency *ca* 10^{-8}/donor cell). Such recombinants were then tested in turn for their donor ability for chromosomal markers. Although the majority of these tested showed meagre ability to transfer chromosome, a few recombinants were found which showed appreciable donor ability, some up to about 1000 times that of the parent R68 plasmid.

The manner by which these variant forms arise is not known. Selection for some regions of the *P. aeruginosa* chromosome, achieved by selecting for particular auxotrophic markers in the recipient, does not produce any such variants amongst the recombinants. By contrast, selection for that region of the chromosome where a second origin site for FP2 has been detected (at about 18′) results in the highest frequency of such efficient donor variants of R68 (Holloway and Watson, in preparation).

One such variant, R68-44 has been studied in some detail. The increased frequency of chromosome transfer evidently does not arise from increased conjugation frequency because R68-44 shows the same frequency of transfer of antibiotic resistance markers (C, N/K, T) as R68 from which it was derived. It requires bacterial recombination function for chromosome transfer to occur. The mechanism by which chromosome transfer is initiated is still under study and it is likely that more than one origin is involved. A variety of regions of the chromosome can be transferred at high frequency and there is not the polarity of transfer frequency with increasing distance from the 0′ region of the map as found with FP2 or FP39. Plate matings in which donor strains carrying R68-44 give high frequencies of marker recovery, but interrupted matings using the same technique found to be satisfactory with PF2 or FP39 donors, result in low frequencies of marker recovery and no clear times of entry for the selected markers. Where linkage values obtained with R68-44 plate matings have been compared to those obtained with FP2, no anomalies have been found. While R68 is highly stable in strain PAO, R68-44 can only be maintained in PAO if cultured in the presence of carbenicillin. Reversion of R68-44 has been observed to a form which can only promote chromosome transfer in strain PAO at the low frequencies of the wild type R68.

It is possible to carry out PAO × PAO, PAT × PAT, PAO × PAT, 8602 × 8602 and 8602 × PAO crosses using R68-44, 8602 (=PAC) being the strain of *P. aeruginosa* used extensively by Clarke and her co-workers for genetic, biochemical and regulatory studies of amidase (Clarke, 1970). This potential breadth of donor ability has not been previously demonstrated with any of the other sex factors characterized in *P. aeruginosa*, and these donor variants of R68 may have the useful role of acting as a genetic bridge between strains of *P. aeruginosa* with particular experimental characteristics and the genetically well known PAO or PAT strains. Genetic and phenotypic elimination of DNA restriction has been achieved in both these strains so that host specificity strains differences need not affect recovery of recombinants in such interstrain matings.

The fact that R68 and R68-44 may be readily transferred from *P. aeruginosa* to other bacteria such as *E. coli* or *Pseudomonas putida* raises the interesting possibility that variants of R68 could be isolated which may mediate intergeneric transfer of chromosomal material.

GENETIC ANALYSIS OF CHROMOSOMAL GENES AFFECTING DRUG SUSCEPTIBILITY

We have examined the chromosomal relationships of certain antibiotic resistance and sensitivity markers. It had been previously shown that a streptomycin resistant determinant (*str*) occurred at about 30′ on the PAO chromosome (Stanisich and Holloway, 1969) and earlier work (Holloway, 1955) had shown that this region of strains PAO and PAT

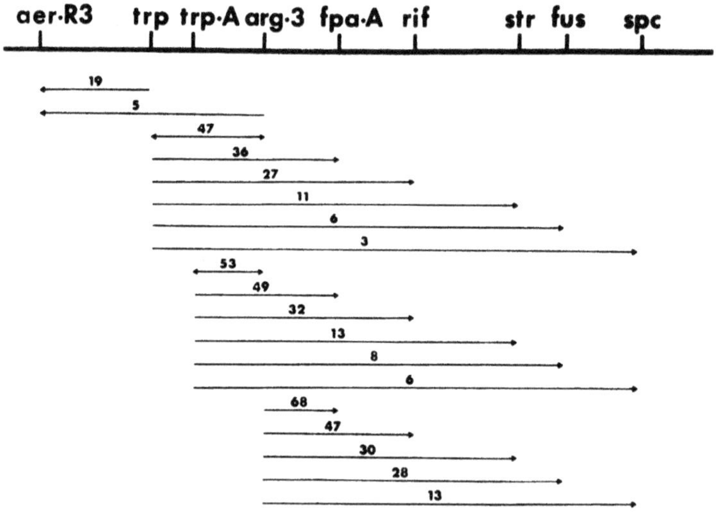

Figure 2. Chromosomal location of markers affecting antibiotic resistance and sensitivity in *P. aeruginosa* strain PAT. The figures indicate the cotransduction frequency (%) obtained using phage F116L. Arrowheads indicate the unselected markers. Marker abbreviations, *aer-R3*, R-type aeruginocin determinant; *arg3*, arginine requirement; *fpaA*, resistance to *p*-fluorophenylalanine; *fus*, hypersensitivity to fusidic acid; *rif*, resistance to rifampicin; *spc*, resistance to spectinomycin; *str*, resistance to streptomycin; *trpA*, tryptophan or anthronilate requirement; *trp*, tryptophan requirement.

were similar with respect to this and certain other markers. A detailed transduction analysis in strain PAT using phage F116L has shown that the markers for resistance to spectinomycin (*spc*) and rifampicin resistance (*rif*) were also located in this region. Further transduction analysis in this strain has led to the map of this region shown in Figure 2 which includes the loci *fus* and *fpa*.

Mutants of *P. aeruginosa* resistant to fusidic acid are difficult to obtain as there is a high native resistance to this agent. However, mutants which are hypersensitive to this agent can be readily obtained. These are inhibited at a level of 2 mgm/ml, compared to the wild type which can survive 10 mgm/ml. The location of this mutation (*fus*) is between *str* and *spc*.

In addition, it has been previously shown for *P. aeruginosa* that mutants resistant to *p*-fluorophenylalanine (fpa-A) are suppressed by certain mutations at the *str* locus indicating a possible functional relationship of these genes (Waltho and Holloway, 1966). Other *fpa* loci at other chromosomal sites were shown to be capable of altering the fidelity of protein synthesis (Dunn and Holloway, 1971). The fpa-A gene is located in the chromosomal region under examination.

The map for strain PAT of these various markers is shown in Figure 2. The relative location of these genes suggests a functional arrangement of genes concerned with RNA and ribosomal function and work is underway to identify the phenotypic lesions of these various mutants.

Mutants resistant to spectinomycin were isolated which did not map in this region, suggesting at least one other genetic site which may produce this phenotype, in agreement with the results of Yamada and Davies (1971) who reported two *spc* loci in *S. typhimurium*.

PLASMID DETERMINED ANTIBIOTIC RESISTANCE IN *P. AERUGINOSA*

Until 1969 there was no substantial evidence that antibiotic resistance in *P. aeruginosa* was other than chromosomally determined, and only limited reports that R factors from Enterobacteriaceae could even be transferred to *P. aeruginosa*. In fact until 1969 the only characterized plasmid for this species was the sex factor FP2 (Holloway, 1969). This situation has now changed dramatically and Table 1 shows the list of known plasmids for *P. aeruginosa*.

Clearly these constitute a heterogeneous genetic group, and as with the plasmids of the Enterobacteriaceae, effective epidemiological studies must be preceded by a workable classification of these various plasmids, in addition to more fundamental biological investigations of such an interesting collection of genetic structures. Accordingly, we have sought to establish taxonomic criteria by which these plasmids can be identified, using those indices which have been of value in other R plasmids, as well as seeking new ones.

The criteria we have used to date are:

(1) Antibiotic resistance pattern
(2) Host range, particularly with respect to the ability to be transferred only to other strains of *P. aeruginosa* as compared to those which can be transferred to more distantly related bacteria particularly Enterobacteriaceae.
(3) Response to bacteriophages including plasmid specific bacteriophages.
(4) Exclusion and incompatibility.

TABLE 1

Plasmids of P. aeruginosa

Plasmid designation	Alternative designation	Location of isolation	Drug resistance profile	Comments	Reference
RP1	R1822 R18	Birmingham	C, K/N, T		1, 2, 3, 4, 5, 6, 8
RP1-1	R18-1	Birmingham	C		5, 9
R91	R9169	Birmingham	C, K/N, T	Only C in some strains of *P. aeruginosa* Transfers chromosome	1, 4, 7
R19		Birmingham	C, K/N, T	Only C in some strains of *P. aeruginosa* Transfers chromosome	7
R68	R6886	Birmingham	C, K/N, T	Transfers chromosome	1, 4, 6
R30		Birmingham	C, K/N, T		6
R74		Birmingham	C, K/N, T		6
R88		Birmingham	C, K/N, T		6
RP8		Glasgow	C, K/N, T		10
R2	R2–72	Japan	C, S, K		11
R38	R38–72	Japan	T, S	Confers Hg resistance	11
R39	R39–72	Japan	T, S	Confers Hg resistance	11
R931		Canada	S, T		12, 13
R679		Canada	S, Su		12, 13
R1162		Canada	S, Su		12, 13
R3108		Canada	S, T, Su		13
R209		Canada	S, Su, G		13
R130		Canada	S, Su, G		13
R716		Canada	S		13
R503		Canada	S		13
R5265		Canada	S, Su		13
R64		Paris	A, C, Su, Gk		14, 15, 16
R40a		Paris	A, K/P, Su		16
pMG1		Atlanta	G, Su, S	Confers Hg resistance	17
pMG2		Atlanta	G, Su, S	Confers Hg resistance	17
FP2		South Africa	—	Confers Hg resistance Transfers chromosome	18
FP39		Melbourne	—	Confers LEU+ phenotype	18
FP5		Japan	—	Confers Hg resistance Transfers chromosome	19

Abbreviations: C, carbenicillin; K/N, kanamycin & neomycin; T, tetracycline; S, streptomycin; A, ampicillin; K/P, kanamycin & paromycin; Su, sulphonamide; G, gentamycin; GK, gentamycin & kanamycin; Hg, mercuric salts; LEU, leucine.

REFERENCES

1. Lowbury *et al*, 1969
2. Sykes & Richmond, 1970
3. Fullbrook *et al*, 1970
4. Stanisich & Holloway, 1971
5. Holloway & Richmond, 1973
6. Chandler & Krishnapillai, 1974a
7. Chandler & Krishnapillai, 1974b
8. Olsen & Shipley, 1973
9. Ingram *et al.*, 1972
10. Black & Girdwood, 1969
11. Kawakami *et al*, 1972
12. Bryan *et al*, 1972
13. Bryan *et al*, 1973
14. Witchitz & Chabbert, 1971
15. Witchitz & Chabbert, 1972
16. Chabbert *et al*, 1972
17. Jacoby, 1974
18. Holloway *et al*, 1971
19. Matsumoto & Tazaki, 1973

Progress has been made in the classification of plasmids of *P. aeruginosa*. Because some can be transferred to *E. coli*, it has been possible to compare them with known groups of enteric plasmids. The ones examined so far fall into two fi⁻ incompatibility groups, group C (Datta and Hedges, 1972) and group P (Datta, Hedges, Shaw, Sykes and Richmond, 1971).

However, Jacoby (personal communication) has recently shown that R plasmids of groups W and N can be transferred to *P. aeruginosa* thus making further comparisons possible. Furthermore, Jacoby (1974) has shown that the plasmid pMG1 determining gentamicin, streptomycin and sulphonamide resistance does not belong to either the P, W, C or Ng groups but belongs to a separate compatibility group which also includes a second independently isolated plasmid, pMG2, with the same resistance pattern as pMG1 and certain plasmids isolated in Canada, R931 (S, T) and R3108 (S, Su, T) (Bryan *et* al., 1973; Bryan, Van Der Elzen and Tsong, 1972).

We have examined a group of R factors isolated in the MRC Industrial Injuries and Burns Unit, Birmingham Accident Hospital, England, and kindly sent to us by Dr E. J. Lowbury. They are listed in Table 2 and all were first identified in *P. aeruginosa* strains with the exception of R74 which was first isolated in a strain of *Klebsiella*. All have been transferred to strain PAO to provide, as far as possible, an isogenic background in which to compare the various R factor properties. In addition we have attempted to show relationships between these R factors and others isolated from different geographical regions. With respect to the Birmingham isolates, they can be placed into three broad groups (I, II and III) on the principle criteria of resistance profile and promiscuity (Table 2).

TABLE 2

Characteristics of R factors isolated in Birmingham, England

Designation	Drug resistance profile in *P. aeruginosa* strain PAO*	Group
R18.1	C	I
R18	C, N/K, T	II
R30	C, N/K, T	II
R68	C, N/K, T	II
R74	C, N/K, T	II
R88	C, N/K, T	II
R19	C	III
R91	C	III

* Abbreviations as for Table 1

(a) Group II R factors can transfer readily between *P. aeruginosa* and enterobacteria including *E. coli*, *Shigella flexneri* and *S. typhimurium*, whereas Group I transfers at low frequency to *E. coli* but readily to *Sh. flexneri*. Group III R factors do not transfer at all to enterobacteria.

(b) The antibiotic resistance patterns follow these groupings based on host range, Group II members having the triple resistance pattern C, N/K, T. The other two groups show resistance to carbenicillin only.

(c) Group II R factors are unstable in *P. aeruginosa* strain PAT, Groups I and III are stable in this strain, and all are stable in strain PAO. Group I is unstable in *E. coli* and *Sh. flexneri*.

(d) Group III R factors are normally repressed for pilus synthesis in contrast to Group II

R factors. Presence of a repressed Group III R factor in the same cell as a Group II R factor did not affect the control of pilus formation by the Group II R factor.

We have examined other phenotypic aspects of these plasmids, to further characterize them, as well as to select other taxonomic criteria for their differentiation.

BACTERIOPHAGE SUSCEPTIBILITY

A variety of bacteriophage have been isolated which are specific for *P. aeruginosa* strains carrying R factors. Olsen and Shipley (1973) described the first of these, PRR1, which was specific for the presence of R1822 (synonymous with RP1 and R18) carrying each of the group II R factors, but no plaques are formed with Group I or Group III plasmids. However, this phage can multiply in such plasmid carrying strains, as determined by an increase in phage titre in liquid medium.

TABLE 3

Host range of plasmid specific phages (Stanisich, 1974b)

Plasmid	Plaque formation		Multiplication in liquid	
	PRR1 PR4	Pf3	PRR1 PR4	Pf3
R18-1, RP1-1	—	—	+	—
R91, R19	—	—	+	—
RP8	(+)	—	+	—
R18, RP1, RP4, R1822,	+	+	+	+
R68, R30, R74, R88	+	+	+	+
R2	(+)	—	+	—
R38, R39, R130, R931, R679, R716, FP2, FP5, FP8, FP39	—	—	—	—

TABLE 4

Plating efficiency of phages on isogenic strains carrying different plasmids (Krishnapillai, 1974)

Plasmid	Phage					
	G101	B3	B39	F116L	D3	E79
R18, R30, R68, R74, R88, RP1, RP4, R1822	10^{-4}	1	1	1	1	1
R18-1, RP1-1, R716	1	1	10^{-6}	1	1	1
R19, R91, RP8, R2, R679	1	1	1	1	1	1
R38, R39	$<10^{-6}$	$<10^{-6}$	$<10^{-6}$	1	$<10^{-6}$	$<10^{-6}$
R130, R931	$<10^{-6}$	$<10^{-6}$	$<10^{-6}$	$<10^{-6}$	$<10^{-6}$	$<10^{-6}$

The method of assay was by spot tests

Three other plasmid specific phages have now been isolated, PR3, PR4 and Pf3 (Stanisich, 1974). PR3 and PR4 are very similar but Pf3 is a filamentous phage and clearly shows a different host range to PRR1, PR3 and PR4.

The ability of these various phages to multiply in *P. aeruginosa* strain PAO when carrying various plasmids is shown in Table 3. Clearly these phages can be used for distinguishing different plasmids.

Other bacteriophages have been shown to be of value in classifying these plasmids, and five groupings of plasmids can be distinguished on this basis (Krishnapillai, 1974) as seen in Table 4.

EXCLUSION AND INCOMPATIBILITY

Compatibility studies have been made with these plasmids, although not all the possible groupings of plasmids have yet been examined. Group I and Group II plasmids have been found to belong to different compatibility groups (Morgan, Saunders, Stanisich and Richmond, 1974). Groups I and II were found to be different compatibility groups to each other and to R2 (Kawakami *et al.*, 1972), R 38 (Kawakami *et al.*, 1972), RP8 (Black and Girdwood, 1969) was compatible with R2 and R38. Determination of these compatibility groups was aided by the isolation of a variant of a Group II plasmid which lacks β-lactamase production (Curtis, Richmond and Stanisich, 1973) and a similar variant of R2.

Thus the compatibility relationships of all the plasmids determining penicillin resistance can be examined. There is no incompatibility or exclusion between groups I, II, III and R2, although exclusion and incompatibility occurs between members of Group II. It could well be that the plasmids grouped together under Group II (Table 2) are really separate isolates of the same plasmid, although there do seem to be some differences between them such as the ability to promote chromosome transfer in *P. aeruginosa*, and ease of transferability from *P. aeruginosa* to other species. Groups II and III show some similarities to each other.

TABLE 5

A provisional classification of drug resistance plasmids of P. aeruginosa

Group P	R18, RP1, RP4, R1822, R30, R68, R74, R88
Group C	R64, R40a
Group I	R18-1, RP1-1
Group III	R91, R19
Group IV	R2
Group V	R38, R39, pMG1, pMG2, R931, R130, R3108
Currently unclassified	RP8, R679, R1162, R716, R503, R5265

It is not clear at this point what taxonomic criteria will provide the best differentiation of a group of plasmids such as this. Undoubtedly DNA hybridisation will be the criterion of choice in the future, but for the present some simpler general laboratory procedures are needed such as the various criteria which are used above. It seems worthwhile on our own published and unpublished data, as well as the published data of other workers, to suggest a provisional classification of some of the known R plasmids of *P. aeruginosa*, and this is shown in Table 5. It is possible that as more data is accumulated, particularly with respect to incompatibility, exclusion and fertility inhibition, changes are likely to occur in these groupings. As yet no close relationship between the R plasmids and the FP plasmids has been established.

THE ROLE OF THE BACTERIAL GENOME IN THE EXPRESSION OF PLASMID GENOMES

It is to be expected that expression of plasmid genes will depend on aspects of the bacterial genome. With the transfer of promiscuous plasmids to a variety of genetic backgrounds, it is likely that the same plasmid will have a variety of modes of expression. As Luria (1968) has pointed out, plasmids may be primarily concerned with the genetic control of structures related to the cell surface, including the cell envelope, components of the cytoplasmic membrane and periplasmic enzymes. Further, it is commonly and reasonably held that plasmids may have a strong physical association with the cell membrane, at least with respect to their replication (Clowes, 1972).

We have obtained and characterized a range of mutants of *P. aeruginosa* in which the structure and function of the cell envelope is altered. It has been shown in *E. coli* that mutants which are indifferent to the lethal action of colicins (the so-called tolerant or *tol* mutants) commonly have alterations in the cell membrane or cell envelope and are pleiotropic for a wide range of functions associated with these structural regions. Mutants of *P. aeruginosa* tolerant to aeruginocins seem to be essentially the same (Holloway *et al.*, 1974). They show alterations in a variety of properties associated with the cell membrane including membrane protein components as examined by acrylamide gel electrophoresis, changes in sensitivity (both decreased and increased) to toxic agents, response to bacteriophage infection including lysogenic response, activity of membrane associated enzymes, and colonial and cellular morphology. Mapping data obtained by both conjugation and transduction indicate that at least three loci are concerned (Holloway *et al.*, 1974; Dodge and Holloway, in preparation).

TABLE 6

Reduced ability of an aeruginocin tolerant mutant to acquire plasmids (Dodge, 1974)

Plasmid*	Acquisition frequency/10^6 donor cells	
	PAO12 (*tol*+)	PAO1701 (*tol*)
R18-1	150	150
R68	30	19
R91	60	1
FP2	100	<1
FP39	800	<1

* See Table 1 for a description of these plasmids.

TABLE 7

A comparison of plasmid donor ability for a tol+ (PAO12) and a tol (PAO1701) strain (Dodge, 1974)

Plasmid	Transfer frequency/10^6 donor cell	
	PAO12 (*tol*+)	PAO1701 (*tol*)
R18-1	400	80
R68	300	400
R91	80	3000

We have sought to investigate whether such mutations affecting the cell membrane and cell envelope have any effect on the behaviour of plasmids. Our results show that there are such effects and in the present context three distinct interactions can be examined.

One aeruginocin-tolerant mutant, PAO1701, was found to have a markedly reduced response to lysogenisation by the temperate phage G101 (Dodge, 1974). It also shows differences from the parent strain in response to deoxycholate and *p*-fluorophenylalanine, typical of a membrane mutant in which permeability had been altered. More important, this mutant shows significant differences in its response to plasmid acquisition and transfer, as seen in Tables 6 and 7.

Some slight reduction in stability of R68, R91 and FP2 occurs in clones of PAO1701 carrying each of these plasmids.

Clearly, PAO1701 transfers the R91 plasmid at a significantly higher frequency than found with the corresponding Tol+ strain. As transfer frequency is an expression of the level of derepression of a plasmid, this finding introduces the notion that the cell membrane may have a role in the regulation of the plasmid genome.

Support for this view comes from the behaviour of the plasmid R38 in a strain carrying a different *tol* mutation to that in PAO1701. Normally R38 can express its full drug resistance pattern in derivatives of strain PAO at both 37°C and 43°C. However, at 43°C, the *tol* mutant PAO1710 carrying R38 is not streptomycin resistant (Table 8).

This phenotype reflects a property of the host, not the R factor because if R38 is enzyme involved is thought to be an adenylase, as R38 confers resistance to both strepto-

TABLE 8

Expression of SM resistance by R38 in tol+ and tol strains
(Mills & Holloway, in preparation)

Host strain	Surviving fraction					
	NA* (37°C)	NA (43°C)	NA+SM+ (37°C)	NA+SM (43°C)	NA+TC (37°C)	NA+TC (43°C)
Tol+ . R38	1	1	3×10^{-1}	3×10^{-1}	1×10^{-1}	1
PAO1710 . R38	1	1	2.5×10^{-2}	1×10^{-8}	1×10^{-1}	2×10^{-1}

* NA, nutrient agar
+ SM, streptomycin

TABLE 9

Inducibility of SM resistance in PAO1710 . R68
(Mills & Holloway, in preparation)

Strain	Previous medium at 37°C	Surviving fraction (colony formation)			
		NA (37°C)	NA (43°C)	SM (37°C)	SM (43°C)
PAO12 . R38	Growth in NB	1	1	3×10^{-1}	3×10^{-2}
PAO12 . R38	Growth in NB+500 μgm SM	1	1	1	1
PAO1710 . R38	Growth in NB	1	1	5×10^{-2}	10^{-9}
PAO1710 . R38	Growth in NB+500 μgm SM	1	1	10^{-1}	10^{-1}

transferred out of PAO 1710 into a *tol*+ strain, that strain then shows the normal pattern of streptomycin resistance characteristic of R38. Further, a range of other *tol* mutants do not demonstrate this property. The effect is also not due to instability of the R38 plasmid during growth at 43°C in PAO1710, in fact R38 is much more stable at 43°C in PAO1710 than in the *tol*+ strain.

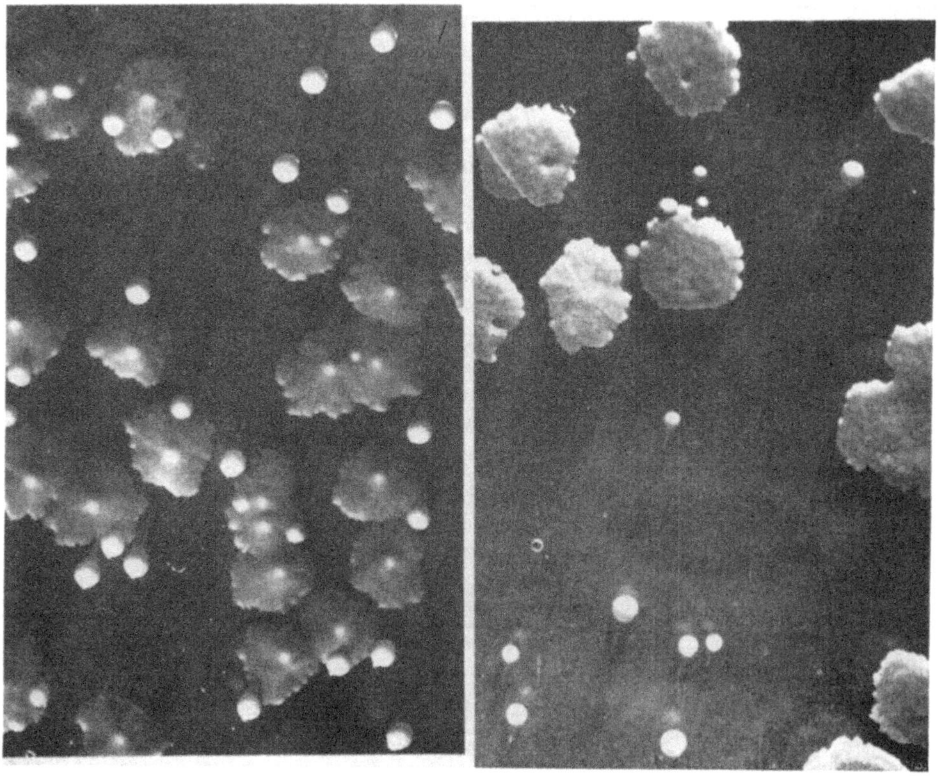

Figure 3. Effect of R plasmids on colonial morphology in *P. aeruginosa*. The left hand photograph shows a mixture of colonies of PAT971 (rough) and PAT900 (smooth) growing on nutrient agar (NA). On the right, a mixture of colonies on NA+250 μgm/ml carbenicillin of PAT971.R91 showing the same rough morphology as PAT971, and PAT971.R30 showing the smooth colonial morphology of PAT900.

The slightly lower ability to form colonies of strains carrying R38 on media containing streptomycin is believed to be due to the non-constitutive nature of the streptomycin degrading enzyme carried by these bacteria. It has been found that with PAO1710. R38 cells, growth can be obtained at 43°C if the cells are grown previously at 37°C in the presence of SM (Table 9).

A possible explanation of these results is that when R38 is carried by PAO1710, the control mechanism of the streptomycin degrading enzyme of R38 is altered from a predominantly constitutive mode to an apparently inducible form and at 43°C, the conditions for induction, possibly involving permeability, do not allow colony formation. The mycin and spectinomycin. Moreover, with the unrelated plasmid R2 (Kawakami *et al.*, 1972) similar, if not quite as extreme, results are obtained. With R2 it is likely that the streptomycin resistance determinant is acting by means of a phosphorylase, showing that the effect of the *tol* mutant is not plasmid specific and lending support to the view that plasmid control involves membrane structure and function.

Plasmids may also have more overt effects on strains carrying them. Strain PAT of *P. aeruginosa* can be mutated to produce a variety of colonially rough forms (Krishnapillai and Fyfe, unpublished). It has been found that the presence of an R plasmid may convert the rough bacterial variant back to the smooth form. Loss of the plasmid through natural instability results in reversion to the colonially rough form. The R plasmid may be contributing to the structural components of the bacterial cell surface, but the mechanism by which it does is unclear. This effect is obtained only by the Group II plasmids (R30, R68, R74, R18). (Figure 3.)

GENETIC MECHANISMS FOR THE EVOLUTION OF PSEUDOMONAS PLASMIDS

Despite the detection of drug resistance plasmids in the middle nineteenfifties and the increasing realization of the importance of the drug resistance properties of *P. aeruginosa* in nosocomial infections, no R plasmid carrying strains of *P. aeruginosa* were detected in isolation from patients until the late 1960's. Does this mean that the evolution of plasmids was such that the ability to become part of the genome of *P. aeruginosa* was delayed or that the plasmid carrying *P. aeruginosa* did not become selected out from the general population of *P. aeruginosa* strains until the appropriate conditions presented themselves? The latter is more likely, but the genetic variation of the isolates obtained in the last few years points to genetic interactions probably involving other plasmids as well as the genome of *P. aeruginosa*. It is therefore important to determine if any of these potential mechanisms of plasmid evolution can occur. Recombination between an R factor and a resident plasmid of *P. aeruginosa* has recently been demonstrated by Stanisich (1974a).

TABLE 10

Genetic analysis of plasmid recombinants
(Stanisich, 1974)

Resistance phenotype of transcipients from C, N, T, Hg donor	Proportion of conjugal transfer of C	Cotransducible markers
C, N, T	6/40	C-N-T
C, N, T, Hg	0/3	C-N-T C-N-T-Hg
C, Hg	0/22	C-Hg
C, T	0/4	C-T

Stanisich and Holloway (1972) described a derivative (PAT900) of the donor strain PAT (FP2) which had lost donor ability and behaves as a recipient in mating. This derivative is still mercury resistant so that either only part of the FP2 has been lost or that there is a second plasmid in PAT which can confer mercury resistance. Mercury resistance in this recipient derivative is not transferable in conjugation but unlike FP2, can be transferred by transduction with phage F116L.

Transfer of the mercury resistance determinant from PAT900 can be accomplished by R factors such as R30 or R18-1, and as the latter does not mobilise chromosome at the frequencies observed for the transfer of the mercury resistance determinant, it can be concluded that this phenotype in PAT900 is determined by a non-transferable extra-chromosomal element.

When C-r transcipients are selected on minimal medium from matings involving PAT.R30^{+} donors and PAO recipients, the majority are found to have coinherited Hg-r. However, acquisition of the latter phenotype is predominantly associated with loss of other R-specified functions such as resistance to N or T, or loss of transfer ability. The description and analysis of such transcipients are shown in Table 10.

It can be proposed that the range of phenotypes seen among the transcipients is due to recombination involving R30 and the mercury-resistance element. This is indeed found to be so, from the transductional analysis of these bacteria which identifies cotransduction of the determinants C-r and Hg-r, a feature not observed in the donor parent.

In view of these transduction results, recombination has clearly occurred between the R30 plasmid and the mercury resistance plasmid and establishes the possibility of such a mechanism in the evolution of Pseudomonas plasmids. Pemberton and Clark (1973) have demonstrated the existence of small, extrachromosomal elements of DNA in strain PAO and no known phenotypic function can presently be ascribed to such elements. The possibility now exists that they may provide a reservoir of genetic variability whenever transferable plasmids establish themselves in this strain.

DISCUSSION

The isolation of drug resistance plasmids has added a new dimension to the genetic study of *P. aeruginosa*. The three genetic effects which provide the most interesting challenge are the extent of genetic variability among the plasmids, their extent of contribution to the phenotype of *P. aeruginosa*, and the ability of some to transfer between *P. aeruginosa* and other unrelated genera.

Clearly the latter has the most immediate epidemiological significance as the potential reservoir for the type of plasmid for nosocomial infections must include almost all the gram negative genera encountered in such situations.

There are suggestions that the expression of certain promiscuous plasmids may differ in particular genera. The demonstration that such plasmids may be able to contribute to the structural components of bacteria as shown with the smooth-rough conversion by type II plasmids, and if this could be shown to be due to specific surface molecular changes it could provide evidence that the selective advantages conferred by these plasmids may lie not only in their drug resistance phenotype.

It remains to be seen whether *P. aeruginosa* occupies a peripheral or focal point in the evolution of these plasmids. It could well eventuate that intergeneric genetic exchange by them plays a dynamic role in bacterial evolution and which when understood at the laboratory level may provide a means of producing bacterial strains with any desired phenotype for experimental, industrial or prophylactic purposes.

One unsolved problem concerns the native resistance of *P. aeruginosa* to such a wide variety of antibacterial agents. It is not known whether this results from structural properties of say the cell envelope or from metabolic mechanisms which protect the action sites in the cell of these agents. It can be anticipated that a genetic approach to this problem through the isolation of hypersensitive mutants to antibacterial substances will prove rewarding.

REFERENCES

BLACK, W. A. and GIRDWOOD, R. W. A. (1969). Carbenicillin resistant *Pseudomonas aeruginosa.* British Medical Journal *iv*, 234.

BRYAN, L. E., VAN DEN ELZEN, H. M. and TSENG, J. T. (1972). Transferable drug resistance in *Pseudomonas aeruginosa.* Antimicrob. Ag. Chemother. **1**, 22.

BRYAN, L. E., SEMAKA, S. D., VAN DEN ELZEN, H. M., KINNEER, J. E. and WHITEHOUSE, R. L. S. (1973). Characteristics of R931 and other *Pseudomonas aeruginosa* R factors. Antimicrob. Ag. Chemother. **3**, 625.

CHABBERT, Y. A., SCAVIZZI, M. R., WITCHITZ, J. L., GERBAUD, G. R. and BOUANCHAUD, D. H. (1972). Incompatibility groups and the classification of fi resistance factors. J. Bact. **112**, 666.

CHANDLER, P. M. and KRISHNAPILLAI, V. (1974a). Phenotypic properties of R factors of *Pseudomonas aeruginosa*: R factors readily transferable between *Pseudomonas* and the *Enterobacteriaceae.* Genet. Res. (in press).

CHANDLER, P. M. and KRISHNAPILLAI, V. (1974b). Phenotypic properties of R factors of *Pseudomonas aeruginosa*: R factors transferable only in *Pseudomonas aeruginosa.* Genet. Res. (in press.)

CLOWES, R. C. (1972). Molecular Structure of Bacterial Plasmids. Bact. Rev. **36**, 361.

CLARKE, P. H. (1970). The aliphatic amidases of *Pseudomonas aeruginosa.* Advan. Microbiol. Physiol. **4**, 179.

CURTIS, N. A. C., RICHMOND, M. H. and STANISICH, V. A. (1973). R factor mediated resistance to penicillins which does not involve a β-lactamase. J. Bact. 79, 163.

DATTA, N., HEDGES, R. W., SHAW, E. J., SYKES, R. B. and RICHMOND, M. H. (1971). Properties of an R factor from *Pseudomonas aeruginosa.* J. Bact. **108**, 1244.

DATTA, N. and HEDGES, R. W. (1972). R factors identified in Paris, some conferring gentamycin resistance, constitute a new compatibility group. Ann. Inst. Pasteur **123**, 849.

DODGE, J. (1974). A study of membrane mutants of *Pseudomonas aeruginosa.* M. Sc. thesis, Monash University, Clayton, Victoria, Australia.

DUNN, N. W. and HOLLOWAY, B. W. (1971). Pleiotropy of *p*-fluorophenylalanine-resistant and antibiotic hypersensitive mutants of *Pseudomonas aeruginosa.* Genet. Res. **18**, 185.

FULLBROOK, P. D., ELSON, S. W. and SLOCOMBE, B. (1970). R factor mediated β-lastamase in *Pseudomonas aeruginosa.* Nature (Lond.) **226**, 1054.

GOVAN, J. R. W. and WILLIAMS, R. J. (1973). Pyocine typing of mucoid strains of *Pseudomonas aeruginosa* isolated from children with cystic fibrosis. Journal of Medical Microbiology **6**, 409.

HOLLOWAY, B. W. (1955). Genetic recombination in *Pseudomonas aeruginosa.* J. Gen. Microbiol. **13**, 572.

HOLLOWAY, B. W. (1969). Genetics of Pseudomonas. Bact. Rev. **33**, 419.

HOLLOWAY, B. W. and RICHMOND, M. H. (1973). R-factors used for genetic studies in strains of *Pseudomonas aeruginosa* and their origin. Genet. Res. **21**, 103.

HOLLOWAY, B. W., KRISHNAPILLAI, V. and STANISICH, V. A. (1971). Pseudomonas genetics. Ann. Rev. Genet. **5**, 425.

HOLLOWAY, B. W., ROSSITER, H., BURGESS, D. and DODGE, J. (1974). Aeruginocin tolerant mutants of *Pseudomonas aeruginosa.* Genet. Res. **22**, 239.

INGRAM, L., SYKES, R. B., GRINSTED, J., SAUNDERS, J. R. and RICHMOND, M. H. (1972). A transmissible resistance element from a strain of *Ps. aeruginosa* containing no detectable extrachromosomal DNA. J. Gen. Microbiol. **72**, 269.

JACOBY, G. (1974). Properties of R plasmids determining Gentamycin resistance by Acetylation in *Pseudomonas aeruginosa.* J. Bact. (in press).

KAWAKAMI, Y., MIKOSHIBA, F., NAGASAKI, S., MATSUMOTO, H. and TAZAKI, T. (1972). Prevalence of *Pseudomonas aeruginosa* strains possessing R factors in a hospital. J. Antibiot. **25**, 607.

KRISHNAPILLAI, V. (1974). The use of bacteriophages for differentiating plasmids of *Pseudomonas aeruginosa.* Genet. Res. (in press).

LOWBURY, E. J. L., KIDSON, A., LILLY, H. A., AYLIFFE, G. A. J. and JONES, R. J. (1969). Sensitivity of *Pseudomonas aeruginosa* to antibiotics: Emergence of strains highly resistant to carbenicillin. Lancet *ii*, 448.

LURIA, S. E. Opening Remarks in Bacterial Episomes and Plasmids. A Ciba Foundation Symposium. Ed. G. E. W. Wolstenholme and M. O'Connor. J. & A. Churchill Ltd., London, 1969.

MATSUMOTO, H. and TAZAKI, T. (1973). FP5 factor, an undescribed sex factor of *Pseudomonas aeruginosa.* Jap. J. Microbiol. **17**, 399.

MORGAN, T., SAUNDERS, J. R., STANISICH, V. and RICHMOND, M. H. (1974). Compatibility relationships between R factors originally isolated from strains of *Pseudomonas aeruginosa.* J. Gen. Microbiol. (abstr.) (in press).

OLSEN, R. H. and SHIPLEY, P. (1973). Host range and properties of the *Pseudomonas aeruginosa* R factor R1822. J. Bact. **113**, 772.
PEMBERTON, J. M. and CLARK, A. J. (1973). Detection and characterisation of plasmids in *Pseudomonas aeruginosa* strain PAO. J. Bact. **114**, 424.
STANISICH, V. A. (1974a). Interaction between an R factor and an Element conferring resistance to Mercuric ions in *Pseudomonas aeruginosa*. Molec. Gen. Genet. **128**, 201.
STANISICH, V. A. (1974b). The properties and host range of male specific bacteriophages of *Pseudomonas aeruginosa*. Genet. Res. (in press).
STANISICH, V. A. and HOLLOWAY, B. W. (1969). Genetic effects of acridines on *Pseudomonas aeruginosa*. Genet. Res. **13**, 57.
STANISICH, V. A. and HOLLOWAY, B. W. (1971). Chromosome transfer in *Pseudomonas aeruginosa* mediated by R factors. Genet. Res. **17**, 169.
STANISICH, V. A. and HOLLOWAY, B. W. (1972). A mutant sex factor of *Pseudomonas aeruginosa*. Genet. Res. **19**, 91.
SYKES, R. B. and RICHMOND, M. H. (1970). Intergeneric transfer of a β-lactamase gene between *Pseudomonas aeruginosa* and *Escherichia coli*. Nature (Lond.), **226**, 952.
WALTHO, J. A. and HOLLOWAY, B. W. (1966). Suppression of fluorophenylalanine resistance by mutation to streptomycin resistance in *Pseudomonas aeruginosa*. J. Bact. **92**, 35.
WITÇHITZ, J. L. and CHABBERT, Y. A. (1971). Résistance transférable à la gentamicine. I Expression du caractère de résistance. Ann. Inst. Pasteur **121**, 733.
WITCHITZ, J. L. and CHABBERT, Y. A. (1972). Résistance transférable à la gentamicine. II Transmission et liaisons du caractère de résistance. Ann. Inst. Pasteur **122**, 367.
YAMADA, T. and DAVIES, J. (1971). A genetic and biochemical study of streptomycin and spectinomycin-resistance in *Salmonella typhimurium*. Molec. Gen. Genet. **110**, 197.

R PLASMIDS DETERMINING GENTAMICIN OR TOBRAMYCIN RESISTANCE IN PSEUDOMONAS AERUGINOSA

G. A. JACOBY
Massachusetts General Hospital, Boston, U. S. A.

These studies were supported by Public Health Service Grant No AI 10237 from the National Institute of Allergy and Infectious Diseases.

Pseudomonas aeruginosa has emerged in the antibiotic era as one of the most feared causes of hospital infection. Its success as a pathogen is due in large part to its resistance to antibiotics, and this, in turn, has provided a stimulus for the development of new anti-pseudomonas drugs. Carbenicillin, gentamicin and, more recently, tobramycin provide the clinician with agents effective against many *P. aeruginosa* strains. However, when carbenicillin (Lowbury et al., 1969) or gentamicin (Shulman, Terry and Hough, 1971) has been used intensively in a setting where cross-infection can occur, such as in a burn unit, carbenicillin or gentamicin resistant *P. aeruginosa* strains have appeared and replaced the originally sensitive populations. Even in a general hospital, such as my own, gentamicin resistant strains have increased in prevalence from 3% in 1972 to currently 9% of all *P. aeruginosa* isolates. Resistance to tobramycin, which is in limited use as an investigational drug, is found in less than 1% of our *P. aeruginosa* strains.

It is well known that carbenicillin resistance in *P. aeruginosa* can be produced by R plasmids (Sykes and Richmond, 1970; Fullbrook, Elson and Slocombe, 1970), and transmissible gentamicin resistance has been found as well (Witchitz and Chabbert, 1971; Bryan et al., 1973; Knothe et al., 1973). In this presentation I would like to describe four kinds of R plasmids that can produce gentamicin or tobramycin resistance in *P. aeruginosa*, to indicate how these plasmids can be classified and how they relate to other known pseudomonas plasmids, and to describe some biologic effects they produce in addition to antibiotic resistance.

Properties of pMG1 and pMG2

The first type of R plasmid was discovered in *P. aeruginosa* strains Capetown No. 18 and Stone No. 130 that originated in burn units in South Africa and Georgia (Weinstein et al., 1971). These strains inactivate gentamicin via gentamicin acetyltransferase I (Mitsuhasi, Kobayashi and Yamaguchi, 1971; Brzezinska et al., 1972). Transmissible gentamicin resistance was detected on mating with *Pseudomonas aeruginosa* strain PU21, a rifampin resistant auxotroph, using rifampin for counterselection (Jacoby, submitted for publication). The R plasmids involved, pMG1 and pMG2, produce resistance to streptomycin and sulfonamides as well as to gentamicin. Although transmissible between many *P. aeruginosa* strains, they are not transferable to *Escherichia coli* or other enterobacteriaceae. They are thus restricted in host range, unlike pseudomonas plasmids of the P group, such as RP1 (Grinsted et al., 1972) or the C group, such as R64 (Witchitz and Gerbaud, 1972; Datta and Hedges, 1972), which can be transferred to *E. coli* for compatibility classification. Their compatibility properties have therefore been tested in *P. aeruginosa* with plasmids originating in pseudomonas and with R plasmids that can be transferred from *E. coli* to *P. aeruginosa*. In addition to P and C-type R factors, these include W-type plasmids, such as R7K (Coetzee, Datta and Hedges, 1972), and N-type R46 (Datta and Hedges, 1971), that can be passed to *P. aeruginosa* and have the same compatibility properties in this genus as in *E. coli* (Jacoby, American Society for Microbiology Conference on Extrachromosomal Elements in Bacteria, 1974).

Table 1 shows the results of crosses between PU21 (pMG1) and R plasmid derivatives of PU1, a prototrophic *P. aeruginosa* strain, with counterselection by omission of isoleucine, leucine and valine, the nutritional requirements of PU21. pMG1 transferred to strains carrying P, C or W-type plasmids at the same frequency as to R plasmid negative PU1. An apparent ten-fold reduction in transfer efficiency was observed with PU1 (R46) as recipient, but this effect appears to be related to decreased expression of gentamicin resistance in the presence of this N group plasmid. When pMG1 transconjugants were tested for retention of the plasmid resident in the recipient, it was apparent that pMG1 was compatible with plasmids of P, C, W or N groups. Furthermore, pMG1 could be transferred independently (data not shown) from such double R factor containing strains. Quite different results were obtained with PU1 carrying R931 or R3108, R plasmids discovered by Bryan and coworkers in streptomycin resistant clinical *P. aeruginosa* isolates from Canada that, like pMG1, are not transmissible to *E. coli* (Bryan et al., 1973). Here there was a 50 to 100-fold decrease in pMG1 transfer frequency. Transconjugants either lost the resident plasmid (20/20 for R3108, 13/20 for R931) or in those R931 recipients that still carried tetracycline resistance, it could be shown that gentamicin resistance and tetracycline resistance were now serially cotransmissible

TABLE 1
Compatibility of pMG1 with other plasmids

Recipient[a]	R plasmid incompatibility group	pMG1 transfer frequency[b]	R^+ $pMG1^+$ / $pMG1^+$[c]
PU1		5×10^{-3}	
PU1 (RP1)	P	1×10^{-2}	20/20
PU1 (R64)	C	7×10^{-3}	20/20
PU1 (R7K)	W	4×10^{-3}	17/20
PU1 (R46)	N	7×10^{-4}	38/40
PU1 (R931)	P-2	1×10^{-4}	7/20
PU1 (R3108)	P-2	4×10^{-5}	0/20
PU1 (CAM)		7×10^{-5}	3/20

[a] PU1 is a prototrophic $FP2^+$ *P. aeruginosa* strain identical to Loutit's OT15 (Loutit, Pearce and Marinus, 1968). PU1 (R^+) derivatives were constructed from donor strains supplied by E. J. L. Lowbury (RP1), Y. A. Chabbert (R64), N. Datta (R7K, R46), L. E. Bryan (R931, R3108) and I. C. Gunsalus (CAM).

[b] The donor strain was PU21 (pMG1). PU21 is a rif^r derivative of OT47 ($ilv\ leu\ str^r$) (Loutit, Pearce and Marinus, 1968). Transconjugants were selected on minimal gentamicin plates after a two hour mating, except with PU1 (R64) where minimal streptomycin plates were used.

[c] The number of PMG1 transconjugants retaining the plasmid resident in the recipient/the number tested.

whichever agent was used for selection and hence, by this genetic test, that hybrid plasmids had been formed. Bryan et al. have reported similar incompatibility results and have assigned R931 and R3108 to the P-2 incompatibility group (Bryan et al., 1973). pMG1 and pMG2, which gave similar results, thus belong to the same group. Table 1 also shows that the transfer efficiency of pMG1 to a PU1 strain carrying CAM, a metabolic plasmid determining enzymes involved in camphor oxidation (Rheinwald, Chakrabarty and Gunsalus, 1973), was reduced 100-fold with loss of CAM in the majority of pMG1 transconjugants. Again genetic tests indicated that, in those recipients still carrying CAM, hybrid plasmids were formed. R931 and R3108 interact similarly with the CAM plasmid, which thus appears to be a member of the P-2 incompatibility group as well.

Pseudomonas plasmids can also be classified by their sensitivity to plasmid specific phages, including PRR1, an RNA phage that lyses strains carrying pili determined by P group plasmids (Olsen and Thomas, 1973; To and Brinton, personal communication), DNA phage PRD1 (Siak, Olsen and Gray, Abstracts of the Annual Meeting of the American Society for Microbiology, 1974), and phages Pf3, PR3 and PR4 (isolated by V. Stanisich) that appear to attach to the cell wall of RP1 containing *P. aeruginosa* strains (Bradley, personal communication). Table 2 shows the sensitivity of plasmid containing

TABLE 2

Phage Sensitivity[a]

Strain	PRR1	PRD1	Pf3	PR3	PR4
PU1	0	0	0	0	0
PU1 (RP1)	+++	+++	+++	+++	+++
PU1 (R64)	0	0	0	0	0
PU1 (R7K)	0	++	0	+	+++
PU1 (R46)	0	0	0	0	0
PU1 (pMG1)	0	0	0	0	0
PU1 (pMG2)	0	0	0	0	0
PU1 (RP1 pMG1)	++	+++	+++	+++	+++

[a] PRR1 and PRD1 were provided by R. Olsen and Pf3, PR3 and PR4 by V. Stanisich. Phage were grown on PU21 (RP1) and spot tested on the indicated strains.

TABLE 3

Properties of pMG1 and related plasmids[a]

Plasmid	Antibiotic resistance	Transmissible to *E. coli*	Resistance to				Inhibition of pyocin production
			Hg^{++}	Phenylmercuric nitrate	UV	Phage	
pMG1	Gm Sm Su	no	R	S	R	R	yes
pMG2	Gm Sm Su	no	R	S	R	R	yes
R931	Sm Tc	no	R	S	R	R	no
R3108	Sm Su Tc	no	R	R	S	R	no
CAM	none	not tested	S	S	R	R	no

[a] Properties of the plasmids in PU21 are indicated. Mercury resistance was tested with brain heart infusion agar plates containing 5×10^{-4} M $HgCl_2$ or 1×10^{-4} M phenylmercuric nitrate and pyocin production by the Gillies and Govan technique (Gillies and Govan, 1966).

PU1 strains to these phages. Note that PRD1, PR3 and PR4 lyse pseudomonas strains carrying W-type R7K as well as P-type RP1. Some N group plasmids in *E. coli* (but not R46) confer PR3, PR4 and PRD1 sensitivity as well (data not shown), so that these phages are neither pilus dependent nor specific for a single plasmid incompatibility group. Strains carrying pMG1 or pMG2 are not sensitive to these phages, confirming the results of compatibility testing that these plasmids do not belong to the P, W or N groups.

The plasmids in this incompatibility group share a number of biologic properties in addition to antibiotic resistance and inability to be transferred to *E coli.* Table 3 shows that pMG1, pMG2, R931 and R3108 all determine Hg^{++} resistance, that for pMG1 and pMG2 at least is due to an inducible system reducing Hg^{++} to metallic Hg° (Summers and Lewis, 1973). R3108 provides resistance to phenylmercuric nitrate as well and causes volatilization of mercury from organic linkage (Silver, personal communication). All except R3108 confer resistance to ultraviolet irradiation. In PU21 pMG1 and pMG2 have the remarkable property of phenotypically inhibiting pyocin production. This effect could be significant for the interpretation of epidemiologic studies based on pyocin typing. For example, pMG2 was detected in a strain from Atlanta where gentamicin resistant, pyocin type 5 organisms largely replaced gentamicin sensitive strains of diverse pyocin types under the selective pressure of gentamicin usage (Shulman, Terry and Hough, 1971). Cross-infection undoubtedly played a major role in the spread of type 5 organisms, but the spread of an R factor, like pMG2, that determines gentamicin resistance and changes the pyocin type could have produced similar results. Thus plasmids can have important effects on pseudomonas typing techniques.

In addition to pyocin typing, typing by phage sensitivity and serological reaction are widely used for *P. aeruginosa*. The immunotype (Fisher, Devlin and Gnabasik, 1969) of strains carrying pMG1 and related plasmids is not altered, but the phage type is affected since PU21 (pMG1), for example, is not lysed by phages 16, F8 and M6 of a standard typing set (Shooter et al., 1966), all of which are active on PU21. This effect has been studied using phages well characterized by Holloway and Bradley (Holloway, 1969; Bradley and Robertson, 1968; Bradley at Pitt, personal communication). Table 4 shows that strains carrying pMG1, pMG2, R931 or CAM failed to propagate phages B3, D3, E79, G101, PB1 or M6. Sensitivity to phages F116, C5, PO4, PP7 and 7S was not changed. R3108 produced a more selective inhibition of phage growth since strains carrying this plasmid are still lysed by E79, PB1 and M6. The mechanism of this

TABLE 4

Effect of plasmids on phage propagation

Strain		Phage[a]					
		B3	D3	E79	G101	PB1	M6
PU21	titer	6×10^{10}	1×10^{10}	4×10^{9}	4×10^{9}	2×10^{10}	9×10^{10}
PU21 (pMG1)	e.o.p.	$<10^{-9}$	$<10^{-9}$	$<10^{-8}$	$<10^{-8}$	$<10^{-9}$	$<10^{-9}$
PU21 (R931)	e.o.p.	$<10^{-9}$	$<10^{-9}$	$<10^{-8}$	$<10^{-8}$	$<10^{-9}$	$<10^{-9}$
PU21 (R3108)	e.o.p.	$<10^{-9}$	$<10^{-9}$	1.0	$<10^{-8}$	1.0	4×10^{-2}
PU21 (CAM)	e.o.p.	$<10^{-9}$	$<10^{-9}$	$<10^{-8}$	$<10^{-8}$	$<10^{-9}$	$<10^{-9}$

[a] DNA phages B3, D3, E79 and G101, provided by B. W. Holloway, and phages PB1 and M6, provided by D. E. Bradley, were grown on PU21 and titered on the indicated strains. Efficiency of plating (e.o.p.) was calculated from the ratio of plaque counts. With undiluted phage lysates, D3 produced clearing on PU21 (R3108) and B3 formed minute plaques on all plasmid^{+} derivatives, but with neither phage could propagation be demonstrated.

alteration in phage growth is not known. Both phages attaching to the cell wall (B3, E79, G101 and PB1) and pilus dependent phage (M6) showed the effect. Phage adsorption could be shown to occur normally. DNA restriction has not been excluded, but no host modified phage have been detected.

Properties of pMG5

The next plasmid I would lide to consider was discovered (Jacoby, in preparation) in *P. aeruginosa* strain GN315 isolated by Mitsuhashi in Japan (Yagisawa et al., 1972). This strain produces kanamycin acetyltransferase and is resistant to antibiotics that are substrates for this enzyme including tobramycin. Tobramycin resistance is readily transmissible to PU21, again using rifampin for counterselection. Sulfonamide resistance is transferred as well. Since kanamycin acetyltransferase does not inactivate all the components of the gentamicin complex, strains carrying this R plasmid, which I will call pMG5, appear gentamicin sensitive.

pMG5 shares a number of properties with pMG1 and related R factors. It is not transmissible to *E. coli*, provides resistance to Hg^{++} and to phenylmercuric nitrate, and inhibits the propagation of phages B3, D3, E79, G101, PB1 and M6. It does not give ultraviolet light resistance or affect pyocin production. Table 5 shows that it indeed belongs to the

TABLE 5

Compatibility of pMG5 with other plasmids

Recipient[a]	R plasmid incompatibility group	pMG5 transfer frequency[b]	R^+ $pMG5^+$ / $pMG5^+$ [c]
PU1		3×10^{-1}	
PU1 (RP1)	P	5×10^{-1}	20/20
PU1 (R64)	C	5×10^{-2}	20/20
PU1 (R7K)	W	2×10^{-1}	20/20
PU1 (R46)	N	2×10^{-1}	20/20
PU1 (pMG1)	P-2	3×10^{-5}	8/20[d]

[a] See Table 1.

[b] The donor strain was PU21 (pMG5), and transconjugants were selected on minimal tobramycin plates after a two hour mating except with PU1 (R64) where BB-K 8 was used for selection.

[c] The number of pMG5 transconjugants retaining the plasmid resident in the recipient/the number tested.

[d] Of these 8 colonies, 1 was streptomycin resistant and gentamicin sensitive, 5 streptomycin sensitive and gentamicin resistant, and 2 resistant to both antibiotics. Colonies of each type could serially transmit their resistance phenotypes to a further recipient whether tobramycin, gentamicin or streptomycin was used for selection.

same incompatibility group as pMG1. pMG5 could be transferred to recipients carrying P, W, C or N group R plasmids at the same frequency as to an R^- strain and could coexist stably with each of these plasmids. Transfer frequency was reduced 10^{-4} to a pMG1 recipient, and the progeny of this cross either had lost pMG1 or contained hybrid plasmids with one or both of the resistance markers from pMG1 linked to pMG5. The formation of such hybrid plasmids appears to be common in crosses between members of the P-2 incompatibility group in these strains and may provide a mechanism for evolving new assortments of plasmid genes.

Fertility inhibition by R64

The next type of plasmid, R64, was discovered by Witchitz and Chabbert (Witchitz and Chabbert, 1971) in a *P. aeruginosa* strain in Paris. It is transmissible to *E. coli*, belongs to the C incompatibility group, and is reported to determine gentamicin adenylyltransferase; an enzyme able to inactivate both gentamicin and tobramycin. R64 also confers carbenicillin resistance and so combines in one plasmid resistance to these three antipseudomonas drugs. Pseudomonas strains carrying R64 are resistant to Hg^{++} but not to ultraviolet radiation or to bacteriophage; they produce pyocin normally.

R64 does have a novel effect not found with pMG1 or related plasmids: it inhibits the fertility of the pseudomonas sex factor FP2 (unpublished observations). FP2 is a plasmid that is self-transmissible at high frequency, can promote transfer of the pseudomonas chromosome at lower frequency (Holloway, 1969,) and also confers Hg^{++} (Loutit, 1970) and phenylmercuric nitrate resistance. The plasmids that have been discussed are all compatible with FP2. In particular R64 has the same transfer frequency to an $FP2^+$ as to an $FP2^-$ recipient, and the two plasmids coexist stably and can be independently transferred from an $R64^+$ $FP2^+$ strain. However, FP2 transfer from such a strain is diminished about 100-fold and chromosomal recombinants are undetectable on mating with an FP^- recipient. R55, another C group plasmid (Witchitz and Chabbert, 1971; Chabbert et al., 1972), has the same effect, which may thus be a general property of C-type plasmids although further R factors need to be tested. Matsumoto and Tazaki have also described interference with FP5 mediated recombinant formation by a pseudomonas R plasmid of unknown incompatibility group (Matsumoto and Tazaki, 1973). Fertility inhibition thus promises to be an additional means of classifying pseudomonas R plasmids.

Host determined resistance of S-a

The fourth type of plasmid was discovered not in *P. aeruginosa* but in *Shigella* in Japan (Watanabe, Furuse and Sakaizumi, 1968). This plasmid, S-a, provides kanamycin resistance in *E. coli* but such a low level of gentamicin resistance (Hedges and Datta, 1971) that the strain appears sensitive by disc testing. Like other W group plasmids, S-a can be transferred to *P. aeruginosa* (unpublished observations). In pseudomonas S-a produces high level resistance to kanamycin, gentamicin and tobramycin. The specific enzymatic mechanism of this resistance is unknown. S-a does not confer resistance to Hg^{++}, ultraviolet radiation or bacteriophage and does not change the pyocin type.

W-type R plasmids have not yet been reported in naturally occurring *P. aeruginosa*. In fact they are rather unstable in pseudomonas. However, I have preliminary evidence that S-a is incompatible with a plasmid that did originate in *P. aeruginosa*, the FP39 sex factor (Pemberton and Holloway, 1973). FP39 is somewhat unstable in the presence of R plasmids of several incompatibility groups but appears to be totally unable to coexist with S-a.

Spectrum of aminoglycoside resistance produced by R plasmids in P. aeruginosa

One reason for studying R plasmids in *P. aeruginosa* is for the information they can provide about the potential for resistance to new aminoglycoside antibiotics as they are introduced clinically. For the biochemically inclined this information is implicit in the substrate specificity of the known inactivating enzymes (Benveniste and Davies, 1973). For the genetically inclined, Table 6 shows the spectrum of resistance determined by these four R plasmids in *P. aeruginosa* toward eleven drugs. Although the R factors can be presented individually as examples of plasmids determining gentamicin resistance

TABLE 6

Spectrum of aminoglycoside resistance

Strain	MIC (μg/ml)[a]										
	SM	KM	NM	PM	LMA	GM	SIS	TOB	GMB	BUT	BB-K 8
P. aeruginosa											
PU1	50	100	50	1,000	100	5	10	1.5	100	50	5
PU1 (pMG1)	10,000	100	50	1,000	100	200	100	5	50	50	10
PU1 (pMG5)	20	1,000	100	500	100	10	50	100	>100	>100	50
PU1 (R64)	20	100	50	500	50	500	>100	100	>100	25	2.5
PU1 (S-a)	10,000	1,000	100	1,000	100	250	>100	100	>100	50	15
E. coli											
J53	<2	<5	<5	<5	5	<0.5	<0.5	<0.5	0.75	1.5	0.75
J53 (S-a)	50	100	<5	<5	10	2.5	10	10	25	1.5	1.5

[a] Minimal inhibitory concentration (MIC) values were determined by agar dilution in Mueller Hinton medium. The concentration of antibiotic is in terms of the free base. Abbreviations: SM, streptomycin; KM, kanamycin; NM, neomycin; PM, paramomycin; LMA, lividomycin A; GM, gentamicin complex; SIS, sisomicin; TOB, tobramycin; GMB, gentamicin B or Sch 14342; BUT, butirosin.

but not tobramycin resistance (pMG1 and pMG2), tobramycin but not gentamicin resistance (pMG5), or resistance to both drugs (R64, S-a), they in fact determine resistance to other aminoglycosides as well, depending on the specificity of the enzymes involved. In particular all four plasmids give resistance to sisomicin and all but pMG1 and pMG2 to gentamicin B (Sch 14342). Only pMG5 significantly enhances resistance to butirosin or BB-K 8 (amikacin). Table 6 also shows the striking difference in the level of resistance produced by S-a in *P. aeruginosa* compared to *E. coli*. Note that the spectrum of resistance produced by S-a in *P. aeruginosa* closely resembles that produced by R64, so that differences in level of resistance may be due to differences in the expression of gentamicin adenylyltransferase in the two organisms.

I would not like to leave the impression that gentamicin and tobramycin resistance in *P. aeruginosa* is always plasmid mediated. There are many resistant pseudomonas strains in which transmissible resistance cannot be found. In some resistance may be determined by chromosomal genes, in others by plasmids that are nontransmissible or inhibited in their transfer functions by a coexisting plasmid, just as certain R plasmids can inhibit the transfer of pseudomonas sex factors. So far plasmid mediated resistance appears to be uncommon in *P. aeruginosa* (Kawakami et el., 1972; Bryan et al., 1973). However, plasmids determining resistance to gentamicin, tobramycin and related drugs are already widely distributed geographically, can recombine with each other to evolve new groupings of resistance markers, and are likely to become increasingly significant as usage of these drugs provides a selective advantage for their spread.

REFERENCES

BENVENISTE, R. and DAVIES, J. (1973): Mechanisms of antibiotic resistance in bacteria. Ann. Rev. Biochem. **42**, 471.

BRADLEY, D. E. and ROBERTSON, D. (1968): The structure and infective process of a contractile *Pseudomonas aeruginosa* bacteriophage. J. Gen. Virol. **3**, 247.

BRYAN, L. E., SEMAKA, S. D., VAN DEN ELZEN, H. M., KINNEAR, J. E. and WHITEHOUSE, R. L. S. (1973): Characteristics of R931 and other *Pseudomonas aeruginosa* R factors. Antimicrob. Ag. Chemother. **3**, 625.

BRZEZINSKA, M., BENVENISTE, R., DAVIES, J., DANIELS, P. J. L. and WEINSTEIN, J. (1972): Gentamicin resistance in strains of *Pseudomonas aeruginosa* mediated by enzymatic N-acetylation of the deoxystreptamine moiety. Biochemistry **11**: 761.

CHABBERT, Y. A., SCAVIZZI, M. R., WITCHITZ, J. L., GERBAUD, G. R. and BOUANCHAUD, D. H. (1972): Incompatibility groups and the classification of *fi*⁻ resistance factors. J. Bacteriol. **112**, 666.

COETZEE, J. N., DATTA, N. and HEDGES, R. W. (1972): R factors from *Proteus rettgeri*. J. Gen Microbiol. **72**, 543.

DATTA, N. and HEDGES, R. W. (1971): Compatibility groups among *fi*⁻ R factors. Nature **234**, 222.

DATTA, N. and HEDGES, R. W. (1972): R factors identified in Paris, some conferring gentamicin resistance, constitute a new compatibility group. Ann. Inst. Pasteur **123**, 849.

FISHER, M. W., DEVLIN, H. B. and GNABASIK, F. J. (1969): New immunotype schema for *Pseudomonas aeruginosa* based on protective antigens. J. Bacteriol. **98**, 835.

FULLBROOK, P. D., ELSON, S. W. and SLOCOMBE, B. (1970): R-factor mediated β-lactamase in *Pseudomonas aeruginosa*. Nature **226**, 1054.

GILLIES, R. R. and GOVAN, J. R. W. (1966): Typing of *Pseudomonas pyocyanea* by pyocine production. J. Pathol. Bacteriol. **91**, 339.

GRINSTED, J., SAUNDERS, J. R., INGRAM, L. C., SYKES, R. B. and RICHMOND, M. H. (1972): Properties of an R factor which originated in *Pseudomonas aeruginosa* 1822. J. Bacteriol. **110**, 529.

HEDGES, R. W. and DATTA, N. (1971): *fi*⁻ R factors giving chloramphenicol resistance. Nature **234**, 220.

HOLLOWAY, B. W. (1969): Genetics of *Pseudomonas*. Bacteriol. Rev. **33**, 419.

KAWAKAMI, Y., MIKOSHIBA, F., NAGASAKI, S., MATSUMOTO, H. and TAZAKI, T. (1972): Prevalence of *Pseudomonas aeruginosa* strains possessing R factor in a hospital. J. Antibiot. **25**, 607.
KNOTHE, H., KRCMÉRY, V., SIETZEN, W. and BORST, J. (1973): Transfer of gentamicin resistance from *Pseudomonas aeruginosa* strains highly resistant to gentamicin and carbenicillin. Chemotherapy **18**, 229.
LOUTIT, J. S. (1970): Investigation of the mating system of *Pseudomonas aeruginosa* strain 1. VI. Mercury resistance associated with the sex factor (FP). Genet. Res. **16**, 179.
LOUTIT, J. S., PEARCE, L. E. and MARINUS, M. G. (1968): Investigation of the mating system of *Pseudomonas aeruginosa* strain 1. I. Kinetic studies. Genet. Res. **12**, 29.
LOWBURY, E. J. L., KIDSON, A., LILLY, H. A., AYLIFFE, G. A. J. and JONES, R. J. (1969): Sensitivity of *Pseudomonas aeruginosa* to antibiotics: emergence of strains highly resistant to carbenicillin. Lancet **2**, 448.
MATSUMOTO, H. and TAZAKI, T. (1973): FP5 factor, an undescribed sex factor of *Pseudomonas aeruginosa*. Japan. J. Microbiol. **17**, 409.
MITSUHASHI, S., KOBAYASHI, F. and YAMAGUCHI, M. (1971): Enzymatic inactivation of gentamicin C components by cell-free extracts from *Pseudomonas aeruginosa*. J. Antibiot. **24**, 400.
OLSEN, R. H. and THOMAS, D. D. (1973): Characteristics and purification of PRR1, an RNA phage specific for the broad host range *Pseudomonas* R1822 drug resistance plasmid. J. Virol. **12**, 1560.
PEMBERTON, J. M. and HOLLOWAY, B. W. (1973): A new sex factor of *Pseudomonas aeruginosa*. Genet. Res. **21**, 263.
RHEINWALD, J. G., CHAKRABARTY, A. M. and GUNSALUS, I. C. (1973): A transmissible plasmid controlling camphor oxidation in *Pseudomonas putida*. Proc. Nat. Acad. Sci. USA **70**, 885.
SHOOTER, R. A., WALKER, K. A., WILLIAMS, V. R., HORGAN, G. M., PARKER, M. T., ASHESHOV, E. H. and BULLIMORE, J. F. (1966): Faecal carriage of *Pseudomonas aeruginosa* in hospital patients. Lancet **2**, 1331.
SHULMAN, J. A., TERRY, P. M. and HOUGH, C. E. (1971): Colonization with gentamicin-resistant *Pseudomonas aeruginosa*, pyocine type 5, in a burn unit. J. Infect. Dis. **124** (Suppl.), S18.
SUMMERS, A. O. and LEWIS, E. (1973): Volatilization of mercuric chloride by mercury-resistant plasmid-bearing strains of *Escherichia coli*, *Staphylococcus aureus*, and *Pseudomonas aeruginosa*. J. Bacteriol. **113**, 1070.
SYKES, R. B. and RICHMOND, M. H. (1970): Intergeneric transfer of a β-lactamase gene between *Pseudomonas aeruginosa* and *E. coli*. Nature **226**, 952.
WATANABE, T., FURUSE, C. and SAKAIZUMI, S. (1968): Transduction of various R factors by phage P1 in *Escherichia coli* and by phage P22 in *Salmonella typhimurium*. J. Bacteriol. **96**, 1791.
WEINSTEIN, M. J., DRUBE, C. G., MOSS, E. L., Jr. and WAITZ, J. A. (1971): Microbiologic studies related to bacterial resistance to gentamicin. J. Infect. Dis. **124** (Suppl.), S11.
WITCHITZ, J. L. and CHABBERT, Y. A. (1971): Résistance transférable à la gentamicine. I. Expression du caractère de résistance. Ann. Inst. Pasteur **121**, 733.
WITCHITZ, J. L. and GERBAUD, G. R. (1972): Classification de plasmides conférant la résistance à la gentamicine. Ann. Inst. Pasteur **123**, 333.
YAGISAWA, M. NAGANAWA, H., KONDO, S., TAKEUCHI, T. and UMEZAMA, H. (1972): 6′-N-Acteylation of 3′,4′-dideoxykanamycin B by an enzyme in a resistant strain of *Pseudomonas aeruginosa*. J. Antibiot. **25**, 495.

GENTAMICIN RESISTANCE IN *PSEUDOMONAS AERUGINOSA* II NON-TRANSFERABLE GENTAMICIN RESISTANCE

L. E. BRYAN, R. HARAPHONGSE, M. S. SHAHRABADI
Department of Medical Bacteriology, University of Alberta, Edmonton, Canada.

In testing clinical isolates of *Pseudomonas aeruginosa* at the University of Alberta Hospital (UAH), in Edmonton, Canada, for susceptibility to aminoglycoside antibiotics by a disc diffusion method (Bauer et al., 1966), we have detected many strains which were resistant to gentamicin as well as to other aminoglycosides. Streptomycin-resistant strains have been examined in detail in the past (Tseng, Bryan and Van Den Elzen, 1972). Most of these strains (about 90%) have MIC's (minimal inhibitory concentration) of from 10 to 200 μg/ml and apparently do not possess R-factors. A few strains had higher MIC values with resistance due to ribosomal mutation (5%) and R-factors (5%). The R-factors in that study were only detected by using *P. aeruginosa* recipients in conjugation mixtures and were either not transferable or inconsistently transferable to *Escherichia coli* K12. Some of these R-factors have been classified as P-2 and P-4 R-factor groups of *Pseudomonas* (Shahrabadi, Bryan and Van Den Elzen, 1974). Thus for streptomycin resistance R-factors were uncommon and the commonest form of resistance was a type which associated with an altered pattern of streptomycin accumulation by cells (Tseng, Bryan and Van Den Elzen, 1972).

In this study we have similarly examined gentamicin, tobramycin and BB-K8-resistant. *P. aeruginosa* for R-factors and have selected 14 gentamicin-resistant strains for a detailed analysis of their mechanism of resistance.

MATERIALS AND METHODS

Organisms. Strains selected from the clinical laboratory of the UAH are shown in Table 1. Strains used in conjugation studies as recipients were rifampin-resistant strains of *P. aeruginosa* 280 met$^-$, 3503, 211, and 1310 and *E. coli* 1306 pro$^-$ met$^-$ nal^r (Bryan et al., 1973). Strain 280 GSK^r is a resistant derivative of strain 280 obtained by subculturing strain 280 in increasing concentrations of gentamicin which is resistant to gentamicin, streptomycin, and kanamycin (Tseng, Bryan, Van Den Elzen, 1972). Strain PS130 (R130), a strain known to produce gentamicin acetyltransferase 1 was obtained from J. Davies, University of Wisconsin, and strain POW 151 known to adenylylate gentamicin, from S. A. Kabins, Michael Reese Medical Center, Chicago.

Media. The media used were trypticase soy (TS) (BBL) broth or trypticase soy agar (TSA, BBL) or Mueller-Hinton agar (BBL).

Mating methods. Broth matings were carried out as previously described (Bryan, Van Den Elzen and Tseng, 1972), using either 5 × 10^8/ml or 5 × 10^7/ml donor bacteria with 5 × 10^8/ml recipient bacteria for 2 hr, overninght and for 2 hr followed by 10 hr of shaking in a shaking water bath at 37C. Selection was by conventional counter-selection with rifampin (100 μg/ml) plus one of the following: gentamicin (1 μg/ml, or 5 μg/ml), streptomycin (16.5 or 33 μg/ml), tobramycin (1 μg/ml, Eli Lilly & Co.) or BB-K8 (2 μg/ml, Bristol-Myers).

Strain characterization. Methods used were as described previously (Bryan, Van Den Elzen and Tseng, 1972).

Minimal inhibitory concentration (MIC). A conventional tube dilution method was used with 2.5 ml of TS broth in an 18 × 125 mm tube. The inoculum size was 10^7 organisms.

Antibiotic inactivation assays. Aminoglycoside acetylating, adenylylating and phosphorylating radioactive assays were carried out by described methods (Benveniste and Davies, 1971; Brzezinska et al., 1972). Cell-free extract was prepared as a 30,000 × g supernatant (S-30) of sonically disrupted cells. The buffer used throughout was 10 mM Tris HCl, 50 mM NH_4Cl, 10 mM $MgCl_2$ and 2 mM dithiothreitol pH 7.8 (buffer A). Microbiological assays contained 0.8 ml of S-30 fraction of cell extract, 40 μg gentamicin sulfate (Schering Corpn.) and 400 μg ATP or Acetyl Coenzyme A (total volume – 1 ml). The test strain was *P. aeruginosa* 280 (MIC for gentamicin 0.25 μg/ml). Control *P. aeruginosa* strains containing acetylating (PS-130), or adenylylating activity (POW 151) were used with each microbiological assay. Assays were considered negative if the test strain was unable to grow at two times the normally inhibitory concentration of antibiotic.

Bioassay of culture filtrates. Overnight cultures containing subinhibitory concentrations of gentamicin (combined ^{3}H-gentamicin and unlabelled gentamicin) for the strain under test were filtered through 0.22 μ membrane filters (Sartorius, British Drug Houses) and gentamicin quantitation was based on radioactive counts of aliquots of the culture filtrate and the known specific activity of the added gentamicin. Filters were determined to contain $< 0.1\%$ of the total radioactivity in the culture. Bioassay of the filtrate was determined using 1.5- and 2-fold serial dilutions of the filtrate in TS broth. These dilutions were inoculated with 4×10^6 cells/ml of *P. aeruginosa* 280 (MIC of gentamicin – 0.25 μg/ml) and incubated at 35C for 18 hr. Residual gentamicin activity was determined using gentamicin standards assayed in an identical manner.

Cell-free amino acid incorporation. The poly-uridylic acid (poly-U) system used was that described by Nirenberg (1963) (specific activity of ^{14}C-phenylalanine – 55mCi//mMole, New England Nuclear). R17 RNA was prepared and used for amino acid incorporation as previously described (Tseng, Bryan, Van Den Elzen, 1972). Preincubation of complete amino acid incorporation mixtures for 30 min at 37C was carried out prior to the addition of R17 RNA and ^{14}C-labelled valine (specific activity – 50 mCi//mMole, Schwarz/Mann). The S-100 (100,000 × g supernatant) fraction and ribosomes were prepared as described. Endogenous protein synthesis was carried out using a non-preincubated deoxyribonuclease-treated S-30 fraction (30,000 × g supernatant) of alumina ground cells (Nirenberg, 1963). Concentrations of reagents were the same as used for R17 RNA directed synthesis with a magnesium concentration of 9 m molar. The S-30 fraction served as a source of mRNA, tRNA, ribosomes and other nonribosomal components required in protein synthesis – see Table 7 for amino acid concentrations.

Purification of tritiated gentamicin. Gentamicin sulfate was supplied as a gift from Schering Corpn. (gentamicin potency 588 μg/mg) and subject to tritium exchange by Amersham-Searle. The product was purified by G-10 chromatography as described by Mahon, Ezer and Wilson, (1973). The purified fraction was determined by paper chromatography (Wagman, Marquez and Weinstein, 1968) to contain less than 0.5% contaminating tritiated material. Final specific activities were from 0.5 Ci/mMole to 0.065 Ci/mMole.

Whole cell incorporation of gentamicin. Incorporation of gentamicin was determined in a similar manner to that described previously for streptomycin (Tseng, Bryan, Van Den Elzen, 1972). Specific activity of ^{3}H-gentamicin in the incorporation mixtures was 0.056 μCi/μg for gentamicin concentrations up to 2 μg/ml and 0.035 μCi/μg for concentrations above that. Measurement of cell-associated radioactivity with sodium azide (0.025%, w/v) was carried out at 37C rather than at 0–4C as described for streptomycin (Tseng, Bryan, Van Den Elzen, 1972).

Extraction of gentamicin. Bacterial cultures were exposed to ^{3}H-gentamicin as for whole cell incorporation studies for 2 hr at 37C. Cells were centrifuged at 4C at 10,000 × g for 10 min and washed 3 times with TS broth. The final cell pellet containing about 1 × × 10″ cells was resuspended in 3 ml of 6% perchloric acid and a 50 μl sample was counted in Bray's scintillation fluid in a liquid scintillation counter (= total count). The mixture was heated in boiling water for 5 min at 100C, cooled and 100 μl was removed and centrifuged in a Beckman microfuge for 1 min; 40 μl of the supernatant was counted as above (= extracted count). For paper chromatography 1 ml of hot acid treated cells was centrifuged at 20,000 × g for 10 min, the supernatant removed by centrifugation (10,000 × g × 10 min) and the supernatant chromatographed as described (Wagman, Marquez and Weinstein, 1968).

Physical studies for extrachromosomal DNA. Analytical CsCl gradient centrifugation of extracted DNA was carried out as previously described (Bryan et al., 1973). Labelling of cells with ^{32}P, isolation of covalently closed circular (CCC) DNA by dye-buoyant density centrifugation and examination of DNA by electron microscopy was done as described (Bryan et al., 1973).

RESULTS AND DISCUSSION

Survey of gentamicin, BB-K8 and tobramycin-resistant isolates of P. aeruginosa from University of Alberta Hospital (UAH) (Edmonton). Sixty-five gentamicin-resistant (broth MIC > 10 μg/ml), 20 tobramycin-resistant (MIC > 10 μg/ml) and 15 BB-K8 resistant (MIC > 20 μg/ml) strains of *P. aeruginosa* were examined for R-factor mediated resistance to these antibiotics. Recipient strains were *P. aeruginosa* strains 280 met$^-$ Rif^r (methione-requiring, rifampin-resistant) 350 Rif^r, and 211 Rif^r and *E. coli* K12 1306 Rif^r (Bryan, et al., 1973). Mating periods were 2 hr, overnight or 2 hr followed by 10 hr of incubation in a shaking water bath at 37C to allow phenotypic expression of resistance (Bryan, Shahrabadi, Van Den Elzen, 1974). No R-factors were detected in this series of strains although we have readily detected R-factors specifying gentamicin resistance in strains obtained from centers outside UAH by these methods. Strains 8803 and 1136 (see Table 1) were further examined by analytical CsCl gradient centrifugation and a band of extra-chromosomal DNA was detected in strain 8803 (Fig. 1) but not strain 1136. Ethidium bromide CsCl gradient centrifugation was carried out using phenol-extracted ^{32}P-labelled DNA from strain 1136 and on unlabelled DNA from strain 8803. Again no extra-chromosomal DNA was detected in strain 1136 but covalently closed circular (CCC) molecules of DNA were purified from strain 8803. The nature of the genetic information specified by the latter DNA is unknown. Attempts to mobilize the DNA with other R-factors (RP4, R679) (Bryan, Shahrabadi, Van Den Elzen, 1974) have been unsuccessful.

Further characterization and selection of strains. Fourteen strains of *P. aeruginosa* (Table 1) were selected to include those strains which varied from moderately resistant to resistant to gentamicin, by Kirby-Bauer (Bauer et al., 1966) testing standards and thus represented the total spectrum of resistance seen at the UAH. Strains 13934, 8803

TABLE 1

Characterization of P. Aeruginosa Strains

Strain	Motility	Growth in air	Catalase	Oxidase	Glucose (acid)	O-F test	Growth at 42°C*	Glucose oxidation	King medium A	King medium B	Growth TPTZ**	Acetamide hydrolysis	Arginine dihydrolysis
8803	+***	+	+	+	+	O + F—	+	+	+	+	—	+	+
1136	+	+	+	+	+	O + F—	+	+	+	+	+	+	+
8476	+	+	+	+	+	O + F—	+	+	+	+	+	+	+
14786	+	+	+	+	+	O + F—	+	+	+	+	+	+	+
11276	+	+	+	+	+	O + F—	+	+	+	+	+	+	+
2148	+	+	+	+	+	O + F—	+	+	—	+	+	+	+
996	+	+	+	+	+	O + F—	+	+	—	+	+	+	+
14043	+	+	+	+	+	O + F—	+	+	+	+	+	+	+
8625	+	+	+	+	+	O + F—	+	+	+	+	+	+	+
13934	+	+	+	+	+	O + F—	+	+	+	+	+	+	+
1444	+	+	+	+	+	O + F—	+	—	+	+	+	+	+
PMU78	+	+	+	+	+	O + F—	+	+	+	+	+	+	+
10804	+	+	+	+	+	O + F—	+	—	+	+	+	+	+

* = Three consecutive transfers.
** = One per cent triphenyltetrazolium chloride in nutrient agar No. 1.
*** + = positive reaction for that test, for King medium + indicates the presence of pigment.

and 14786 (see Table 1) were obtained from the same patient who was under treatment with topical gentamicin for *P. aeruginosa* infection of an extensive burn. All of these strains were of the same pyocine type (Table 2). Five different pyocine types occurred

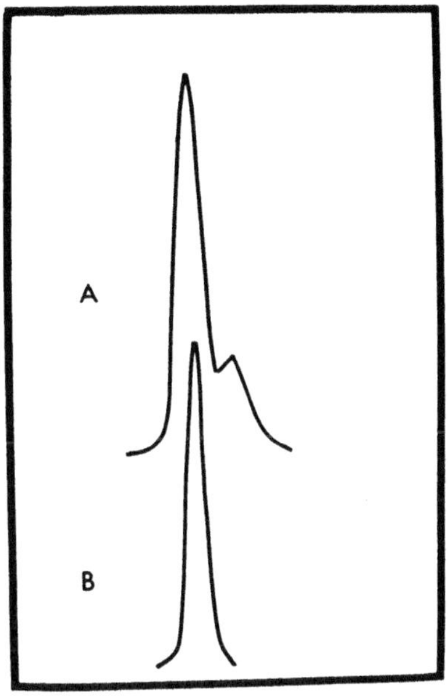

Fig. 1. CsCl density gradient centrifugation of *P. aeruginosa* strains 8803 (A.) and 1136 (B.).

in the 14 strains (Table 2) and six different hospital wards were represented. Clinical cases included burns, a renal infection, an infection of a decubitus ulcer, an infected skin graft and other skin infections.

Strains were characterized as *P. aeruginosa* by previously described criteria (Bryan, Van Den Elzen, Tseng, 1972). The results of those studies are shown in Table 1.

Minimal inhibitory concentrations (MIC) of gentamicin and other aminoglycosides.

MIC values ranged from 6.25 μg/ml to 100 μg/ml using TS broth as the growth medium (Table 3). MIC values have been shown to be dependent on the medium in use and specifically the divalent cation concentration (Gilbert et al., 1971). The divalent cation concentration of broth is less than equivalent agar containing medium. Thus MIC values in excess of 6–10 μg/ml with a broth system would be considered clinically resistant. Strains were also grown in the presence of subinhibitory concentrations of gentamicin for 7 days and their MIC values re-determined. From the results in Table 3 it is clear that most of the strains undergo an increase in resistance and all would be considered clinically resistant. These strains did not change growth characteristics as do strains in which gentamicin resistance is developed by serial sub-culture with increasing gentamicin concentration in the medium (Weinstein et al., 1971). Strain 280 GSK^r is an example of a laboratory induced gentamicin-resistant strain (Tseng, Bryan, Van Den Elzen, 1972) in that its generation time is in excess of 2 hr, it no longer grows on minimal medium and is of lowered animal virulence. The strains shown in Table 3 exhibited none of these characteristics following serial transfer in medium with 5 μg/ml gentamicin. Thus the increase in resistance level would seem of importance in that exposure to subinhibitory gentamicin levels *in vivo* might have a similar effect.

Table 3 also illustrates the resistance pattern of these strains to tobramycin, sisomicin, and BB-K8. Only 2 of the gentamicin-resistant strains are clearly sensitive to tobramycin with the remainder being resistant although more susceptible to tobramycin than genta-

TABLE 2

Pyocine Typing of Gentamicin-Resistant P. aeruginosa

Strain	Type	Sub-Type
8803	16	
1136	3	
8476	3	
14786	16	
11276	3	
2148	NT	
966	3	
14043	16	
8625	16	
13934	16	
1444	1	
PMU78	1	c
10804	12	c

Pyocine typing was by the method of Govan and Gillis (1969).

TABLE 3

MIC Values of Several Aminoglyside Antimicrobial Agents for P. aeruginosa Gentamicin-Resistant Strains

	MIC (μg/ml)				
	Gentamicin		Tobramycin	BB K8	Sisomicin
	Untreated	Pretreated*			
8803	100	200	50	>100	100
1136	50	100	25	>50	50
8476	50	100	25	>50	50
14786	50	100	25	>50	50
11276	50	100	25	>50	50
2148	25	50	12.5	37,5	25
966	25	100	12.5	37.5	25
14043	25	100	12.5	37.5	25
8625	25	100	12.5	37.5	25
13934	25	100	12.5	37.5	25
1444	12.5	50	6.25	25	12.5
10804	6.25	25	0.75	25	1.5
PMU78	6.25	20	0.75	12.5	0.75
280**	0.25	—	0.375	0.75	0.15
3503**	0.6	—	0.3	0.6	0.3

* Pre-treated cells were grown in the presence of gentamicin (5 μg/ml) for 7 serial transfers.

** Strains considered susceptible to gentamicin.

micin. Thirteen of 14 strains have an MIC of 20 μg/ml for BB-K8 and are considered clinically resistant. Thus most of the strains are relatively resistant to all four newer aminoglycosides and, in addition, have been found resistant to streptomycin and kanamycin.

Antibiotic inactivating enzymes. Microbiological and radioactive assays were used to examine cell extracts from each of the 14 strains as well as control strains containing gentamicin-acetylating or adenylylating activity and a known susceptible strain (Table 4).

TABLE 4

Gentamicin Adenylylating or Acetylating Activity in Strains of P. aeruginosa

Strain	Assay***			
	Acetylating		Adenylylating	
	Microbiological	Radioactive	Microbiological	Radioactive
8803*	—	—**	—	—**
1136*	—	—	—	—
8476	—		—	
14786	—		—	
11276	—	—	—	—
966	—		—	
14043	—		—	
8625	—		—	
13934*	—	—	—	—
1444	—		—	
PMU78	—	—	—	
10804	—		—	
PS130 (R130)	+	+**	—	—
280 (R130)	+	+	—	—
POW151	—	—	+	+**
280	—	—	—	—
3503	—	—	—	—

* Also contain no detectable kanamycin acetylating or phosphorylating activity, streptomycin phosphotransferase, or gentamicin phosphorylating activity.

** $<1 \times 10^2$ protein = —
$>6 \times 10^3$ cpm/mg protein = +.

*** See methods for details of assay procedure.

No evidence of either activity was detected in any of the 14 strains although control strains were positive and we have detected both acetylating and adenylylating activity in strains of *P. aeruginosa* from other sources (Bryan, Shahrabadi, Van Den Elzen, 1974). Strains 8803, 13934 and 1136 were examined and found to be negative for gentamicin phosphorylation. This form of modification has been previously detected although it does not cause strains to be gentamicin-resistant (Davies et al., 1969). Strains 8803, 1136 and 13934 were examined for streptomycin phosphotransferase and kanamycin and BB-K8-acetylating activity but none was detected. Thus known forms of enzymatic activity which could account for the resistance to gentamicin, tobramycin or BB-K8 were not detected in any strain examined.

Cell-free amino acid incorporation studies. Tables 5 and 6 outline results of cell-free amino acid incorporation studies using as messenger RNA either R17 RNA (Table 5)

TABLE 5

Effect of Gentamicin and Tobramycin on the Incorporation of ^{14}C Phenylalanine in a Cell-Free System directed by Poly U*

Strain	μμmoles incorporated in 25 μl mixture – gentamicin	μμmoles incorporated in 25 μl mixture + gentamicin	Δ μμmoles	% inhibition	μμmoles incorporated in 250 μl mixture + tobramycin	Δ μμmoles	% inhibition
8803**	58.1	25.1	33.0	56.7	24.6	33.5	57.6
1136	73.9	23.7	50.4	67.8	29.7	44.1	59.7
8476	256.7	148.9	107.8	42.0			
14786	266.9	107.0	159.8	59.8			
11276	80.9	31.5	49.4	61.0	26.0	54.8	67.7
2148	69.9	22.5	47.4	67.8			
966	99.6	30.8	68.8	69.0			
14043	311.5	104.5	207.0	66.4			
8625	87.0	19.3	67.6	77.7	12.3	74.7	85.8
13934	67.1	20.9	46.2	68.8	16.1	50.9	75.8
1444	225.4	88.8	136.6	60.5			
PMU78	33.6	16.9	16.7	49.6	15.5	18.1	53.8
10804	47.7	32.7	15.0	31.5	14.8	32.9	69.0
3503	225.1	88.5	136.6	60.5			
280	502.5	192.1	310.3	61.8	180.2	322.3	64.1
8803RS**	78.8	36.7	42.1	53.4			

* Concentration of gentamicin and tobramycin = 1 μg/ml.

** Ribosomes were obtained from the strains listed and the S-100 fraction (see Methods) from strain 280 for all experiments except for the repeat listing of strain 8803 (noted as 8803RS) where both ribosomes and S-100 fractions were obtained from strain 8803.

TABLE 6

Effect of Gentamicin C Complex on the Incorporation of ^{14}C Valine in a Cell-Free System Directed by R17/RNA*

Strain	μμmoles incorporated in 250 μl —gentamicin	μμmoles incorporated in 250 μl +gentamicin	Δ μμmoles	% inhibition
8803**	19.8	6.6	13.1	66.4
1136	73.4	46.3	27.1	36.9
8476	47.2	25.9	21.2	45.1
14786	48.4	27.3	21.0	43.0
11276	36.1	21.1	14.9	41
2148	84.4	37.3	47.1	55.7
966	93.1	43.7	49.3	53.0
14043	68.4	38.6	29.8	43.5
8625	83.8	38.1	45.7	54.5
13934	28.4	10.5	17.8	62.8
1444	26.2	16.2	10.0	38.5
280	35.9	16.4	19.5	54.2
8803RS**	75.2	41.4	33.8	55

* gentamicin – 10 μg/ml.

** Ribosomes were obtained from the strains listed and the S-100 (see Methods) fraction for all experiments from strain 280 except for the repeat list of strain 8803 (noted as 8803RS) where ribosomes and S-100 fractions were obtained from strain 8803.

or polyuridiylic acid (Poly U) (Table 6). In these studies the non-ribosomal components of the system were obtained as 100,000 × g supernatants (S-100) (see methods) from the gentamicin susceptible strain 280. Ribosomes were prepared from the respective resistant strains. In both the R17 RNA and Poly-U directed systems the ribosomes from resistant strains were generally as susceptible to gentamicin and in selected cases to tobramycin as ribosomes from strain 280.

In order to examine the possibility that non-ribosomal components of protein synthesis were resistant to gentamicin, an S-100 fraction from the resistant strain 8803 was used with ribosomes from strain 8803 in both the Poly-U directed (Table 5) and in the R-17 RNA directed (Table 6) systems. Finally the effect of gentamicin on ^{14}C-labelled amino acid incorporation by endogenous RNA from strain 8803 was also examined (Table 7).

TABLE 7

Effect of Gentamicin and Streptomycin on Incorporation of Amino Acids in a Cell Free System Directed by Endogenous Messenger RNA

Strain	MIC of gentamicin (μg/ml)	Experiment number	dpm* of ^{14}C amino acids incorporated** No antibiotic	+ gentamicin	% inhibition	+ streptomycin	% inhibition
8803	100	1	596	84	84	67	89
		2	714	285	60	128	75
		3	688	146	79	111	84
3503	0.6	1	786	196	75	102	87
		2	1096	428	61	340	69
		3	542	195	64	189	65

* dpm-disintegrations per minute, counting efficiency – 60%.

** Each assay (0.7 ml) contains 0.5 ml of deoxyribonuclease-treated extract from alumina ground cells (Nirenberg, 1963) as a source of mRNA, tRNA, ribosomes and other non-ribosomal component of protein synthesis; 0.7 nmoles of all amino acids (12 of the 20 amino acids were labelled with ^{14}C each with a specific activity of 50 μCi/μmole; Schwarz/Mann); ATP, ATP-generating system, and other reagents were added as described (Tseng, Bryan, Van Den Elzen). Magnesium concentration was 9 mMolar. Concentrations of gentamicin were 10 μg/ml and streptomycin 20 μg/ml. Sample size was 50 μl and incubation was at 34C for 30 min. Control tubes (zero time) dpm were 100–150 and have been substracted from the values given above.

In each of these circumstances gentamicin inhibited amino acid incorporation, confirming the susceptibility of all components of the system to gentamicin. We did not detect a stimulatory effect of gentamicin on amino acid incorporation with endogenous RNA as reported by Milanesi and Ciferri (1969) using cell-free extract from *Escherichia coli*. Table 7 also shows the effect of streptomycin on the endogenous amino acid-incorporating system. Strain 8803 was resistant to streptomycin while strain 3503 was sensitive. Clearly, as expected, streptomycin inhibits amino acid incorporation. Thus, no evidence of resistance of the protein-synthesizing system to gentamicin in resistant strains could be demonstrated in the strains tested.

Chromatography of extracellular and extracted gentamicin. Purified gentamicin (Schering Corpn.) was subjected to tritium labelling by Amersham-Searle. The product of this procedure was purified and characterized as described in the Methods. The paper chromatography profile of labelled gentamicin was similar to that shown in Fig. 2.

The profile corresponded exactly to ninhydrin positive spots of chromatographed purified unlabelled gentamicin. Cells of all 14 resistant strains were individually exposed to ^{3}H-gentamicin (40 μg/ml, specific activity – 3 × 10^4 dpm/ug) for 2 hr at 37C in TS broth and the cell-free supernatant obtained after filtration (0.2 μ filter) was examined by bio-assay and paper chromatography. No detectable loss of biological activity nor change in the chromatograph pattern occurred with any of these strains when compared to assays containing standard gentamicin.

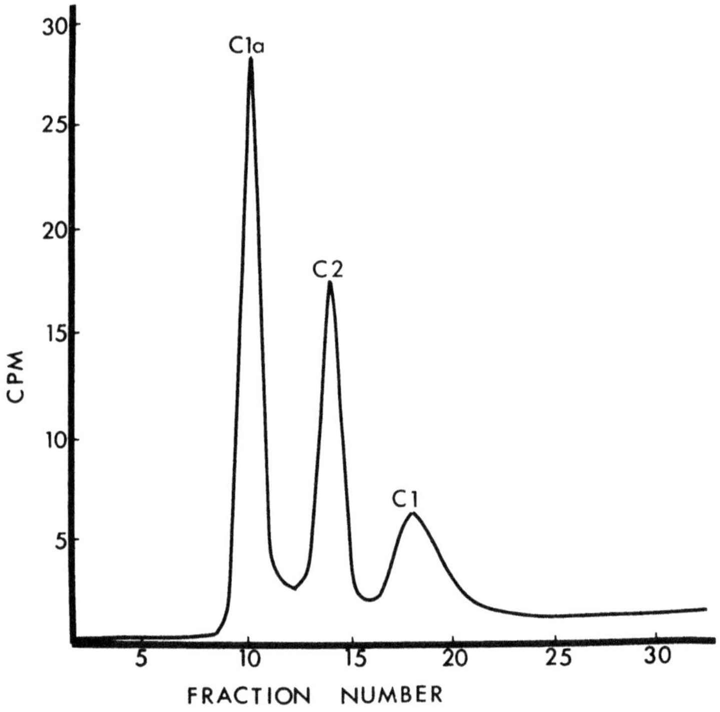

Fig. 2. Paper chromatography of hot acid treated (100C for 5 mins) purified ^{3}H-gentamicin. Following development the chromatogram was divided into 1 × 2.5 cm strips and these were counted using a liquid scintillation system. The three peaks corresponded to those obtained with untreated purified ^{3}H-gentamicin and with standard unlabelled gentamicin (detected by reaction with ninhydrin).

Gentamicin was extracted by a hot acid method from washed cells of strain 3503 and strain 8803 after exposure of these strains in TS broth at 37C for 2 hr to 0.625 and 100 μg/ml gentamicin respectively (see Methods). Treatment of gentamicin with hot acid did not change the chromatographic profile of gentamicin (Fig. 2) or the biological activity of the drug. Greater than 99% of radioactivity was recovered from cells of both strains. Bioassay of the recovered filtered extracted gentamicin showed that relative to gentamicin standards 35% of the gentamicin activity was lost. However, a comparison of the biological activity of the extracts obtained from the resistant (strain 8803) and sensitive (strain 3503) strains showed equal activity. Similarly the paper chromatographic profile of the latter extracts (Fig. 3) was the same and was similar to gentamicin standards though less C_2 component of the gentamicin complex was recovered perhaps due to lower cellular incorporation of that gentamicin component.

The preceding studies indicate that there is no detectable evidence of inactivation of gentamicin by a resistant strain. The loss of potency of gentamicin after hot acid extraction is felt to be due to gentamicin binding to cellular components and is the same for gentamicin extracted from either sensitive of resistant strains.

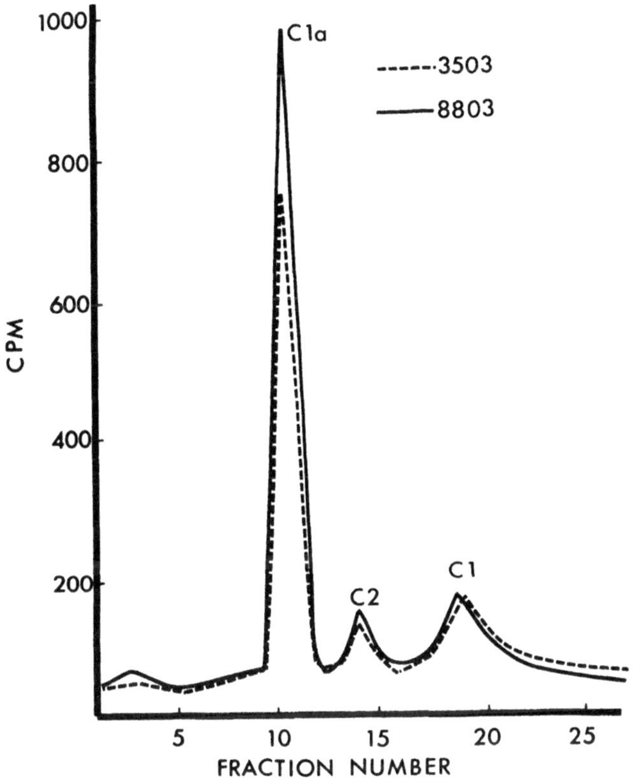

Fig. 3. Paper chromatography of hot acid extracted gentamicin from strains 8803 and 3503. Chromatograms were treated as given in Fig. 2.

Relationship of whole cell uptake of gentamicin to gentamicin resistance.

When cells of *P. aeruginosa* strain 3503 are exposed to ^{3}H-gentamicin two phases of gentamicin accumulation occur (Fig. 4). The primary phase is rapid and does not significantly increase with time. Primary uptake does increase with gentamicin concentration being directly proportional up to at least 100 μg/ml gentamicin in the uptake medium (not shown). A secondary phase of gentamicin accumulation is observed after about 15 min providing the concentration of gentamicin is approximately equal to or greater than the MIC of gentamicin for the strain. Fig. 4 illustrates the primary and secondary uptake for strain 3503 at 0.625 μg/ml gentamicin. Secondary uptake can be eliminated by exposing cells to 0.025% (w/v) sodium azide (Fig. 4) or by 4C (not shown). A similar situation has been described for streptomycin (Dubin, Hancock, Davis, 1963:

Tseng, Bryan, Van Den Elzen, 1972). The secondary, or azide sensitive phase of gentamicin uptake is not seen with strain 1136 at 0.625 μg/ml gentamicin (Fig. 5). In fact secondary uptake begins for strain 1136 at about 50 μg/ml gentamicin (Table 8) which is the MIC of gentamicin for that strain as well as the concentration of gentamicin which reduced incorporation of ^{14}C-proline into acid insoluble cell materials (presumably protein) by 50%.

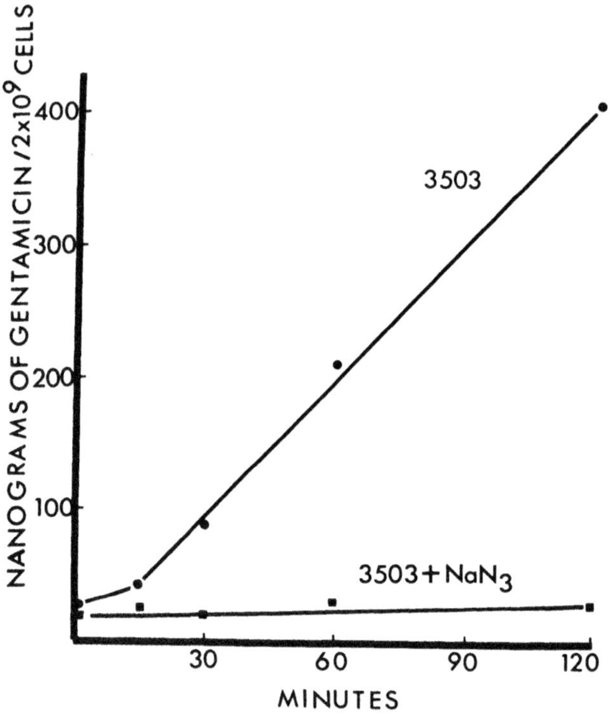

Fig. 4. Whole cell uptake of ^3H-gentamicin by strain 3503 with or without 0.025% sodium azide (NaN_3). Cells were exposed to 0.625 μg/ml gentamicin (+ or — NaN_3) at 37C in TS broth for 60 mins and then washed three times with TS broth. The final pellet was resuspended in 1.2 ml TS broth, treated with ultra-sound and 0.3 ml counted in a liquid scintillation system.

Table 8 demonstrates the relationship between secondary gentamicin uptake and MIC of gentamicin for several strains of *P. aeruginosa*. For each of the five strains examined uptake begins at or about the MIC of gentamicin for that strain. Additional evidence of the clear correlation of MIC and secondary gentamicin uptake is seen in Table 9. Secondary uptake of strain 13934 was determined before and after seven daily transfers in 5 μg/ml gentamicin in TS broth. This resulted in an increase of MIC of from 25 μg/ml to 100 μg/ml gentamicin and a corresponding elimination of uptake of gentamicin at 30 μg/ml gentamicin in the uptake medium. The strain did not change generation time and was able to grow on minimal medium after the seven daily transfers.

Two explanations of the relationship of MIC and secondary gentamicin uptake may be made – either inhibition of protein synthesis and cell growth is the result of secondary gentamicin uptake or is the cause of such uptake. Streptomycin has been shown capable of producing leakage of cell membranes and secondary uptake of that drug has been ascribed to such leakage which occurs as a result of inhibition of protein synthesis (Dubin, Hancock and Davis, 1963). Our own studies do not currently support that hypothesis (manuscript in preparation). We have shown secondary uptake, although reduced in magnitude, still occurs in a strain showing high-level streptomycin resistance, due to a mutation affecting ribosomes and thus occurs in the absence of inhibition of protein synthesis. In this strain secondary uptake is initiated at the MIC of the parent

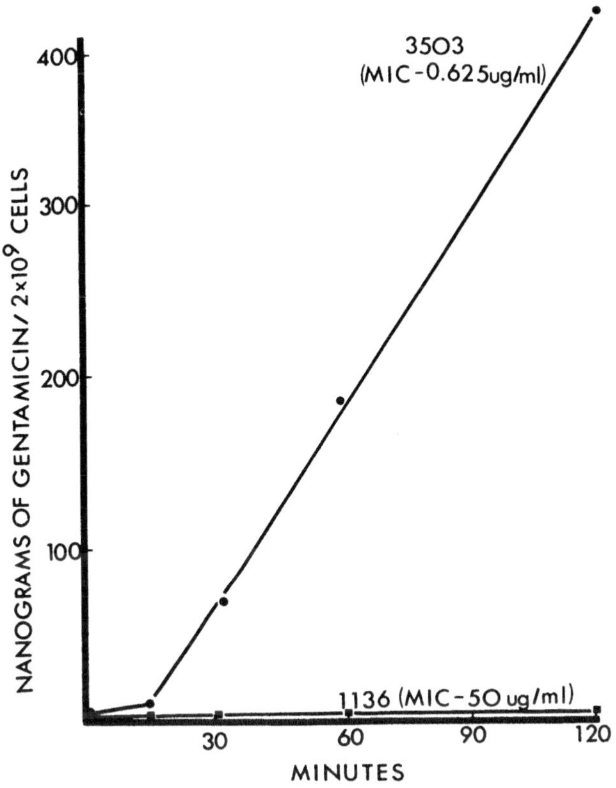

Fig. 5. Whole cell uptake of ^{3}H-gentamicin by strains 3503 and 1136 at 0.625 μg/ml gentamicin. Uptake was performed as given in Fig. 4. Values given for uptake were determined by subtracting uptake in the presence of 0.025% NaN_3 from that without NaN_3.

and not the mutant strain and is not associated with leakage of intracellular nucleotides. The introduction of R-factors specifying gentamicin acetylation or streptomycin phosphorylation produces similar results. Thus for streptomycin resistance due to a mutation affecting the ribosome or streptomycin phosphorylation and gentamicin resistance due to gentamicin acetylation, inhibition of protein synthesis, MIC and membrane leakage are dissociated from secondary streptomycin or gentamicin uptake (manuscript in preparation). We feel that primary uptake is extra-membranous as exposure of susceptible strains to gentamicin or streptomycin concentrations several times the MIC in the presence of 0.025% sodium azide or 4C is associated with primary drug uptake without any significant loss of cell viability (manuscript in preparation).

TABLE 8

Whole Cell Uptake of Gentamicin by Strains of P. aeruginosa at various Gentamicin Concentrations

Strain	MIC (μg/ml)	Gentamicin concentration (μg/ml)									
		0.3125	0.625	1.25	2.5	5.0	10.0	25.0	50.0	75.0	100
3503	0.625	11.0[a]	*170.0*[b]	220.0	420	905.0	2727.4				
1310	1.6	2.67	26.7	*320.2*[b]	672.2	1380.0					
8803	100	0	0	0	0	0	0	0	0	0	*820.7*[b]
1136	50.0	0	0	0	0	0	0	0	*471.1*[b]		
113934	25.0	0	0	0	0	0	0	*132.1*[b]			

[a] The gentamicin uptake was calculated by subtracting the amount of gentamicin bound in the presence of 0.025% NaN_3 (primary uptake) from the gentamicin bound in the absence of 0.025% NaN_3 and given as nanograms of gentamicin per 2×10^9 cells.

[b] Values represent gentamicin uptake at or about the MIC.

TABLE 9

Whole Cell Uptake of Gentamicin by P. aeruginosa 13934 before and after daily Transfer in Gentamicin (5 μg/ml) for 7 Days

Strain	MIC (μg/ml)	Gentamicin uptake* ng/10^9 cells	% decrease of uptake
13934	25	97.6	
13934**	100	0	100

* Uptake at 30 μg/ml gentamicin. Uptake was assessed as described in Table 8.

** Strain 13934 transferred daily in TS broth containing 5 μg/ml gentamicin for 7 days.

The absence of any data supporting inactivation of gentamicin or demonstrating the insensitivity to gentamicin of any component of cellular protein synthesis is suggestive that gentamicin resistance may be due to failure of gentamicin to penetrate into bacterial cells. This concept is further supported by the close relationship of the MIC and secondary uptake of gentamicin. Our studies to be reported elsewhere indicate that such a relationship is not due to membrane leakage resulting from cellular effects produced at the MIC. The strains reported here are also resistant to several other aminoglycoside antibiotics and no evidence of inactivation of streptomycin, BB-K8 or tobramycin has been detected in several strains. Until the mechanism of gentamicin entry into bacterial cells is better understood, a final conclusion concerning the mode of resistance in these bacterial strains cannot be made. However, the weight of evidence at the present time clearly favours some form of defective permeability to gentamicin as the explanation of the gentamicin resistance.

ACKNOWLEDGMENTS

The study reported here was supported by Medical Research Council of Canada grant MT. 4350.

REFERENCES

BAUER, A. W., KIRBY, W. M., SHERRIS, J. C., and TURCK, M. (1966). Antibiotic susceptibility testing by a standardized single-disk method. Amer. J. Clin. Pathol. **45**: 493–496.

BENVENISTE, R. and DAVIES, J. (1971). Enzymatic acetylation of aminoglycoside antibiotics by *Escherichia coli* carrying an R-factor. Biochemistry **10**: 1787–1796.

BRYAN, L. E., VAN DEN ELZEN, H. M. and TSENG, J. T. (1972). Transferable drug resistance in *Pseudomonas aeruginosa*. Antimicrob. Ag. Chemother. **1**: 22–29.

BRYAN, L. E., SEMAKA, S. D., VAN DEN ELZEN, H. M., KINNEAR, J. E. and WHITEHOUSE, R. L. S. (1973). Characteristics of R931 and other *Pseudomonas aeruginosa* R-factors. Antimicrob. Ag. Chemother. **3**: 625–637.

BRYAN, L. E., SHAHRABADI, M. S. and VAN DEN ELZEN, H. M. (1974). Gentamicin resistance in *Pseudomonas aeruginosa*. I. R-factor specified resistance. Antimicrob. Ag. Chemother. (In Press).

BRZEZINSKA, M., BENVENISTE, R., DAVIES, J., DANIELS, P. J. L. and WEINSTEIN, J. (1972). Gentamicin resistance in strains of *Pseudomonas aeruginosa* mediated by enzymatic N-acetylation of the deoxystreptamine noiety. Biochemistry. **11**: 761–765.

DAVIES, J., BENVENISTE, R., KVITEK, K., OZANNE, B. and YAMADA, T. (1969). Aminoglycosides: Biologic effects of molecular manipulation. J. Infect. Dis. **119**: 351–354.

DUBIN, D. T., HANCOCK, R., DAVIS, B. D. (1963). The sequence of some effects of streptomycin in *Escherichia coli*. Biochim. Biophys. Acta. **74**: 476–489.
GILBERT, D. N., KUTSCHER, E., IRELAND, P., BARNETT, J. A. and SANFORD, J. P. (1971). Effect of concentration of magnesium and calcium on the *in vitro* susceptibility of *Pseudomonas aeruginosa* to gentamicin. J. Infect. Dis. **124**: 537–545.
GOVAN, J. R. W. and GILLIES, R. R. (1969). Further studies in the pyocine typing of *Pseudomonas pyocyanea*. J. Med. Microbiol. **2**: 17–25.
MAHON, W. A., EZER, J. and WILSON, T. W. (1973). Radioimmunoassay for measurement of gentamicin in blood. Antimicrob. Ag. Chemother. **3**: 585–589.
MILANESI, G. and CIFERRI, O. (1966). Studies on the mechanism of action of gentamicin. Effect on protein synthesis in cell-free extracts of *Escherichia coli*. Biochemistry. **5**: 3926–3935.
NIRENBERG, M. W. (1963). Cell-free protein synthesis by messenger RNA. *In* P. C. Colowick and N. O. Kaplan (Eds.) *Methods in Enzymology*. Vol. 6, pp. 18–23. Academic Press, New York.
SHAHRABADI, M. S., BRYAN, L. E. and VAN DEN ELZEN, H. M. (1974). Further properties of P-2 R-factors of *Pseudomonas aeruginosa* and their relationship to other plasmid groups. Antimicrob. Ag. Chemother. (In Press.)
TSENG, J. T., BRYAN, L. E. and VAN DEN ELZEN, H. M. (1972). Mechanisms and spectrum of streptomycin resistance in a natural population of *Pseudomonas aeruginosa*. Antimicrob. Ag. Chemother. **2**: 136–141.
WAGMAN, G. H., MARQUEZ, J. A. and WEINSTEIN, M. J. (1968). Chromatographic separation of the components of the gentamicin complex. J. Chromatog. **34**: 210–215.
WEINSTEIN, M. J., DRUBE, C. G., MOSS, E. L. Jr. and WAITZ, J. A. (1971). Microbiological studies related to bacterial resistance to gentamicin. I. Infect. Dis. **124**: 511–517.

BROAD HOST RANGE DRUG RESISTANCE PLASMID ASSOCIATED CELL WALL ALTERATIONS AND THEIR RELATIONSHIP TO PHAGE SENSITIVITY

R. H. OLSEN

Department of Microbiology, The University of Michigan, Ann Arbor, U.S.A.

This investigation was supported by Public Health Service research grant AI 07533 from the National Institute of Allergy and Infectious Diseases.

Drug resistance factors originating in hospital isolates of *Pseudomonas aeruginosa* have been shown to transfer these determinants to other strains of *P. aeruginosa* and also to mobilize the transfer of chromosomal determinants in several instances (Stanisich and Holloway, 1971). In addition, the transfer of multiple drug resistance to unrelated species from *P. aeruginosa* has been reported by several laboratories by Bryan *et al.*, (1972), Datta *et al.*, (1971), Fulbrook *et al.*, (1970), Grinsted *et al.*, (1972), Smith and Armour, (1966), Sykes and Richmond, (1970) and our own laboratory (Olsen and Shipley, 1973). The list of species capable of maintaining broad host range drug resistance plasmids among Gram-negative bacteria is extensive. It includes bacteria distinctly unrelated when compared on the basis of their physiological activities or nucleic acid composition. Accordingly, we considered that the possible expression of plasmid specified phage receptors in bacterial species having otherwise disparate properties might greatly facilitate the isolation of plasmid specific phages plating on any host containing a given plasmid and capable of expression of plasmid specified phage receptors. We have been testing this possibility in our laboratory recently and have isolated several phages of varied types whose host range extends beyond related Gram-negative bacteria as does the drug resistance plasmid required for adsorption of the phages.

Our work has focused on two considerations regarding drug resistance plasmid dependent phages.

1. How may plasmids change cells resulting in phage sensitivity?
2. What may be concluded regarding the origin of plasmid genes required for a productive phage infection?

Many surface and subsurface components of Gram-negative bacteria are known to function as phage adsorption sites. They include nonessential surface layers (capsule or slime), cellular appendages (flagella or pili) as well as structural components of the cell wall. Among the cell wall components, lipoprotein and lipopolysaccharide (LPS) components have been shown specific for the adsorption of a particular phage. In fact, altered phage sensitivity reflecting mutations in the synthesis of LPS components has been useful in the description of these cell structures (for a recent review see Rapin and Kalckar, 1971). In addition, changes in LPS composition have been shown to affect other properties of the cell in addition to an alteration of phage sensitivity. Tamaki *et al.*, (1971) showed a relationship between LPS content and phage adsorption or antibiotic resistance in *Escherichia coli* K12. A similar change in permeability to antibiotics has been demonstrated by Schlect and Westphal (1970) for *Salmonella typhimurium* subsequent to genetic alterations in its LPS. Plasmid containing bacteria often exhibit many properties associated with the presence of alterations in LPS structure when compared with isogenic non-plasmid containing strains. For example, the presence of plasmids has been shown to result in altered colicin sensitivity patterns, increased detergent sensitivity and resistance to heavy metals (Novick, 1969). Drug resistance plasmids have also been shown to effect changes in permeability of *E. coli* to rifampicin and other antibiotics (Riva *et al.*, 1972).

Thus, from even a cursory perusal of the LPS and plasmid literature, it is clear that plasmids may cause direct changes in the phenotypic properties of cells that resemble changes heretofor associated with alterations in the outer wall of bacteria not containing plasmids. In view of the correspondence of these observations. we have considered the possibility that plasmid directed changes in cell wall may also be expressed as the deposition of wall specific phage receptor sites. This report describes the behavior of several phages whose infectivity depends on the presence of a broad host range drug resistance plasmid.

Plasmid dependent phages whose adsorption is dependent upon the production of pili associated with the presence of a plasmid have been described previously. F pili produced by the *E. coli* sex factor have been shown to function as the adsorption sites for a large number of independently isolated RNA phages. Similarly, filamentous single stranded DNA containing phages have been isolated adsorbing to the tips of F pili. In most instances, these *E. coli* sex factor specific phages will also adsorb to pili produced by F related drug resistance factors designated fi^+ factors. Thus, in this instance, the presence of the drug resistance factor and its associated pili in a bacterium confers phage sensitivity. A filamentous phage specific for type I pili produced by plasmids belonging to the I incompatibility group of plasmids also has been described by Lawn *et al.*, (1967). Another plasmid dependent phage, Ike, has been isolated by Khatoon and Iyer which is specific for plasmids of incompatibility group N (1971). We have recently reported the isolation of an RNA phage specific for RP1 or RP4, members of the P incompatibility group of drug resistance plasmids (Olsen and Thomas, 1973).

PROCEDURES AND RESULTS

Plasmid specific phages were isolated from filtered sewage. For this, a broth culture of plasmid containing bacteria at a cell density of approximately 10^8/ml was added to broth and filtered sewage in the proportion of 1 : 1 : 1. This was incubated overnight at 37°C under static conditions, centrifuged and the supernate filtered. These filtered supernates were then plated by the overlay technique using unrelated bacteria containing the same plasmid as used for initial sewage enrichment as bacterial indicators. Plaques were then picked and purified by the usual procedures. In most instances, either *E. coli* or *P. aeruginosa* was used as the primary enrichment host or for plaque isolations. These bacterial strains have been described in previous reports from our laboratory (Olsen and Shipley, 1973; Olsen and Thomas, 1973). The phages isolated by this procedure were initially grouped on the basis of their sensitivity to ribonuclease (RNase) inhibition of infectivity. One of the phages isolated by the procedure, an RNase sensitive phage designated PRR1, is unique in that its broad host range resembles that observed for the RP1 plasmid required for the production of its pilus receptors. A summary of the host range of PRR1 is shown in Table 1. As we have previously reported (Olsen and

TABLE 1

PRR1 host range on RP1 containing bacteria

RP1 indicator strain	Strain	Lytic reaction
A.		
Pseudomonas aerigunosa	PA067	Forms plaques of variable clearness
	PAT904	
Pseudomonas fluorescens	PF015.4	
Pseudomonas putida	PPO13.12	
Escherichia coli	B	
	C	
	K12	
	0111 : B4	
Salmonella typhimurium	LT2	
Vibrio cholerae	596B	
B.		
Actinetobacter calcoaceticus	ACJ1	No individual plaques; does elicit inhibition zones when spotted on surface inoculated bacterial lawns
Neisseria perflava	NPM	
Azotobacter vinelandii	AVM	
Proteus mirabilis	PM17	
C.		
Shigella boydii	SBM	No phage susceptibility but conjugal RP1 donor

TABLE 2

DNA phage PRD1 host range on RP1 containing bacteria

RP1 indicator strain	Lytic reaction
A. Strains forming plaques	
P. aeruginosa PA067	Plaque morphology variable with the isolate tested
PAT904	
P. fluorescens PP015.4	
P. putida PP013.12	
E. coli B	
C	
K12	
S. typhimurium LT2	
P. mirabilis PM17	
V. cholerae 596B	
A. calcoaceticus ACJ1	
B. Strains not forming plaques	
A. vinelandii AVM	No plaques or inhibition of bacterial growth
N. perflava NPM	
S. boydii SBM	

Thomas, 1973) this phage does not infect non-plasmid containing (R^-) strains. The host range of PRR1, based on the inhibition of surface lawns spotted with concentrated phage suspension, is also less than that observed for the RP1 plasmid. However, this may merely reflect the fragility or masking of RP1 pili in some bacteria since PRR1 insensitive RP1 containing hosts can still act as low frequency donors in conjugal mating. We have also tested the susceptibility of bacteria containing plasmids of the A, C, F (fi+), J, W and T incompatibility groups to PRR1. They were insensitive by plating criteria and also were unable to show a phage titer increase in broth experiments. Thus, the behavior and specificity of PRR1 for the P group plasmids parallels that of the male specific for F related plasmid containing bacteria although PRR1 is distinctly unrelated to the male specific phages.

Several phages were isolated using RP1 containing bacteria which could be distinguished from the foregoing on the basis of their RNase insensitivity, plaque morphology and growth characteristics. We have studied one of these in some detail and it is the subject of another report (Olsen, Siak and Gray, 1974). This phage, designated PRD1, has not been observed to plate on R⁻ bacteria tested in our laboratory and thus requires the presence of a plasmid for its infectivity. The host range of PRD1 among RP1 containing bacteria tested is shown in Table 2. The host range of PRD1 is somewhat less extensive than that observed previously for RNA phage PRR1. The PR1 containing bacteria shown in section B, Table 2, can act as conjugal donors of RP1 but have never been observed to propagate phage PRD1 in broth or semisolid plating medium. Thus, apparently not all strains are capable of expression of the PRD1 receptor site, or alternatively, it is masked by the presence of other cell wall components.

TABLE 3

Host range of DNA phage WRD1

Bacterial strain	Plasmid	Lytic reaction
E. coli J53	none	—
	R388	+
	Rsa	+
K12	none	+
C	none	+
B	none	—
	R388	—
	Rsa	—
S. typhimurium LT2	none	—
	R388	—
	Rsa	—
P. aeruginosa PA067	none	—
	R388	—
	Rsa	—
PAT904	none	—
	R388	+
	Rsa	+

+ = phage sensitive; — = insensitive

We have also examined the PRD1 phage sensitivity of *E. coli* strains containing plasmids of other incompatibility groups. Of the incompatibility groups tested, representatives of groups N and W indicated susceptibility to phage PRD1. However, one plasmid containing *E. coli* strain of group W, J53(R7K), showed no susceptibility to PRD1. Therefore, within a recognized incompatibility group, variable expression of the PRD1 receptor site was observed. Of further interest was the observation that bacteria containing W or N group plasmids required 0.85 percent NaCl in tryptone glucose yeast extract medium to demonstrate PRD1 susceptibility whereas P group plasmid containing bacteria plated PRD1 on this medium in the absence of NaCl. Therefore, PRD1 plating on bacteria containing either, P N or W group plasmids indicate that these otherwise unrelated plasmid groups may contain functionally equivalent or identical genetic information for the specification of the PRD1 receptor. Observations of PRD1 phage adsorbed to P, N or W group plasmid containing bacteria using electron microscopy show PRD1 adsorbed to the cell wall of these bacteria and not pili as observed for PRR1 phage. Thus, these plasmids are altering the cell wall of susceptible bacteria resulting in the occurrence of the PRD1 receptor site.

We have also isolated a third type of plasmid dependent phage, based on its host range and plating behavior, which we have designated WRD1. This phage was initially isolated from sewage using bacteria containing the W group plasmid, R388. The plating behavior of WRD1 is unlike the foregoing and unique in several respects. It will productively infect some R^- strains of *E. coli* although with other strains the presence of a W group plasmid is required for plating. Furthermore, *E. coli* B, *S. typhimurium* LT2, or *P. aeruginosa* PAO strains containing R388 or Rsa have not been observed to plate WRD1 although the W group plasmids are stably maintained in these bacteria and express the expected resistance determinants. Plating behavior of WRD1 corresponding to these observations is shown in Table 3. As indicated in foregoing observations on PRD1 phage, the W group of plasmids also specify the receptor for phage PRD1. However, W plasmid strains insensitive to WRD1 shown in Table 3 do plate PRD1 phage. Thus, based on plating behavior alone, the WRD1 and PRD1 phage receptors specified by the W group plasmids are clearly different. Of perhaps more fundamental interest, is the observation that some R^- *E. coli* strains show susceptibility to WRD1. This observation may imply that W incompatibility group plasmids contain genes identical or functionally homologous to those normally part of the *E. coli* K12 or C strain chromosome.

Several considerations are pertinent to the interpretation of data shown in Table 3.

1. WRD1 phage receptors may be masked by the presence in some bacterial strains of capsule or slime.
2. The WRD1 phage receptor may be part of the LPS component of the cell wall and masked in bacteria having O antigens as part of their LPS.
3. "Rough" bacteria strains, *i.e.*, O antigen deficient bacteria, may not have LPS components compatible with the deposition of the WDR1 receptor.

The inability of *S. typhimurium* LT2 containing either R388 or Rsa plasmids to plate WRD1 indicates that the WRD1 receptor may be masked by the O antigen component of their cells wall LPS. The corollary of this tentative view, namely, that WRD1 phage is rough specific (*i.e.*, adsorbs to O antigen deficient bacteria) is supported by its observed infectivity for known rough *E. coli* strains. (For an extensive discussion of the relationship between phage susceptibility and LPS composition see Rapin and Kalckar, 1971). However, *E. coli* B is also an O antigen deficient bacterium. In this instance, the lack of WRD1 susceptibility and hence absence of the W plasmid receptor may reflect an LPS composition incompatible with the deposition of the WRD1 receptor. In this regard, *E. coli* B is known to lack galactose as part of its LPS core structure (Rapin and Kalckar, 1971). *E. coli* K12, on the other hand, is known to contain galactose as part of its LPS core structure (Rapin and Mayer, 1966). It follows from this observation that the appropriate metabolite pool of nucleosidesugars is available in *E. coli* K12 bacteria for transferase activity resulting in the synthesis of galactose containing LPS core structure which, if not masked by the presence of O antigens, can serve as the WRD1 receptor. The implication, of this surmise, then, is that, the W group plasmids contain genes associated with the addition of galactose to LPS core polysaccharide structure. Although this function is normally present in most *E. coli* K12 or C strains, it may be provided, when absent, by W plasmid genes as perhaps with *E. coli* J53, or *P. aeruginosa* PAT904. A precondition for this proposed addition of galactose would be a LPS core structure compatible with the addition of W plasmid specified galactose additions. The later instance is apparently not the case with either *E. coli* B or *P. aeruginosa* PAT cores. In any case, WRD1 phage susceptibility is clearly plasmid dependent with some bacterial strains and may represent an instance of gene pickup during the evolution of these plasmids of functions usually associated with chromosomal *E. coli* genes.

DISCUSSION

Preliminary observations in our laboratory on the relationship between plasmid specified cell components and phage sensitivity have indicated considerable diversity in the nature of the plasmid specified phage receptors. With the RNA phage, PRR1, this relationship is clear. The P incompatibility group plasmids specify the synthesis of pili required for PRR1 adsorption and the occurence of these P specific pili is apparently restricted to the P group of plasmids. The drug resistance plasmid specification of cell wall changes resulting in the adsorption of the DNA phage PRD1 provides another example of the nature of plasmid modifications of cellular structure. In this instance, PRD1 plating allows a direct manifestation of plasmid dependent alteration of cell structure heretofore associated with the presence of some drug resistance plasmids (*e.g.*, changes in permeability, sensitivity to detergents, etc.). Also of particular interest, with regard to PRD1 phage susceptibility, is the observation that plasmids of three different incompatibility groups, P, N and W, have in common the genetic information for the synthesis and expression of PRD1 receptors although these groups do not share many other properties. The host range of the P and W group plasmids is broad and includes diverse Gram-negative bacteria unrelated by host phenotypic traits commonly used for this purpose, whereas the host range of the N group of plasmids is confined to some genera of the *Enterobacteriaceae* (Datta and Hedges, 1972).

The elucidation of the nature of the WRD1 phage receptor may, in the final analysis, prove to be the most interesting of plasmid associated changes in cellular structure. In this instance, a cell wall component, whose general composition and properties are predictable, based on precidents from the extensive studies on *S. typhimurium* LPS, may be resolved to the extent that a specific change in cellular structure may be associated with the activity of a single or few gene functions carried by the plasmid. Although it has been generally known for some time that the presence of plasmids *per se* in bacteria alter a wide range of properties usually associated with cell wall functions, the description of the PRD1 and WRD1 phage receptors should facilitate a more clear description of the relationship between structure and function changes in cell wall properties prompted by the acquisition of a plasmid. In the case of the WRD1 phage receptor site specified by the W group plasmids, it seems that the W plasmid gene(s) concerned with WRD1 adsorption perform a function normally present in some R^- strains. Thus, implied by this observation, is the possibility that the origin of these genes has been the *E. coli* chromosome.

REFERENCES

Bryan, L. E., Van Den Elzen, H. M. and Tseng, J. R. (1972). Transferable drug resistance in *Pseudomonas aeruginosa*. Antimicrob. Ag. Chemother. 1, 22.

Datta, N., Hedges, R. W., Shaw, E. J., Sykes, R. B. and Richmond, M. H. (1971). Properties of an R factor from *Pseudomonas aeruginosa*. J. Bacteriol. 108, 1244.

Datta, N. and Hedges, R. W. (1972). Host ranges of R factors. J. Gen. Microbiol. 70, 453.

Fulbrook, P. D., Elson, S. W. and Slocombe, B. (1970). R factor mediated β-lactamase in *Pseudomonas aeruginosa*. Nature 226, 1054.

Grinsted, J. J., Saunders, R., Ingram, L. C., Sykes, R. B. and Richmond, M. H. (1972). Properties of an R factor which originated in *Pseudomonas aeruginosa* 1822. J. Bacteriol. 110, 529.

Khatoon, H. and Iyer, R. V. (1971). Stable coexistence of R*fi*$^-$ factors in *Escherichia coli*. Can. J. Microbiol. 17, 669.

Lawn, A. M., Meynell, G. G., Meynell, E. and Datta, N. (1967). Sex pili and the classification of sex factors in the *Enterobacteriaceae*. Nature, 216, 342.

Novick, R. P. (1969). Extrachromosomal inheritance in bacteria. Bacteriol. Revs. 33, 210.

Olsen, R. H. and Shipley, P. (1973). Host range and properties of the *Pseudomonas aeruginosa* R factor R1822. J. Bacteriol. 113, 772.

Olsen, R. H. and Thomas, D. D. (1973). Characteristics and purification of PRR1, an RNA phage specific for the broad host range *Pseudomonas* R1822 drug resistance plasmid. J. Virol. 12, 1560.

Olsen, R. H., Siak, J. and Gray, R. H. (1974). Characteristics of PRD1, a plasmid dependent broad host range DNA phage. J. Virol. (in press).

Rapin, A. M. C. and Mayer, H. (1966). Complex polysaccharides in different strains of *Escherichia coli* K-12. Ann. N.Y. Acad. Sci. 133, 425.

Rapin, A. M. C. and Kalckar, H. M. (1971). The relation of bacteriophage attachment to lipopolysaccharide structure *in* Microbial toxins, G. Weinbaum, S. Kadis and S. J. Ajl (Eds.). Academic Press Inc., New York.

Riva, S., Fietta, A. M., Silvestri, L. G. and Romero, E. (1971). R factors-determined changes in permeability of *E. coli* towards rifampicin and other antibiotics. *in* Bacterial plasmids and antibiotic resistance, V. Krcméry, L. Rosival and T. Watanabe, (Eds.) Avicenum, Czechoslovak Medical Press, Prague.

Schlect, S. and Westphal, O. (1970). Untersuchungen zur typisierung von Salmonella-R-formen 4. Mitteilung: Typisierung von *S. minnesota-R*-mutanten mittels Antibiotica. Z. Bakteriol. 213, 356.

Smith, D. H. and Armour, S. E. (1966). Transferable R factors in enteric bacteria causing infection of the genito-urinary tract. Lancet *ii*, 15.

Stanisich, V. A. and Holloway, B. W. (1971). Chromosome transfer in *Pseudomonas aeruginosa* mediated by R factors. Genet. Res. 17, 1.

Sykes, R. B. and Richmond, M. H. (1970). Intergeneric transfer of β-lactamase gene between *Ps. aeruginosa* and *E. coli*. Nature 226, 952.

Tamaki, S., Sata, T. and Matsuhashi, M. (1971). Role of lipopolysaccharides in antibiotic resistance and bacteriophage adsorption of *Escherichia coli* K-12. J. Bacteriol. 105, 968.

EPIDEMIOLOGICAL TRACING OF *PSEUDOMONAS AERUGINOSA* BY BACTERIOCIN PRODUCTION

R. A. BOBO, L. F. JONES, E. T. THOMAS and J. J. FARMER, III.
Department of Clinical Pathology, University of Alabama Medical Center, Birmingham, U. S. A., Department of Microbiology, University of Alabama, Alabama, U. S. A.[1]

Nosocomial infections caused by *Pseudomonas aeruginosa* constitute a major problem in many hospitals (Bobo et al., 1965, Cooke et al., 1970, Falcao et al., 1972, Fierer et al., 1967, Govan and Gillies, 1969, Myerowitz et al., 1971). Exposure to *P. aeruginosa* is a serious, potential danger to many members of today's hospital population, particularly the premature infant, the immunosuppressed, and those patients whose normal host defense mechanisms are altered.

P. aeruginosa will survive and often multiply in such diverse environments as resuscitators, humidifiers, incubators, milk formulas, breast pumps, washbasins, water faucets, sinks, sponges, drains, detergents and disinfectant solutions and personnel (Bassett et al., 1965, Cartwright and Hargrawe, 1970, Cole et al., 1971, Cooke et al., 1970, Falcao et al., 1972, Fierer et al., 1967, Rees, 1970, Renate and Marget, 1964, Shooter et al., 1969, Zabransky and Day, 1969).

Currently epidemiologists use several methods to compare or "fingerprint" isolates: bacteriocin produetion (Farmer and Herman, 1969, Gillies and Govan, 1966, Govan and Gillies, 1969, Zabransky and Day, 1969) bacteriocin sensitivity (Gillies and Govan, 1966, Osman, 1965), bacteriophage susceptibility (Sutter et al., 1965), antibiotic succeptibility, agglutination in antisera so sonatic antigens (Farmer, 1970), or simply gross phenotypic properties. Combinations of these have been used successfully, but no stadardized, method presently exists.

To monitor successfully the epidemiology of this organism, strain or isolate relationships must be positively established. Presently isolate relationships have been studied primarily using bacteriocin production as an epidemiological marker and employing the scrape and streak method (Gillies and Govan, 1966, Govan and Gillies, 1969, Zabransky and Day, 1969). It is a simple method and easily performed in most laboratories; however, the disadvantages are as follows: lack of reproducibility, the occurrence of untypable strains, the difficulty of stimulating bacteriocin production on an agar surface and the subjectivity of interpretation. In Figure I the results are easily read, however, in Figure 2 the pattern would be a subjective interpretation.

An alternative to the scrape and streak method in studying the epidemiology of *P. aeruginosa* infections is the typing of isolates by bacteriocin production in which the bacteriocins are p3oduced in a broth (Farmer and Herman, 1969, Jones et al., 1973)

[1] This research was supported in part by research grant CC 00592 from the Center for Disease Control, Atlanta, Georgia U.S.A.

Using this method, large numbers of isolates can be tested, subjectivity in interpretation of results is reduced or eliminated, and the isolates produce bacteriocin in a liquid medium without the requirements of mitomycin C or ultraviolet ray stimulation.

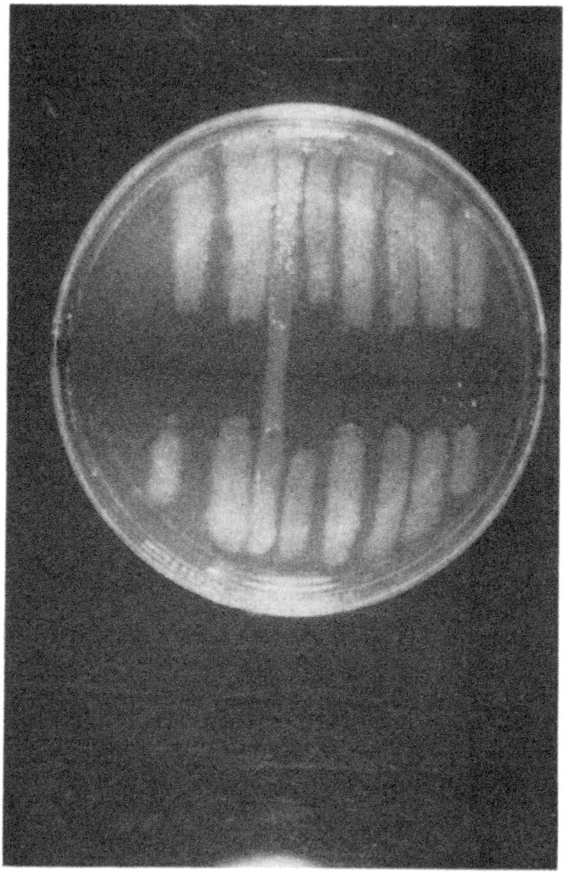

Fig. 1

Several studies have been conducted using this more sensitive bacteriocin typing technique (Bobo et al., 1973, Falcao et al., 1972, Farmer and Herman, 1969, Jones et al., 1973). This report describes the important features of this new technique application in determining the precise epidemiological relationship among 23 isolates of *P. aeruginosa* recovered from a variety of sources during an outbreak that occurred in a high risk nursery in the United States and produced septicemia in 9 infants.

Ordinarily when *P. aeruginosa* is cultured in Trypticase Soy Broth most of the growth is restricted to the surface where a pellicle is usually formed. However, the same medium minus glucose but supplemented with 1% potassium nitrate, gives uniform growth throughout the broth and the surface pellicle is reduced or absent. This new medium is available as bacteriocin production medium 81 (Jones et al., 1973).

In the routine typing of *P. aeruginosa*, bacteriocin production is done as follows: a single colony from a 24 hr. culture of the strain to be typed is picked and transferred into medium 81. The cultures are incubated in a 32°C water bath for 24 hours during which time bacteriocins are produced. One ml of chloroform is added to each tube, the mixture shaken vigorously for 10 seconds, allowed to stand at room temperature

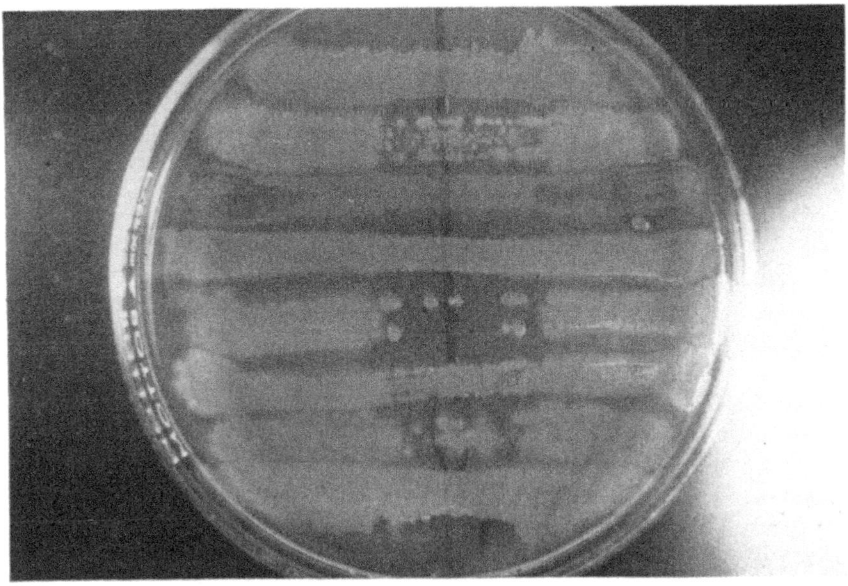

Fig. 2

for 10 minutes, and then shaken again. The residual chloroform is evaporated by leaving the caps loose for several hours at room temperature or overninght in the refrigerator. These suspensions can be centrifuged to remove cellular debris, but this is not essential for typing. Each lysate is then placed in individual droppers of the "Accu-Drop" bacteriophage-bacteriocin applicator (Sylvana Corp., Milburn, N. J.).

TABLE 1

Code for Reporting Bacteriocin Production Patterns★

3 Bacteriocin Reactions	*Representation*
+++	1
++—	2
+—+	3
—++	4
+——	5
—+—	6
——+	7
———	8

★A strain with a bacteriocin production pattern + — + — + + — — — would be coded "348" (+ — + + 3, — + + 4, — — — = 8). If the total number of bacteriocin reactions is not evenly divisible by 3, a second (+ — = A, + — = B, — + = C, — — = D) and third mnemonic (+ = E, — = F), can be used to code the remaining reactions.

Bacterial lawns of indicator strains or unknown strains are prepared by removing growth from a Trypticase Soy agar plate and preparing a standardized suspension by inoculating a tube of TBS without glucose and incubating overnight at 37°C (Fig. 3). The suspension is then diluted 1 : 1,000 in medium 81. Two ml of the dilution is flooded

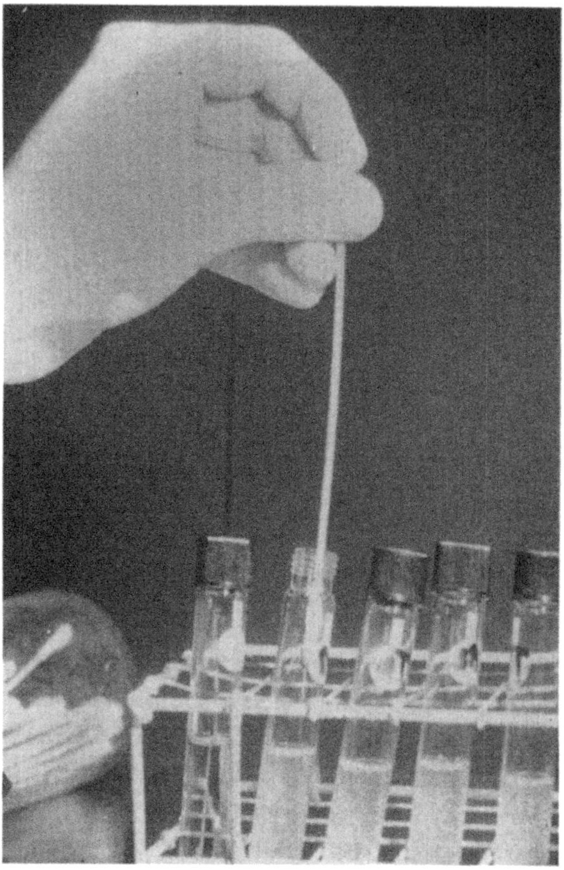

Fig. 3

onto a TSA plate and all excess liquid immediately removed using an automatic pipette shown in Figure 4. Figure 5 shows the placing of a lysate in a single applicator of the Accu-Drop applicator which as manufactured applies 24 drops in a single operation, thus greatly facilitating bacteriocin typing. The Accu-Drop applicator is shown in Figure 6 as assembled for operation to drop lysates for strains to be tested on indicator strain lawns. Figure 7 shows a typical bacteriocin typing operation with the lysates in the Accu-Drop applicator and the indicator strain lawns in the petri dishes. Bacteriocins are dropped onto the indicator strain lawns and incubated at 35–37°C for 24 hr. Figure 8 illustrates a bacteriocin typing pattern which can be converted using a reporting code (Table 1) (Farmer, 1970).

The outbreak to be described occurred in two episodes, separated by a period of approximately five weeks, in a high risk nursery in the United States and resulted in septicemia in 9 infants. Table 2 shows the pertinent clinical features of the nine cases studied in the U. S. outbreak. Seven of the nine infants were premature by both weight

Fig. 4

and gestational age. Two births were associated with prolonged rupture of membrane and two infants were delivered by Casesarean sections. Perinatal asphyxia requiring endoctrachael intubation occurred in eight of nine babies. Hyaline membrane disease

TABLE 2

Pertinent Features of 9 Cases of P. aeruginosa Sepsis

Prenatal Asphyxia	8
Immaturity and Prematurity	7
Pneumonia	6
Hyaline Membrane Disease	3
Meningitis	1
Peritonitis	1
Outcome	
Death	8
Survivors	1

was diagnosed in three infants and meningitis and peritonitis in one newborn. All infants were nursed in isolettes with high humidity and oxygen. Eight of the 9 infants studied expired. The surviving infant was gestationally mature, required only bulb suctioning at delivery, had one of four blood cultures positive for *P. aeruginosa*, and remained in an isolette with oxygen therapy for only ten hours. Trachael cultures from eight infants yielded *Pseudomonas*; thus the respiratory tract appeared to be the portal

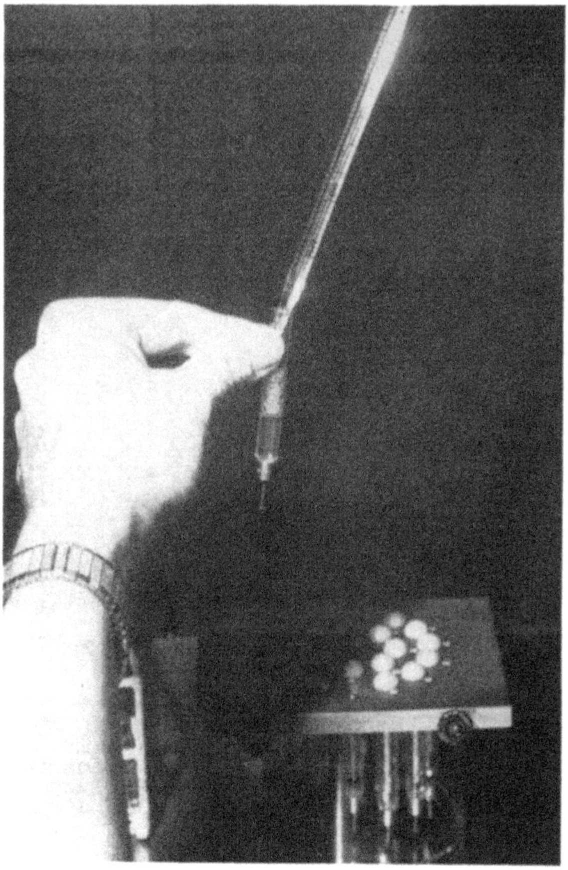

Fig. 5

of entry. A culture survey of all other infants occupying the same nursery and of the nursery environment, but not nursery personnel yielded the additional isolates shown in Table 3.

TABLE 3

P. aeruginosa Isolates from High Risk Nursery

Source	Number
Blood	10 (9 infants)
Suction Bottles	6
Suction Bottle Tubing	5
Oxygen Bottle Tubing	1
O-SYL Solution	1

Typing of the outbreak isolates by pyocin production was determined by the method of Farmer and Herman (1969) and using the 60 indicators listed in Table 4. Twenty-one isolates showed identical bacteriocin production patterns. Each of the 60 bacteriocin

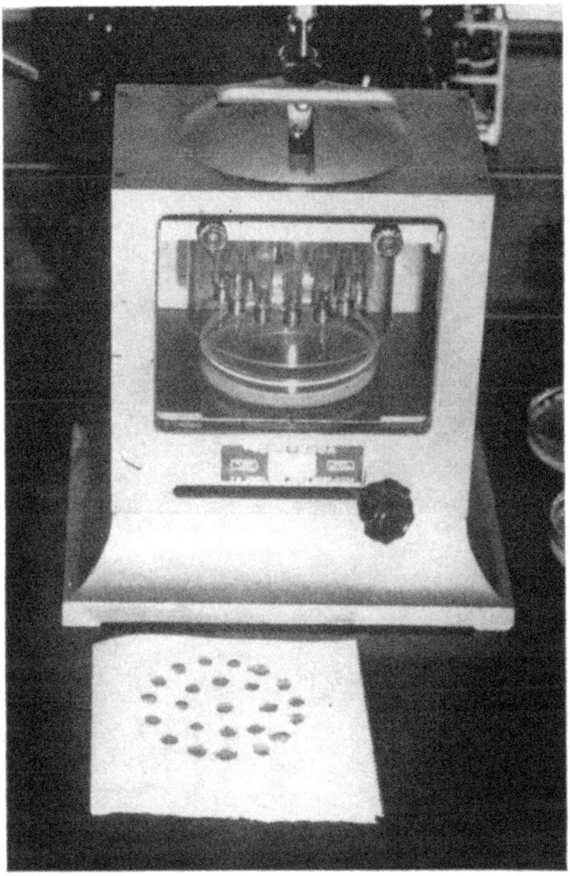

Fig. 6

TABLE 4

Pyocin Production Patterns of the 23 Outbreak Isolates

	Indicators of				
	Farmer & Herman	Gillies	Govan & Gillies	Zabransky & Day	Griffith
21 Isolates Blood, Suction Btls Oxygen Btls, Tubings	613 846 133	11 A	8 A	167 3	66 B
Isolate 1 Suction Btl	851 635 767	16 B	2 C	324 7	22 B
Isolate 1 0-SYL Solution	847 868 677	52 C	8 B	267 7	88 B

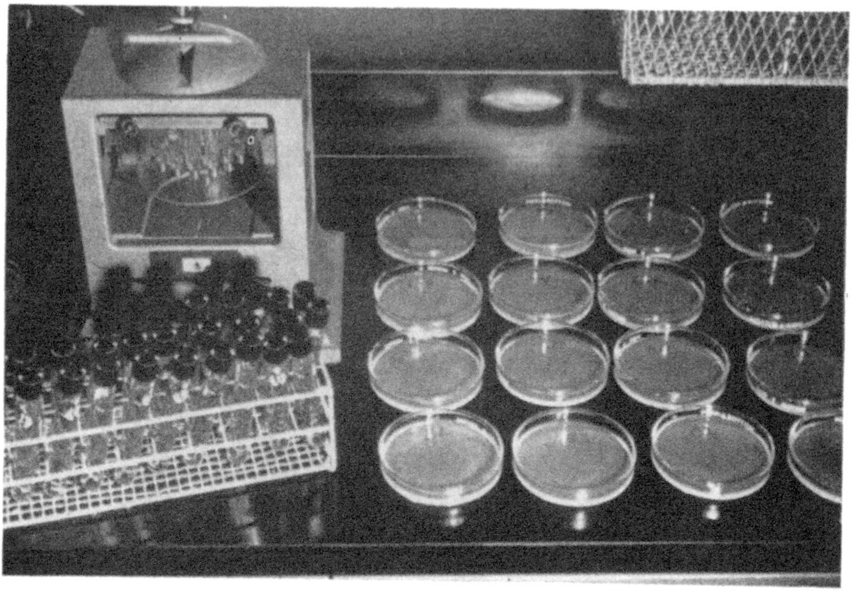

Fig. 7

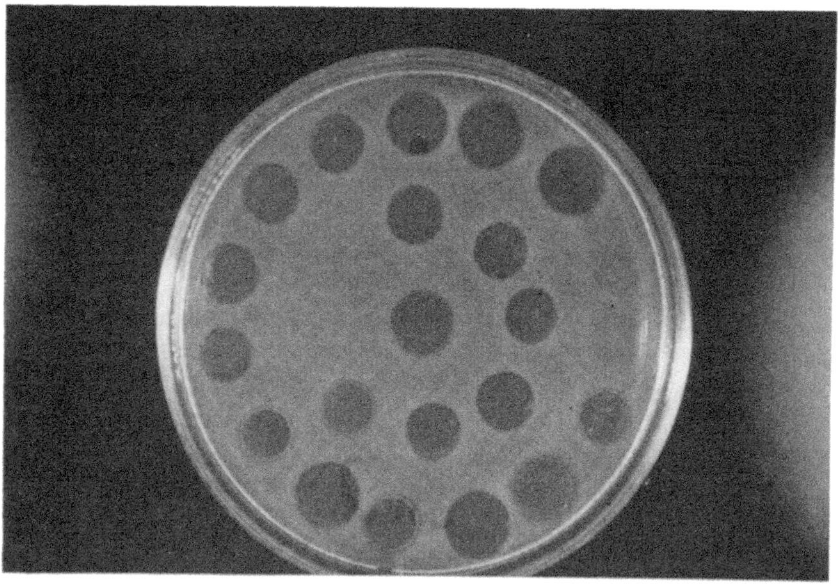

Fig. 8

reactions were strikingly identical in both size and shape of inhibition zones. Additionally, none of the 21 isolates produced a bacteriocin against any of the others. However, isolate 23 from the 0 - Syl solution and isolate 22 from a suction bottle tubing showed different bacteriocin production patterns.

Serotyping of the isolates by both slide and tube agglutination with antisera to the 13 0 antigens of the Habs schema showed that each of the 20 isolates possessing identical bacteriocin production patterns strongly agglutinated in antisera to 0 antigens 7 and 8; thus, serologically they were also identical (Table 5). The two isolates with different bacteriocin production patterns possessed different 0 antigens: 0 - Syl isolate 23 was 0 2, 5, and suction bottle tubing isolate 22 was 0 6.

TABLE 5

Serotypes of the 23 Outbreak Isolates

	Habs Somatic Agglutination
21 Isolates Blood Suction Btls Oxygen Btls Tubing	0 7, 8
Isolate 1 Suction Btl	0 6
Isolate 1 O-SYL Solution	0 2, 5

The absolute identity of the 21 isolates in their bacteriocin production patterns and serotypes convinced us that the outbreak was due to a single strain of *P. aeruginosa* which had spread throughout the nursery. Preliminary antibiotic susceptibilities showed that all strains were resistant by disk to tetracycline, cephalothin, chloramphenicol, kanamycin and ampicillin; and susceptible to gentamycin, polymyxin B, and carbenicillin. However, 5 outbreak strains were resistant to $>$ 1,000 μg/ml of sulfadiazine whereas the remaining 16 were susceptible to this agent. This difference in antibiotic susceptibility prompted us to look for additional phenotypic differences. We defined four phenotypic groups based on differences such as antibiotic susceptibility, pigment and phage production, hemolysis of sheep blood, and formation of autoplaques. These differences were reproducible and are summarized in Table 6.

TABLE 6

Phenotypic Differences Among the 21 Isolates of the Outbreak Strain

Property tested	Phenotype A (10 Isolates)	Phenotype B (4 Isolates)	Phenotype C (2 Isolates)	Phenotype D (5 Isolates)
1. Zone size in mm with 250 μg sulfadiazine disk, 48 hr.	23	16	23	0
2. As above, but sulfathiazole	27	16	23	11
3. Pigment on TSA	Green	Light Green	Dark Green	Blue Green
4. Fluroscein on MH	—	—	—	+
5. Autoplaques on MH	—	—	+	+
6. Zone size (mm) of hemolysis on SBA after 72 hr.	2	3.5	1	1
7. Prdouced phage which lysed strains of Phenotype D	+	+	+	—

When old TSA slants of phenotype D (all 5 isolates) were streaked on TSA plates, two colony types were usually observed, the wild type and a mutant which was easily distinguishable. The mutant colony was picked, purified, and then tested for the properties listed in Table 6. The mutant was identical to phenotype B in all properties tested. To rule out any possible chance of contamination, three of us repeated this experiment independently and obtained the same results, namely, dissociation of phenotype D into phenotype B. Thus, dissociation in the infants and environment probably resulted in the difference in phenotype. This explains why incorrect epidemiological results were obtained when typing by bacteriocin sensitivity, antimicrobial susceptibility and gross phenotypic properties were used as epidemiological markers. Although typing *P. aeruginosa* using bacteriocin sensitivity is a convenient marker for tentative conclusions, its lack of stability may limit its usefulness. We have also observed changes in bacteriocin sensitivity patterns of isolates in other outbreaks we studied (Falcao et al., 1972).

Until recently determining the epidemiology of *P. aeruginosa* infections has been due to limitations in methodology. Zierdt and Schmidt demonstrated that dissociation mutants from the same strain of *P. aeruginosa* fluctuate in their sensitivity to bacteriophage thus making results from phage typing confusing and difficult to interpret (Zierdt and Schmidt, 1964). Additionally serological typing suffers because most strains fall into relatively small numbers of groups. Also the sensitivity of this technique alone is inadequate for epidemiological purposes, but can be used in combination with other methods as we have shown in this report.

Recently a simplified method has been devised for typing *P. aeruginosa* by bacteriocin production which eliminates mechanical agitation and mitomycin C. induction of bacteriocins (Jones et al., 1973). Bacteriocins are produced as isolates grow overnight in

TABLE 7

Reaction of the 18 ALA Indicator Strains Against 100 Hospital Strains of P. Aeruginosa

ALA indicator	Former designation	Reaction for each indicator that were: Clearly positive	Clearly negative	Questionable[α] (— or +)	Due to phage activity
1	NIH 7	42	52	1	5
2	NIH 14	50	46	0	4
3	NIH 15	47	49	2	2
4	NIH 20	52	47	0	1
5	Gillies T_4	58	33	2	7
6	Brazil isolate	42	55	2	1
7	Brazil isolate	49	49	0	2
8	Mayo 2	56	40	2	2
9	Mayo 9	77	29	2	2
10	Griffith 5	36	60	4	0
11	Griffith 7	73	22	5	0
12	Bethesda isolate	60	37	0	3
13	Bethesda isolate	72	23	1	4
14	Bethesda isolate	54	42	4	0
15	SH isolate	46	52	2	0
16	Gillies T_{10}	48	50	2	0
17	Mayo 5	74	26	0	0
18	Bethesda isolate	51	48	1	0

[α] These reactions are the ones most likely to lead to misreading errors.

TSB without glucose plus 1 % potassium nitrate. All bacteriocin indicators currently in use work with the method, but some indicators give clearer results than others. Because of these reaction differences a new set of bacteriocin indicator strains was established.

Two hundred fifty isolates of *P. aeruginosa* were screened for use as possible indicators, and of this number 172 were eliminated. The remaining 78 indicators were evaluated against 100 different strains isolated from clinical specimens or from hospital environments. The resulting 7,800 reactions were analyzed using a computer program designed to select the most useful indicators for differentiating these 100 strains. Table 7 shows the 18 indicator strains chosen, their former designations, and the reactions of the 100 different strains. These indicator strains are noted as AlA strains to designate their origination. Of 1,800 test results, 96.5% were clearly positive or clearly negative. Only 1.7% were reactions classified as questionable, and only 1.8% were due to phage activity. The new set of indicator strains includes 11 strains of *P. aeruginosa* used in other indicator sets.

Classification of positive reactions is shown in Table 8. In addition, any of the above zones can be surrounded by an area of partial clearing, perhaps due to lysozyme or other

TABLE 8

Examples of Symbols Used in Representing Different Zones of Inhibition

Symbol	Meaning
+	Complete clearing due to pyocin
$+^{50}$	Clear zone with 50 colonies in it
+d	Clear zone, but larger than original drop (d is for diffusion)
+w	Weak inhibition, completely overgrown with colonies
∅ cf	Confluent lysis by bacteriophage
∅ 15	15 bacteriophage plaques

cell wall digesting enzymes, which was defined to be a metabolic effect, and the symbol "M" was placed after the original designation. For example, a zone with 50 colonies in it, but also surrounded by this partial clearing, would be designated as "$+^{50}$M". When the same strain was tested 10 times in the same experiment, the results were identical with respect to size and shape of zones.

When the reactions of the 100 strains were converted to the six digit codes, 92 of the strains showed completely different codes. In Table 9 is shown how two strains can be differentiated when all 18 reactions are the same. If differences are apparent in the size, shape, or degree of inhibition of the zone, it is a very good indication that the strains are genetically and epidemiologically distinct. The two isolates were recovered from

TABLE 9

Use of Differences in Zone Size and Shape in Differentiating Two Strains with the Same 18 Reactions (Scored only as + or —)

Isolate	Pyocin Production Pattern	Reaction against indicator:			
		ALA 9	ALA 10	ALA 12	ALA 16
Brooke 1	62 1111	∅ cf	+	+m	+d
SH 1	62 1111	+	$+^{50}$	$+^{200}$	+

different patients in hospitals separated by over 1,000 miles, so they are epidemiologically distinct. Additionally, careful examination of the individual zones of inhibition showed that there were other major differences between the two strains.

TABLE 10
Differentiation of Strains with the Same 18 Reactions (Scored only as + or —)

Group	Strains	Pyocin production pattern	Conclusion when zones were compared
I	NIH K	38 7412	Different
	Brazil isolate	38 7412	
	Bethesda isolate	38 7412	
II	NIH M	37 7111	Same
	Brazil isolate	37 7111	
III	NIH B	61 1224	Different
	Brazil isolate	61 1224	
IV	Brooke isolate	62 1111	Different
	SH isolate	62 1111	
V	Brazil isolate	61 3124	Different
	Bethesda isolate	61 3124	
VI	Brooke isolate	62 1724	Different
	SH isolate	62 1724	
	Brazil isolate	62 1724	

In Table 10 data show that all strains with the same code could be differentiated, except for strains in Group II. These two strains are epidemiologically distinct and appear to have identical reactions due to chance. No additional markers were evaluated to differentiate these strains as the purpose of this study was the evaluation of the new Alabama set of indicator strains. The new typing method was used to compare the 23 isolates from the U. S. nursery outbreak and gave results which permitted identical epidemiological conclusions.

Because of the high degree of antibiotic resistance and ubiquitous distribution in the hospital environment, *P. aeruginosa* poses a serious, potential threat to many hospital patients. The need for a simple, reliable, and reproducible technique for comparing in-house isolates is evident. During our epidemiological studies on *P. aeruginosa* outbreaks, we have evaluated a number of epidemiological markers. Techniques based on antibiotic susceptibility, bacteriophage susceptibility, and bacteriocin sensitivity all have as a common factor, the cell surface of the unknown strain. The phenomenon of dissociation of *P. aeruginosa* is well known and dissociation can bring about surface changes in the dissociated strains which affect the receptivity to any test substances (Holloway, 1969, Zierdt and Schmidt, 1964). Bacteriocin production and serological typing have been the best and most stable markers, and for this reason we have sought to improve bacteriocin production. Antisera for serological typing are not easily obtainable at present, although in the next few years this situation should change. Certainly the logical first step in any typing system for P. *aeruginosa* is the determination of the 0 antigen. However, many institutions will need a more sensitive marker to differentiate strains with the same 0 type, and at present, bacteriocin production appears to be one of the best supplementary markers.

The procedure described in this report was designated with simplicity in mind, and to be extremely sensitive in differentiating strains. The new set of 18 indicator strains was chosen from 250 strains which were evaluated. Eleven indicators in the new set had been used previously as bacterocin indicators. All indicators were chosen with the aid of a computer program which analyzes bacterocin reactions and chooses only the best ones for routine typing.

The methods and data presented should not be considered as final. It is tempting to propose it as a standardized method, but as time will bring about improvements and the discovery of other indicator strains, we have resisted the temptation. When antisera becomes commercially available for typing *P. aeruginosa*, the roles of serological typing and bacteriocin production may be better defined.

ACKNOWLEDGEMENTS

I thank Eleanor J. Newton, Linda Farmer, Evan T. Thomas, and J. P. Zakanycz for their excellent work in these studies.

REFERENCES

Bassett, D. C. J., Thompson, S. A. S., and Page, B. (1965). Neotal infections with *Pseudomonas aeruginosa* associated with contaminated resuscitation equipment. Lancet 1: 781–784.

Bobo, R. A., Newton, E. J., Jones, L. F., Farmer, L. F., and Farmer, J. J., III. (1973). Nursery outbreak of *Pseudomonas aeruginosa*: epidemiological conclusions from five different typing methods. Appl. Microbiol. 25: 414–420.

Cartwright, R. Y., and Hargrave, P. R. (1970). *Pseudomonas* in ventilators. Lancet 1: 40.

Cole, A. P., Thom, A. R., and Watrasiewicz, K. (1971). Infant carriers of *Pseudomonas aeruginosa*. Lancet 2:1155.

Cooke, E. M., Shooter, R. A., Farrell, S. M. O., and Martin, D. R. (1970). Faecal carriage of *Pseudomonas aeruginosa* by newborn babies. Lancet 2: 1045–1046.

Falcao, D. P., Mendonca, C. P., Scrassolo, A., De Almeida, B. B., Hart, L., Farmer, L. H., and Farmer, J. J., III. (1972). Nursery outbreak of severe diarrhea due to multiple strains of *Pseudomonas aeruginosa*. Lancet 2: 38–40.

Farmer, J. J., III, and Herman, L. G. (1969). Epidemiological finger-printing of *Pseudomonas aeruginosa* by the production of and sensitivity to pyocin and bycteriophage. Appl. Microbiol. 18: 760–765.

Farmer, J. J., III. (1970). Mnemonic for reporting bacteriocin and bacteriophage types. Lancet 2: 86.

Fierer, J., Taylor, P. M., and Gezon, H. M. (1967). *Pseudomonas aeruginosa* epidemic traced to delivery room resuscitators. N. Engl. J. Med. 276: 991–996.

Gillies, R. R., and Govan, J. R. W. (1966). Typing of *Pseudomonas pyocyanea* by pyocine production. J. Pathol. Bacteriol. 91: 339–345.

Govan, J. R., and Gillies, R. R. (1969). Further studies on the pyocin typing of *Pseudomonas pyocyanea*. J. Med. Microbiol. 2:17–25.

Habs, I. (1957). Untersuchungen über die 0-Antigene von *Pseudomonas aeruginosa*. Z. Hyg. Infekt. 144: 218–228.

Holloway, B. W. (1969). Genetics of *Pseudomonas*. Bacteriol. Rev. 33: 419–443.

Jones, L. F., Pinto, B. V., Thomas, E. T., and Farmer, J. J., III. (1973). Simplified method of producing pyocins from *Pseudomonas aeruginosa*. Appl. Microbiol. 26: 120–121.

Myerowitz, R. L., Mediros, A. A., and O'Brien, T. F. (1971). Recent experience with bacillemia due to gram-nega.

Osman, M. A. M. (1965). Pyocin typing of *Pseudomonas aeruginosa*. J. Clin. Pathol. 18: 200–202.

REES, T. A. (1970). Bacteria in suction machines. Lancet 1: 240.
RENATE, P., and MARGET, W. (1964). *Pseudomonas aeruginosa* infections (pyocyaneous sepsis) in infancy. Ger. Medical Monthly. 9: 119–123.
SHOOTER, R. A., COOKE, E. M., GAYA, H., KUMAR, P., PATEL, N., PARKER, M. T., THOM, B. T., and FRANCE, D. R. (1969). Food and medicaments as possible sources of hospital strains of *Pseudomonas aeruginosa*. Lancet 1: 1227–1229.
SHOOTER, R. A., FAIERS, M. C., COOKE, E. M., BREADEN, A. L., and O'FARRELL, S. M. (1971). Isolation of *Escherichia coli*, *Pseudomonas aeruginosa*, and *Klebsiella* from food in hospitals, canteens, and schools. Lancet 2: 390–392.
SUTTER, V. L., HURST, and FENNEL, J. (1965). A standardized system for phage typing *Pseudomonas aeruginosa*. Health Lab. Sci. 2: 7–16.
WAHBA, A. H. (1963). The production and inactivation of pyocins. J. Hyg. 61: 431–441.
ZABRANSKY, R. J., and DAY, F. E. (1969). Pyocin typing of clinical strains of *Pseudomonas aeruginosa*. Appl. Microbiol. 17: 293–296.
ZIERDT, C. H., and SCHMIDT, P. J. (1964). Dissociation in *Pseudomonas aeruginosa*. J. Bacteriol. 87: 1003–1010.

TRANSMISSIBLE HIGH-LEVEL RESISTANCE TO GENTAMICIN IN PSEUDOMONAS AERUGINOSA

BY E. PAPACHRISTOU AND P. KONTOMICHALOU

Athens University School of Medicine, Department of Clinical Therapeutics, „Alexandra" Hospital. Athens, Greece

All clinical *Pseudomonas aeruginosa* strains isolated in clinical practice in Greece were sensitive to gentamicin during the first two years (1970–71) of its use. Since 1972 gentamicin has become a very popular drug, because of its broad-spectrum activity and high potency.

The aim of this study was to find out whether resistance to gentamicin would appear in *Pseudomonas* and to investigate the nature of this resistance. Tobramycin, which has not yet been introduced in clinical practice in Greece, was also included.

MATERIALS AND METHODS:

The material of the study consists of hundred and thirty strains of *Pseudomonas aeruginosa*, of which 101 were all *Pseudomonas aeruginosa* strains (one from each patient), isolated in the Department of Therapeutics from November 1, 1972 to February 28, 1974. Five other strains were randomly selected out of the *Pseudomonas aeruginosa* strains isolated at the same hospital in 1971, and 24 strains were isolates of 1972 from the Athens Accident Hospital. Eighty three of the 130 strains were from urine.

All the strains tested belonged to the aeruginosa group, identified by conventional methods. Their sensitivity to gentamicin and tobramycin and sometimes kanamycin was examined by the tube-dilution method in brain heart infusion broth with an inoculum of 10^4 bacteria per ml. The results of the sensitivities to these aminoglycosides were confirmed by performing in parallel disc sensitivity tests and cultures in ditch-plates, in MacConkey agar. Gentamicin-resistant bacteria were further tested for transferability of their resistance by conjugation to a rifampicin resistant clone of *E. coli* K12F$^-$RC85. Mixed cultures were started from 1 ml donor, 4.5 ml recipient and 4.5 ml fresh complete broth. They were incubated for 3 hrs and centrifuged, and their sediment was transferred to membrane filters on the surface of nutrient agar and incubated overnight. The cells were washed from the membranes and dilutions were inoculated to minimal media with lactose or MacConkey agar incorporating rifampicin and other antibiotics. In isogenic transfers with K12 derivatives the mixed cultures were incubated for 3hrs only.

This work was supported by grants from:

1) The Greek Ministry of Culture and Civilization
2) The Wellcome Trust
3) The Ebirikion Foundation

RESULTS

The sensitivities of the strains are shown in Tables 1 and 2. Strains with an MIC exceeding 6 μg/ml of the aminoglycosides were considered to be resistant. Table 1 shows that the percentage of resistant strains was 30% and 28.4% for gentamicin and tobramycin respectively. The cross-resistance between the two aminoglycosides was almost complete.

TABLE 1

Cumulative resistance to gentamicin and tobramycin in clinical strains of Pseudomonas aeruginosa

Antibiotic	Number of strains tested	Resistant strains (MIC >6 μg/ml)	
		No.	%
Gentamicin	130	39	30.0
Tobramycin	130	37	28.4

TABLE 2

Sensitivity of the 130 Ps. aeruginosa strains by the tube dilution method to the two aminoglycosides

MIC	Gentamicin			Tobramycin		
	MIC μg/ml	MBC μg/ml	Number of strains	MIC μg/ml	MBC μg/ml	Number of strains
<6 μg/ml	<0.25	<0.25	74	<0.25	<0.25	81
	<0.25	0.5–2	7	<0.25	0.5–1	4
	0.5–4	1–8	10	0.5–4	0.5–16	8
Total			91			93
>6 μg/ml	8–32	32–64	3	8–32	16–64	34
	62–250	500–1000	6	62–250	125–250	3
	500–1000	500–2000	30	500–1000		0
Total			39			37

Table 2 shows that out of the 91 strains sensitive to gentamicin 74 had a MIC and MBC lower than 0.25 μg/ml, compared to 81 out of 93 for tobramycin. Most of the remaining strains resistant to gentamicin, i.e., 30 out of 39, were highly resistant with a MIC of 500–1000 μg/ml, whereas of the strains resistant to tobramycin 34 out of 37 had a lower level of resistance, 8–32 μg/ml, with only 3 strains with a MIC of 62–250 μg/ml. This difference in the resistance of *Pseudomonas* strains to gentamicin and tobramycin was demonstrated only by the tube-dilution method, whereas by discs of 30μg and ditch plates it was less pronounced.

According to sources, gentamicin-resistant strains were found exclusively in one of the two hospitals, where the percentages of those resistant increased from 21.4% in 1972 to 41.6% in 1974 (Table 3). Most of the resistant strains of this hospital were isolated from urine: 49.3% of the *Pseudomonas aeruginosa* from urine were gentamicin-resistant compared to 6.4% of the strains from all other sources.

Transferability of the gentamicin resistance to *E. coli* $K12F^-$ was tested in all 39 gentamicin-resistant strains, with selection made by gentamicin, kanamycin and streptomycin for strains resistant to the aminoglycosides and, in addition to them, by carbenicillin for strains resistant to this drug as well. As shown in Table 4 positive intergeneric

TABLE 3

Distribution of gentamicin resistant Ps. aeruginosa according to sources and year of isolation

Year	Hospital	Number of strains tested	Resistant to gentamicin	
			Number of strains	%
1971	Th*	5	0	0
1972	Ac	24	0	0
1972	Th	14	3	21.4
1973	Th	75	30	40.0
1974	Th	12	5	41.6

*Th = Department of Therapeutics
Ac = Accident Hospital

TABLE 4

Intergeneric transfer of Gentamicin–Tobramycin resistance-determinants

Ps. aeruginosa donors		Selection by	Conjugal transfer to *E. coli K12 F^-Rif^r*	
Resistance pattern*	Number of strains		Positive	%
Cb Gm Tb Km Sm	10	Cb	10	100
		Km or Sm	8	80
		Gm	6	60
Gm Tb Km Sm	29	Gm or Km or Sm	0	0

* Concerning only the aminoglycosides and carbenicillin. For symbols see Table 5.

transfer occurred with all Pseudomonas donors, which were resistant to both carbenicillin and the aminoglycosides. For these strains selection was achieved in 100% by carbenicillin, in 80% by kanamycin and streptomycin and in 50% by gentamicin. The frequencies of these transfers were as a rule of the order of 10^{-6} for selection by carbenicillin and 10^{-7} to 10^{-8} for selection by the aminoglycosides. In contrast, all transfer experiments with donors lacking resistance to carbenicillin were negative.

The resistance patterns of the *E. coli* $K_{12}RC85$ transconjugants were subsequently tested. With selection by any of the drugs used, for plasmids derived from two donors, transconjugants were encountered carrying in single colonies either all eleven determinants for resistance to the aminoglycosides (gentamicin, tobramycin, kanamycin and streptomycin), the b-lactam drugs (carbenicillin, benzypenicillin, ampicillin and cephaloridin), to chloramphenicol, tetracyclines and sulfanamides, or parts of these resistance markers (pPK 183 and pPK 237, Table 5). It was most common to find in individual clones, infected with the R- plasmids of each Pseudomonas, a variety of segments of this pattern, with the whole pattern reconstructed in most cases for each donor. Resistance determinants, single or in pairs, were missing in varying proportions. The gentamicin and tobramycin resistance genes were found transferred in 62% of isolated clones tested, they were transferred always together with other resistance determinants and have never been segregated from each other. The carbenicillin marker was present in 92% of the K_{12} transconjugants.

TABLE 5

Resistance patterns in R plasmids originating from Pseudomonas

R plasmids	Resistance determinants*	In F^-K12 transconjugants
pPK183, pPK237	Gm Tm Km Sm Cb Pc Ap Cp Cm Tc Su	In single or from many colonies
pPK148, pPK170, pPK175 pPK179, pPK192 pPK207	Gm Tm Km Sm Cb Pc Ap Cp Cm Tc Su	From many colonies
pPK133	Gm Tm Sm Cb Pc Ap Cp Cm Tc Su	From many colonies
pPK141	Km Sm Cb Pc Ap Cp Cm Tc Su	From many colonies

* Gm = Gentamicin, Tm = Tobramycin, Km = Kanamycin, Sm = Streptomycin, Cb = Carbenicillin, Pc = Benzylpenicillin, Ap = Ampicillin, Cp = Cephaloridin, Cm = Chloramphenicol, Tc = Tetracycline, Su = Sulfonamides

TABLE 6

*Frequencies of transfers of R plasmids originating from Ps. aeruginosa in E. coli K12 isogenic crosses (F^-XF^+)**

Plasmids		Selection by		
Designation	Resistance pattern**	Carbenicillin	Gentamicin	Kanamycin
p PK170–4	Gm Km		1×10^{-1}	1×10^{-1}
p PK179	Cb Gm Km	1.2×10^{-4}	1.2×10^{-3}	1.2×10^{-3}
p PK192	Cb Gm Km	1×10^{-3}	5×10^{-4}	5×10^{-4}
p PK237	Cb Gm Km	4×10^{-4}	2.5×10^{-2}	1.2×10^{-2}

* Frequencies per donor.
** Concerning carbenicillin, gentamicin and kanamycin only.

Interspecies transfers of these plasmids in K_{12} derivatives were made from F^- transconjugants to Hfr and F^+ recipients, for identification of the fi class of the plasmids. There were two plasmids of fi^+ class and the remaining were fi^-. The carbenicillin and gentamicin resistance markers had in these transfers different transfer frequencies, with the gentamicin plasmid being in some cases derepressed for transfer (Table 6).

The levels of the transferable resistance to the two aminoglycosides (gentamicin-

TABLE 7

Levels of expression of extrachromosomal resistance to gentamicin–tobramycin in Pseudomonas and E. coli K12 host. MICs in μg/ml.

Plasmids	Pseudomonas aeruginosa		E. coli K12 F⁻RC85	
	Gentamicin	Tobramycin	Gentamicin	Tobramycin
p PK 148	500	16	16	16
p PK 170	500	32	32	32
p PK 179	64	16	32	16
p PK 183	125	16	32	16
p PK 192	125	8	16	16

tobramycin), as shown in Table 7, were in Pseudomonas and *E. coli* K12 host different only for gentamicin. The high level resistance to gentamicin of *Pseudomonas* was not found in *E. coli* K12 transconjugants, in which the resistance to both aminoglycosides was expressed at the same levels of 16–32μg/ml. Resistance to kanamycin and neomycin was as a rule present in all K12 transconjugants tested, carrying the gentamicin resistance genes.

DISCUSSION

The epidemiology of gentamicin resistance in *Pseudomonas* varies considerably in different countries, due mainly to the date of introduction of the drug in clinical practice and the extent of its use. In Greece gentamicin became available on the market in 1970 when other countries had already 6 or 7 years of clinical experience and had published reports from hospitals with 100% resistant *Pseudomonas* (Stone and Kolb, 1971). Such cases were almost exlusively from topical application of the antibiotic (Stone and Kolb, 1971, Yourassowski et al., 1971). The variability in the emergence of gentamicin resistance in *Pseudomonas* is reflected in the fact that several surveys from different countries, or from different hospitals in the same country, still report low percentages or even an absence of gentamicin resistance (Damaso and Moreno-Lopez 1973, Karney et al., 1973, Lockwood and Lawson, 1973, Duncan, 1974). In Greece, in a recent Symposium on *Pseudomonas*, the percentage of contemporary strains resistant to gentamicin was higher in our report than in three other hospitals (Papachristou and Kontomichalou, 1974, Maniatis et al., 1974, Demenaga-Papapetropoulou et al., 1974, Skevakis et al., 1974). Most of our *Pseudomonas*, of which 92% were from urine, were highly resistant to gentamicin. Similar strains from urine have emerged in other countries as well (Knothe et al., 1973).

High-level gentamicin resistance and its spread in a hospital unit are often attributed to the episomal nature of this resistance (Tanaka, 1970, Christol et al., 1971, Witchitz and Chabbert, 1971). Transfer of plasmids carrying gentamicin-resistance determinants in Pseudomonas has been demonstrated *in vitro in* a few cases, occurring mostly in interspecific crosses (Bryan et al., 1973). The intergeneric transfer of hish-level gentamicin resistance from *Pseudomonas* to *E. coli* and to other *Enterobacteriaceae* has been described by Witchitz and Chabbert (1971), whereas other authors obtained negative results from such attempts (Khan et al., 1970, Knothe et al., 1973). The levels of the transferable gentamicin resistance in this study were higher than those found by other investigators.

As both kanamycin and neomycin markers were present in all K12 clones tested, carrying the gentamicin-tobramycin marker, we could assume that more than one enzyme inactivating the deoxystreptamine group of aminoglycosides were present, that is a gentamicin-adenylating enzyme and a kanamycin-phosphorotransferase (Davies, 1972, Witchitz, 1972).

The positive transfer of plasmids with gentamicin-inactivating enzymes, only in the presence of penicillinase plasmids in the *Pseudomonas* donors, could suggest that mobilisation of the gentamicin-resistance plasmids was due to the penicilinase-plasmids sex factors.

The carbenicillin and gentamicin-resistance markers, as well as other resistance markers (results not shown here), when entering K12 transconjugants, behaved afterwards, as a rule, as independent self-transmissible replicons. It is difficult to accept that these plasmids, independent in K12, could be in *Pseudomonas* in cointegrated form.

In the intergeneric transfers, a variety of segregants were also detected in K12 transconjugants from each donor and among them were some gentamicin-kanamycin epiclones lacking the carbenicillin resistance markers. Hence, in these strains mobilization of the gentamicin marker might not have been due only to the penicillinase plasmids. However, there was no difference in the phenotype of the resistance to the aminoglycosides in the group of the ten *Pseudomonas* donor strains with positive and the group of 29 strains with negative intergenetic transfer, suggesting that the nature of their gentamicin-kanamycin resistance might not be different. Thus, if the presence of the penicillinase plasmids is not the only mechanism permitting intergeneric transfers, it is not easy to understand the complete obstruction of such transfers in the absence of penicillinase plasmids in this latter group of Pseudomonas.

Our report on these findings is considered to be preliminary since many aspects, for which experiments are already in progress, need further elucidation.

SUMMARY

Resistance to gentamicin and tobramycin was found in 30% and 28.4% respectively of the 130 strains of *Pseudomonas aeruginosa,* isolated in our laboratory in 1971 and from 1972 to 1974. Resistance to gentamicin was as a rule of a high level, 500–1000 μg/ml. The resistance determinants to gentamicin and tobramycin were conjugally co-transferred to *E. coli* K12 derivatives together with resistance genes for other or unrelated antibiotics, most commonly with penicillinase plasmids. In selection by gentamicin, transfer frequencies were low, 10^{-7} or 10^{-8} per donor.

Resistance to the above two aminoglycosides has never been segregated in transfers and was always accompanied by resistance to kanamycin. There was a host specificity in the expression of transmissible resistance only for gentamicin. In isogenic transfers in *E. coli* K12 derivatives the gentamicin-tobramycin plasmids had independent transfers and in some cases they were derepressed.

REFERENCES

Bryan, L. E., Van den Elzen, H. M. and Tseng, J. T. (1972): Transferable drug resistance in *Pseudomonas aeruginosa.* Antimicrob. Ag. Chemother. 1: 22–29.

Christol, D., Bure, A., Boussougant, Y. et Witchitz, J. (1971): Evolution de la résistance â la gentamicine. Pres. Med. 79: 467–470.

Damaso, D. and Moreno-Lopez, M. (1973): Study on in vitro activity of the antibiotics tobramycin and gentamicin against *Pseudomonas aeruginosa* clinical strains. J. Antibiot. 16: 233–237.

Davies, J. (1972). Bacterial resistance to antimicrobial agents. Triangle, 11: 131–135.

DEMENAGA-PAPAPETROPOULOU, M., STRATAKOU, E. and KOUMENTAKOU, E. (1974): Sensitivity to carbenicillin and gentamicin of *Pseudomonas aeruginosa* strains isolated in the last two years. Proc. 6th Nat. Congr. Bacteriol. Athens. In press.

DUNKAN, J. B. R. (1974): Susceptibility of 1.500 isolates of *Pseudomonas aeruginosa* to gentamicin, carbenicillin, colistin and polymyxin B. Antimicrob. Ag. Chemother. 5: 9–15.

KARNEY, W., HOLMES, K. K. and TURCK, M. (1973): Comparison of five aminocyclitol antibiotics in vitro against Enterobacteriaceae and Pseudomonas. Antimicrob. Ag. Chemother. 3: 338–342.

KNOTHE, H., KRČMÉRY, V., SIETZEN, W. and BORST, J. (1973): Transfer of gentamicin resistance from *Pseudomonas aeruginosa* strains highly resistant to gentamicin and Carbenicillin, Chemotherapy, 18: 229–234.

KONTOMICHALOU, P., PAPACHRISTOU, E. and LEVIS, G. (1974): TEM-like penicillinases in contemporary plasmids of Enterobacteriaceae and Pseudomonas. Comparison with b-lactamases of the preantibiotics Klebsiella. 2nd Intern. Symp. on Antib. Resist. Smolenice, Czechoslovakia.

LOCKWOOD, W. R. and LAWSON, L. A. (1973): Studies on the susceptibility of 150 consecutive clinical isolates of *Pseudomonas aeruginosa* to tobramycin, gentamicin, colistin, carbenicillin and five other antimicrobials. Antimicrob. Ag. Chemother. 4: 281–284.

MANIATIS, A., PEROGAMBROS, A. and ARSENI, A. (1974): Trends in drug resistance of Pseudomonas in the last decade. Proc. 6th Nat. Congr. Bacteriol. Athens. In press.

PAPACHRISTOU, E. and KONTOMICHALOU, P. (1974): Resistance of *Pseudomonas aeruginosa* to antibiotics with specific antipseudomonal activity. Proc. 6th Nat. Congr. Bacteriol. Athens. In press.

SKEVAKIS, L., KAVADIAS, N., AHIMASTOS, A. and PAVLATOU, M. (1974): Sensitivity of *Pseudomonas aeruginosa* to gentamicin. Proc. 6th Nat. Congr. Bacteriol. Athens. In press.

STONE, H. H. and KOLB, L. D. (1971): The evolution and spread of gentamicin-resistant Pseudomonas. J. Trauma, 11, 586–589.

TANAKA, N. (1970): Biochemical studies on gentamicin resistance. J. Antibiot. (Japan) 23: 469–471.

UMEZAWA, H., YAGISAWA, M., MATSUHASHI, Y., NAGANAWA, H., YAMAMOTO, H., KONDO, S., TAKENCHI, T. and CHABBERT, Y. A. (1973): Gentamicin acetyl transferase in *Escherichia coli* carrying R factor. J. Antibiot. 16: 612–614.

WITCHITA, J. L. (1972): Plasmid-mediated gentamicin resistance not associated with Kanamycin resistance in Enterobacteriaceae. J. Antibiot. 15: 622–624.

WITCHITZ, J. L. et CHABBERT, Y. A. (1971): Résistance transférable à la gentamicine. I. Expression du caractère de résistance. Ann. Inst. Past. 121: 733–742.

YOURASSOWSKY, E., SCHOUTENS, E. and BEUMER, J. (1971): *Pseudomonas aeruginosa* with exceptionally high resistance to gentamicin in hospital wards. Susceptibility and phage typing Rev. Eur. Etud. Clin. Biol. 16: 427–929.

DEMONSTRATION OF R PLASMIDS FROM *PSEUDOMONAS AERUGINOSA* CONFERRING GENTAMICIN AND CARBENICILLIN RESISTANCE

[1]V. KRČMÉRY, H. SAGAI, K. HASUDA, H. KAWABE, S. IYOBE [2]H. KNOTHE AND S. MITSUHASHI

Department of Microbiology, Gunma University, Maebashi, 371, Japan and ([2]) *Institute of Hygiene, University of Frankfurt, F. R. G.*

Knothe et al., 1973 isolated in Frankfurt urologic clinic strains of *Pseudomonas aeruginosa* from the urine, which were highly resistant to a series of aminoglycoside antibiotics including gentamicins, kanamycin, streptomycin, to carbenicillin and to several other antibiotics (see Table 1). It seemed interesting to demonstrate the presence of R plasmids in these strains and to elucidate the enzymatic background of the resistance, mainly to aminoglycoside drugs.

MATERIALS AND METHODS

Bacterial strains. Eleven strains of *Pseudomonas aeruginosa* isolated by H. Knothe in Frankfurt clinics were used. Three auxotrophic strains of *P. aeruginosa* without sex

TABLE 1

Drug resistance of Pseudomonas aeruginosa strains used

Strain	Level of resistance[a]) (mcg/ml)									
	TC	DM	SM	SA	KM	LV	CPC	GM	DKB	AK
Ps50	50	200	≥800	≥6400	100	50	≥200	100	0.8	3.2
Ps79	50	100	≥800	≥6400	50	50	200	3.2	0.4	1.6
Ps80	50	100	≥800	≥6400	400	50	≥200	100	0.4	3.2
Ps110	≥400	50	≥800	≥6400	100	≥100	50	100	50	0‘8
Ps125	50	100	≥800	≥6400	100	50	≥200	100	1.6	3.2
Ps133	100	200	≥800	400	100	50	≥200	≤3.2	0.8	—
Ps135	100	100	400	400	100	50	≥200	≤3.2	1.6	—
Ps138	≥400	100	≥800	≥6400	400	≥100	100	100	100	3.2
Ps140	50	100	≥800	≥6400	50	50	≥200	100	0.8	3.2
Ps141	50	100	≥800	≥6400	50	50	≥200	100	0.8	1.6
Ps142	50	100	≥800	≥6400	50	50	≥200	100	1.6	0.8

[a]) Minimum inhibitory concentration.

Overnight culture in peptone water was diluted 100-fold in fresh peptone water and one loopful of the diluted culture was spotted in HIA plates containing serial two-fold dilutions of each drug. After 18 hr of incubation at 37 °C, the growth of bacteria was examined and the minimum inhibitory concentration was noted.

[1] Present adress: Research Institute of Hygiene, Bratislava, Czechoslovakia.

factor FP, ML4262 (*trp⁻ his⁻ ilv⁻ met⁻ rif*^{r}) ML4561 (*trp⁻ his⁻ ilv⁻ arg⁻ leu⁻ rif*^{r}) and ML4600 (*trp⁻ his⁻*) were used. ML4600 was a transductant of ML4262 *rif*^{s} obtained with F116 phage propagated on M88 (*met*$^+$ *ilv*$^+$).

Drugs. Tetracycline (TC), chloramphenicol (CM), streptomycin (SM), sulfanilamide (SA), kanamycin (KM), lividomycin (LV), gentamicin C complex (GM), 3′, 4′ — dideoxykanamycin B(DKB), amikacin (AK), carbenicillin (CPC) and rifampicin (RF) were used.

Media and conjugation experiments. Two ml of recipient and 0.5 ml of donor cultures were mixed and gently shaken at 37 °C. After 2 hr of incubation, the mixture was spread on the nutrient agar containing one of selective drugs and RF, or on A-glucose agar supplemented with appropriate amino acids and drug.

Inactivation systems of aminoglycoside antibiotics. Phospholylation or adenylylation was examined by system 1) and acetylation was done by system 2). The system 1) contained the enzyme, ATP, $MgCl_2$ and drug. The system 2) contained the enzyme, ATP, magnecium acetate, Co-enzyme A and drug. The reaction mixture was incubated at 37 °C and the residual activity of drug was determined by paper disk method.

Tetracycline uptake determination. TC uptake was determined by the method reported by Inoue (Inoue at al.).

Assay of β-lactamase. β-Lactamase was determined iodometrically (Perret, 1954).

RESULTS AND DISCUSSION

1. Transfer of gentamicin resistance

As can be seen from Table 1, eight out of eleven strains studied had a high-level gentamicin resistance, and 9 out of 11 strains were highly resistant to carbenicillin.

Three strains were able to transfer R plasmids to recipients. They were Ps110, Ps138 and Ps 142 strains. Both Ps110 and Ps138 strains carried R plasmids resistant to (TC. SM.SA.KM.LV.GM.DKB). They were designated as R_{ms146} and R_{ms147}, respectively. Their transfer frequencies were 7.4×10^{-4} (R_{ms146}) and 4.8×10^{-6} (R_{ms147}).

TABLE 2

Drug resistance mechanisms of R factors

	R146 (TC.SM.SA.KM. LV.GM.DKB)	R147 (TC.SM.SA.KM. LV.GM.DKB)	R148 (SM)	R149 (SM.SA.GM.CPC)
TC	per	per	—	—
SM	ade or pho	ade or pho	ade or pho	ade or pho
KM	pho	pho	—	—
LV	pho	pho	—	—
GM	ace	ace	—	ace
DKB	ace	ace	—	—
CPC	—	—	—	β-lactamase

per, change of permeability; ade, adenylylation; pho, phosphorylation; ace, acetylation.

The patterns of resistance transfer were the same in these R plasmids, even when ex-conjugants were selected for any of the 6 drugs or to any of the three recipient strains.

On the other hand, Ps142 transferred (SM) resistance at a high frequency (5.3×10^{-1}) and (SM.SA.GM.CPC) resistance at a low frequency (2.0×10^{-6}). The ex-conjugant which received only (SM) resistance, ML4262 R_{ms148} (SM), could transfer the (SM) resistance at a frequency of 1.4×10^{-1} to the second recipient, ML4600. The almost all of the other exconjugants which received (SM. SA.GM.CPC) resistance also transferred (SM) resistance at a high frequency and (SM.SA.GM.CPC) resistance at a low frequency. We could isolate ML4262 R_{ms149} (SM.SA.GM.CPC) from exconjugants which transferred only (SM.SA.GM.CPC) resistance at a low frequency (1.6×10^{-6}) to ML4600 but did not transfer (SM) resistance. Furthermore, we isolated ML4262 R_{ms150} (SA.CPC) from ML4262 R_{ms149}, which had lost (SM.GM) resistance. This R_{ms150} could co-exsist with R_{ms148} in a same cell. Results of these experiments indicate that the original strain Ps142 posesses two types of R plasmids, R_{ms148} and R_{ms149} in a cell, called hetero-R state (Table 3, 4, 5).

TABLE 3

Transfer of drug resistance from Ps110 and Ps138

Donor	Recipient	Transfer frequency[a)]	Drug-resistance transferred
Ps110	ML4561	7.1×10^{-4}	TC, SM, SA, KM, LV, GM, DKB
	ML4262	3.1×10^{-4}	TC, SM, SA, KM, LV, GM, DKB
Ps138	ML4561	1.0×10^{-5}	TC, SM, SA, KM, LV, GM, DKB
	ML4262	4.8×10^{-6}	TC, SM, SA, KM, LV, GM, DKB

a) Transfer frequency was expressed by the ratio of the number of ex-conjugants per that of donor cells. The selection was made for the transfer of SM resistance.

TABLE 4

Transfer of drug resistance from Ps142

Donor	Recipient	Selection	Transfer frequency	Resistance transfered
Ps142	ML4561	SM	5.3×10^{-1}	SM
		GM	1.5×10^{-6}	SM, SA, GM, CPC
		CPC	2.0×10^{-6}	SM, SA, GM, CPC
ML4262 R^{+}[a)]	ML4600	SM	1.0×10^{-1}	SM
		GM	3.0×10^{-3}	SM, SA, GM, CPC
		CPC	2.3×10^{-3}	SM, SA, GM, CPC

a) ML4262 R^{+} was an ex-conjugant of ML4262 which acquired (SM.SA.CM.CPC) resistance from Ps142.

TABLE 5

Transfer frequency of R factors derived from Ps142 to ML4600 with or without R factors

Donor	Recipient (ML4600 carrying)	Selection	Transfer frequency
ML4262 R148	—[a)]	SM	1.4×10^{-1}
$(SM)^{+}$	R150 (SA.CPC)	SM	2.6×10^{-1}
	R150 (SA.CPC)	GM	2.0×10^{-6}
ML4262 R150	—	CPC	1.8×10^{-5}
$(SA, CPC)^{+}$	R148(SM)	CPC	1.6×10^{-5}

a) ML4600 without R factor
R148 was an R factor carrying only SM resistance and its transfer frequency was high markers.
R149 carried (SM.SA.GM.CPC) resistance and its transfer frequency was low.
R150 was a segregant obtained from an R149.

It should be noted that exconjugant ML4262 R_{ms148} was resistant to SM but susceptible to spectinomycin, while ML4262 R_{ms149} was resistant to both drugs, indicating the difference of biochemical mechanisms of SM resistance in both plasmids.

2. Enzymatic degradation of antibiotics coded by 4 plasmids

The biochemical mechanisms of drug resistance conferred by the four plasmids are given in Table 2 and will be described in detail elsewhere.

REFERENCES

Knothe, H., Krčméry, V., Sietzen, W. and Borst, J. (1973). Chemotherapy 18: 229–234.

Inoue, M., Hashimoto, H. and Mitsuhashi, S. (1969). Progress in antimicrobial and anticancer chemotherapy (Univ. of Tokyo Press), 433–439.

Perret, C. J. (1954). Nature 174: 1012–1013.

V.

ECOLOGY, EPIDEMIOLOGY AND NOSOCOMIAL PROBLEMS OF ANTIBIOTIC RESISTANCE

CHAIRMEN: R. E. DIXON (U.S.A.) and L. ROSIVAL (CZECHOSLOVAKIA)

OPENING LECTURE

EPIDEMIOLOGY OF DRUG RESISTANCE IN HOSPITALS

R. E. DIXON

Chief, Hospital Infections Branch, Bacterial Diseases Division, Center for Disease Control, Atlanta, Georgia, U.S.A.

In the past decade, great concern has been expressed about the emergence of antimicrobial drug-resistant bacteria as a cause of human disease. Examples of this emergence are numerous. Many strains of *Neisseria gonorrhoeae* were found to be resistant to low-dose of sulfonamides shortly after these agents became widely available. For a period, low-dose penicillin was effective in treating gonorrheal infection, but, by the mid-1950s, it was apparent that *N. gonorrhoeae* was also becoming more resistant to penicillin; and, in the United States, we are currently seeing a progressive increase in the resistance of these organisms. Only recently we learned of 3 cases of *Hemophilus influenzae* meningitis caused by an ampicillin-resistant strain (1) — a finding not previously documented. Pneumococci resistant to tetracycline were first noted in the United States in 1964 (2), 16 years after that antimicrobial was first marketed in the United States. A widespread epidemic of typhoid fever in Mexico was caused by a strain of *Salmonella typhi* resistant *in vivo* and *in vitro* to chloramphenicol (3), and this resistance was conferred by a transferable resistance factor (4). These shifts were presaged by the development of resistance in some organisms soon after antimicrobials were introduced. Several mechanisms have accounted for this development of resistance. For some organisms, antimicrobial-induced mutations conferred high-level drug resistance. Other strains seemed to be selected by the availability of antimicrobials: penicillinase-producing staphylococci, probably present long before penicillin became available therapeutically, had a selective advantage after the use of penicillin became widespread, and resistant strains have become quite prevalent around the world. Still other strains — particularly gram-negative organisms in the *Family Enterobacteriaceae* — have been shown capable of acquiring antimicrobial resistance through interbacterial transfer of genetic information.

Drug resistance of gram-negative bacilli is especially worrisome. Between July 1972 and June 1973, aerobic gram-negative bacilli accounted for 67 percent of community-acquired and 68 percent of nosocomial (hospital-acquired) infection isolates from hospitalized patients in 8 community hospitals participating in the Center for Disease Control's (CDC's) Comprehensive Hospital Infections Project (CHIP). Many of these organisms were resistant to 1 or more antimicrobial agents.

Two gloomy themes permeate the infectious disease literature. One proposes that gram-negative infections are becoming increasingly common, and the other suggests that gram-negative organisms, even compared with those seen several years ago, are becoming more resistant to available antimicrobials. Many investigators correlate these apparent changes with the use — and misuse — of antimicrobial agents. They contend that widespread antimicrobial use either results in replacement of sensitive strains with resistant ones or induces drug resistance in those organisms remaining. These harmful effects of antimicrobials have been often cited in support of efforts to assure their more rational use (5). Bulger and associates noted that resistant staphylococci (6) and gram-negative organisms (7) were isolated relatively less frequently as more rational antimicrobial usage was instituted in a single hospital. However, not all share this view. Professor R. E. O. Williams has said that "the secular changes that we have seen have not been wholly, or even mainly, due to alterations in the use of antibiotics... there are indications that the rise in numbers of cases of gram-negative bacteremia is very largely due to changes in the population and in our ability to recognize the condition, rather than to the backwash of antibiotic therapy" (8). Thus, at this point, it seems appropriate to ask 3 questions: 1. What is the evidence that gram-negative infections and particularly drug-resistant infections are increasing; (2) if these trends do exist, what evidence supports the contention that the trends are adversely affected by antibiotic usage; and (3) if a problem exists, how important is it?

Considering the complexity of these questions, it is not surprising that arguments are based more upon anecdote and inference than upon controlled study. Five types of evidence are available: (1) the effects of antimicrobials on microorganisms *in vitro*; 2. the effects of antimicrobials on an individual host — either human or animal; 3. extrapolations from epidemic investigations; 4. longitudinal studies of the incidence of patient disease; and 5. longitudinal studies of susceptibility reactions from clinical bacterial isolates. The first effect is discussed by others at this symposium, and I shall limit my remarks to the last 4.

EFFECTS OF ANTIMICROBIALS UPON HOST CARRIAGE OF MICROORGANISMS

It is well established that the resident microbial flora of a host changes when it is exposed to an antimicrobial agent either systemically or topically. Sensitive strains are either eradicated or markedly reduced in number, and resistant strains are isolated in greater numbers. The proliferation of resistant strains may result from exogenous acquisition which is not prevented by the normal, inhibitory effects of the normal host flora; from overgrowth of such organisms already present in low numbers but, because of the presence of the antimicrobial, now having a selective advantage; or from induction of resistance in a previously sensitive strain by the antimicrobial itself.

In addition to these well-known and well-documented changes, we must consider two additional less well-known effects of antimicrobial usage. First, an individual need not receive an antimicrobial agent himself to be affected directly by that agent. In an institutional setting particularly, where transmission of organisms among persons occurs with unfortunately high frequency, antimicrobial therapy of other persons may influence their flora and, because of secondary spread of microorganisms, may then result in infection with ar resistant strain by a person not directly exposed to the antimicrobial agent.

Second, antimicrobial therapy may influence the ability of a host to transmit a particular organism. Bennett and associates (9) attempted to control endemic staphylococcal skin disease in a institution for the retarded by artifically colonizing a sample of residents with a supposedly avirulent strain of *Staphylococcus aureus* (the 502-A strain). This strain was resistant to tetracycline. After nasal installation of 502-A, apparent nasal carriage of the organism fell to low levels by 8 weeks after exposure in the absence of tetracycline therapy. However, when tetracycline was subsequently administered, many previously treated individuals could again be documented to carry the strain although they had negative cultures before treatment. This phenomenon is called recall. More striking was the apparent spread of this tetracycline-resistant strain to previously unexposed patients in the institution. At the time of initial nasal installation of 502-A, few patients in the untreated control group acquired carriage by secondary spread. However, when tetracycline therapy was given, such spread to the control group did occur.

Thus when evaluating the epidemiology of drug resistance, we must consider not only the immediate effects of antimicrobial therapy upon the host who receives the drug, but we must also be concerned with the effects of such therapy upon the untreated.

EPIDEMICS CAUSED BY MULTIPLE-RESISTANT PATHOGENS

The importance of multiply drug-resistant bacteria is dramatically illustrated by epidemics within hospitals. Numerous such outbreaks have been thus far described (10–13) and have required, on occasion, limitation of antimicrobial usage to achieve control (14). Examples of 3 such out breaks recently investigated by CDC epidemiologists illustrate the diversity in the epidemiology of nosocomial infections caused by resistant strains.

In the first example, 9 patients hospitalized in the rehabilitation unit of a university-affiliated community hospital acquired *Proteus rettgeri* urinary tract infection (UTI) in a 12-month period. The epidemic strain was resistant to all antimicrobials routinely tested (ampicillin, carbenicillin, cephalothin, colistin, gentamicin, kanamycin, nalidixic acid, nitrofurantoin, streptomycin, sulfonamides, and tetracycline) and had, in addition, a negative indole reaction. The first patient with disease was admitted from another hospital with UTI caused by the epidemic strain. She had an indwelling urinary catheter, and she subsequently shared the same room as other patients with catheters. The epidemiologic investigation showed that patients with *P. rettgeri* UTI were the major reservoir for the organism and that infection was transmitted from infected patients with catheters to uninfected patients who were also catheterized by personnel providing urinary catheter care. The strain did not appear to reside in the patients' gastrointestinal tracts, nor was it isolated from the inanimate environment. Furthermore, antimicrobial therapy appeared to play little direct role either in preventing infection or in making patients more susceptible to infection. Despite numerous attempts, presence of an R factor was not demonstrated in CDC laboratories.

In the second outbreak, increased use of gentamicin was temporally associated with the appearance of a multiply resistant epidemic strain. In this epidemic, 130 patients in a large municipal hospital developed nosocomial infection with another *P. rettgeri* strain that was resistant to gentamicin as well as numerous other antimicrobials. By October 1971 (Figure 1), the epidemic strain ("highly resistant PR infection," listed at the top of the figure) appears to begin to replace the more sensitive *P. rettgeri* strains that had been present in the hospital (shown at the bottom of the figure), and, by early 1972, the resistant strain had become epidemic. The increased incidence of infection with the gentamicin-resistant strain followed the increased use of gentamicin in the hospital (Figure 2). Resistance transfer factor was not isolated from this epidemic strain either, nor did other organisms seem to acquire this resistance pattern.

Finally, 36 surgical patients in a large university-affiliated federal hospital developed infection with a multiply-resistant type 21 *Klebsiella penumoniae* strain. Nineteen patients had significant clinical illness, and 4 patients died as a result of the infection. Each of the epidemic strains was resistant to ampicillin, cephalosporin, tetracycline, chloramphenicol, kanamycin, and gentamicin and was sensitive to colistin, nitrofurantoin, and BBK8. The epidemic strain was not isolated from a common source in the inanimate environment; instead, the epidemiologic investigation strongly suggested that infection was transmitted

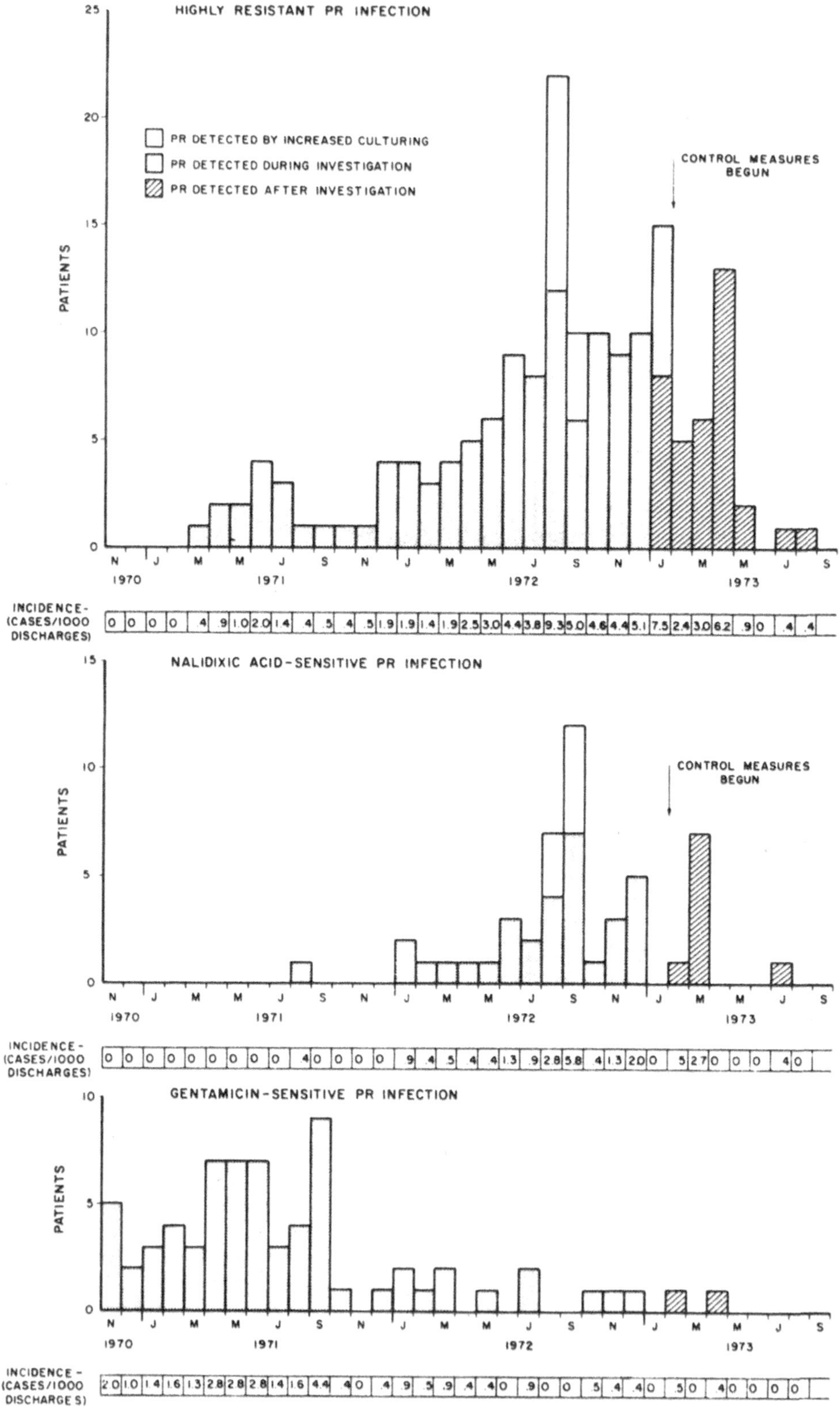

HIGHLY RESISTANT PR INFECTION
PR DETECTED BY INCREASED CULTURING
PR DETECTED DURING INVESTIGATION
PR DETECTED AFTER INVESTIGATION
CONTROL MEASURES BEGUN
PATIENTS
1970
1971
1972
1973
INCIDENCE-(CASES/1000 DISCHARGES)
NALIDIXIC ACID-SENSITIVE PR INFECTION
CONTROL MEASURES BEGUN
PATIENTS
INCIDENCE-(CASES/1000 DISCHARGES)
GENTAMICIN-SENSITIVE PR INFECTION
PATIENTS
INCIDENCE-(CASES/1000 DISCHARGES)

from infected patient to susceptible patient on the hands of ward personnel. Gastrointestinal colonization appeared to be the primary reservoir for the epidemic strains. Although an R factor carrying the resistance determinants was demonstrated *in vitro*, no other species acquired these resistances; and, since all of the epidemic strains of Klebsiella tested have the same serotype, it is uncertain whether the R factor was paramount in determining the course of the epidemic.

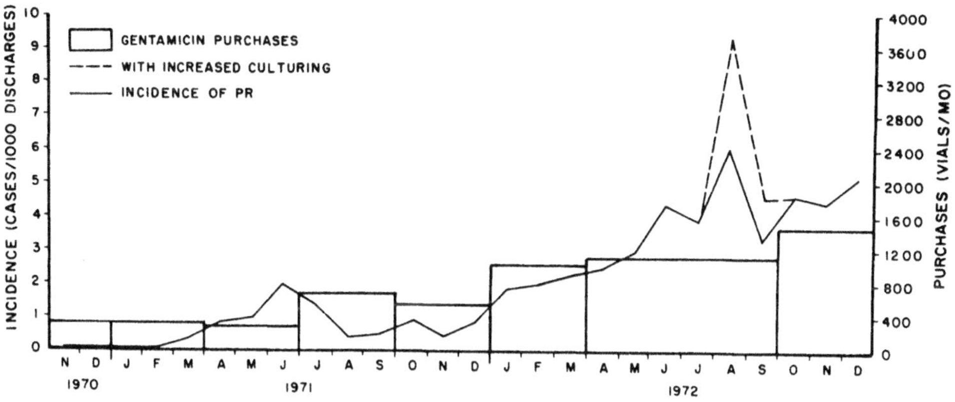

Figure 2. Monthly incidence of highly resistant *Proteus rettgeri* infection and monthly Gentamicin purchases. The dotted line reflects the increase in incidence resulting from widespread culturing of patients during part of the epidemiologic investigation.

I have briefly summarized these three outbreaks to make several probably obvious points. Nosocomial infections caused by multiply-resistant organisms are not solely R factor problems. Antimicrobial use may influence which organisms are prevalent in the hospital, and by R factor selection and enhancement, they strongly influence the suceptibilities of these strains. However, other aspects of nosocomial infection epidemiology are also important in determining whether patients become ill. These other aspects include the virulence of the microorganism, the susceptibility of the host population, and the mode of transmission of the infecting pathogen.

CHANGES IN THE INCIDENCE OF INFECTIOUS DISEASES

On the basis of published reports, there is little doubt that the types of microorganisms responsible for infections are changing in relative importance, especially in the hospitalized patient. In 1959, Rogers (15) reviewed deaths on the medical services at New York Hospital in 1938–40 compared with 1957–58 and found that death due to pneumococci, group A streptococci, and tuberculosis had been virtually eliminated in the latter period while they were quite common in the former. By the late 1950s, however, he found that fungal and enteric bacterial infections had markedly increased as a cause of death. He postulated that increased use of antimicrobials had strongly influenced these findings.

◀ *Figure 1.* Patients with resistant *Proteus rettgeri* infection, by month of first positive culture, in a municipal hospital. Top figure: cases of disease with the highly resistant strain; middle figure: cases of disease with Nalidixic acid-sensitive variant; Bottom figure: cases of disease with other Proteus strains.

Dr. Maxwell Finland at the Boston City Hospital has also repeatedly commented upon changes in the pathogens responsible for disease in hospitalized patients (16). He, too, has correlated these changes with the availability and use of antimicrobials. In the most recent publication (17) from Finland's group, McGowan and co-workers examined the incidence of bacteremia, primarily in surgical patients; in 8 years between 1935 and 1969 selected to assess potential influences of new antimicrobial agents, they found a 10-fold stepwise increase in the incidence of gram-negative bacillary bacteremia from 0.9 cases per 1,000 admissions in 1935 to 8.7 cases per 1,000 in 1969. Despite the increasing availability of potent antimicrobial agents, the case-fatality ratios were the same in 1969 (49 percent) as in 1935 (50 percent) and have persisted at about that level for the entire time. In contrast to the marked increases in the incidence of gram-negative bacteremia, the incidence of gram-positive bacteremia fluctuated markedly in the period and only increased 1.5-fold in the 34-year period. As in Rogers's study, these changes were attributed, in part, to the influences of antibiotics.

In contrast to information on many infections of public health importance such as smallpox, tuberculosis, measles, or diphtheria where surveillance systems provide information derived from very large non-hospitalized populations, broadly based studies of infections seen in hospitals have not been initiated until recently. As a result, data are available primarily from single institutions where local changes in patient population, types of medical care, or laboratory technique may markedly influence the apparent trends in disease incidence. None of the studies showing an increased incidence of gram-negative infections are adequately controlled for other changes in periods compared. Therefore, it might be argued that the apparent increase in gram-negative infections has resulted from other advances in medical therapy whereby an older population or a population with substantially diminished resistance to infection is proportionally more often hospitalized now than previously. Epidemic or hyperendemic nosocomial infection problems, not characteristic of other institutions' experiences, might markedly influence data obtained from small samples of patients in single institutions. Finally, other therapeutic advances not directly related to the availability of antimicrobials may play a paramount role in determining patient susceptibility to disease. As a result, it is indeed possible to argue that the seemingly increased occurrence of gram-negative infection is more apparent than real and that if a current hospitalized population were standardized so as to be comparable to populations hospitalized 10, 20, or even 50 years ago, there might be no actual increased risk of disease. Without large-scale studies involving large populations, it is difficult to say with certainty what role these other factors have played in the epidemiology of drug resistance. But no matter what the reasons, it seems quite clear that there has been both an absolute and relative increase in these infections over the past several decades, and that increase, in itself, is a cause for concern. However, it seems equally clear that we cannot attribute these changes solely — or even primarily — to the effects of antimicrobials without an evaluation of the other factors playing a role in that increase.

CHANGES IN ANTIMICROBIAL SUSCEPTIBILITIES OF CLINICAL ISOLATES

In addition to there being an increased incidence of infection with organisms likely to be inherently resistant to antimicrobials, it is also often argued that shifts from relative sensitivity to resistance have occurred within bacterial species. This has certainly occurred with some organisms such as *Staphylococcus aureus*, *N. gonorhoeae*, and *Neisseria meningitidis* (18). Fewer data are available about the long-term trends of gram-negative bacilli in this regard, but the available data suggest that these organisms too have become more resistant over time. The evidence supporting this contention comes primarily from longitudinal studies of clinical isolates in individual institutions.

Finland has recently reviewed his extensive experience with a wide range of pathogens isolated at the Boston City Hospital (19). From data collected over 2 decades, he has confirmed the observations of others about the increased resistance of *S. aureus*, gonococci, and meningococci. He also found that common gram-negative bacilli have become increasingly resistant, and he attributed that shift to "excessive use of antibiotics; the use of multiple, highly active, and broad-spectrum agents; the use of excessive dosages; and particularly their widespread use of prophylaxis." He thereby implied that patients treated with antibiotics or having contact with others so treated are at greater risk of infection with a resistant strain than are patients not so exposed. And the results obtained by Eickhoff (20) of Finland's group seem to support that view. In that study, the sucepti-bilities of 206 nosocomial strains and 64 community-acquired strains of *Tribe Klebsielleae* were compared for sensitivity to chloramphenicol, tetracycline, kanamycin, streptomycin, cephalothin, and polymyxin B. For each antimicrobial, nosocomial strains exhibited a higher resistance than did community strains (20).

More recent data obtained by CDC in its CHIP hospitals suggest that the differences between nosocomial and community isolates are not so striking. The 8 community hospitals that participated in CHIP between 1970 and 1973 were chosen to reflect the infection experience of the majority of hospitals providing medical care in the United States. They are average-size community hospitals, located in communities of different types across the United States, serving different patient populations, having different physical facilities and medical care structures, and not serving as major teaching hospitals for medical schools.

In each CHIP hospital, surveillance for both community-acquired and nosocomial infections was carried out by trained infection control nurses who used standard definitions to determine when an infection was present and to differentiate nosocomial from community-acquired infection (21). The quality of surveillance was constantly monitored throughout the project by CDC epidemiologists, who performed frequent on-site prevalence surveys seeking every case of infection. These quality-control checks demonstrated that the infection control nurses were more than 70 percent effective in finding infections; for serious infections and for infections documented by positive bacteriolologic cultures, they were even more efficient. Laboratory competency was also assessed in the project. Each hospital's laboratory used acceptable techniques for isolation and identification of potential pathogens, and each used a standard modification of the high potency agar disk diffusion technique for antimicrobial susceptibility testing (22). Microbiology laboratories in CHIP hospitals participated in a proficiency testing program specifically designed for the project, and they agreed with CDC laboratories in the identification of unknowns in over 90 percent of samples and in susceptibility test results in over 96 percent of tests.

Between July 1972 and June 1973, the last complete year of the project, 9,169 bacterial isolates were reported to be responsible for nosocomial and community-acquired infection. All isolates were from patients with active infection, and the reported pathogens were considered to be responsible for disease on the basis of the established criteria (21).

Figure 3 shows the results of susceptibility testing of 809 *S. aureus* and 189 *Staphylococcus epidermidis* strains; these isolates were almost equally divided between nosocomial (520) and community-acquired (478) infections that had, insofar as could be determined, no association with hospitalization. The susceptibility patterns for nosocomial and community strains of *S. aureus* are remarkably similar; although nosocomial isolates are more resistant for 7 of 12 tested antimicrobials, the percentage differences are minimal.

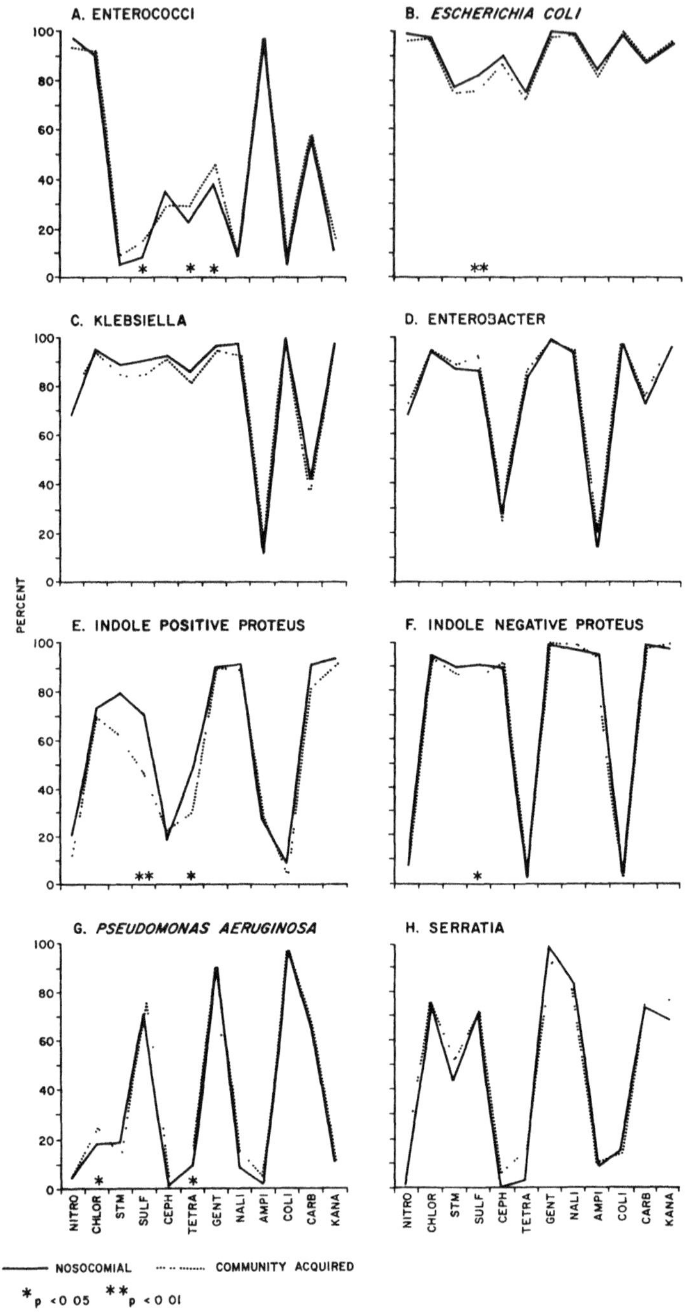

Figure 3. Percentage of staphylococcal clinical disease isolates susceptible to 12 antimicrobials; Nosocomial vs. Community-Acquired, CHIP Hospitals, July 1972 — June 1973.

Similarly, for *S. epidermidis*, the patterns are not remarkably different, but, again, community strains appear slightly more sensitive than the nosocomial strains. For streptomycin and erythromycin, the differences are statistically significant at less than the 5 percent level (Chi-square, 2 tail test).

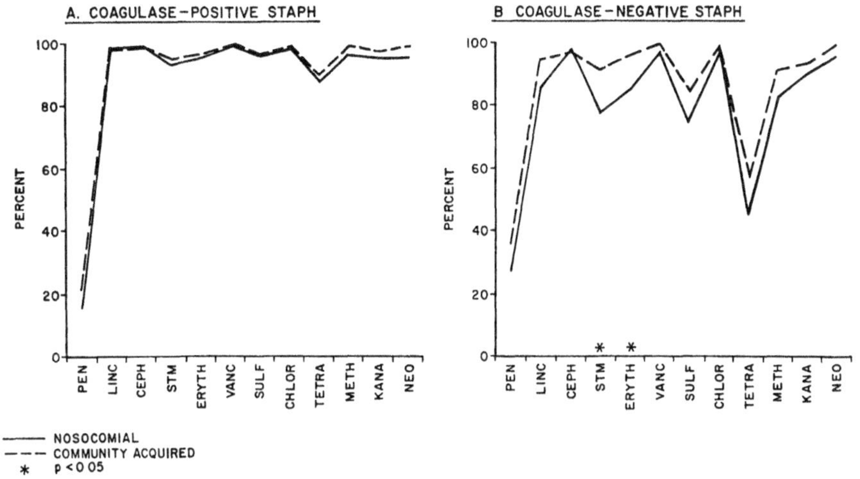

Figure 4. Percentage of enterococcal and gram-negative clinical disease isolates susceptible to 12 antimicrobials: Nosocomial vs. Community-Acquired, CHIP Hospitals, July 1972 — June 1973.

We were surprised however when we compared susceptibility patterns of nosocomial and community isolates of other organisms. Figure 4 illustrates the comparative susceptibility patterns of 8 other groups of organisms: *Pseudomonas aeruginosa*, enterococci, and 6 genera in the Family *Enterobacteriaceae*. (We include enterococci with gram-negative organisms because we routinely test them against the gram-negative battery of antimicrobial agents). Of the 96 combinations of organism with antimicrobial, the community-acquired isolates were more susceptible in 45 (47 percent), the nosocomial isolates were more susceptible in 40 (42 percent), and susceptibilities were comparable in 11 (11 percent). Furthermore, although differences were seen in 85 of the 96 combinations, the differences were usually minor. In fact, only 10 of the differences are statistically significant; community-acquired strains are more sensitive in 5, and nosocomial strains are more sensitive in the other 5.

When the organisms or the antimicrobials are evaluated independently, one still sees few striking differences. Only *Escherichia coli* and Indole-positive proteus showed significant differences when the reactions of community and nosocomial strains are compared for the entire test battery of antimicrobials. Most strikingly, for both organisms, the nosocomial isolates showed greater overall sensitivity (*E. coli*: $p < 0.01$; Chi-square, 2 tail test and indole-negative proteus; $p < 0.01$; Chi-square, 2 tail test). Furthermore, only 2 of the 12 antimicrobials routinely used for these organisms showed significant differences between susceptibilities of nosocomial and community strains. Community-acquired isolates were more susceptible to kanamycin (79 percent) than were nosocomial isolates (74 percent; $p < 0.001$; Chi-square, 2 tail test), while the converse is true with sulfonamides where community isolates were more sensitive (74 versus 71 percent; $p < 0.01$, Chi-square, 2 tail test).

How can these data be reconciled with those obtained by Eickhoff and co-workers (20)? First, the studies were conducted in different types of institutions. Second, different methods were used in the two studies. In Eickhoff's studies, a sampling of organisms from a single institution over a short time period were evaluated; in contrast data in the CHIP study came from 8 widely differing hospitals and represented a relatively complete sample of clinically significant nosocomial and community-acquired pathogens for an entire year. As a result, an epidemic in any single institution or community would have had little impact upon the overall results. In this regard, Eickhoff's data appear somewhat biased in that a multiply-resistant strain of *K. pneumoniae*, serotype 21, was hyperendemic in the hospital during the period of study. As a result, the nosocomial strains evaluated may not have been representative of normal nosocomial flora. Furthermore, those investigators grouped together various members of *Tribe Klebsielleae* in presenting their results. Since *Serratia marcescens*, for example, is more often nosocomial than community-acquired and has greater intrinsic resistance to antimicrobials, a relative over-representation of this organism in the nosocomial group could have contributed to the observed greater resistance among nosocomial pathogens.

Surveillance of nosocomial and community-acquired infections has not been conducted long enough by CDC to evaluate long-term changes in susceptibilities of any of the organisms discussed here. However, the comparisons of community and nosocomial isolates suggest that the conventional thoughts regarding changes in susceptibility patterns need to be reconsidered, for these findings can only be reasonably explained by one of three explanations. First, the data from CHIP may be misleading because infection control personnel did not effectively differentiate community and nosocomial sources for the pathogens reported, because identifications or susceptibility tests were inaccurate in CHIP hospitals, or because CHIP hospitals were most unusual and did not reflect the experiences of other institutions. None of these explanations seems reasonable since quality control evaluations of both surveillance and laboratory performance by CDC showed consistent high-quality performance in participating hospitals and since participating hospitals were selected only because of their interest in infection control and were unlikely to be an otherwise biased population.

As a second explanation, it is possible that antibiotic usage plays little role in the overall development of antimicrobial resistance, for, despite the expressed commitment to infection control manifest by CHIP hospitals, antimicrobial usage was widespread in these hospitals. Such usage was evaluated throughout the study. Findings in the pilot phases (23) are representative. In that period, Scheckler and co-workers found that 30.6 percent of patients whose charts were reviewed by CDC epidemiologist during prevalence surveys received antimicrobials at some time during hospitalization. Over 60 percent of treated patients had no recorded evidence of any infection. However, this explanation seems unlikely, for, as noted earlier, there is substantial evidence from *in vitro*, *in vivo*, and epidemic investigations that antimicrobial usage influences the occurrence of infection and the microorganisms responsible for disease.

Finally, it remains possible that antimicrobial usage does influence susceptibility patterns of gram-negative and other enteric organisms. If this be true, one must postulate that patients in the community have either extensive contact with antimicrobials or substantial exposure to microorganisms that have, themselves, been in contact with antimicrobials. Although antibiotic use is widespread among outpatients, it seems unlikely that this alone could explain the resistance patterns seen. On the other hand, almost all persons in the United States have contact with foods derived from animals that have been routinely fed broad-spectrum antimicrobials, and resistance of community-acquired pathogens might be enhanced by this mechanism. Alternatively, R factors might be widely spread throughout the community because of dissemination from hospitals or through the intermediary of animal food products.

Of course, other hypotheses might be advanced, but at the risk of sounding too speculative, I think it is indeed probable that antimicrobial agents do influence the susceptibilities of infecting microorganisms. Furthermore, data from the CHIP study suggest that community as well as nosocomial pathogens can be so influenced. It is perhaps not a coincidence that problems with multiply-resistant shigellae and salmonellae, which had such a profound influence upon work with R-factor-mediated resistance, occurred in the community — not in the hospital.

CONCLUSIONS

Tentative answers to the questions initially posed can now be offered although the data upon which these conclusions must be based are incomplete and in some respects contradictory. The shift in the types of organisms responsible for sporadic disease acquired in the hospital or requiring hospitalization clearly appears to be real. Although most data documenting these shifts come from single institutions where local community or hospital factors may bias results so that they are not representative of the general experience, the unanimity with which observers have noted similar trends suggests that those observations are valid. It is less certain whether individual species responsible for infection are themselves becoming more resistant. But again, most data do suggest such a shift.

Next, is there any evidence that antimicrobial usage has influenced resistance patterns within bacterial species? For many microorganisms, such as *S. aureus* and *N. genorhoeae*, there is strong associational evidence suggesting such an influence. For the gram-negative enteric bacteria, the data are less firm, but epidemics caused by multiply-resistant strains occur with some frequency and often coincide with antimicrobial usage patterns that appear to have played a causative role. In light of the well established mechanisms whereby antimicrobial exposure may result in enhanced resistance, it seems highly unlikely that antimicrobial usage plays no role.

Finally, one must ask how important multiply-resistant microorganisms are. Here too, data are insufficient to say with certainty. On occasion, such organisms cause major epidemic or hyperendemic problems. But fortunately we are not yet faced with the situation where the bulk of infections are caused by these strains. While the failure of multiply-resistant microorganisms to take over as the major pathogens responsible for disease is heartening, that failure should not cause us to be unconcerned about the problem. These organisms continue to be a major threat. Thus, continued research is absolutely required in order to understand better the causes of drug resistance and to develop new methods to treat such infections. Along with that research, it is also most important that we carefully study the epidemiology of drug resistance so that we can know what the problems are and understand the factors that influence these problems.

ACKNOWLEDGEMENTS

The epidemiologic investigations reported herein were conducted by James O. Lindsey, M.D., Richard A. Kaslow, M.D., and Robert C. Aber, M.D., Hospital Infections Branch, Bureau of Epidemiology, Center for Disease Control. Mr. Don C. Mackel, M.S., M.P.H., Mr. Roger Anderson, B.S., M.S., Miss Kaye Wachsmuth, B.S., and Mrs. Connie

Phillips, B.S., M.S., assisted in the laboratory investigations reported. I appreciate the assistance provided by John V. Bennett, M.D., Director, Bacterial Diseases Division, Bureau of Epidemiology, CDC, in the preparation of this manuscript.

REFERENCES

1. Center for Disease Control: Ampicillin-resistant *Hemophilus influenzae* meningitis-Maryland Georgia. Morbidity and Mortality Weekly Rep 23: 77–78, Mar 2, 1974.
2. SHAEDLER, R. W., CHOPPIN, P. W., ZABRISKIE, J. B. (1964). Pneumonia caused by tetracyclineresistant pneumococci. N Engl. J Med 270: 127–129.
3. Center for Disease Control: Follow-up on chloramphenicol-resistant *Salmonella typhi*-Mexico. Morbidity and Mortality Weekly Rep 22: 159, May 5, 1973.
4. GANGAROSA, E. J., BENNETT, J. V., WYATT, C., et al. (1972). An epidemic-associated episome? J Infect Dis 126: 215–218.
5. SIMMONS, H. E., STOLLEY, P. D. (1974). This is medical progress? Trends and consequences of antibiotic use in the United States. JAMA 227: 1023–1028.
6. BULGER, R. J., SHERRIS, J. C. (1968). Decreased incidence of antibiotic resistance among *Staphylococcus aureus*. A study in a university hospital over a 9-year period. Ann Intern Med 69: 1099 — 1108.
7. BULGER, R. J., LARSON, E., SHERRIS, J. C. (1970). Decreased incidences of resistance to antimicrobial agents among *Escherichia coli* and *Klebsiella-Enterobacter*. Observations in a university hospital over a 10-year period. Ann Intern Med 72: 65–71.
8. WILLIAMS, R. E. O.: Changing perspectives in hospital infection. In Proceedings of the International Conference on Nosocomial Infections. Chicago, American Hospital Association, 1971, pp. 1–10.
9. BENNETT, J. V., SHULMAN, J. A., ROSENSTEIN, B. J., et al. (1968). Staphylococcal interference studies. Amer. J Epidemiol 88: 410–421.
10. ADLER, J. L., SHULMAN, J. A., TERRY, P. M. et al. (1970). Nosocomial colonization with kanamycinresistant *Klebsiella pneumoniae* types 2 and 11 in a premature nursery. J Pediatr 77: 376–385.
11. SELDEN, R., LEE, S., WANG, W. L. L. et al. (1971). Nosocomial klebsiella infections: intestinal colonization as a reservoir. Ann Intern Med 74: 657–664.
12. MARTIN, C. M., IKARI, N. S., ZIMMERMAN, J., et al. (1971). A virulent nosocomial *Klebsiella* with transferable R factor for gentamicin: emergence and suppression. J Infect Dis 124: S24–S29.
13. FRANCO, J. A., EITZMAN, D. V., BAER, H. (1973). Antibiotic usage and microbial resistance in an intensive care nursery. Amer J Dis Child 126: 318–321.
14. PRICE, D. J. E., SLEIGH, J. D. (1970). Control of infection due to *Klebsiella aerogenes* in aneurosurgical unit by withdrawal of all antibiotics. Lancet 2: 1213–1215.
15. ROGERS, D. E. (1959). The changing pattern of life-threatening microbial disease. N Engl J Med 261: 677–683.
16. FINLAND, M. (1970). Changing ecology of bacterial infection as related to antibacterial therapy. J Infect Dis 122: 419–431.
17. MCGOWAN, J. E. Jr., BARNES, M. W., FINLAND, M. (1974). Host-pathogen-drug interactions in surgical patients with bacteremia. Surg Gynecol Obstet 138: 50–54.
18. GILL, F. A., HOOK, E. W. (1965). Changing patterns of bacterial resistance to antimicrobial drugs. Amer. J Med 39: 780–795.
19. FINLAND, M. (1972). Changing patterns of susceptibility of common bacterial pathogens to antimicrobial agents. Ann Intern Med 76: 1009–1036.
20. EICKHOFF, T. C., STEINHAUER, B. W., FINLAND, M. (1966). The *Klebsiella-Enterobacter-Serratia* division. Biochemical and serological characteristics and susceptibility to antibiotics. Ann Intern Med 65: 1163–1179.
21. Center for Disease Control: Outline for Surveillance and Control of Nosocomial Infections. U.S. Department of Health, Education, and Welfare, Atlanta, Georgia, Reprinted May 1974.
22. Center for Disease Control: Antimicrobial susceptibility testing. National Nosocomial Infections Study Quarterly Report, Fourth Quarter 1972, Issued April 1974, pp 26–34.
23. SCHECKLER, W. E., BENNETT, J. V. (1970). Antibiotic usage in 7 community hospitals. JAMA 213: 264–267.

STUDIES ON THE DRUG RESISTANCE OF *S. AUREUS* AND *E. COLI* AGAINST ANTIBIOTICS

H. OTAYA, S. MACHIHARA, M. YOSHIMURA, M. YAMAMOTO

Shionogi & Co., Ltd. Osaka. Japan

1. The processes of the susceptibility survey in Japan.
2. Differentiation of drug sensitive- and resistant-strains by minimal inhibitory concentration of drug and annual change of sensitivity of the bacteria to various drugs.
3. Difference of sensitivity of isolates from various clinical specimens.

We have been investigating the drug resistance of clinical strains of *S. aureus* and *E. coli* isolated in main hospitals all over Japan, annually since 1965. Our main objective was to investigate the annual trend of bacterial resistance to antibiotics and sulfonamide frequently used in hospitals in Japan. In this paper, we will report about three items, firstly on the processes of susceptibility survey in Japan, secondly on the grouping concentration of drug sensitive- and resistant-strains by the moving pooling method and

TABLE 1

Record of supplied isolate

Strain; *S. aur.* (coagulase positive) *E. coli.* Streptococcus, beta.			
Prefecture. ________ No. ______	Patient; _____ years. in-patient out-patient ♀| ♂| Name of disease;	Name of hospital Name of supplier	Isolated, date.
Name of strain	Pathogenic bacterium;	Specimens;	
	1. definite 2. probable 3. possible	pus, sputum, pharyngeal mucus, urine, bile, blood, spinal fluid, faeces, others.	
The result of susceptibility at your lab. method of test, (name of disc.) Drugs; Erythromycin, Penicillin-G, Aminobenzyl-PC, Kanamycin, Streptomycin, Chloramphenicol, Tetracyclin, Cephalothin, Cephaloridin, Cephalexin, Cephazolin, Sulfonamid, Others.			

annual change of sensitivity of bacteria to various drugs. Finally, on the difference of sensitivity of isolates from various clinical specimens.

The detail of each items are described as follows;

1. *The processes of the susceptibility survey in Japan.*

METHODS

Sampled hospitals and isolates.

Hospitals and bacterial strains were selected by a statistical sampling method according to population and number of general hospitals which have more than 100 beds in each Prefecture.

TABLE 2

No. of isolates in each district. May, 1971 — June, 1972

	Population	No. of hospitals	S. aureus		E. coli	
			No. hosp.*	No. isol.	No. hosp.*	No. isol.
A	28,354	462	49	124	82	246
B	11,815	190	30	72	46	131
C	12,375	219	15	32	32	86
D	50,974	947	97	219	152	455
Tot	103,518	1,818	191	447	312	918

* Sampled hospitals

TABLE 3

Test of single correlation coefficients for the items

** $\alpha = 0.01$
* $\alpha = 0.05$

S. aureus

population	**	**	**
	No. of hospitals	**	**
		No. of sampling hospitals	**
			No. of isolates

E. coli

population	**	**	**
	No. of hospitals	**	**
		No. of sampling hospitals	**
			No. of isolates

During 2 years, 1970–1971, the number of hospitals and isolates in each district is shown in Table 2 and the result of significance test in Table 3. From the result, the trend of sensitivity of bacteria all over Japan may be estimated.

The processes of the susceptibility test of isolates.

Figure 1 shows the processes; a specimen was isolated from patient examined by physician and then sent to clinical laboratory for examination. We ask the physician for the isolates which were selected by a random sampling method, then those isolates are

The processes of the susceptibility survey in Japan

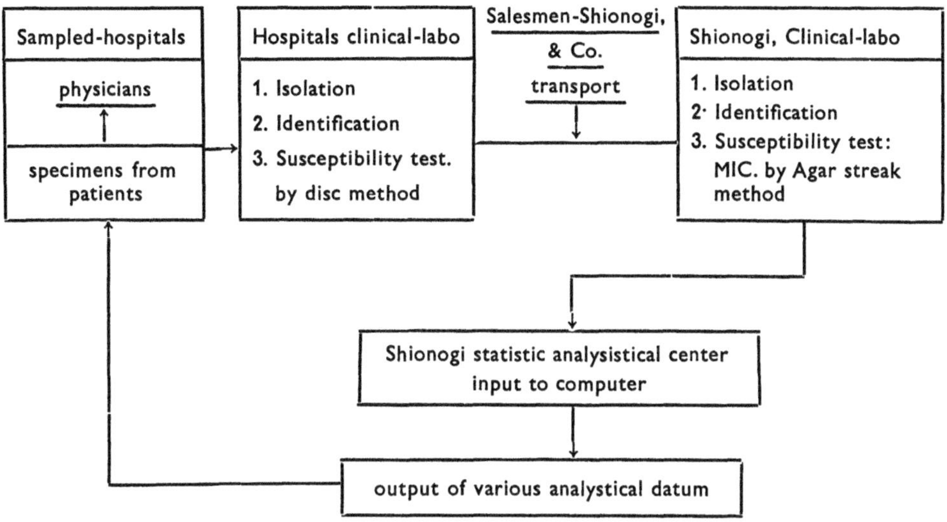

Fig. 1

brought to our clinical laboratory by our salesmen. In our laboratory the isolates were tested by the standard method.[1]) The record of supplied isolates is shown in Table 1. The results of test are sent to our Computer center for mathematical and statistical analysis, for input to computer and various data are made for the respective purposes, some of them are informed to the sampled hospitals immediately and others are published in a journal.

2. *Differentiation of drug sensitive- and resistant-strains* by minimal inhibitory concentration and annual change of sensitivity of the bacteria to various drugs.

MATERIALS AND METHODS

Strains.

Coagulase positive strains of *S. aureus* and *E. coli* isolated from various clinical specimens during 1965–1971 were used. Period of experiments was divided as follows:

1965: June 1965 – May 1966
1966: June 1966 – May 1967
1967: June 1967 – May 1968
1968: June 1968 – May 1969
1969: June 1969 – May 1970
1970: June 1970 – May 1971
1971: June 1971 – May 1972

Clinical specimens

The number of *S. aureus* and *E. coli* isolated during 1965–1971 from various clinical specimens. Every year both bacteria were isolated from various clinical specimens at a similar rate. About 70% of *S. aureus* were originated from pus, 10% from pharyngeal mucus and 7% from sputum. In *E. coli* about 70%, 10% and 8% were originated from urine, pus and faeces, respectively.

Sensitivity test of *S. aureus* isolated in 1971 is partly under investigation.

Method of sensitivity test (MIC test).

MIC values of drugs in isolates were determined by streak culture on agar using two-fold serially diluted agar (Modified Müller-Hinton medium)[1, 2]. For MIC of sulfonamide in *E. coli*, Sauton medium was used from 1965 through 1968 and then changed to Sauton-agar with streak culture.

ANTIBIOTICS TESTED

(a) Antibiotics examined for 7 years:

Tetracycline (TC), Chloramphenicol (CP), Streptomycin (SM), Kanamycin (KM), Penicillin G (PC-G), Cephalothin (CET), Cephaloridine (CER), Erythromycin (EM) and Sulfamethoxazol (SMX).

(b) Antibiotics examined for 3 or 2 years:

Aminobenzylpenicillin (AB-PC), Gentamycin (GM), Cephaloglycin (CEG), Cephalexin (CEX), Cephazolin (CEZ), Methylchlorisoxazol-Penicillin (MCI-PC), Carbobenzylpenicillin (CB-PC).

Determination of MIC value to differentiate the isolated into sensitive and resistant strains.

For the evaluation of MIC values of drugs, following conditions should be considered.

i) Variation of scale of hospital and of area where the strains tested were isolated[2].
ii) Variation of characters of strains due to difference of clinical specimens from which the strains were isolated[2, 3].
iii) Variation of quality of the medium due to difference of the date produced.
iv) Variation of technique of the investigator[4].

To minimize experimental errors due to variation of quality of the medium and variation of technique of the investigator, MIC of standard strains for both *S. aureus* and *E. coli* were examined together with the isolated strains. *S. aureus* 209P and *E. coli* mutaflora were tested on control strains. These control strains were examined once for about 50–100 experiments.

Fig. 2 and 3 showed the variations of MIC values of various drugs in the control strains of *S. aureus* and *E. coli*. Due to differences of experimental conditions employed, considerable variations of MIC values were observed in most antibiotics studied.

In a previous paper[1], we have simply classified the isolates into three groups according to the degree of MIC values of each antibiotics. They were groups of high sensitivity, intermediate sensitivity and low sensitivity. In this grouping, however, specificity of each antibiotics was not considered. In the present study, to determine the MIC value for grouping of the isolates into sensitive- and resistant-groups, the following method for the determination was adopted.

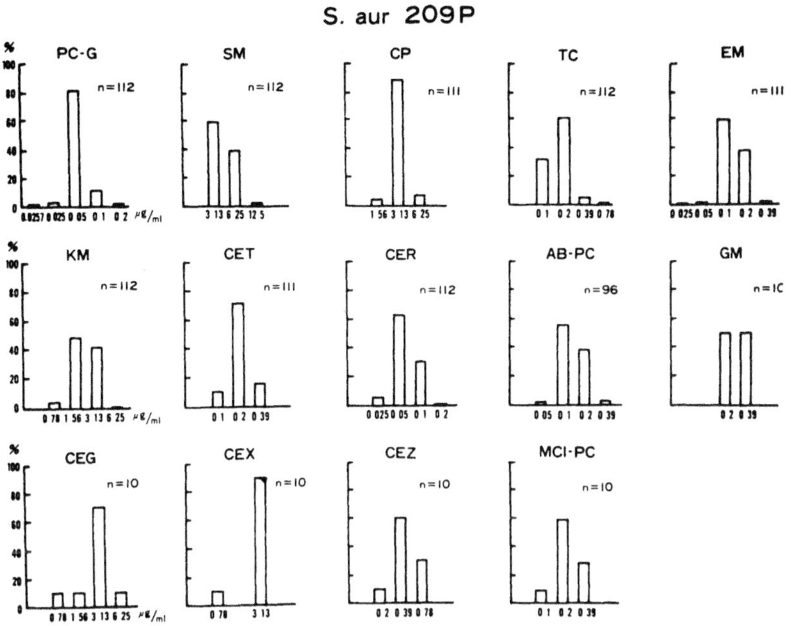
S. aur 209P
%
PC-G
n=112
μg/ml
SM
n=112
CP
n=111
TC
n=112
EM
n=111
KM
n=112
CET
n=111
CER
n=112
AB-PC
n=96
GM
CEG
n=10
CEX
n=10
CEZ
n=10
MCI-PC
n=10

Fig. 2

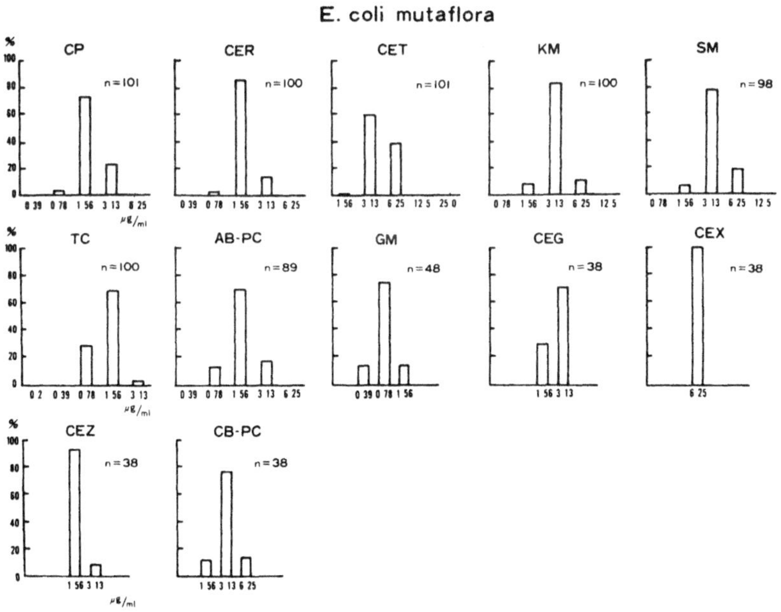
E. coli mutaflora
%
CP
n=101
μg/ml
CER
n=100
CET
n=101
KM
n=100
SM
n=98
TC
n=100
AB-PC
n=89
GM
n=48
CEG
n=38
CEX
n=38
CEZ
n=38
CB-PC
n=38

Fig. 3

MIC values of isolates were pooled for either 2 or 3 years and were plotted in a single figure. Then the MIC value of the lowest point of the curve between peaks obtained was determined as the grouping value for grouping of isolates into sensitive and resistant groups.

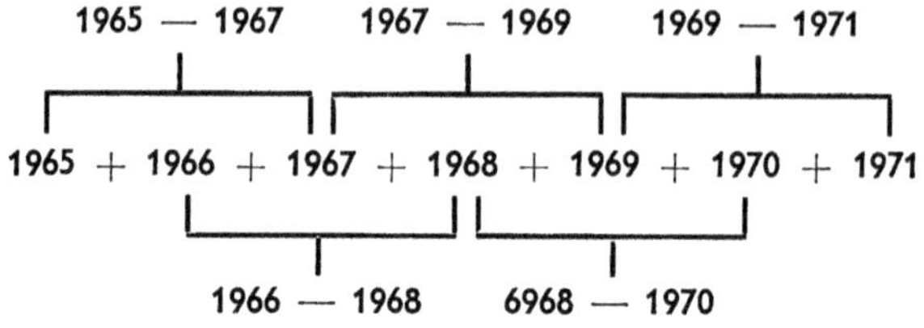

Fig. 4. Model of the moving pooling method

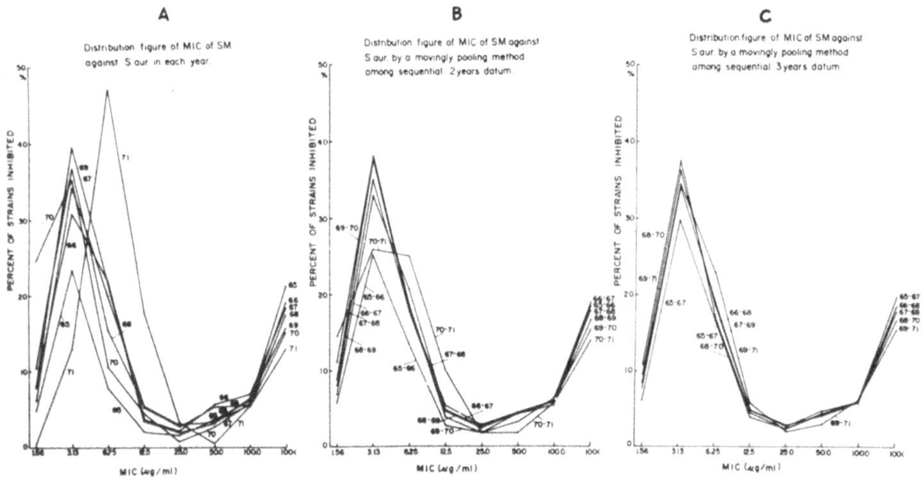

Fig. 5

In Fig. 4, a typical model of pooling the MIC values of isolates for 3 years was presented. In this paper we designated this method as the moving pooling method.

RESULTS

Differentiation of sensitive and resistant strains by the grouping value.

a) *S. aureus*

Fig. 5-A showed annual distribution of MIC values of SM. If we determine the lowest point of the curve as the MIC value of SM for grouping of isolates into sensitive and resistant groups, the grouping value for 1965 should be either 12.5 µg/ml or 25 µg/ml. For 1970 and 1971, the grouping values should be 25 µg/ml and 50 µg/ml, respectively. On the other hand, as shown in the Fig. 5-B, if we pooled the MIC values in isolates for 2 years the grouping value for all samples should be 25 µg/ml. Fig. 5-C, showed the results of pooling the MIC values in isolates for 3 years. These figures clearly demonstrated that the grouping value for all samples should be 25 µg/ml.

b) *E. coli*

Similar experiment was carried out with *E. coli*, the results being presented in Fig. 6. Annual distribution of MIC values of SM showed that the grouping values for 1970 and 1971 were 25 μg/ml. Pooling the MIC values in isolates for 2 years also showed that the grouping values for 1970 and 1971 were 25 μg/ml. However, the grouping values obtained by pooling the MIC values in isolates for 3 years were 12.5 μg/ml. The above findings indicated that for determination of the MIC value of SM for grouping of the isolates into sensitive- and resistant-groups, pooling the MIC values for 2 years in *S. aureus* and for 3 years in *E. coli* were necessary.

Determination of the grouping value by pooling the MIC values for more than 3 years was also carried out both in *S. aureus* and *E. coli*, but no shift of the grouping values were observed.

Distribution of MIC values of various antibiotics were studied and MIC values of each antibiotics for grouping of isolates into sensitive and resistant groups were determined. The grouping values thus determined were summarized in Table 4.

TABLE 4

Grouping values of drug to differentiate the isolates into sensitive and resistant groups

S. aureus					*E. coli*				
Drug	MIC ≦	Peak MIC			Drug	MIC ≦	Peak MIC		
CET	1.56	0.39	2*	3**	CET	50.00	12.50	2*	3**
CER	0.78	0.20	2	3	CER	12.50	3.13	2	3
CEX	12.50	3.13	2	2	CEX	50.00	12.50	2	2
CEG	12.50	3.13	2	2	CEG	25.00	6.25	2	2
CEZ	3.13	0.39	3	2	CEZ	12.50	1.56	3	2
KM	12.50	1.56	3	3	KM	25.00	6.25	2	3
GM	1.56	0.20	3	3	GM	6.25	1.56	2	3
SM	25.00	3.13	3	3	SM	12.50	6.25	1	3
AB-P	0.39	0.10	2	3	AB-P	25.00	3.13	3	3
PC-G	0.20	0.05	2	3	CB-P	50.00	12.50	2	2
MCI-P	0.78	0.20	2	2	TC	12.50	3.13	2	3
CP	12.50	6.25	1	3	CP	25.00	6.25	2	3
SMX	125.00	31.30	2	3	SMX	3.90	0.50	3	3
TC	3.13	0.39	3	3					
EM	0.78	0.20	2	3					

* Fold index, from peak MIC to resistant MIC.

** Sequential years for a moving pooling method.

Also, some of the results of MIC values of various antibiotics determined by the same method were presented in Fig. 7–14. For both *S. aureus* and *E. coli*, the grouping values determined were two or four times higher concentrations than peak MIC values, which represented MIC values of highest numbers of isolates.

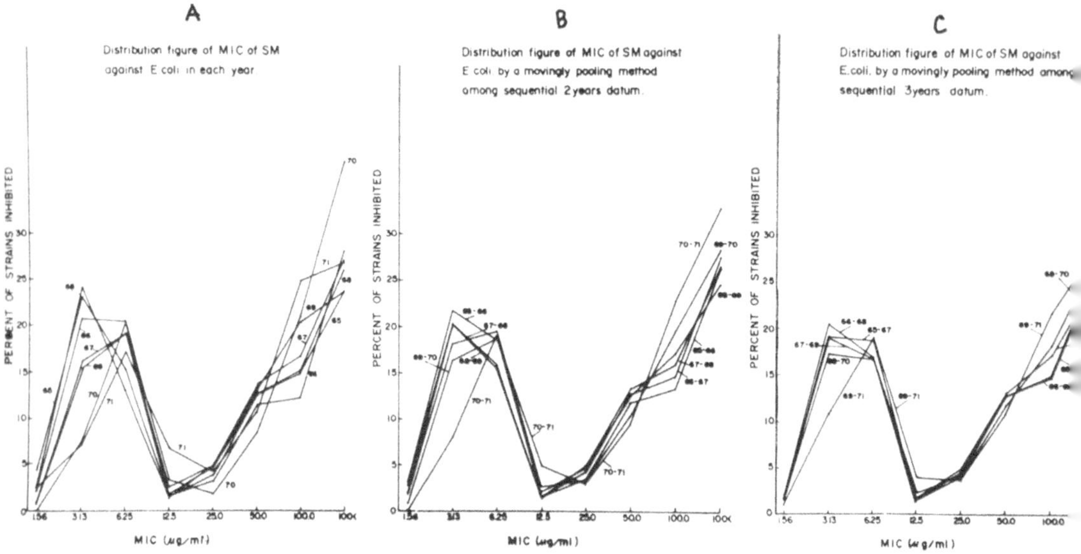

Fig. 6

ANNUAL CHANGE OF SENSITIVITY OF THE BACTERIA TO VARIOUS ANTIBIOTICS

a) *S. aureus*

Table 5 shows the annual change of sensitivity of *S. aureus* to various antibiotics. The grouping values were determined by the moving pooling method for 2 or 3 years. Test of significance of the values shown in Table 5 was presented in Table 6.
In PC-G, SM, CP and TC, increase of sensitivity was observed, while in EM, GM and KM, sensitivity decreased gradually. No significant change was observed in CET, CER and SMX.

b) Similar determination was made in *E. coli* and the results were presented in Table 7. No drug showed increase of sensitivity in *E. coli*. Decrease of sensitivity was observed in CP, TC, KM and CET. A little decrease of sensitivity was observed in CER and SMX. In AB-PC, though the period of the experiment was not long, showed decrease of sensitivity. In GM, though the experiment was interrupted once, no change of sensitivity was observed. Test of significance of the values shown in Table 7 was presented in Table 8.

TABLE 5

*Annual change of sensitive strains of S. aureus to various drugs**

Drug	P											
	65+66 (A)		66+67 (B)		67+68 (C)		68+69 (D)		69+70 (E)		70+71 (F)	
	N	S	N	S	N	S	N	S	N	S	N	S
PC-G	2056	13.5%	2304	14.3%	2432	14.8%	1905	15.7%	1046	18.1%	449	19.4%
SM	2054	67.0	2321	68.2	2439	69.0	1905	70.2	1046	73.3	449	76.6
CP	2061	84.9	2313	86.3	2439	86.8	1904	87.8	1045	89.2	449	87.1
TC	2064	54.9	2311	60.4	2429	60.5	1905	60.1	1046	62.0	449	68.6
EM	2070	70.4	2315	69.1	2433	66.9	1905	63.8	1046	61.2	449	59.7
KM	2041	94.2	2302	91.2	2432	90.7	1903	89.7	1045	88.4	449	88.2
SMX	2060	44.1	2297	48.1	2417	54.5	1901	50.9	1039	38.5	437	25.9
CET	1540	99.7	2299	99.4	2426	99.5	1905	99.7	1046	99.4	449	99.3
CER	1541	99.1	2297	98.7	2424	98.7	1905	99.2	1046	99.1	449	97.1
GM			1601	99.0							449	96.7
ABPC							1903	16.8	1045	18.7	448	19.6
MIPC											449	98.9
CEG											449	99.8
CEX											449	95.1
CEZ											449	98.7

Note: N: Number of strains examined S: Percent of sensitive strains
* The grouping values of each drugs were determined by the moving pooling method for 2 or 3 years

TABLE 6

*Test of significance of the experiment in Table 5**

Drug	A-C	A-D	A-E	A-F	B-D	B-E	B-F	C-E	C-F	D-F
PC-G	—	*	**	**	.	**	*	*	*	*
SM	.	*	**	**	.	**	**	**	**	**
CP	*	**	**	—	.	*	—	*	—	—
TC	**	**	**	**	—	—	**	—	**	**
EM	**	**	**	**	**	**	**	**	**	.
KM	**	**	**	**	.	**	*	*	.	—
SMX	**	**	**	**	*	**	**	**	**	**
CET	—	—	—	—	—	—	—	—	—	—
CER	—	—	—		.	—	*	—	*	**
GM							**			
ABPC										.
MIPC										
CEG										
CEX										
CEZ										

Note: ** P = 0.01 * P = 0.05 . P = 0.1 — Not significant
* By the exact probability method by R. A. Fischer

TABLE 7

*Annual change of sensitive strains of E. coli to various drugs**

Drug	P											
	65+66 (A)		66+67 (B)		67+68 (C)		68+69 (D)		69+70 (E)		70+71 (F)	
	N	S	N	S	N	S	N	S	N	S	N	S
SMX	1008	35.8%	1363	34.6%	1635	34.5%	1473	37.0%	1107	35.3%	901	31.0%
CP	1036	55.8	1424	54.5	1690	53.6	1512	54.1	1122	50.5	922	44.9
CER			1410	93.6	1685	91.8	1512	90.1	1122	85.8	922	82.6
CET	912	95.6	1408	94.9	1688	94.4	1512	94.7	1122	92.4	922	88.3
GM			1259	99.6							901	99.7
KM			1421	95.7	1688	95.0	1512	93.6	1122	90.7	922	88.6
SM	938	43.3	1407	39.2	1690	38.1	1512	38.0	1122	32.1	922	27.3
TC	1023	45.2	1418	40.8	1684	40.1	1512	40.8	1122	36.4	922	34.6
ABPC							1500	82.0	1122	77.0	922	76.8
CBPC											922	78.0
CEG											922	90.2
CEX_1											922	95.8
CEZ											922	92.0

Note: N: Number of strains examined S: Percent of sensitive strains

* The grouping values of each drugs were determined by the moving pooling method for 2 or 3 years

TABLE 8

Test of significance of the experiment in Table 7

Drug	A-C	A-D	A-E	A-F	B-D	B-E	B-F	C-E	C-F	D-F
SMX	—	—	—	*	—	—	*	—	*	**
CP	—	—	**	**	—	*	**	•	**	**
CER					**	**	**	**	**	**
CET	—	—	**	**	—	**	**	*	**	**
GM							—			
KM					**	**	**	**	**	**
SM	**	**	**	**	—	**	**	**	**	**
TC	**	*	**	**	—	*	**	*	**	**
ABPC										**
CBPC										
CEG										
CEX										
CEZ										

Note: ** P = 0.01 * P = 0.05 • P = 0.1 — Not significant

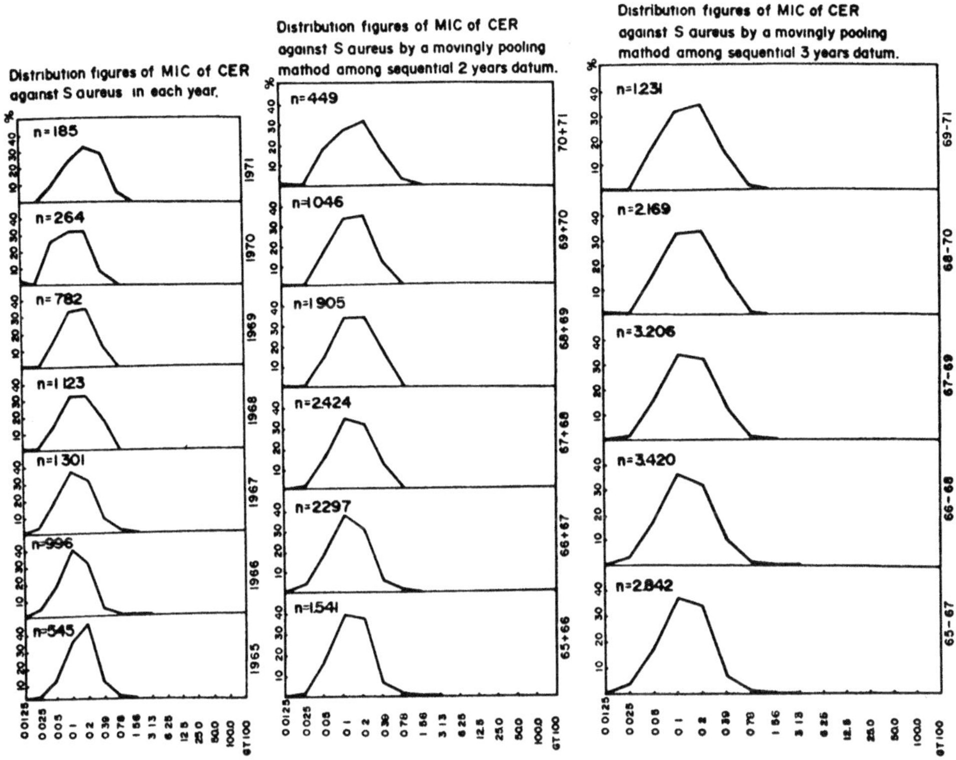
Distribution figures of MIC of CER against S aureus in each year.
n=185
1971
n=264
1970
n=782
1969
n=1123
1968
n=1301
1967
n=996
1966
n=545
1965
Distribution figures of MIC of CER against S aureus by a movingly pooling mathod among sequential 2 years datum.
n=449
70+71
n=1046
69+70
n=1905
68+69
n=2424
67+68
n=2297
66+67
n=1541
65+66
Distribution figures of MIC of CER against S aureus by a movingly pooling mathod among sequential 3 years datum.
n=1231
69-71
n=2169
68-70
n=3206
67-69
n=3420
66-68
n=2842
65-67
%
0.0125
0.025
0.05
0.1
0.2
0.39
0.78
1.56
3.13
6.25
12.5
25.0
50.0
100.0
GT100

Fig. 7

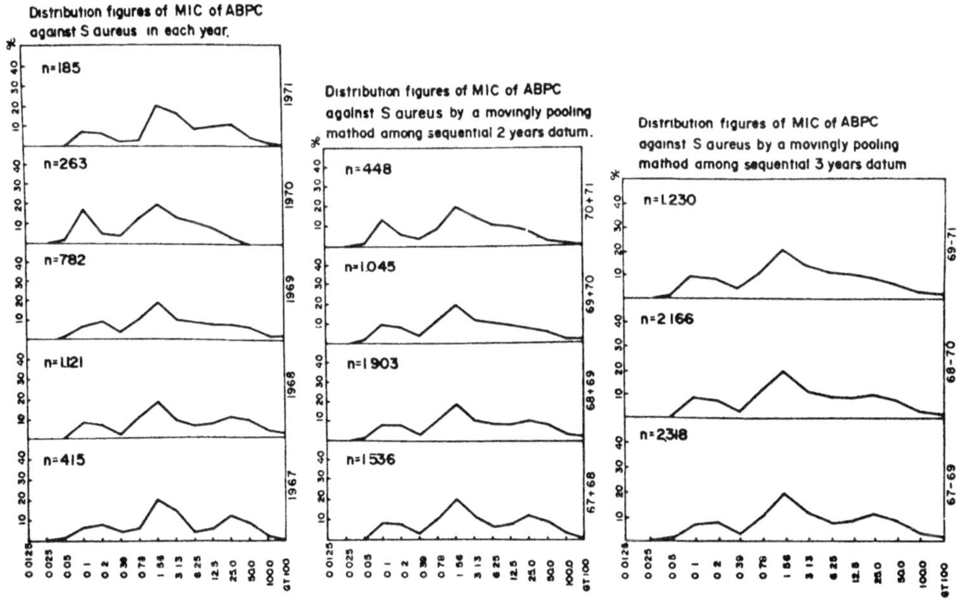
Distribution figures of MIC of ABPC against S aureus in each year.
n=185
1971
n=263
1970
n=782
1969
n=1121
1968
n=415
1967
Distribution figures of MIC of ABPC against S aureus by a movingly pooling mathod among sequential 2 years datum.
n=448
70+71
n=1045
69+70
n=1903
68+69
n=1536
67+68
Distribution figures of MIC of ABPC against S aureus by a movingly pooling mathod among sequential 3 years datum
n=1230
69-71
n=2166
68-70
n=2318
67-69
%
0.0125
0.025
0.05
0.1
0.2
0.39
0.78
1.56
3.13
6.25
12.5
25.0
50.0
100.0
GT100

Fig. 8

Fig. 9

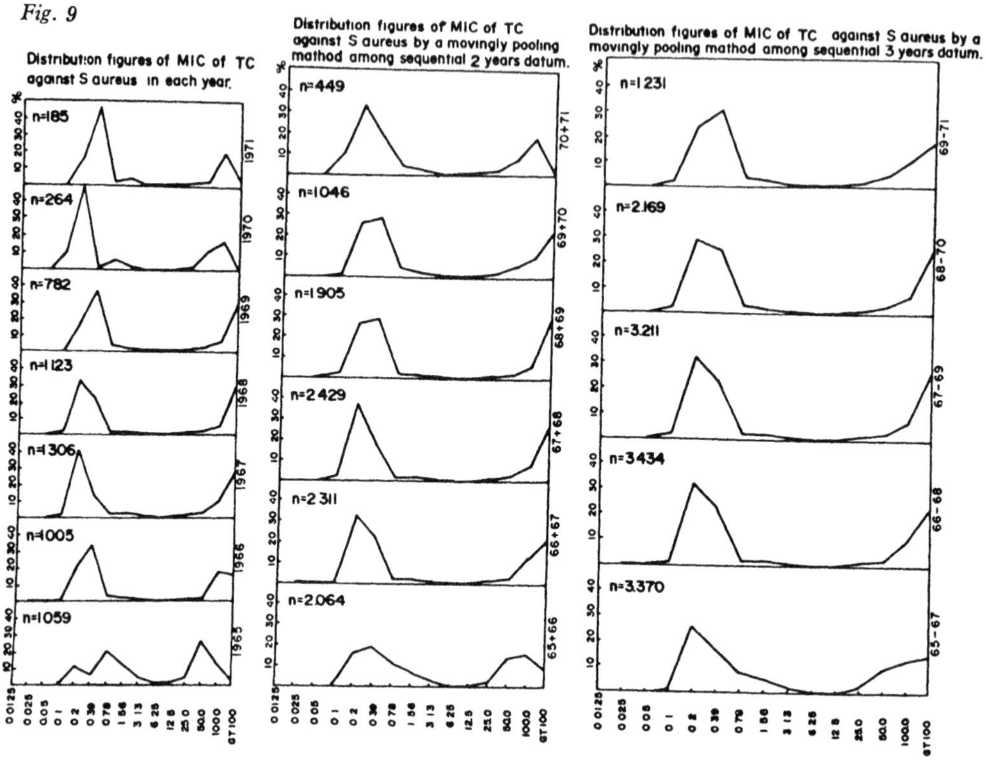
Distribution figures of MIC of TC against S aureus in each year.
%
n=185
n=264
n=782
n=1123
n=1306
n=1005
n=1059
1971
1970
1969
1968
1967
1966
1965
Distribution figures of MIC of TC against S aureus by a movingly pooling mathod among sequential 2 years datum.
n=449
n=1046
n=1905
n=2429
n=2311
n=2064
70+71
69+70
68+69
67+68
66+67
65+66
Distribution figures of MIC of TC against S aureus by a movingly pooling mathod among sequential 3 years datum.
n=1231
n=2169
n=3211
n=3434
n=3370
69-71
68-70
67-69
66-68
65-67
10
20
30
40
0.0125
0.025
0.05
0.1
0.2
0.39
0.78
1.56
3.13
6.25
12.5
25.0
50.0
100.0
GT 100

Fig. 10

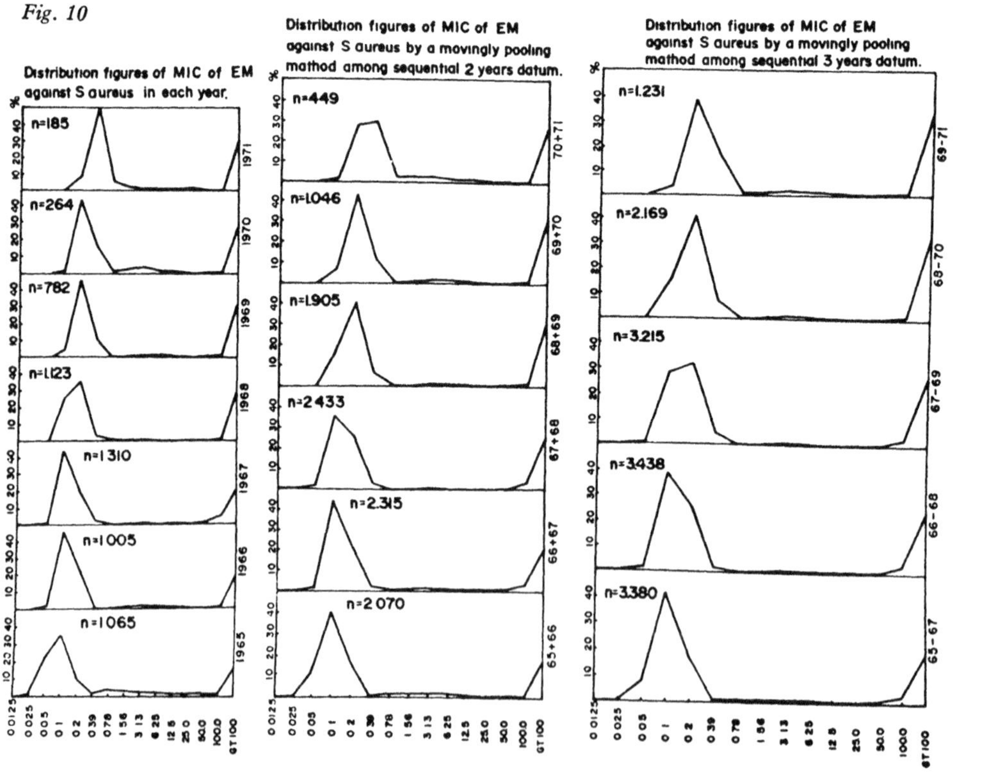
Distribution figures of MIC of EM against S aureus in each year.
%
n=185
n=264
n=782
n=1123
n=1310
n=1005
n=1065
1971
1970
1969
1968
1967
1966
1965
Distribution figures of MIC of EM against S aureus by a movingly pooling mathod among sequential 2 years datum.
n=449
n=1046
n=1905
n=2433
n=2315
n=2070
70+71
69+70
68+69
67+68
66+67
65+66
Distribution figures of MIC of EM against S aureus by a movingly pooling mathod among sequential 3 years datum.
n=1231
n=2169
n=3215
n=3438
n=3380
69-71
68-70
67-69
66-68
65-67
10
20
30
40
0.0125
0.025
0.05
0.1
0.2
0.39
0.78
1.56
3.13
6.25
12.5
25.0
50.0
100.0
GT 100

Fig. 11

Distribution figures of MIC of CET against E coli in each year

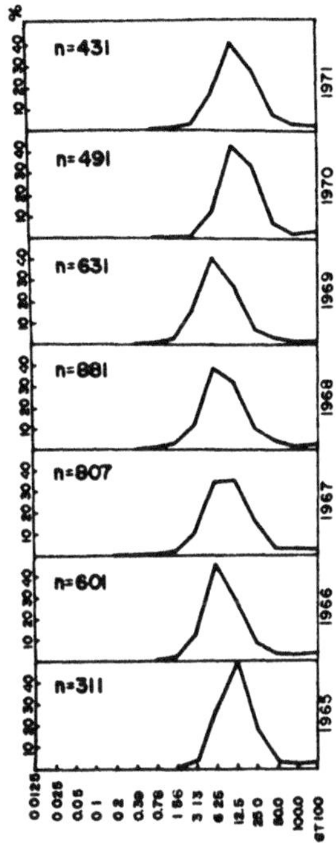

Distribution figures of MIC of CET against E coli by a movingly pooling mathod among sequential 2 years datum.

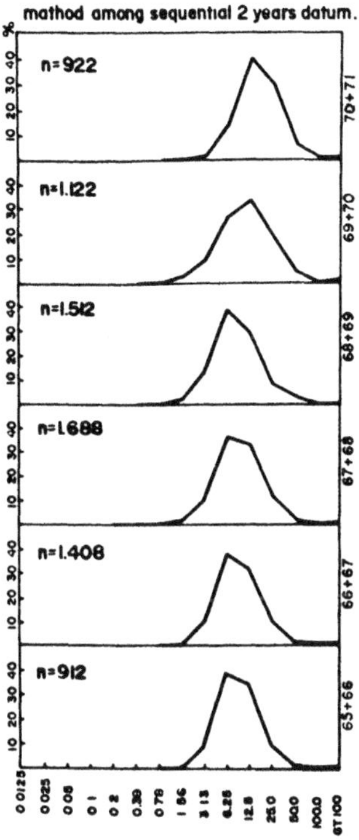

Distribution figures of MIC of CET against E coli by a movingly pooling mathod among sequential 3 years datum.

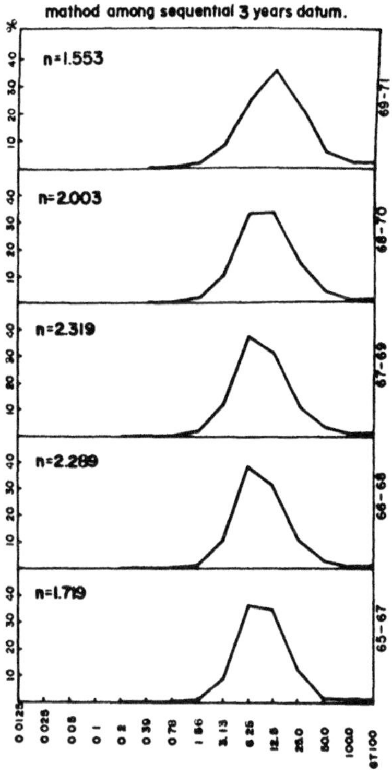

Fig. 12

Distribution figures of MIC of KM against E coli in each year

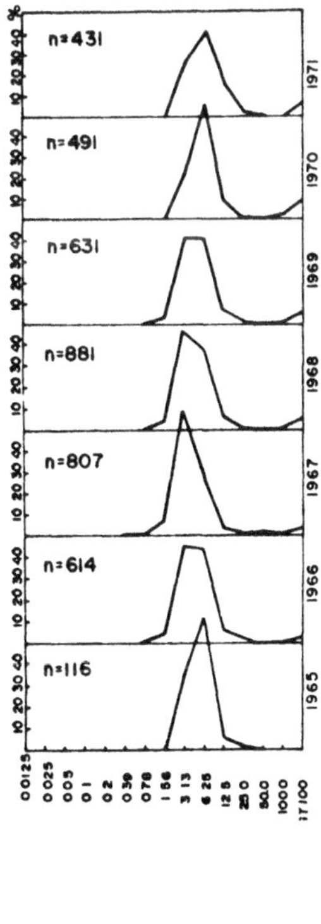

Distribution figures of MIC of KM against E coli by a movingly pooling mathod among sequential 2 years datum

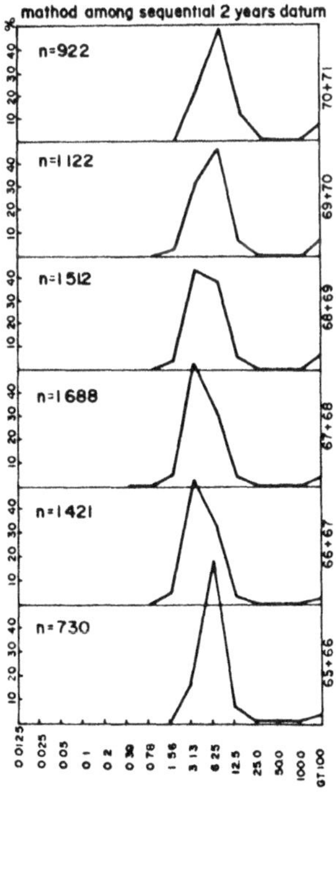

Distribution figures of MIC of KM against E coli by a movingly pooling mathod among sequential 3 years datum.

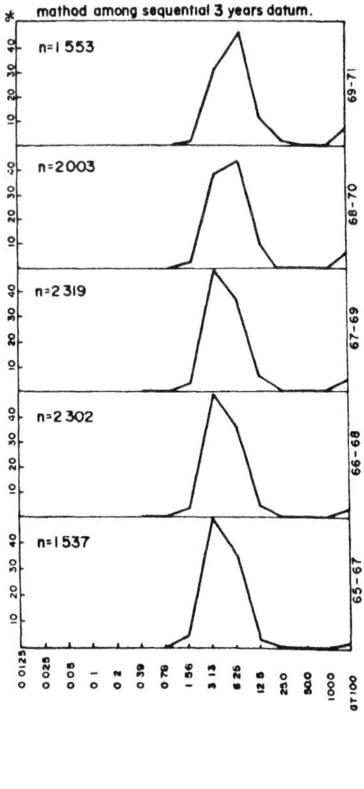

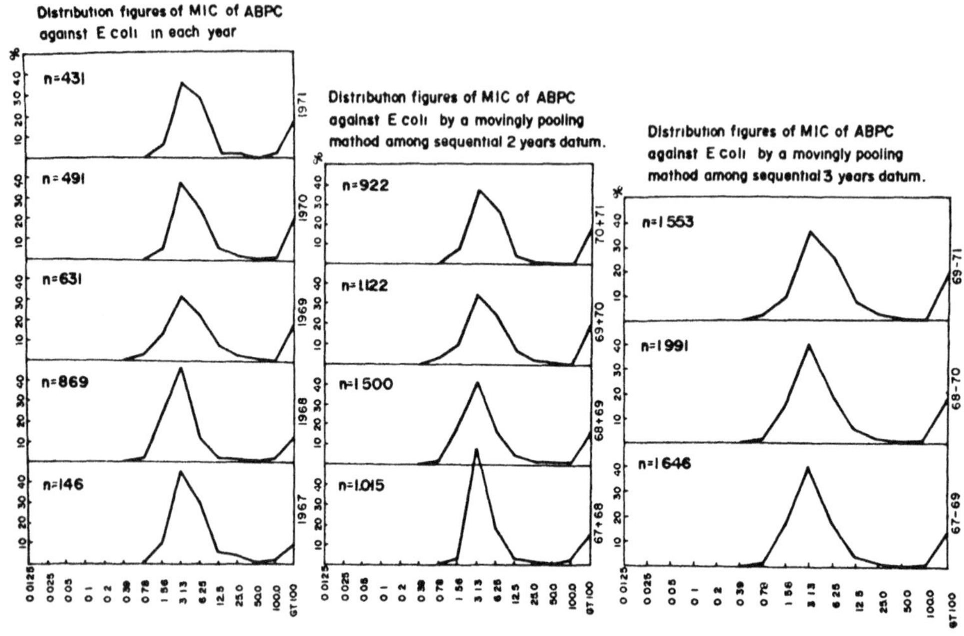
Distribution figures of MIC of ABPC against E coli in each year
Distribution figures of MIC of ABPC against E coli by a movingly pooling mathod among sequential 2 years datum.
Distribution figures of MIC of ABPC against E coli by a movingly pooling mathod among sequential 3 years datum.

Fig. 13

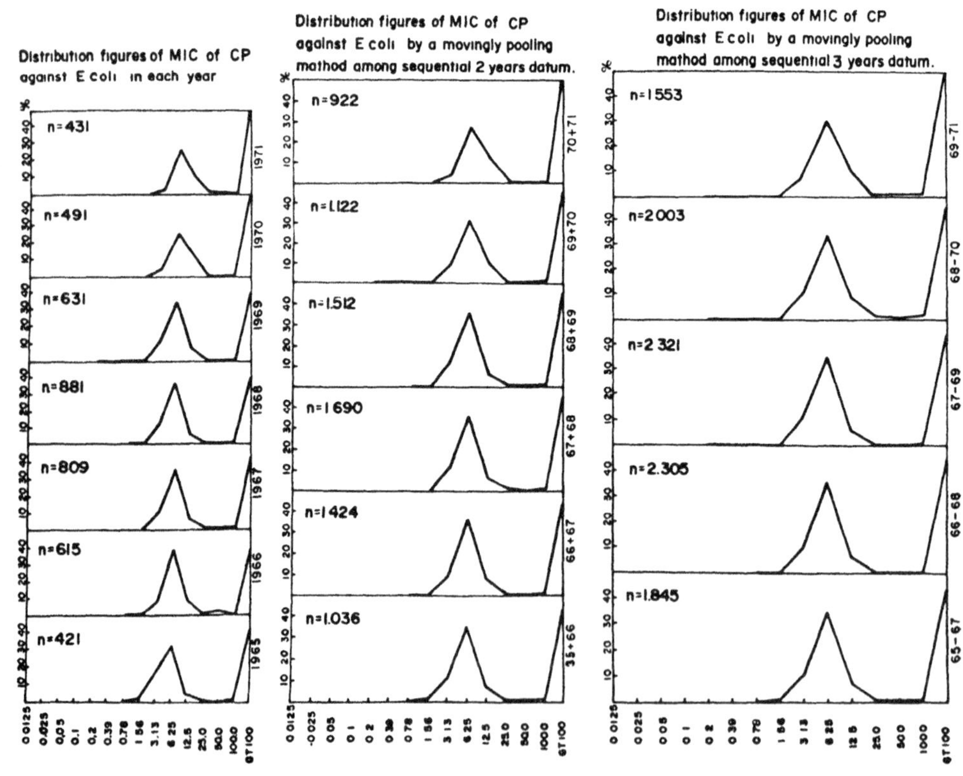
Distribution figures of MIC of CP against E coli in each year
Distribution figures of MIC of CP against E coli by a movingly pooling mathod among sequential 2 years datum.
Distribution figures of MIC of CP against E coli by a movingly pooling mathod among sequential 3 years datum.

Fig. 14

3. Difference of sensitivity of isolates from various clinical specimens.

MATERIALS AND METHOD

Strains

Strains isolated from various clinical specimens during 1969–1971 were used. In 1970, strains of *S. aureus* were isolated from only pus.

Method of analysis.

S. aureus isolated from pus, sputum and pharyngeal mucus, and *E. coli* isolated from urine and bile were analyzed according to the following items;

1. Source of isolates.
2. Isolates from out- and in-patients.
3. Isolates from source of infections.
4. Isolates from districts.
5. Scale of hospitals.

RESULTS

The results of analysis to 5 items are shown in Fig. 15–27, and Table 9–11. Difference of sensitivity of various antibiotics of *S. aureus* and *E. coli* isolated from various clinical specimens due to difference of source of the isolates was studied.

Number of strains studied in this paper was about 82% in *S. aureus* and about 70% in *E. coli* of all isolated strains.

TABLE 9

Staphylococcus aureus isolated from various clinical specimens from June 1965 through May 1971

Year	1965	1966	1967	1968	1969	1970	1971	Total
Number of strains	994	942	1226	1077	751	262	185	5437
Specimen	%	%	%	%	%	%	%	%
Pus	71.0	71.0	67.4	71.2	71.2	100.0	65.4	71.0
Pharyngeal mucus	10.0	9.6	12.3	10.9	11.0		20.5	10.6
Sputum	8.0	6.7	7.5	10.6	5.5		13.0	7.7
Urine	6.0	7.0	7.0	2.6	6.4			5.3
Blood	1.1	1.3	0.7	0.6	0.9			0.8
Chest fluid	0.7	0.7		0.4	0.3			0.5
Ascites	0.6	0.8	0.7	0.2	0.7			0.5
Faeces	0.4	0.2	1.0	0.7	0.4			0.6
Spinal fluid	0.6	0.2	0.6	0.4	0.3			0.4
Bile	0.4	0.2	0.1	0.7	0.1			0.4
Skin piece	0.3	0.1	0.4					0.1
Joing fluid	0.1	0.4		0.4	0.4			0.3
Others	0.4	1.4	2.1	1.4	2.9			1.2

S. aureus isolated from pus specimens
1969-1971, number of strains 908
* 1970-1971, number of strains 383

Fig. 15

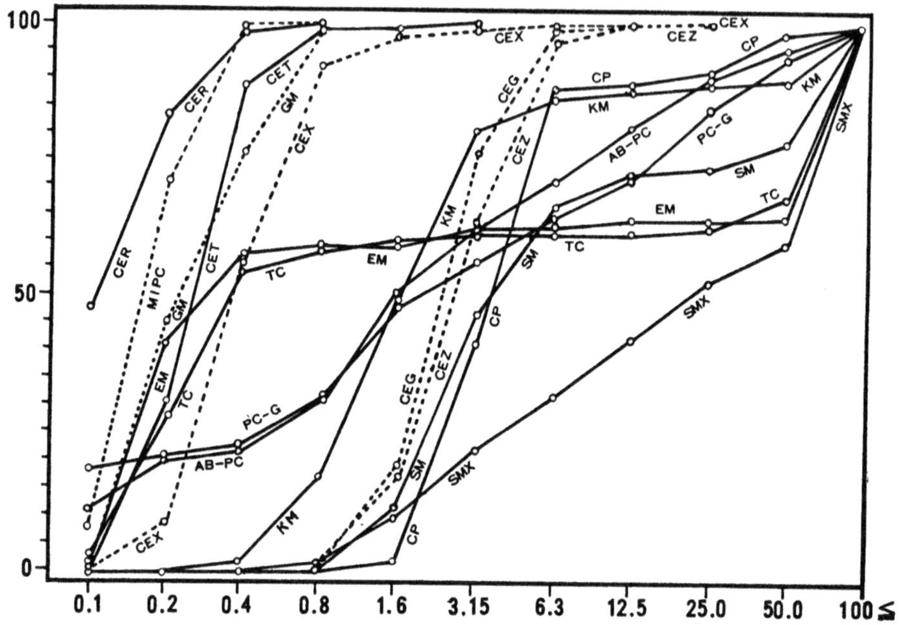

S. aureus isolated from sputum specimens
number of strains 65.
* 1970-1971, number of strains 24.

Fig. 16

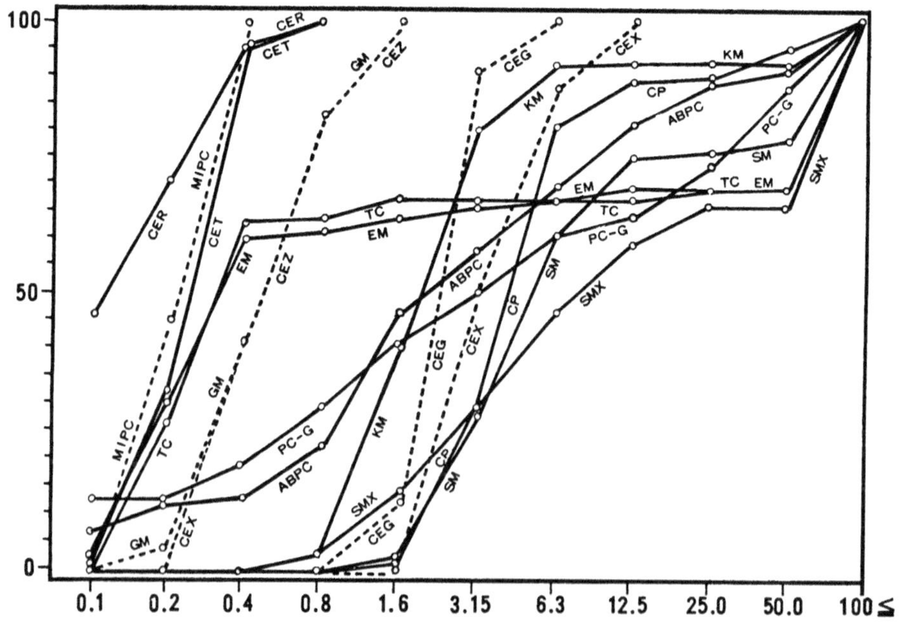

S. aureus isolated from pharyngear mucus specimens
1969–1971, number of strains 121.
* 1970–1971, number of strains 38.

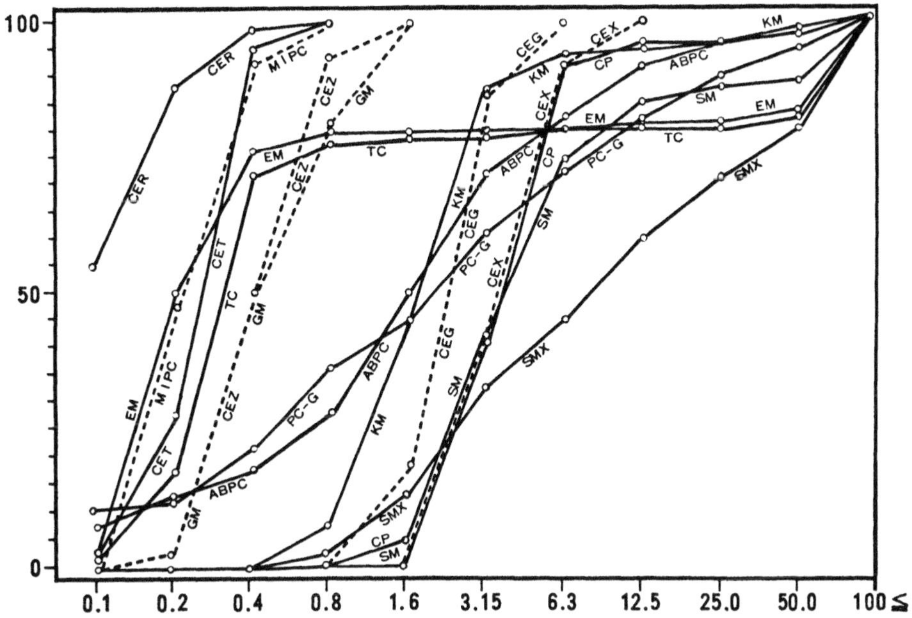

Fig. 17

E. coli isolated from from urine specimens
1969–1971, unmber of strins 1152.
* 1970–1971, number of strains 712.

Fig. 18

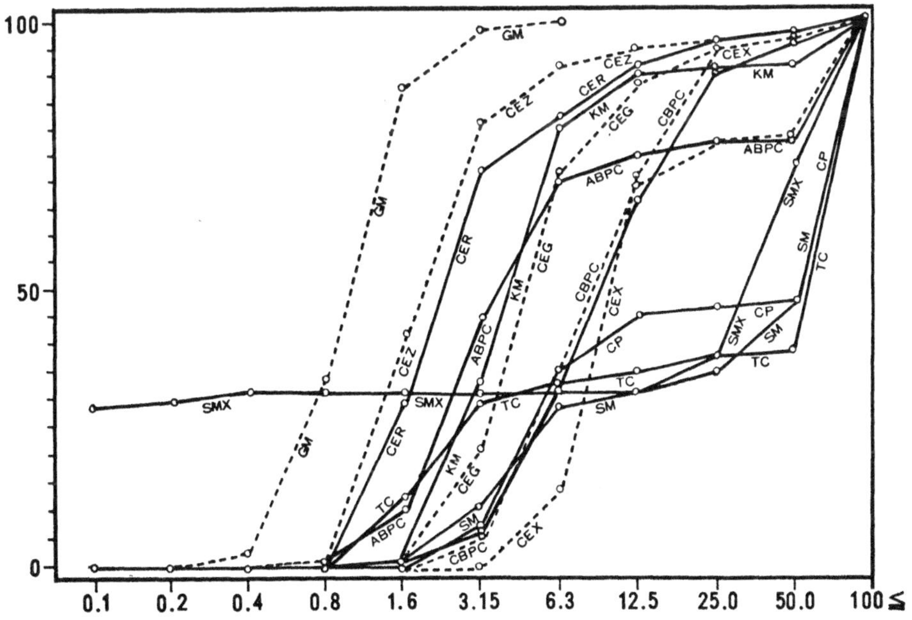

E. coli isolated from pus specimens
1969-1971, number of strains 145.
* 1970-1971, number of strains 86.

Fig. 19

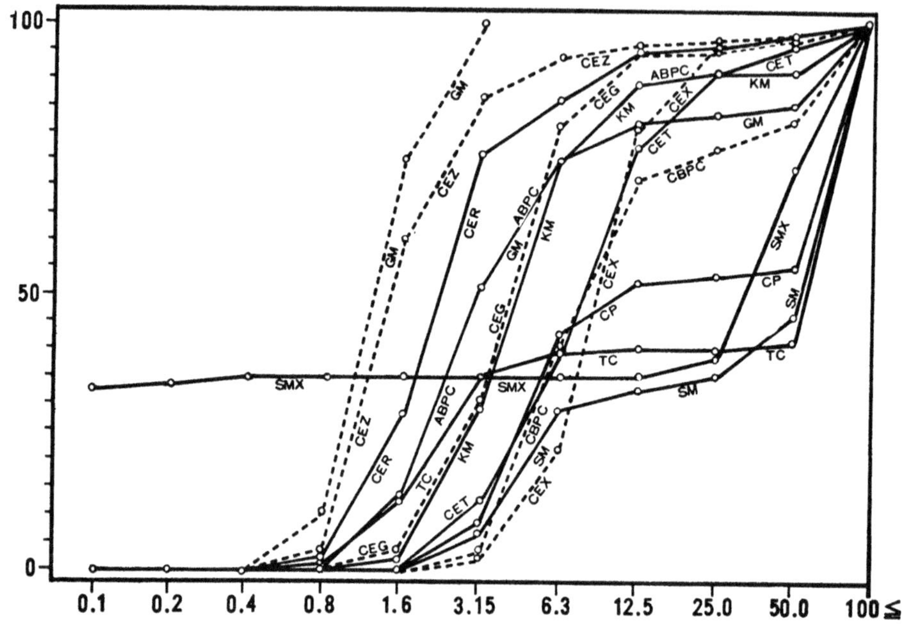

E. coli isolated from bile specimens
1969-1971, number of strains 36.
* 1970-1971, number of strains 20.

Fig. 20

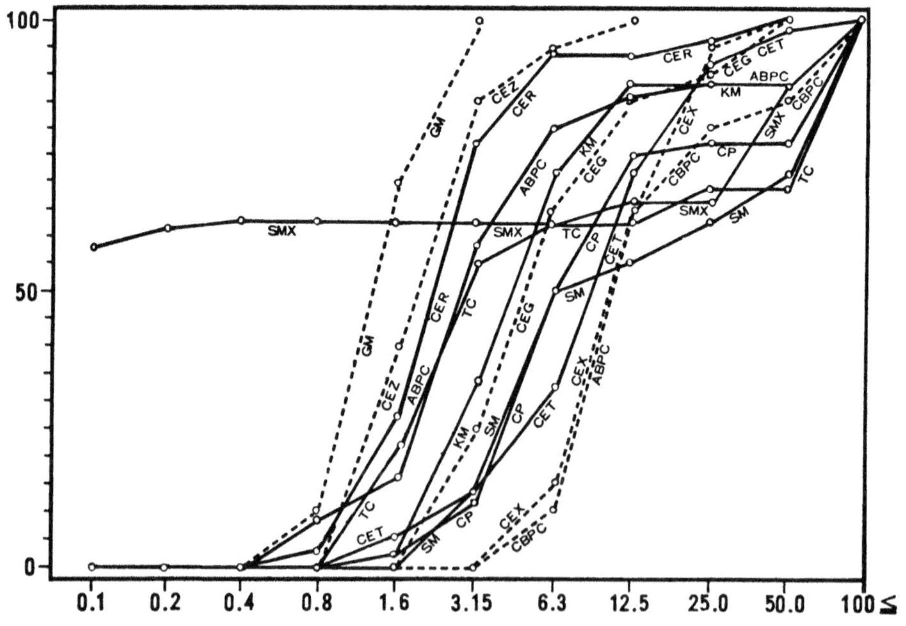

E. coli isolated from expected urine specimens *Fig. 21*
1969-1971, number of strains 399.
* 1970-1971, number of strains 210.

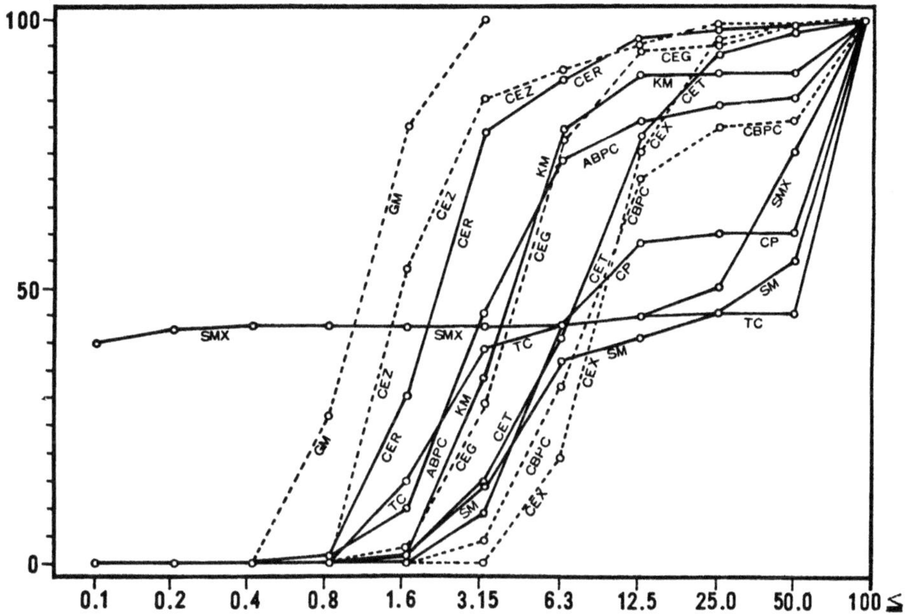

S. aur. *Fig. 22*
Infection of skin and soft tissue, number of strain 233.
1969-1971
* number of strain 113, 1969-1971

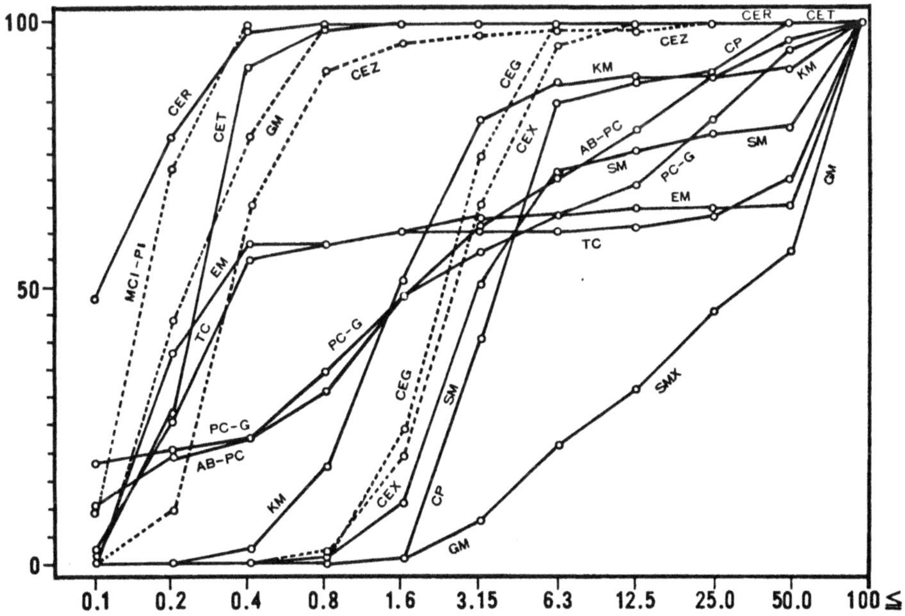

S. aur.
Otorhinological infection, number of strain 156.
1969–1971
* number of strain 69, 1970–1971

Fig. 23

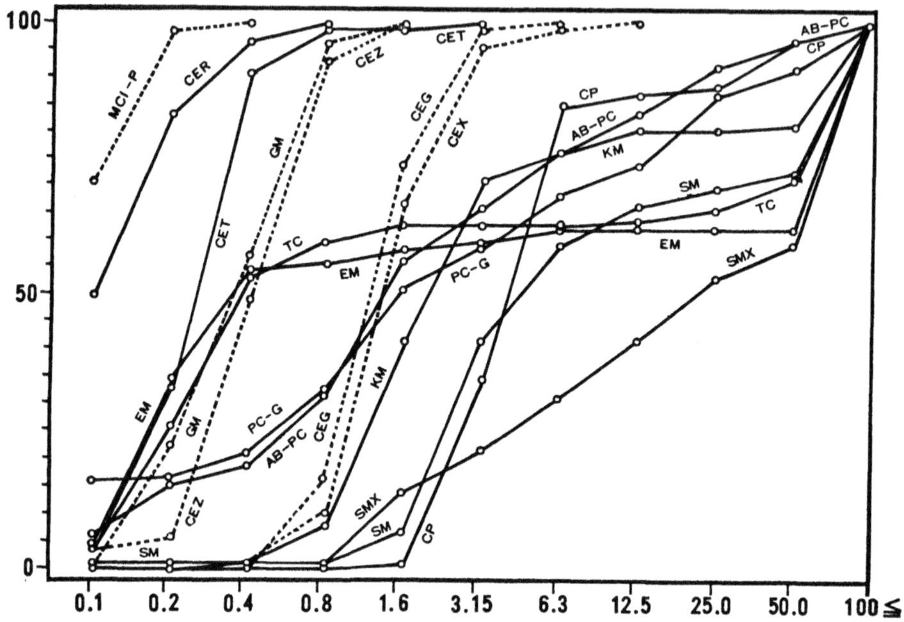

S. aur.
Infections of Respiratory tract, number of strain 110.
1969–1971
* number of strain 42, 1970–1971

Fig. 24

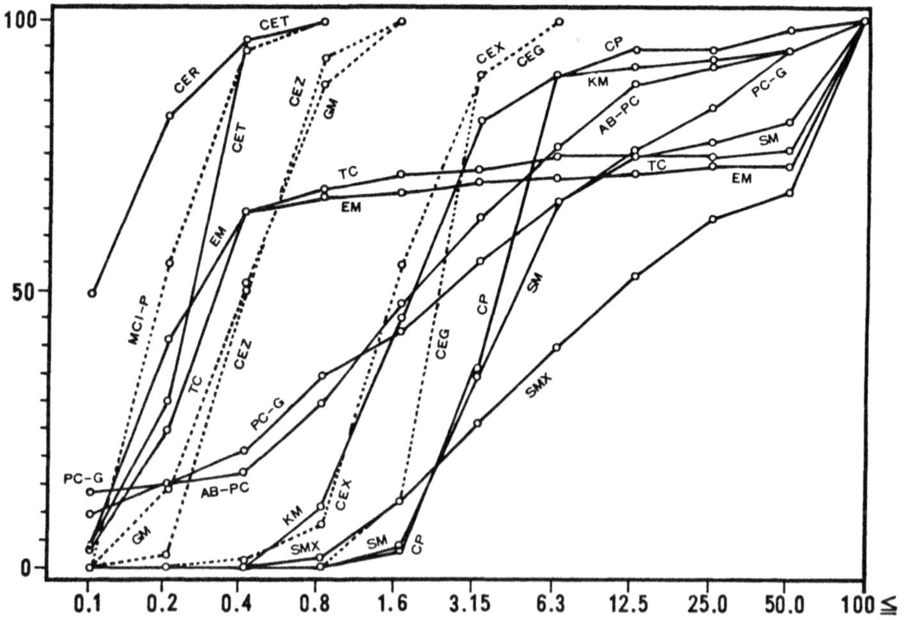

E. coli. *Fig. 25*

Infections of urinary tract, number of strain 736, 1969–1971.

* number of strain 456, 1970–1971

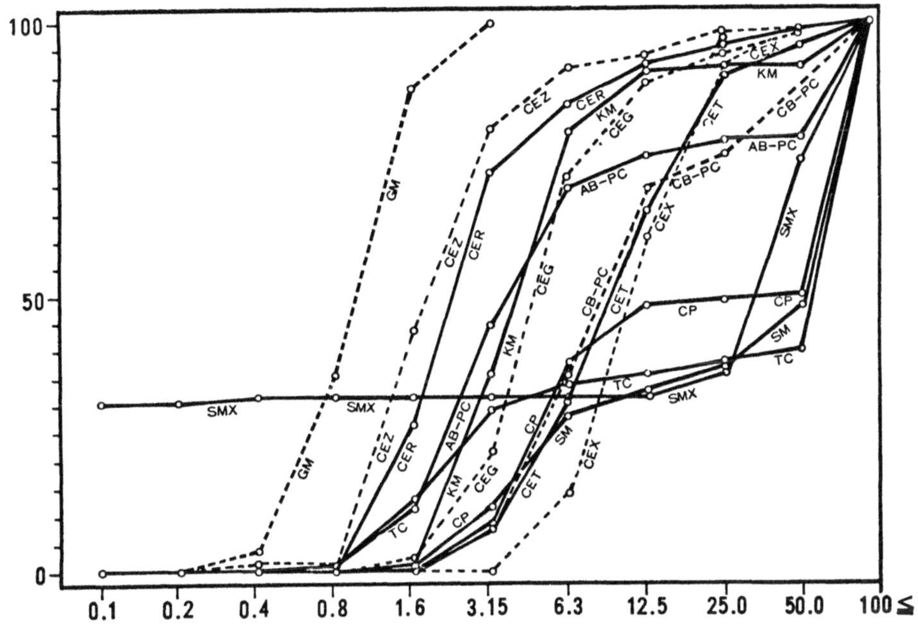

* MIC value of SMX must be 10 fold.

TABLE 10

Escherichia coli isolated from various clinical specimens from June 1965 through May 1971

Year	1965	1966	1967	1968	1969	1970	1971	Total
Number of strains	396	553	776	852	599	483	426	4085
Specimen	%	%	%	%	%	%	%	%
Urine	66.0	67.6	70.0	75.0	73.8	80.0	76.8	72.8
Pus	12.0	12.8	12.0	9.2	9.8	8.1	11.0	10.6
Faeces	9.3	12.0	9.7	6.5	9.2	5.0	1.4	7.8
Bile	5.6	4.0	2.6	3.2	2.7	2.5	1.9	3.1
Sputum	1.3	1.1	1.5	2.7	1.8	2.7	1.9	1.9
Pharyngeal mucus	2.8	0.9	1.7	1.5	1.0	0.6	2.8	1.5
Blood	0.5	0.2	0.8	0.2	0.3	0.2	1.2	0.5
Ascites	0.3	0.2	0.3	0.4	0.5	0.2	0.7	0.3
Chest fluid	0.3	0.2	0.1			0.2		0.1
Others	1.5	0.5	1.4	0.6	0.8	0.2	1.9	0.9

TABLE 11

The kinds of infections, diseases and number of isolates

Infections	Name of diseases	Strains: Number	Strains: Percent
Infection of skin and soft tissue. *Pus.*	Carbuncle	61 (32)	26 (28)
	Fruncle	41 (18)	18 (16)
	Infections of soft tissue	32 (9)	14 (8)
	Impetigo	13 (6)	6 (5)
	Phlegmone	11 (6)	5 (5)
	Others	75 (42)	32 (29)
	Total	233 (113)	100 (100)
Otorhinological. infection *Pus.*	Oitis media	125 (55)	80 (80)
	Oitis externa	12 (6)	8 (9)
	Sinuitis	0 (2)	0 (3)
	Others	19 (6)	12 (10)
	Total	156 (69)	100 (100)
Infections of Respiratory tracts. *Pus, Sputum, Pharyngeal mucus.*	Tonsillitis	26 (11)	24 (26)
	Pneumonia	14 (7)	13 (17)
	Pharyngitis	112 (5)	11 (12)
	Angina	5 (4)	5 (10)
	Others	53 (5)	47 (12)
	Total	110 (42)	100 (100)
Infections of Urine tract. *Urine.*	Cystitis	393 (239)	53 (52)
	Pyelonephritis	191 (125)	36 (27)
	Urine tract inf.	83 (50)	11 (11)
	Others	69 (42)	9 (9)
	Total	736 (456)	100 (100)
Cholecystitis. *Bile.*	Cholecystitis	13 (8)	52 (61)
	Others	12 (5)	48 (39)
	Total	25 (13)	100 (100)

( )... 1970–71

These strains studied were assumed to be a causative agent of the clinical symptoms.

In *S. aureus,* isolates from pus were most resistant and isolates from pharyngeal mucus were most sensitive to various antibiotics tested. The results that no significant differences were observed in *S. aureus* between isolated from out-patients and from in-patients were unexpected. On the other hand, as was expected, sensitivity of isolated from in-patients to antibiotics in *E. coli* was lower than that of isolates from out-patients. This was especially eminent in CER, AB-PC, CEG and CEZ, which were introduced recently in clinical usage.

It is noteworthy that *S. aureus* isolated from infections of skin and soft tissues were more sensitive to KM and other Aminoglycosides than those from otorhinological infections. It is also interesting that *E. coli* isolated from infections of the urinary tract were more resistant than those isolated from infections of bile ducts. It was found that number of sensitive strains of *S. aureus* to PC-G and AB-PC in district A, where populations were large and hospitals of large scale were located, was much less than in other districts. This may be due to that these antibiotics are frequently used in district A. In *E. coli*, no significant difference was observed among each district.

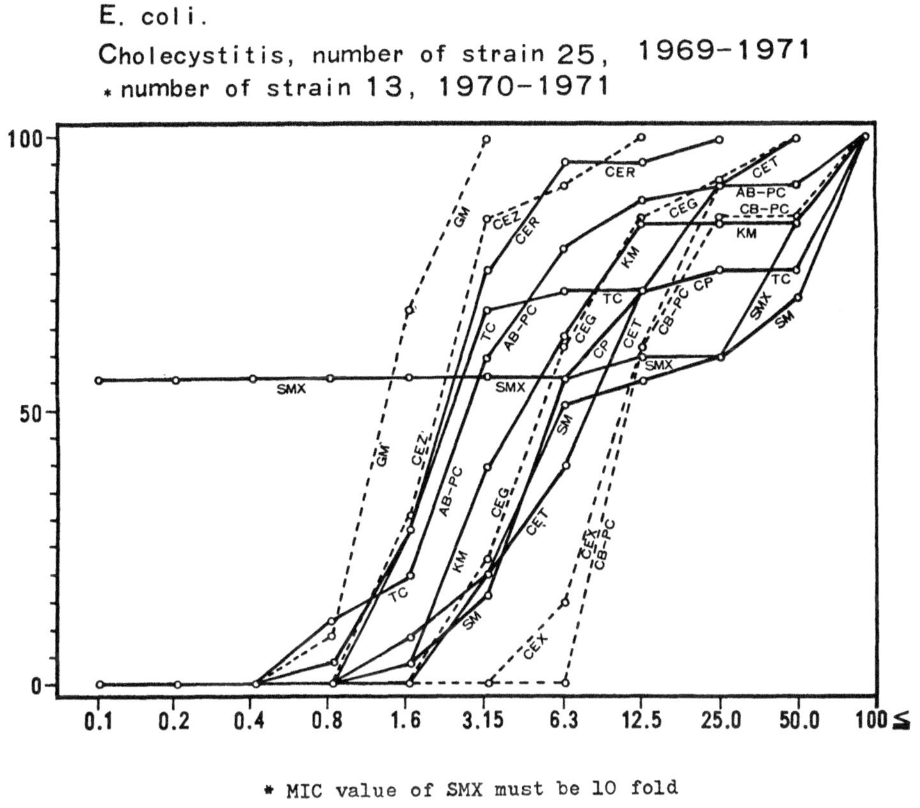

Fig. 26

CONCLUSION

MIC values of various antibiotics in *S. aureus* and *E. coli* isolated from clinical specimens for grouping the isolates into sensitive and resistant strains were determined. To minimize the effect of various variable conditions, the grouping values were determined by the moving pooling method for 2 or 3 years. The values were determined on the basis of MIC values of the drugs in *S. aureus* and *E. coli* under investigation. Annual change of sensitivity of the bacteria to various drugs found in this study was summarized in Table 12.

In the studies on the isolates from various clinical specimens, generally slight difference of sensitivity to various antibiotics was observed in *S. aureus*, while in *E. coli* almost no significant difference of sensitivity to various antibiotics was observed.

This may be mainly due to the existence of R-factor and other resistant plasmids among *E. coli* isolated from various clinical specimens. It will be imaginable that these resistant plasmids are distributing extensively in Japan.

TABLE 12

Annual change of sensitivity of S. aureus and E. coli

	Strains	
	S. aureus	*E. coli*
Group 1. Increase of Sensitivity	PC-G, AB-PC, SM, CP, TC	None
Group 2. No change of Sensitivity	CET, CER*	GM, SMX*
Group 3. Decrease of Sensitivity	EM, GM, KM, SMX	CP, CET, CER, KM, SM, TC, AB-PC

* Almost no change.

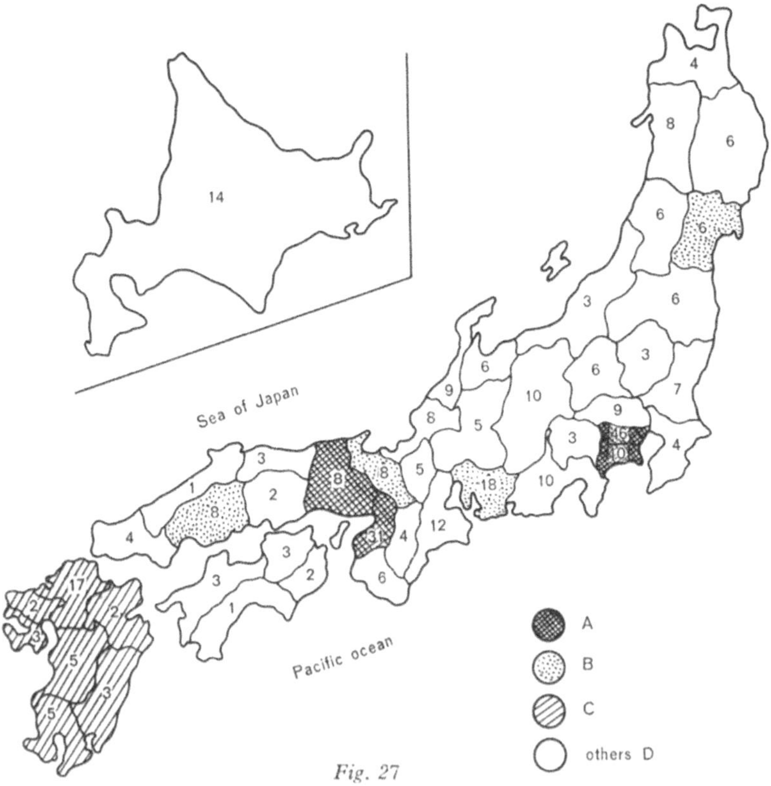

Fig. 27

ACKNOWLEDGEMENT

We express our sincere gratitude to those who have kindly supplied us with the strains for this study. We also thank Prof. N. Kosakai of Juntendo University, School of Medicine for his guidance, Drs. T. Iwaki, H. Nishimura and S. Okamoto of Shionogi & Co., Ltd. for their help, and Messrs. Minamide and Kawakubo for their technical cooperation.

REFERENCES

OTAYA, H., OKAMOTO, H., INOUE, E., ADACHI, Y., MACHIHARA, S. & YOSHIMURA, M.: Studies on the drug resistance of Staphylococcus and *Escherichia coli* against antibiotics II. General tendency of Resistant Strains. J. Antibiotics. 23: 324–335, 1970.

OTAYA, H. & INOUE, E.: Studies on the drug resistance of Staphylococcus and *Escherichia coli* against antibiotics and sulphonamide (I). Mod. Med. 22: 2545–2568, 1965 (in Japanese).

OTAYA, H., OKAMOTO, S., INOUE, E., ADACHI, Y., MACHIHARA, S. & YOSHIMURA, M.: Studies on the drug resistance of Staphylococcus and *Escherichia coli* against antibiotics. Bacterial Plasmids and Antibiotic Resistance. Ist International Symposium Smolenice, Czechoslovakia 63–69, 1971.

KOZAKAI, N.: Proceeding of newdrug Symposium. 16–17. XIX General Meeting of the East-Japan Branch Society of Chemotherapy. November 1972.

FORMATION OF F'TET PLASMIDS AND FERTILITY INHIBITION IN F+ PROTEUS MIRABILIS STRAINS

H. TSCHÄPE

Institute of Experimental Epidemiology, Wernigerode, G.D.R.

Nearly all wild-type strains of *Proteus mirabilis* are resistant to the tetracyclines, except minocycline. The determinant for this resistance cannot be transferred conjugationally to other bacteria or eliminated by ethidium bromide, dodecyl-sulfate, rifampicin or proflavin, even with a following penicillin-, ampicillin, or flavomycin-screening. This may account for a chromosomal nature rather than for a non-transferable plasmid (Tschäpe, 1973). In addition, no satellite DNA band could be detected in density profiles of DNA prepared from several F^-, R^- *P. mirabilis* strains (Hofemeister, pers. comm.; see Clowes, 1972).

Biochemical and kinetic data demonstrated tetracycline resistance of *P. mirabilis* to be

TABLE 1

The formation of tet-recombinants in F^+ Proteus mirabilis PG193 X E. coli K12 CSH-2 crosses

No.	Genotypes	Recombinant genotypes	Numbers[1])
F6	tra^+ fin^- comFI	tra^+ fin^- comFI tet^+	8
		tra^+ fin^+ comFI tet^+	13
RKM	tra^+ fln^- comFII $kan(neo)^+$	no recombinants	0
FSH-2	tra^+ fin^- comFI his^+ rfb^+	no recombinants	0
FSP-8	tra^+ fin^- comFI proA/B	tra^+ fin^+ comFI tet^+	12
FSA-8	tra^+ fin^- comFI $argE(argG)^+$	no recombinants	0
R6d8	tra^+ fin^+ comFII chl str mer sul	tra^+ fin^+ comFII tet^+	35
		tra^+ fin^+ comFII chl str sul mer tet	15
RH517	tra^+ fin^+ comFIII chl str sul mer amp kan(neo)	tra^+ fin^+ comFIII tet^+	6

[1]) 50 different clones of each strain were proved.

The methods for conjugation of F^+, F', or R^+ plasmids to *P. mirabilis* strains and the isolation of recombinants are described by Tschäpe (1973) and Tschäpe and Böhme (1972). The F' plasmids FSH-2, FSP-8, and FSA-8 were isolated following intergeneric crosses of Hfr *S. typhimurium* X *E. coli* K12 (Hacker and Tschäpe, to be published). The R-plasmids are isolated from *E. coli* wild-type strains.

similar to or identical with the tetracycline resistance determined by most of the naturally occurring R-plasmids, i.e. the $Rtet_A$ types (Hofemeister, 1970).

When the Sex-plasmid F6 was introduced into different *P. mirabilis* strains (such as PG 193, PG 273, PM1, different wild-type strains) tet-recombinants have been obtained with a stable linkage of the F and the tet region. In a $recA^-$ *P. mirabilis* strain, however, such recombinant F'tet plasmids have not been found (Tschäpe, 1973; Tschäpe and Böhme, 1972). Therefore, the integration of the F-plasmid into and a wrong excision from the host chromosome but not an unspecific mobilisation leading to the F'tet types, was discussed to be background of this formation. Alternatively, repliconization may also lead to F prime formation, particularly when intergeneric crosses were proved (see Low, 1972; Clark, 1971). This is the fact in the *P. mirabilis* × *E. coli* system used.

In order to distinguish between repliconization, excision or other unspecific mechanisms Hfr *P. mirabilis* strains have to be isolated. F^+ *P. mirabilis* strains, however, were shown to lose very rapidly their ability for pili formation, indicated by a negative F-phage spot test and their reduced fertility in crosses. This is why Hfr *P. mirabilis* strains could as yet not be detected.

TABLE 2

Lysis ability of the F-specific phage fr on different plasmids containing E. coli, Proteus mirabilis and Salmonella typhimurium strains

	E. coli K12		*Proteus mirabilis*			*S. typhimurium*	
	CSH-2	W677	PG-193	PM1	Pm2089	A70	LT 24
F	ol	ol	—	—	ol	ol	—¹)
F'tet-1	ol	ol	—	—¹)	ol	ol	—
FSH-2	ol	ol	—	—	ol	ol	—
FSP-1	ol	ol	—²)	—¹)	ol	ol	—¹)
FSA-8	ol	ol	—²)	—¹)	ol	ol	—¹)
RB9	ol	ol	—	—	ol	ol	—
RKM	ol	ol	—	—	ol	ol	—
F^-	—	—	—	—	—	—	—

Explanation: ol = opaque lysis
— = no lysis
¹) = not tested
²) = pro^-, arg^- variants of PG 193 were proved
RKM: tra^+ fin^- comFII kan(neo)
RB9: tra^+ fin^- comFII chl str sul mer tet
FSH-2: tra^+ fin^- comFI rfb^+ his^+
FSP-8: tra^+ fin^- comFI pro
FSA-8: tra^+ fin^- comFI $argE(argG)^+$
F'tet-1: tra^+ fin^- comFI tet^+
CSH-2: met^- pro^- F^- R^-
PG193: his^- F^- R^- tet^+ Fin^-
PM1: F^- R^- tet^+ Falkow strain) Fin^+
Pm2089: F^- R^- tet^- Fin^- (communicated by P. A. M. Guinee)
A70: F^- R^- Fin^- (communicated by E. S. Anderson)
LT 24: F^- R^- Fin^+ leu^- his^-

The determination of the lysis ability of the F-specific phage fr ($< 10^9$) was performed by spot test on the prospective strain (Tschäpe 1973). All strains were additionally tested for their fr-propagation ability.

The loss of fertility of F+ *P. mirabilis* strains may be due either to a segregation of the F-plasmid itself or to a fertility inhibition of the F-plasmid governed by the host cell, like in many *Shigella* and *Salmonella strains* (Grindley et al., 1971; Tschäpe, unpublished results).

As can be demonstrated in Table 1, Fin⁻ plasmids undergo fertility inhibition when they are introduced into *P. mirabilis* PG193, and PM1. On the other hand, all Fin⁻ plasmid containing *P. mirabilis* 2089 cells are sensitive to the F-specific phage fr.

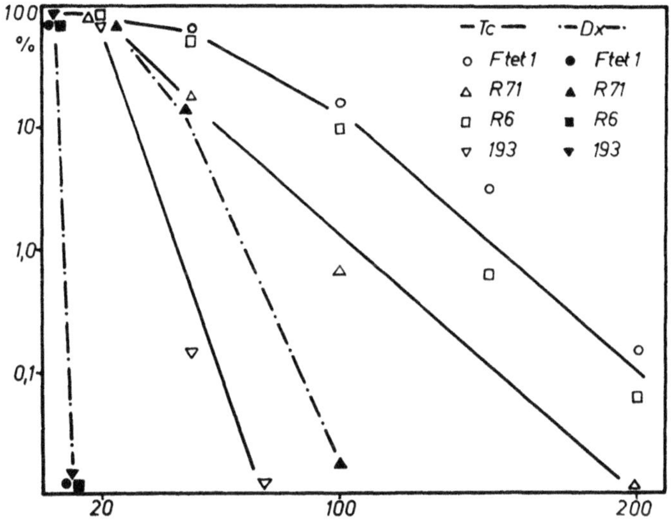

Fig. 1. Inactivation of various R+, F+, or F′ *E. coli* and *Proteus mirabilis* strains by different concentrations of Oxytetracycline (produced by VEB Jenapharm) and Doxycycline (produced by Pfizzer).

F′tet-1: tra+, comFI, fin⁻, tet+
R71: tra+, com9 (fi⁻) tet_B^+, chl+, str+ sul+, amp+
R6: tra+, comFII, fin+, tet_A^+, chl+, str+, sul+, mer+
All plasmids were proved in *E. coli* K12 (CSH-2)

193: *Proteus mirabilis* PG193 carrying the chromosomal tet determinant.

Since the Fin-property prevents the isolation of Hfr *P. mirabilis* strains (which should be the precursor state of the F′tet formation) F′plasmids with different chromosomal regions (e.g. his, pro, arg) were proved to have the ability for tet recombinant formation. F′plasmids (such as F′his) integrate into the chromosome predominantly at those sites which are homologous to the corresponding "prime"-region (e.g. the "his" region). (Campbell, 1969; Scaife and Gross, 1963; Pittard and Adelberg, 1964). According to Campbell's model, only neighboured chromosomal regions can be substituted or added to the integrated F-plasmid genome. Therefore, the tet formation was expected to occur only with such types of F′plasmids that integrate into the host chromosome near to the tet-region.

The results of these experiments are summarized in Table 2.

In addition to the F6 plasmid cannot only the FSP8 plasmid can form tet+ recombinants, while FSH-2, RKM and FSA-8. A phenomenon interesting to note is the loss of the pro-region when FSP-8 undergo tet-recombination.

As demonstrated in Table 2, two types of recombinants were detected in F6 crosses: One type is fin⁻ (F'tet-1), the other one is fin⁺ (F'tet-32). Both types belong to the FI compatibility set and are similar as regards all other properties tested.

Genetic different recombinant types of F6 with *P. mirabilis* PG193 or PM1 have not been found. Similar results have been obtained with FSP-8.

R-plasmids carrying tet-deletions were previously shown to pick up the tet determinant in *P. mirabilis* (Tschäpe and Böhme, 1972). In addition to this result, Table 2 shows that R6d8 was found to form one recombinant type carrying the tet determinant in addition to the other markers and a second type carrying the tet marker instead of the others. This might be explained by observations that the tra-region (RTF) segregates from the R-region in *P. mirabilis* (Cohen and Miller, 1972; see Clowes, 1972).

All F'tet plasmids similar to wild-type strains of *P. mirabilis* and most of the naturally occurring R-plasmids do not determine the resistance to minocycline (doxycycline) (Fig. 1). Therefore, they have been classed into the tet_A group of R-plasmids. Mutations toward a doxycycline resistance have not been detected among several clones proved.

Summarizing, the above-listed data do not account for either excisive recombination or repliconization but they make unspecific processes such as marker rescue very unlikely. Certain recent findings indicate that repliconization is the genetic background of F'tet formation in *P. mirabilis* (to be published).

REFERENCES

CAMPBELL, A. M. (1969). Episomes. Harper and Row, New York.

CLARK, A. J. (1971). Toward a metabolic interpretation of genetic recombination of E. coli and its phages. Ann. Rev. Microbiol. 25: 437.

CLOWES, R. C. (1972). Molecular structure of bacterial plasmids. Ann. Rev. 36: 361.

COHEN, S. N. and MILLER, C. A. (1970). Nonchromosomal antibiotic resistance in bacteria. III. Isolation of the discrete transfer unit of the R factor R1. Proc. Nat. Acad. Sci. 67: 510.

HOFEMEISTER, J. (1970). Mechanismus, Expressivität und Spezifität der R222 Resistenzwirkung in *Proteus mirabilis*. und *E. coli*. Dissertation 1970, Halle.

GRINDLEY, N. D. F., ANDERSON, E. S., SMITH, H. R. and GRINDLEY, J. N. (1971). The effect of *S. typhimurium* on derepressed mutants of F-like factors. Genet. Res. 17: 89.

LOW, K. B. (1972). *Escherichia coli* K12 F-prime factors, old and new. Bact. Rev. 36: 587.

PITTARD, J. and ADELBERG, E. A. (1964). Gene transfer by F' strains of *Escherichia coli* K12. IV. Genetics 49: 995.

TCHÄPE, H. (1973). "tet gene pick up" durch ein F-Plasmid in *Proteus mirabilis*. Zschr. Allg. Mikrobiol. 13: 693.

TCHÄPE, H. and BÖHME, H. (1972). Recombinations between different R(f) and R(i) plasmids and between R-plasmids and the chromosome in *E. coli* and P. mirabilis in Bacterial Plasmids and Antibiotic Resistance ed. V. Krčméry, L. Rosival and R. Watanabe. Avicenum Med. Press. Prague, p. 217.

CHARACTERIZATION ON THE BASIS OF PHAGE RESTRICTION OF R PLASMIDS DERIVED FROM E. COLI STRAINS

HEDDA MILCH, MARIAN GYENES and G. HÉRMÁN
National Institute of Public Health, Budapest, Hungary

The significance of phage restriction and modification caused by plasmids in bacteriologic diagnostics was recognized primarily in the case of *Salmonella* phage typing (Anderson, 1966, Guinée et al. 1967, Tschäpe et al., 1970). Primarily on the basis of grouping R-plasmids by genetic methods, and on the basis of fi character (Datta et al., 1966, Watanabe et al., 1964) phage restriction (Anderson, 1966, Anderson a. Lewis, 1965, Watanabe et al., 1964) and incompatibility (Datta a. Hedges, 1971, Hedges a. Datta, 1971), the route of R-plasmids may be followed from bacteria to bacteria and from the organism carrying bacteria it can be tracked to the environment. In our investigations, we aimed to characterize R-plasmids carried by *E. coli* strains, according to their effect in causing phage restriction.

R-plasmid carrying *E. coli* strains play a significant role in the spread of R-plasmids, and their identification may consequently facilitate the discovery of R-plasmid sources in nature.

MATERIALS AND METHODS

Examined strains: 1. Pathogenic and probably pathogenic *E. coli* strains of a definite serogroup, isolated from pathological material (enteritis). 2. Probably non-pathogenic *E. coli* strains derived from patients (cultivated next to *Shigella* strains). 3. Strains of *E. coli* isolated from foodstuffs and from surface water.

Determination of sensitivity to antibiotics:

Prior to crosses, the following ten antibiotics have been used by the disc method: ampicillin, streptomycin, tetracycline, chloramphenicol, neomycin, kanamycin, colistin, polymyxin B, nalidixic-acid, gentamycin.

Conjugation: by the method of Watanabe (1964).

Recipient strains: K12 HfrH lac^- nal^r and K12 F^- (WI-A2) lac^-nal^r. The HfrH lac^- strain has been derived from lac^+ strain at our laboratory, whilst the K12 lac^- strain is from Sturtevant (1969).

Determination of fi character has been performed on the basis of lysing HfrH recombinant strain by MS2 (Strauss, Sinsheimer, 1963) and f2 (Loeb, Zinder, 1961) male specific RNA phages.

Phages used for phage restriction test:

02 female specific phage (Cuzin, 1965); T1–T7, P1 and λ phages, 28 phages isolated at our laboratory for phage typing of *E. coli* strains, C1–C28 (11).

Test method for phage restriction:

After the acquisition of R-plasmid, the spot test was used to investigate K12 HfrH and K12 F^- strains. The phages were dropped undiluted and in the highest dilution by which the phage still exhibits confluent lysis with plasmid-free HrfH and K12 F^- strains. In case this method produced a weaker lysis in relation to the recipient strain, the EOP value of the examined phage was determined by the strain in question. The

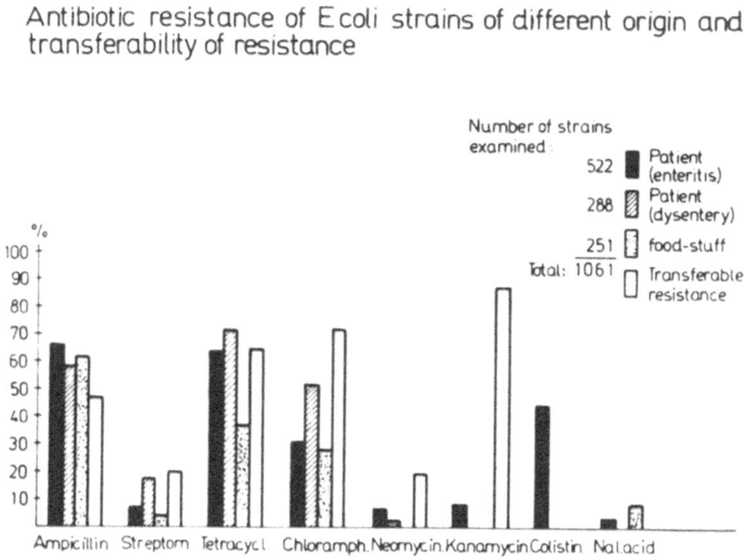

Fig. 1

extent of the restriction has been determined on teh basis of the relative EOP values.

Medium: After conjugation, the strains were streaked on EMB agar, which contained (50 γ/ml) nalidixic acid and the required inhibitor concentration of the investigated antibiotics (tetracycline 50 γ/ml, ampicillin 500 γ/ml, chloramphenicol 25 γ/ml).

RESULTS

1. Of the 522 *E. coli* strains derived from patients, 73.2% were resistant to one or more antibiotics, whilst 30% of the strains were resistant to 3–7 antibiotics.
2. 45.4% of *E. coli* strains (288 strains have been tested) existing as an accompanying flora to *Shigella* strains were resistant to one or more antibiotics, whilst 15,2% showed resistance to three or more antibiotics. The antibiotic resistance markers of *E. coli* and *Shigella* strains cultivated in the same medium were compared in 107 cases. In nearly 30% of these cases, antibiotics resistance markers of *E. coli* and *Shigella flexneri* or *Shigella sonnei* strains have been identical. In these identical cases, resistance was due to the R-factor in 78% of them.
3. Of the 251 *E. coli* strains isolated from foodstuffs or water, 27% were resistant to one or more antibiotics; in this investigated group, only 1.5% of the strains were resistant to three or more antibiotics.

Fig. 1. shows the antibiotic resistance markers of *E. coli* strains from all the three investigated groups. The first column indicates pathogenic *E. coli* strains of patients, whilst column 2 shows the resistance markers of *E. coli* strains existing next to *Shigella*, and column 3 those of *E. coli* strains derived from foodstuffs and water respectively. Values are in %, concerning all the resistant strains. Column 4 shows the percentage of transferable resistance.

Of the pathogenic *E. coli* strains, tetracycline and ampicillin resistance markers occurred in over 60% of the cases, (63.6% and 66.7% respectively), occurrence of colistin resistance was 45.5%, chloramphenicol resistance 30.6%, streptomycin resistance 17.8%, kanamycin resistance 9.4%, neomycin resistance 7.6%, nalidixic-acid resistance 2.4%, and polymyxin as well as gentamycin resistance was under 1%.

"Accompanying" *E. coli* strains were tetracycline resistant in 71.7% of the cases, ampicillin resistant in 51.9%, streptomycin resistant in 6.1%, neomycin resistant in 1.5%.

TABLE 1

Characterization of R factors derived from serologically different E. coli strains according to fi character and phage restriction

Strain	"O" antigen	Antibiotic resistance pattern	R fi	Phage restriction in	
				Hfr H	K_{12}
R 376*	0 : 4	**A S T C**	—	T_3, T_4, T_5, 14, 23, 24	—
R 228*		**A T**	—	14, 23, 24	—
R 19*	0 : 6	**A** S **T C** N **K** Co	—	T_3, T_4, T_5, 14, 23, 24	T_3, T_5, 14
R 253*		**T C**		T_3, T_4, T_5, 14, 23, 24	—
R 139*	0 : 55	**A** S **C** N **K** Co	—	14	14
R 121*		A S **C K**	—	14	14
R 92*		A S **T C** N **K** Co	+	12	6
R 147*		A S T **C** N **K**	+	—	—
C 160*	0 : 111	A S **T C**	—	—	—
C 275*		A **T C K**	—	—	—
C 162*		**S T C**	—	—	—
C 134*	0 : 148	**T**	—	—	—
C 454**	0 : 149	C **T** Ka	+	14	T_3, T_4, 14
C 436**		C **T** Ka	—	14	T_3, T_4, 4, 14
C 359***		**T A S**	—	14	T_3, 14
C 440**		C T K **A**	+	—	P_1
C 242***		**C S T Ka**	+	—	—
C 400***	0 : ?	**T**	+	—	14
C 443**		**S T Ka**	+	—	14
C 184***		**T**	—	—	P_1, T_3
C 180***		**T**	—	P_1	P_1
C 181***		**T**	—	P_1, 23	P_1

A, S... = transferable antibiotic marker
* = patient
** = intestine of pig
*** = food stuff

A = ampicillin
S = streptomycin
T = tetracycline
C = chloramphenicol
N = neomycin
K = kanamycin
Co = colistin

Of the strains derived from foodstuffs and surface water, tetracycline resistance marker occurred in only 36.8% of the cases, ampicillin resistance in 61.8%, chloramphenicol resistance in 27.9%, streptomycin resistance in 4.4%, nalidixic-acid resistance in 8.8% and the other resistance markers under 1%.

According to marker frequency the antibiotics resistance transfer is as follows: kanamycin was transferable in 88.8% of the cases, chloramphenicol in 75%, tetracycline in 65.8%, and streptomycin and neomycin resistance in 20%. In 74% of the cases, the resistance was proved to be due to plasmids.

Grouping of R-plasmids on the basis of fi character

Of the 65 plasmids investigated, 57 were of fi^- and 8 of fi^+ character.

It is remarkable, in spite of the small number of tests, that of the 8 fi^+ R-plasmids 4 are derived from *E. coli* strain isolated from foodstuffs, and 2 were isolated from pig intestines. It is a matter of further investigations to determine whether the *E. coli* strains derived from foodstuffs carry animal-originated R-plasmids on the basis of fi character, and also whether R-plasmids isolated from human diseases are generally different from these R-plasmids (Table 1).

Characterization of R-plasmids on the basis of phage restriction

Table 1 demonstrates the grouping of *E. coli* strains of different origin and of different serogroup, in accordance with the fi character and phage restriction, on the basis of certain emphasized strains that have been investigated. Restrictions indicated in the Table were carried out by the spot test. Transferable antibiotic markers are indicated separately. The R-plasmid of 04 and 06 strains isolated from infant enteritis having various antibiotic resistance patterns displayed a nearly similar restriction. By investigating the two types of recipient strains we have obtained different restriction patterns. Resistance markers of 055 strains isolated at the same hospital were different, but exhibited a restriction with the same phage. Of serogroup 0111 strains, two strains were fi^+, with one R-plasmid exhibiting a restriction. R-plasmids of other 0111 strains did not provide any restriction. The strain belonging to the serogroup 0148 producing enterotoxin and isolated by Skerman et al., (1972) had shown a tetracycline resistance marker, with the fi^- R-plasmid causing no restriction. Of two pig-pathogenic strains of serogroup 0149 having identical antibiotic markers one was fi^+ and the other fi^-, whilst in HfrH strain they have caused an identical phage restriction.

We have indicated an R-plasmid having tetracycline resistance marker and causing similar phage restriction, from two types of foodstuffs (sausages).

According to a report by Pitton and Anderson (1970), the female specific 02 phage (Cuzin) is capable of distributing strains. The authors have described a fi^+ factor, which provided restriction with 02 phage. In the course of our investigations (which were performed by spot test only, with the exception of a few examples) we have found ten fi^- and three fi^+ factors who restricted phage 02. This observation is verified in a report published by Jonsson et al. (9), who said that on the basis of 02 phage restriction R factors may be classified into three main groups, viz. two types of fi^+ groups and the fi^- groups.

Determining the extent of phage restriction caused by plasmids, on the basis of EOP

Table 2 demonstrates R-plasmid resistance determinants of *E. coli* strains of various origin, the transferred resistance determinants, the fi character, and the relative EOP values. Investigations conducted so far indicate that R-plasmids of strains derived from

TABLE 2

Phage restriction in E. coli HfrH and E. coli K_{12} by "R" plasmids of E. coli strains of different origin.

Donor strain	„O" antigen	Origin of strains	Recipient strain	Relative efficiency of plating: Phages P_1	T_3	T_4	T_5	4	14	23	24	Res. det. of donor	Res det. of recombinant	R fi
C 440	nt ?	intestine of pig	E. coli K12	$0{,}1.10^{-4}$	1	1	1	1	1	1	1	CTKA	A	+
C184	0 : ?	sausage	E. coli K12	$0{,}1.10^{-4}$	$0{,}6.10^{-4}$	1	1	1	1	1	1	T	T	—
C359	0 : ?	river water	E. coli K12	1	$0{,}6.10^{-4}$	1	1	1	1.10^{-2}	1	1	CTAS	TAS	—
C454	0 :149	intestine of pig	E. coli K12	1	$0{,}8.10^{-2}$	1.10^{-1}	1	1	1.10^{-1}	1	1	CTK	T	+
C436	0 :149	intestine of pig	E. coli K12	1	$0{,}8.10^{-2}$	1.10^{-1}	1	$4{,}1.10^{-6}$	$0{,}2.10^{-2}$	1	1	CTK	T	—
C180	0 : ?	sausage	E coli HfrH	$1{,}5.10^{-3}$	1	1	1	1	1	1	1	T	T	—
C181	0 : ?	sausage	E coli HfrH	$1{,}5.10^{-3}$	1	1	1	1	1	$5{,}3.10^{-4}$	1	T	T	—
R376	0 :4	enteritis	E coli HfrH	1	$0{,}4.10^{-2}$	$1{,}1.10^{-3}$	$0{,}4.10^{-2}$	1	$4{,}1.10^{-5}$	$1{,}7.10^{-6}$	2.10^{-3}	CTS	CTS	—
R19	0 :6	enteritis	E coli HfrH	1	$0{,}4.10^{-4}$	$2{,}2.10^{-3}$	$0{,}4.10^{-5}$	1	$0{,}4.10^{-6}$	$0{,}8.10^{-5}$	$0{,}6.10^{-6}$	CT	CT	—
R253	0 :6	enteritis	E coli HfrH	1	$0{,}2.10^{-4}$	$0{,}8.10^{-2}$	$0{,}4.10^{-6}$	1	$0{,}4.10^{-6}$	$0{,}7.10^{-5}$	$0{,}6.10^{-5}$	CT	CT	—

infant enteritis cause the restriction of identical phages, and that they were nearly identical also on the basis of the EOP values. The phage restriction of the R-plasmids of *E. coli* strains isolated from pig intestines and sausages was proved to be identical or nearly identical also on the basis of EOP and differed considerably from restriction characteristics of R-plasmids of human *E. coli* strains causing enteritis.

The examinations conducted so far indicate that fi^- and fi^+ R-plasmids may be classified according to their phage restriction characteristics not only by the phages T, P_1 and 02, but also by the *E. coli* phages 4, 6, 12, 14, 23, 24 isolated by us. In restriction tests, the *E. coli* C strain was not found to be suitable, since it did not provide lysis with several of the phages tested. The HfrH strain has been proved to be convenient for phage restriction investigations. In future we shall carry out investigations to verify, by examining a larger volume of material, that the spread of R-plasmids may be tracked on the basis of phage restriction patterns, and to verify or preclude the relations among R plasmids isolated from various sources.

REFERENCES

ANDERSON, E. S. (1966). Influence of the △ transfer factor on the phage sensitivity of Salmonellae. Nature, 212, 795.

ANDERSON, E. S., LEWIS, M. J. (1965). Drug resistance and its transfer in *Salmonella typhimurium.* Nature (Lond.) 206, 579.

CUZIN, F. (1965). Un bacteriophage spécifique du type sexuel F^- d' E. coli K 12. C.R. Acad. Sci. 260, 6482.

DATTA, N., LAWN, M., MEYNELL, E. (1966). The relationship of F type piliation and F phage sensitivity to drug resistance transfer in R^+F^- *E. coli* K 12. J. gen. Microbiol. 45, 365.

DATTA, N. and HEDGES, R. W. (1971). Compatibility Groups among fi^- R factors. Nature, 234, 222.

GUINÉE, P. A. M., SCHOLTENS, R. Th. and WILLEMS, H. M. C. C. (1967). Influence of Resistance Factors on the phage types of *Salmonella panama.* Antonie van Leeuwenhoek, 33, 30.

GUINÉE, P. A. M. and WILLEMS, H. M. C. C. (1967). Restriction and modification of phage 47 and Lambda by R-factors. Antonie van Leeuwenhoek 33, 397.

HEDGES, R. W., DATTA, N. (1971). fi^- R factors giving Chloramphenicol Resistance-Nature 234, 220.

JONSSON, M., RUTBERG, L., TURNEWALL, G. (1972). Transferable resistance to antibiotics in Gram negative bacteria isolated in a hospital for infectious diseases. Scand. J. Inf. Dis. 4, 209.

LOEB, T., ZINDER, N. D. (1961). A bacteriophage Containing RNA. Proc. nat. Acad. Sci (Wash.) 47, 282.

MILCH, H., and GYENES, M. (1972). Subdivision and Correlation Studies of Serologically Grouped *E. coli* strains by Phage Typing, Colicinogeny, Lysogeny and Biochemical Test. Acta Microbiol. Acad. Sci. hung. 19, 213.

PITTON, J. S., ANDERSON, E. S. (1970). The inhibitory action of trasnfer factors on lysis of *E. coli* K 12 by phages μ 2 and ∅ 2. Genet. Res., Camb., 16, 215.

SKERMAN, F. J., FORMAL, S. B., FALKOW, S. (1972). Plasmid-Associated Enterotoxin Production in a Strain of *E. coli* Isolated from Humans. Infection and Immunity 5, 622.

STRAUSS, J. H., SINSHEIMER, R. L. (1963). Purification and properties of bacteriophage MS_2 and of its ribonucleinacid. J. Molec. Biol. 7, 43.

STURTEVANT, A. B., FEARY, T. W. (1969). Incidence of Infections Drug Resistance Among Lactose-Fermenting Bacteria Isolated from Raw and Treated Sewage, Appl. Microbiol. 18, 918.

TSCHÄPE, H., RISCHE, H. (1970). Veränderungen von Lysotypen durch R-Faktoren und deren epidemiologische Bedeutung bei *Sh. sonnei.* Zbl. Bakt. I. Abt. Orig. 214, 91.

WATANABE, T. (1964). Selected methods of genetic study of episome-mediated drug resistance in bacteria. Meth. med. Res. 10, 202.

WATANABE, T., NISHIDA, H., OGATA, C., ARAI, T., SATO, S. (1964). Episome-mediated transfer of drug-resistance in Enterobacteriaceae. VII. Two types of naturally occurring R-factors J. Bacteriol. 88, 716.

STUDIES ON R-PLASMIDS DERIVED FROM SHIGELLA FLEXNERI STRAINS

VERA LÁSZLÓ and I. RIMANÓCZY
National Institute of Public Health, Budapest, Hungary

The increasing number of drug-resistant bacteria, and mainly that of bacteria carrying R factors presents a serious problem all over the world. We therefore wanted to determine the ratio of R-factor carrying *Sh. flexneri* strains in our country.

Epidemiological practice raised another problem: strains of different phage-types were isolated during the same outbreak. We wanted to study the role of R-factors in the alterations of phage-types and phage-restriction, due to R factors.

MATERIAL AND METHODS

We examined 1712 *Sh. flexneri* strains isolated from patients in Hungary in 1972–1973. Resistance to chloramphenicol, streptomycin, oxytetracycline, neomycin and polymixin B was tested by the disc method. Drug-resistant *Sh. flexneri* strains were crossed with *E. coli* Hfr-nalidixicacid resistant strain, using the method of Tschäpe.

The 18^{hr}-broth cultures of the donor- and recipient strains were centrifuged, the sediments were mixed (ratio 1:1) and plated to the surface of an agar plate. After 5^{hr} incubation (37°C), the plates were washed with broth, and plated to eosin-methylen-blue agar medium, containing 50 γ/ml nalidixicacid + 30 γ/ml chloramphenicol or 30 γ/ml oxytetracycline. The phage-sensitivity of the isolated clones was examined with a phage-set containing 28 phages isolated and elaborated for *E. coli* phage-typing by Milch (1972).

The R factor carrying HfrH strains were crossed with a drug sensitive *Sh. flexneri* strain of variant X, phage type 3 – using the above-mentioned method. The strain was not lysogenic and did not produce colicine. Deoxycholate-citrate agar plates-containing chloramphenicol or oxytetracycline was used as a selective medium. The drug-resistant *Sh. flexneri* strains were examined by a phage-set, containing 19 phages, elaborated for phage-typing of *Sh. flexneri* strains in our laboratory (László et al., 1973).

Fi type was determined by male-specific phages (MS_2 and f_2) using the method of Watanabe and Fukasawa (1962). Visible lysis was assessed on the R factor carrying HfrH strain. To determine the efficiency of plating *Sh. flexneri* phage-set and the agar-layer method were used.

RESULTS AND DISCUSSION

Table 1 shows the antibiotic resistance of *Sh. flexneri* strains isolated from patients in our country in 1972–1973. Out of the 1712 strains 488 were resistant to 1–4 antibiotics (28.5%), with the following distribution:

9.7% were resistant to one antibiotic, 10.8% to two antibiotics, 7.9% to three and four antibiotics respectively.

TABLE 1

Antibiotic resistance of Shigella flexneri strains in Hungary in 1972–1973

Resistant to		Number of strains in years		Total		
		1972	1973	number		%
1 antibiotic	Sm	12	29	41	167	9,8
	Cm	40	14	54		
	Tc	18	51	69		
	Nm	2	1	3		
2 antibiotics	Sm-Cm	4	3	7	185	10,8
	Sm-Tc	1	2	3		
	Sm-Pm	—	1	1		
	Sm-Nm	—	1	1		
	Cm-Tc	80	92	172		
	Tc-Pm	—	1	1		
3 antibiotics	Sm-Cm-Tc	67	40	107	115	6,7
	Sm-Tc-Pm	—	3	3		
	Sm-Cm-Nm	—	2	2		
	Cm-Tc-Pm	—	3	3		
4 antibiotics	Sm-Cm-Tc-Nm	3	14	17	21	1,2
	Sm-Cm-Tc-Pm	—	4	4		
Total		227	261		488	

Sm = streptomycin
Cm = chloramphenicol
Tc = tetracycline
Nm = neomycin
Pm = polimyxin

In a previous study (Egler et al., 1968) conducted in 1967–68 we found that 3.5% of the strains were resistant to 3–6 antibiotics. (The number of examined strains was 1336.)

As concerns the antibiotic-markers, Cm-Tc and Sm-Cm-To were the most frequent in 1972 and 1973. Among the strains resistant to one antibiotic, the Cm-marker was more frequent in 1972, the Tc marker in 1973.

The frequency of R-factors among the resistant strains was examined.

All of the multiple-resistant strains examined in 1967–68 carried R factors; 81.2% of the resistant strains examined in 1972–1973 carried R factors. The number of examined strains was 373.

The determinants Cm-Tc occurred most frequently, amounting to 56.4%, the frequency of Sm-Cm-Tc determinants respectively, was 14.5% (Table 2). There was a difference in the frequency of the Cm and Tc determinants in the two years under study.

The Tc determinant was carried by 5, Cm determinant by 26 R factors in 1972. The occurrence of the two determinants was almost identical in 1973 (Watanabe et al., 1968).

Examination of the fertility inhibition properties of R-factors showed that 83.8% proved to be fi⁻, and 16.2% fi⁺. This finding differs from that of Yoshikawa (1971) because the R factors of *Shigella* strains isolated in Japan were fi⁺ (122 out of 124 strains).

TABLE 2

Distribution of R-factors according to resistance determinants and fi⁻ character

Resistance determinants	Number of R-factors in year				Total number
	1972		1973		
	fi⁺	fi⁻	fi⁺	fi⁻	
Sm	—	4	2	2	8
Cm	2	24	6	12	44
Tc	—	5	4	13	22
Sm-Cm	—	8	—	2	10
Sm-Tc	—	—	1	3	4
Cm-Tc	17	68	13	73	171
Sm-Cm-Tc	—	31	4	9	44
Total	19	140	30	114	303

There was another difference between our results and the data reported in literature. R factors of fi⁺ type determining Cm-resistance were found to be common, but Cm-resistance controlled by fi⁻ R factor occurred rarely. Watanabe et al. (1964) reported in 1964 the absence of fi⁻ R factors, conferring Cm-resistance, they found it only in one *Shigella* strain in 1968 (1968). Fi⁻ R factors carrying the Cm determinant were isolated from *Aeromonas* strains (Aoki et al., 1971).

Examining enteric bacteria, fi⁻ R factors transferring Cm marker could not be isolated in Great Britain by Hedges and Datta (1972). Shaw et al. (1972) reported a fi⁻ R factor determining Cm-acetyl-transferase (Shaw et al., 1972).

In our examinations fi⁻ R factors carried Cm determinants fairly frequently, 36 R-factors possessed Cm determinants. 141 Cm-Tc and 40 Sm-Cm-Tc markers.

There seem to be several differences between the data found in Europe, Japan and the U.S.A. For example, many of the R factors isolated in Europe were known to carry a colicinogenic factor (Siccardi, 1966) but such R factors have never been isolated in Japan (Watanabe and Arai, 1968). Many of the R factors isolated in Japan carry a chloramphenicol marker, but those in Europe and the U.S.A. do not (Yoshikawa et al., 1971).

These data suggest that the existence and dominance of different strains and R factors is due to the influence of various ecological factors.

Epidemiological practice raised another problem. Strains, at different phage-types were isolated during the same outbreak.

Phage restriction is important also from the practical, epidemiological viewpoint. Anderson, (1966) studying fi⁻ R factors of *S. typhimurium*, pointed out that the acquisition of plasmids could alter the phage types of the strains. Similar observations were published on other bacteria, for example *Salmonella*, *Sh. sonnei* (Guiné et al., 1967, Tschäpe and Rische, 1970, Tschäpe et al., 1974). Many of the naturally occurring fi⁻ R factors isolated in Japan cause host-controlled restriction and modification of various phages, while fi⁺ R factors usually do not have such functions (Yoshikawa et al., 1971, Watanabe et al., 1964, Arber and Morse, 1965). Studies on R factors isolated in Britain have subsequently shown that some fi⁺ R factors also cause restriction and modification (Bannister and Glover, 1968). Our examinations proved that, both fi⁺ (in 53%) and fi⁻ R factors (in 44%) could cause phage-restriction.

TABLE 3

Efficiency of Plating (e.o.p.) of Sh. flexneri Phages on Sh. flexneri 232, Carrying fi⁻ R Factors

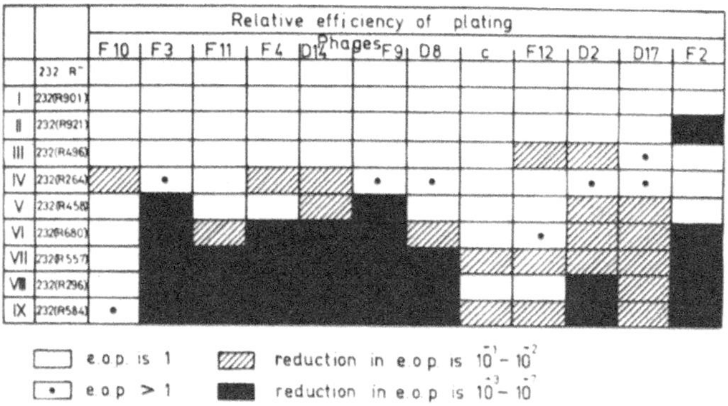

Because of the changes of phage-types mediated by R factors we thought it desirable to characterize R factors on the basis of their restrictive capacity.

First we used different strains as recipients. In the course of the studies the *Sh. flexneri* strain, No. 232, proved to be the best suited because of its broad lytic-spectra. Out of 19 type phages this strain was lysed by 16 phages.

After transferring the R factors into this strain, E.O.P. was determined using our *Sh. flexneri* phage-set.

Table 3 demonstrates the relative E.O.P. values, with 12 phages. (In the case of four phages the E.O.P. values amounted to 1 in all of the recombinants, and these phages were therefore omitted.)

R factors could be divided into 9 types, on the basis of E.O.P.

To make the table simpler, no numbers are indicated where E.O.P. is 1 and the squares are left empty; where the decrease of E.O.P. is $10^{-1}-10^{-2}$ the squares are hatched; and where they are $10^{-3}-10^{-7}$, the squares are dark.

The I. type of R factors comprised those which did not restrict phages.

II. type: one phage was restricted,
III. type: two phages were restricted to a small degree,
IV. type: three phages were restricted to a small degree,
V. type: two phages were restricted to a higher, and three to a smaller degree,
VI. type: 5 phages were restricted to a higher, 4 to a smaller degree,
VII. type: 7 phages were restricted to a higher, 4 to a smaller degree,
VIII. type: 8 phages were restricted to a higher, 1 to a smaller degree,
IX. type: 8 phages were restricted to a higher, 3 to a smaller degree.

The Table shows that the E.O.P. was higher in the case of phages with certain recombinants than with the R^- recipient. Probably, the recipient strain possessed a restrictive capacity which was repressed by the introduced R factor. This problem needs further examina tion.

The presence of R factors caused modification in plaque — morphology, too.

This grouping is stable only in the given recipient strain, in other recipients the degree and pattern of restriction may change. This fact supports the theory that the expression

TABLE 4

Alteration of phage types of Shigella flexneri strains caused by R-factors

Outbreak	Antibiotic-res determinant	Sero-type	Phage type	a	F10	F3	F11	F4	D14	F9	D8	c	F5	F6	D19	F12	D2	D17	F2	b	f	d	Type of R-factor
1	—	4a	81	■	□	□	□	□	□	□	■	□	□	□	□	□	□	□	■	□	□	□	II
	Cm Tc	4a	83	■	□	□	□	□	□	□	■	□	□	□	□	□	□	□	□	□	□	□	
2	—	4a	89	□	□	□	□	□	□	□	■	□	□	□	□	□	□	□	■	□	□	□	II
	Cm Tc	4a	90	□	□	□	□	□	□	□	■	□	□	□	□	□	□	□	□	□	□	□	
3	—	2a	16	■	■	■	■	■	■	■	■	■	■	■	■	□	■	■	□	■	□	□	V
	Sm Cm	2a	50	■	■	□	■	■	■	□	■	■	■	■	■	□	■	■	□	■	□	□	
4	—	3a	31	■	■	■	■	■	■	■	■	■	■	□	■	□	■	■	□	□	□	□	VII
	Cm Tc	3a	66	■	■	□	□	□	□	□	■	□	■	□	■	□	■	□	□	□	□	□	
5	—	2a	71degr	■	□	■	■	■	■	■	■	■	■	■	■	□	■	■	□	□	□	□	VII
	Cm Tc	2a	76	■	□	□	□	□	□	□	■	□	■	■	■	□	■	■	□	□	□	□	

□ no-lysis
■ lysis

of the phage-restrictive effect of R factors depends on the recipient strain, too. Despite this finding, the R-factor types were found helpful in evaluating the alterations in the phage-types.

The next Table (Table 4) shows some examples of alterations of phage-types in five pairs of strains derived from five different outbreaks.

After R factor acquisition the phage types changed from

89 to 90,
81 to 83,
31 to 66,
71 to 76,
16 to 50.

In all of these cases R factor transfer was followed by phage-restriction — and the restrictive types of R factors were useful in establishing the source of the infection.

Analysing the phage-typing results, resistance to antibiotics and occurrence of R factors of *Sh. flexneri* strains in 1972 and 1973, it has been found that the frequently occurring strains of serotypes 2a and 3a included phage types of broad lytic spectra. (Serotype 2a phage type 14, 16, 71; serotype 3a 5, 18, 19, 31, 36, 38). Most of these strains were sensitive to antibiotics, they did not carry R factors, or the carried R factor had no restrictive capacity.

There were bacteria of other phage types, with limited phage sensitivity. These strains were resistant to antibiotics and carried R factors of restrictive capacity (serotype 2a) phage type 58, 65, 69, 73, 74, 76;

serotype 3a:

phage type 55, 58, 61, 64, 66, 68.

Knowing the restriction types of R factors, it is possible, in the case of R factor-carrying strains possessing limited lytic-spectra, to deduce the original, antibiotic sensitive strains of broad lytic spectra.

REFERENCES

ANDERSON, E. S. (1966). Influence of the △ transfer factor on the phage sensitivity of Salmonellae. Nature, London 208, 1016.

AOKI, T., EGUSA, S., OGATA, Y., WATANABE, T. (1971). Detection of resistance factors in fish pathogens *Aeromonas liquefaciens*. J. gen. Microbiol. 65, 343.

ARBER, W., MORSE, M. L. (1965). Host specificity of DNA produced by E. coli. VI. Effects on bacterial conjugation. Genetics 51, 137.

BANNISTER, D. and GLOVER, S. W. (1968). Restriction and Modification of Bacteriophages by R^+ strains of *Escherichia Coli* K 12. Biochem. biophys. Res. Commun. 30, 735.

EGLER, L., LÁSZLÓ, V. G., MILCH, H., SCHMIDT, M. (1968). Studies on R factors. Paper at Symposium of Hungarian Microbiological Society.

GUINÉE, P. A. M., SCHOLTENS, R. Th. and WILLEMS, H. M. C. C. (1967). Influence of Resistance Factors on the phage types of *Salmonella panama*. Antonie van Leeuwenhoek, 33, 30.

GUINÉE, P. A. M. and WILLEMS, H. M. C. C. (1967). Restriction and modification of phage 47 and Lambda by R-factors. Antonie van Leeuwenhoek, 33, 397.

HEDGES, R. W., DATTA, N. (1971) fi^- R Factors giving Chloramphenicol Resistance. Nature, 234, 220.

LÁSZLÓ, V. G., MILCH, H., HAJNAL, A. (1973). Phage typhing of *Shigella flexneri*. Acta Microbiol. Acad. Sci. Hung. 20, 135.

MILCH, H., and GYENES, M. (1972). Subdivision and Correlation Studies of Serologically Grouped *E. coli* Strains by Phage Typing, Colicinogeny, Lysogeny and Biochemical Test. Acta Microbiol. Acad. Sci. Hung. 19, 213.

SHAW, W. V., SANDS, L. C., DATTA, N. (1972). Hybridization of variants of Chloramphenicol acetyltransferase specified by fi^+ and fi^- R factors. Proc. Nat. Acad. Sci. USA 69, 3049.

SICCARDI, A. G. (1966). Colicin resistance associated with resistance factors in *E. coli*. Genet. Res. 8, 219.

TSCHÄPE, H. and RISCHE, H. (1970). Veränderungen von Lysotypen durch R-Faktoren und deren epidemiologische Bedeutung bei *Sh. sonnei*. Zbl. Bakt. I. Abt. Orig. 214, 91.

TSCHÄPE, H., DRAGAS, ANA-ZLATA, RISCHE, H., KÜHN, H. (1974). Conversion of Phage Types of *S. typhimurium* due to Different R-Plasmids. Zbl. Bakt. Hyg., I. Abt. Orig. A 226, 184.

WATANABE, T. and FUKASAWA, T. (1962). Episome-mediated transfer of drug resistance in Enterobacteriaceae. IV. Interactions between resistance transfer factor and F factor in *E. coli* K 12. J. Bacteriol., 83, 727.

WATANABE, T., NISHIDA, H., OGATA, C., ARAI, T. and SATO, S. (1964). Episome-mediated transfer of drug-resistance in Enterobacteriaceae. VII. Two types of naturally occurring R factors. J. Bacteriol. 88, 716.

WATANABE, T., ARAI, T. (1968). Changes in sensitivity to colicine and UV irradiation conferred by R factors. Japan. J. Bacteriol. 23, 271.

WATANABE, T., FURUSE, C. and SAKAIZUMI, S. (1968). Transduction of Various R Factors by Phage P1 in *E. coli* and by Phage P22 in *S. typhimurium*.. J. Bacteriol. 96, 1791.

YOSHIKAWA, M., and AKIBA, T. (1962). Studies on transferable drug resistance in bacteria. 4. Suppression of plaque formation of phages by the resistance factor. Japan J. Microbiol. 6, 121.

YOSHIKAWA, M., NAGASHIMA, Sh. and MATSUSHIMA, Sh. (1971). Genetical Distinction of R Factors Derived from Shigellae and Salmonellae. Japan. J. Microbiol. 15, 425.

PROPERTIES OF R FACTORS FROM *BORDETELLA BRONCHISEPTICA*

N. TERAKADO and S. MITSUHASHI
National Veterinary Assay Laboratory, Kokubunji, Tokyo, Japan
and School of Medicine, Gunma University, Maebashi, Japan

The transferable drug resistance (R) factors (Mitsuhashi, 1971) derived from family *Enterobacteriaceae* have been extensively studied, although an increasing number of R factors have been characterized from *Pseudomonas aeruginosa*. We have recently demonstrated R factors from *Bordetella bronchiseptica* strains isolated from nasal cavities of young pigs (Terakado et al., 1973). Further studies indicate that the R factors are the main cause of drug resistance in *B. bronchiseptica* (Terakado et al., 1974). This organism is a frequent inhabitant of the respiratory tracts and belongs to family *Brucellaceae* (Breed et al., 1957). So, we have much interest in the relationship between the R factors derived from *B. bronchiseptica* and those from *Enterobacteriaceae* and *P. aeruginosa*. In this paper, the results of our recent studies on the R factors from *B. bronchiseptica* are reported.

EPIDEMIOLOGY OF R FACTORS FROM B. BRONCHISEPTICA

Drug resistance of *B. bronchiseptica* strains isolated from pigs in Japan from 1969 to 1972 were examined. Drug resistance was determined by agar dilution method described previously (Terakado et al., 1974). Of 304 strains tested, 71 (23%) were found to be resistant to either one or more of the following three drugs, streptomycin (SM, more than

TABLE 1
Drug resistance patterns of Bordetella bronchiseptica strains isolated from pigs[a)]

Type of drug resistance[b)]	No. of strains isolated in 1969	1970	1971	1972	Total	(%)
Triple (SM.SA.APC)	4	7	41	9	61	71 (23)
Double (SM.SA)	1	.	.	.	1	
Single (SM)	5	.	.	.	5	
(SA)	4	.	.	.	4	
Sensitive	89	104	37	3	234	(77)
Total	103	111	78	12	304	(100)

a) Drug resistance was determined by an agar dilution method described prviously (Terakado et al., 1974).

b) Abbreviations; SM, streptomysin; SA, sulfonamide; APC, aminobenzyl panicillin.

800 μg/ml), sulfonamide (SA, more than 1,600 μg/ml) and aminobenzyl penicillin (APC, 200 μg/ml). The results are shown in Table 1. Triple (SM.SA.APC)-resistant strains were isolated most frequently and followed by single (SM or SA)-resistant and double (SM.SA)-resistant strains in that order.

Next, the incidence of R factors in the resistant strains was examined. The detection of R factors in *B. bronchiseptica* was done by mixed cultivation method described previously (Terakado et al., 1974). The results are shown in Table 2. Of 71 drug-resistant strains, 61 (86%) were found to carry R factors which were capable of conjugal transfer and all of these R factors had the triple (SM.SA.APC)-resistant markers.

TABLE 2

Incidence of R factors among resistant strains of Bordetella bronchiseptica isolated from pigs[a)]

Pattern of drug resistance[b)]	No. of strains tested	No. of R^+ strains (%)	Resistance pattern of R factor
SM.SA.APC	61	61 (100)	SM.SA.APC
SM.SA	1	0	—
SM	5	0	—
SA	4	0	—
Total	71	61 (86)	

a) Detection of R factors was done by a mixed cultivation using *E. coli* ML1410 as a recipient strain (Terakado et al., 1974).

b) Abbreviation of drugs is the same as in Table 1.

PROPERTIES OF THE R FACTORS FROM B. BRONCHISEPTICA

Drug-resistant marker. As described above, all of the R factors derived from *B. bronchiseptica* had the triple (SM.SA.APC)-resistant markers, and no other resistant markers were detected.

Fi (fertility inhibition) type. The *fi* character of the R factors was examined by the method reported previously (Terakado et al., 1974). All of 61 R factors tested did not reduce the plating efficiency of the male specific f2 phage and were identified as fi^- type.

Phage restriction. The phage restriction by the R factors was examined by the double layer method (Adams, 1959), using *Escherichia coli* 1100, F^- nal^r (nalidixic acid resistant) as an indicator strain. In this experiment, phages λ, ∅80, P1, P2, T1, T3, T6, T7, W31

TABLE 3

Phage restriction by R factors derived from Bordetella bronchiseptica[a)]

No. of R factor tested	Phage restriction	Phage									
		λ	∅80	P1	P2	T1	T3	T6	T7	W31	BF-23
61	Yes	0	0	0	0	0	0	0	0	0	0
	No	61	61	61	61	61	61	61	61	61	61

a) Restriction was recorded as No when the plating efficiency of the phages was the same (or within 10^{-1}) as on the R^- host.

and BF-23 were employed. As shown in Table 3, all 61 R factors tested did not reduce the plating efficiency of the phages, indicating that the R factors from *B. bronchiseptica* did not confer the character of phage restriction on their host bacteria.

Macarbomysin (MM) sensitivity. It was reported by Mitsuhashi et al. (1970) and Iyobe et al. (1971) that MM could preferentially inhibit *E. coli* strains carrying R factors; the R

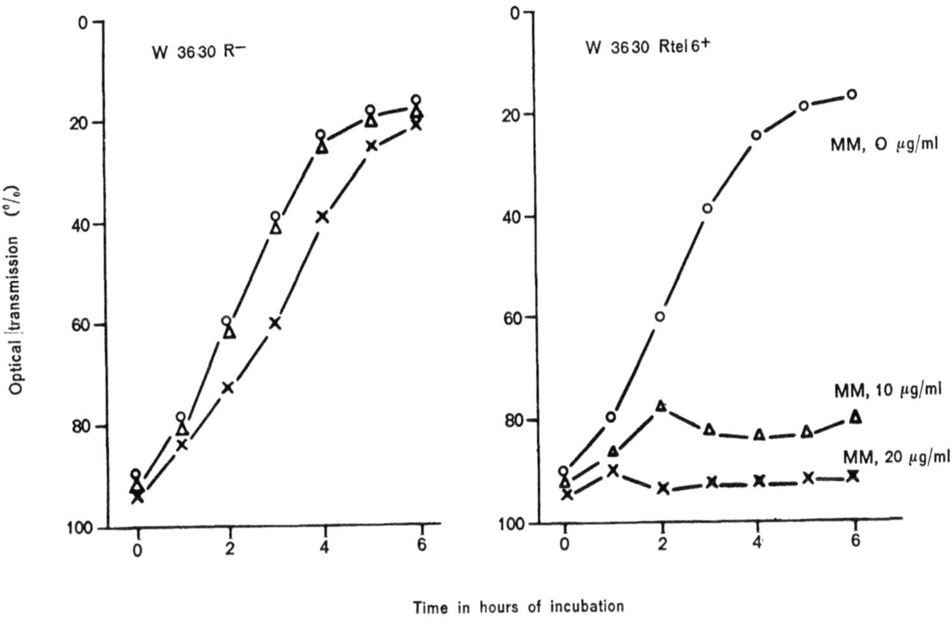

Fig. 1. Effect of macarbomysin (MM) on the growth of W3630 with or without Rte16. In each test, 0.05 ml of overnight culture of W3630 was added to 10 ml of pennasay broth with or without MM. Cultivation was done with shaking at 37 C. Bacterial growth was examined electrophotometrically at 660 mµ.

factors being able to confer MM sensitivity on their host cells when infected. In this experiment, *E. coli* W3630, F^- *mal*$^-$ was used as a host strain of R factors to be tested. The method to determine MM sensitivity was described previously (Mitsuhashi et al., 1970). As shown in Fig. 1, the growth of W3630 R^- was not inhibited by 10 µg/ml of MM and was slightly inhibited by the addition of 20 µg/ml of MM. In contrast, the growth of W3630 R^+ was completely inhibited by 20 µg/ml of MM and was also markedly inhibited by 10 µg/ml of MM. Thus all 61 R factors from *B. bronchiseptica* were concluded to confer MM sensitivity on their host cells when infected. Fig. 1.

Compatibility. The test of compatibility was done by the method reported by Coetzee et al. (1972) and Chabbert et al. (1972). An Rte16 was used as a representative R factor, and the standard R factors for compatibility test were kindly supplied by N. Datta. *E. coli* ML1410, F^- *met*$^-$ *nal*^{r} was used as a donor strain of Rte16 and ML1410-*rfp*^{r} (rifampicin resistant), i.e., a mutant of ML1410, was used as a recipient strain. The results are shown in Table 4. When Rte16 was transferred to the recipient carrying RP4, the exclusion of the resident RP4 was always observed. When the selection was done by both SM and kanamycin (KM), the exconjugants carrying both R factors could be obtained

TABLE 4

Incompatibility of Rte16 with various R factors

R factor[a])	Recipient[b]) carrying		Selectors[c])	Transfer frequency[d])	Characters of purified exconjugant cells
	—		SM+RFP	3.3×10^{-4}	
			APC+RFP	3.0×10^{-4}	
	RP4	(P)[e])	SM+RFP	1.3×10^{-4}	20/20 Rte16 only
			SM+KM	1.1×10^{-7}	10/10 both R factors, but unstable[f])
	R40a	(C)	SM+RFP	3.9×10^{-4}	20/20 both R factors, but unstable
	RA1	(A)	SM+RFP	4.2×10^{-4}	19/20 both R factors 1/20 Rte16 only
Rte16	R386	(F1)	SM+RFP	4.1×10^{-4}	20/20 both R factors
	R100	(F11)	APC+RFP	1.2×10^{-3}	20/20 both R factors
	R391	(J)	SM+RFP	3.2×10^{-4}	20/20 both R factors
	R387	(K)	APC+RFP	1.0×10^{-3}	20/20 both R factors
	R446b	(M)	APC+RFP	1.4×10^{-4}	18/20 both R factors 2/20 Rte16 only
	RN3	(N)	APC+RFP	5.3×10^{-4}	20/20 both R factors
	Rtsl	(T)	SM+RFP	8.6×10^{-5}	20/20 both R factors
	R6K	(X)	SM+RFP	2.0×10^{-4}	20/20 both R factors

[a]) *E. coli* ML1410 was used as a donor strain of Rte16.

[b]) *E. coli* ML1410-rfp^r was used as a recipient strain.

[c]) Abbreviation of drugs: RFP, rifampicin; KM, kanamycin; others, see footnote in Table 1.

[d]) It was expressed as the number of exconjugants per that of donors after mixed culture at 37 C (or 30 C for Rts1) for 2h.

[e]) Figures in parenthesis indicate the name of compatibility groups of each R factor.

[f]) Unstable; high frequent loss of R factors observed in drug-free medium.

at a very low frequency. But both R factors could not stably coexist and Rte16 was mainly eliminated from the exconjugants carrying both R factors after one or two subcultures in the drug-free medium. On the contrary, the frequency of spontaneous loss of Rte16 was about less than 2–3% of the host cells carrying only Rte16 under the same condition, indicating that Rte16 was incompatible with RP4 (compatibility group P). It was also found that Rte16 was also incompatible with R40a (compatibility group C). When Rte16 was transferred to R40a$^+$ recipient, no exclusion of R40a was observed, but both R factors could not stably coexist; Rte16, in many cases, being lost from the exconjugants with frequencies of 70–90% after 10 subcultures in the drug-free medium. From these results, it could be concluded that Rte16 belongs to two compatibility groups P and C and is compatible with the R factors of groups A, F1, F11, J, M, N, T and X.

Host range of Rte16. In transfer experiments, *E. coli* ML1410 was used as a donor strain of Rte16 and rifampicin-resistant mutants of various species of bacteria were used as recipients. Mixed culture was done by the method described previously (Terakado et al., 1973). When the frequency of transfer was very low, the mixed culture for 24h was used. In each mixed culture, the death of the mating partner was not seen. As shown in Table 5, Rte16 was conjugally transmissible to *B. bronchiseptica*, *E. coli*, *Citrobacter freundii*, *Salmonella typhimurium* and *Yersinia enterocolitica* at a frequency of 10^{-4} to 10^{-7}, but did not to *Shigella flexneri*, *Sh. sonnei*, *Proteus mirabilis*, *P. vulgaris*, *P. rettgeri*, *Klebsiella pneumoniae* and *P. aeruginosa*. No segregation of the resistant markers was detected when transferred. It was found that the level of SM-resistance of Rte16 was very high (more than 800 μg/ml) in *B. bronchiseptica*, but rather low (50–100 μg/ml) in *E. coli*, *C. freundii*, *S. typhimurium* and *Y. enterocolitica*. The level of APC-resistance was also low (25–100 μg/ml) in the latter four species, but SA-resistance was similarly high (more than 1,600 μg/ml) among all strains tested. It was also found that Rte16 was fairly unstable in *C. freundii* but stable in *B. bronchiseptica*, *E. coli*, *S. typhimurium* and *Y. enterocolitica*.

TABLE 5

Transfer of Rte16 to various bacteria

Mixed culture[a]		Transfer frequency[b]: Time in mixed culture		Drug resistance which was transferred			
R factor[c]	Recipient[d]	2h	24h	Resistance pattern	Level[e] of resistance to: SM	SA	APC
	B. bronchiseptica 147	10^{-4}	10^{-2}	SM.SA.APC	>800	>1,600	200
	E. coli ML1410-*rfp*ʳ	10^{-4}	10^{-2}	SM.SA.APC	50	>1,600	50
	C. freundii GN587	10^{-6}	10^{-3}	SM.SA.APC	50	>1,600	100
	S. typhimurium LT-2	10^{-6}	10^{-3}	SM.SA.APC	50	>1,600	100
	Y. enterocolitica 7–3	10^{-7}	10^{-4}	SM.SA.APC	200	>1,600	25
Rte16	*Sh. flexneri* YSH-1	$<10^{-8}$	$<10^{-8}$	—	—	—	—
	Sh. sonnei 13389	$<10^{-8}$	$<10^{-8}$	—	—	—	—
	P. mirabilis GN3167	$<10^{-8}$	$<10^{-8}$	—	—	—	—
	P. vulgaris GN3231	$<10^{-8}$	$<10^{-8}$	—	—	—	—
	P. rettgeri GN3213	$<10^{-8}$	$<10^{-8}$	—	—	—	—
	K. pneumoniae GN402	$<10^{-8}$	$<10^{-8}$	—	—	—	—
	P. aeruginosa M1292	$<10^{-8}$	$<10^{-8}$	—	—	—	—

a) Mixed culture was done by the method described previously (Terakado et al., 1973).
b) It was expressed as the number of exconjugants per that of donors.
c) *E. coli* ML1410 was used as a donor strain of Rte16.
d) All of the recipient strains used were rifampicin-resistant mutants.
e) Minimum inhibitory concentration, μg/ml.

β-lactamase specified by Rte16. Yamagishi et al. (1969) and Sawai et al. (1970) reported two types of β-lactamase whose synthesis was specified by R factors. These β-lactamases differ from each other in their physicochemical, enzymological and immunological properties, and were designated type 1 and 11 β-lactamase, respectively. Therefore, we tried to clarify the properties of the β-lactamase mediated by Rte16. *E. coli* W3630 was used as a donor strain of Rte16 and the assay procedures for the enzyme were the same as that of Yamagishi et al. (1969) and Sawai et al. (1970). The results are shown in Table 6, suggesting that Rte16-β-lactamase apparently differs from those of RGN14- and RGN238-β-lactamase. The Rte16-β-lactamase was found to be markedly active against phenethicillin (PE-PC) and cloxacillin (MCI-PC) but not against cephaloridine (CER). In contrast, the RGN14-β-lactamase was not active against PE-PC and MCI-PC but active against CER. Furthermore, marked differences were also observed between these two enzymes in their molecular weight, isoelectric point and effect of inhibitors. The molecular weight of Rtel6- was very large (46,000), and this enzyme was not inhibited by sodium chloride. On the other hand, there were similarlities between the two enzymes specified by Rte16 and RGN238, in their optimal pH, isoelectric point and effect of inhibitors. However, Rte16-β-lactamase differed markedly from RGN238-β-lactamase in molecular weight and optimal temperature. In addition, there were some differences between these two enzymes in their activities against APC, PE-PC and 6-aminopenicillanic acid (6-APA). It is concluded, therefore, that the two enzymes are different from each other.

TABLE 6

Summary of the comparison of R-mediated β-lactamases with types 1 and 11 β-lactamase

Properties [b]	R factor [a]		
	Rte16	RGN14 (type 1)	RGN238 (type 11)
Substrate specificity			
PC-G	100	100	100
APC	131	115	450
PE-PC	449	33	150
6-APA	100	87	260
MCI-PC	310	2	300
CER	23	130	36
Molecular weight	46,000	20,600	25,400
Optimal pH	7.0–7.5	6.5–7.0	7.6
Optimal temperature	40 C	45 C	30 C
Isoelectric point	8.3	5.1	8.3
Effect of inhibitors			
Sodium chloride	+	—	+
p-CMB	—		—

[a]) The results for RGN14 and RGN238 are from Yamagishi et al. (1969) and Sawai et al. (1970).

[b]) Abbreviations; PC-G, benzyl penicillin; APC, aminobenzyl penicillin; PE-PC, phenethicillin; 6-APA, 6-aminopenicillanic acid; MCI-PC, cloxacillin; CER, cephaloridine; p-CMB, p-chloromercuribenzoate.

DISCUSSION

The results presented in this paper show that the R factors are the main cause of drug resistance in *B. bronchiseptica*, and that the R factors derived from *B. bronchiseptica* have the quite homogeneity in their genetical properties, such as resistant markers, fertility inhibition, phage restriction and macarbomycin sensitivity. It should be emphasized that all of them had the triple (SM.SA.APC)-resistant markers and no other resistant markers were detected. Although the high isolation frequency of R factors is due to selection by a wide use of drugs, it is of interest that the resistant markers of the R factors detected in *B. bronchiseptica* do not reflect the kinds of drugs used on livestock in Japan. Tetracycline (TC), SM and SA are widely used but the use of APC is very rare, because this drug has not been permitted for animal use up to the end of 1972 in Japan. Comparing with many R factors detected from *Enterobacteriaceae* and *P. aeruginosa* strains, the R factors of *B. bronchiseptica* can be characterized by the lack of TC-resistant marker but by the presence of APC-resistant marker.

Incompatibility experiments seem to be a reliable method for the classification of R factors. Several classes (or compatibility groups) have been defined; the members of the same group being incompatible with each other but can stably coexist with R factors of other groups (Chabbert et al., 1972, Datta and Hedges, 1971, 1972, Hedges et al., 1973). Since Rte16 derived from *B. bronchiseptica* is incompatible with both RP4 and R40a, it belongs to two compatibility groups P and C. The reason why Rte16 is incompatible with the two R factors of different groups, is not clear. A possibility of the presence of two R factors in one host cell, that is hetero-R state (Tanaka et al., 1967), can be excluded, because the triple (SM.SA.APC)-resistant markers of Rte16 was transmissible "*en bloc*" by either conjugation or transduction with phage P1.

Both RP4 and R40a were originally detected in *P. aeruginosa* strains and had an exceptionally wide host range. Datta et al. (1971) and Datta and Hedges (1972a, b) reported that the R factors of compatibility groups P and C could be transferred to *P. aeruginosa* as well as to *E. coli*. However, Rte16 from *B. bronchiseptica* was not transmissible to *P. aeruginosa*, although this R factor belonged to compatibility groups P and C. It was also found that Rte16 gives low level of resistance to APC when presented in *E. coli*. In contrast, the level of APC-resistance mediated by RP4 and R40a was very high (more than 1,600 μg/ml), indicating that the origin of APC-resistant gene or Rte16 may be different from these two R factors.

Comparison of the β-lactamase Rte16 with RGN14 (type 1) and RGN238 (type 11)--β-lactamase indicates that these three enzymes are different from each other. The marked difference was observed in their molecular weight. The molecular weight of Rte16-β-lactamase was very large (46,000) comparing with those (20,600–25,400) of the latter two types. Also in substrate specificity and other properties, there were some differences among these three enzymes. It is strongly suggested that the β-lactamase specified by Rte16 belongs to one type of enzyme different from the other two types.

Dale and Smith (1971) reported that the β-lactamase R1818 (now termed R46) has the large molecular weight (44,000). Furthermore, many similarities were also observed between two enzymes specified by Rte16 and R1818 in their substrate specificity, optimal pH, isoelectric point and effect of inhibitors. However, R1818 was originally detected in *S. typhimurium* and belonged to compatibility group N (Datta and Hedges, 1971), indicating that the two R factors may be different from each other in the origin of their APC-resistant genes.

ACKNOWLEDGEMENTS

We are grateful to Dr. N. Datta, Royal Postgraduate Medical School, for furnishing the set of standard R factors in compatibility experiments.

REFERENCES

ADAMS, M. H. (1959). Bacteriophages. Interscience Publishers.

BREED, R. S., MURRAY, E. G. D. and SMITH, N. R. (1957). Bergey's manual of determinative bacteriology. 7th ed., Williams & Wilkins Co., Baltimore.

CHABBERT, Y. A., SCAVIZZI, M. R., WITCHITZ, J. L., GERBAUD, G. R. and BOUANCHAUD, D. H. (1972). Incompatibility groups and the classification of *fi*⁻ resistance factors. J. Bacteriol. 112, 666.

COETZEE, J. N., DATTA, N. and HEDGES, R. W. (1972). R factors from *Proteus rettgeri*. J. Gen. Microbiol., 72, 543.

DALE, J. W. and SMITH, J. T. (1971). The purification and properties of the β-lactamase specified by the resistance factor R-1818 in *Escherichia coli* and *Proteus mirabilis*. Biochem. J., 123, 493.
DATTA, N. and HEDGES, R. W. (1971). Compatibility groups among *fi*$^-$ R factors. Nature (London). 234, 222.
DATTA, N. and HEDGES, R. W. (1972 a). Host ranges of R factors. J. Gen. Microbiol., 70, 453.
DATTA, N. and HEDGES, R. W. (1972 b). R factors identified in Paris, some conferring gentamicin resistance, constitute a new compatibility group. Ann. Inst. Pasteur., 123, 849.
DATTA, N., HEDGES, R. W., SHAW, E. J., SYKES, R. B. and RICHMOND, M. H. (1971). Properties of an R factor from *Pseudomonas aeruginosa*. J. Bacteriol., 108, 1244.
HEDGES, R. W., DATTA, N., COETZEE, J. N. and DENNISON, S. (1973). R factors from *Proteus morganii*. J. Gen. Microbiol., 77, 249.
IYOBE, S., MITSUHASHI, S. and UMEZAWA, H. (1971). Relationship between sex-pili formation and macarbomysin sensitivity in *Escherichia coli*. J. Bacteriol., 108, 946.
MITSUHASHI, S. (1971). Transferable drug resistance factor R. University of Tokyo Press.
MITSUHASHI, S., IYOBE, S., HASHIMOTO, H. and UMEZAWA, H. (1970). Preferential inhibition of the growth of *Escherichia coli* strains carrying episomes. J. Antibiot., 23, 319.
SAWAI, T., TAKAHASHI, K., YAMAGISHI, S. and MITSUHASHI, S. (1970). Variant of penicillinase mediated by an R factor in *Escherichia coli*. J. Bacteriol., 104, 620.
TANAKA, T., HASHIMOTO, H., NAGAI, Y. and MITSUHASHI, S. (1967). Drug resistance of enteric bacteria. XI. Isolation of Shigella strains in hetero-R state. Jap. J. Microbiol., 11, 155.
TERAKADO, N., AZECHI, H., NINOMIYA, K. and SHIMIZU, T. (1973). Demonstration of R factors in *Bordetella bronchiseptica* isolated from pigs. Antimicrob. Ag. Chemother., 3, 555.
TERAKADO, N., AZECHI, H., NINOMIYA, K., FUKUYASU, T. and SHIMIZU, T. (1974). The incidence of R factor in *Bordetella bronchiseptica* isolated from pigs. Jap. J. Microbiol., 18, 45.
YAMAGISHI, S., OOHARA, K., SAWAI, T. and MITSUHASHI, S. (1969). The purification and properties of penicillin β-lactamases mediated by transmissible R factors in *Escherichia coli*. J. Biochem., 66, 11.

SOME PROBLEMS OF TRANSMISSIBLE RESISTANCE IN PRACTICE

V. CHALOUPECKÝ
Institute of Hygiene and Epidemiology, Prague, Czechoslovakia

Our work aimed to provide District Hygiene Stations with a standard method for estimating R-factor transmission frequency as well as with the most suitable recipient strain. It happens only too often that the estimation of the frequency with which R-factors are transmitted is done after overnight or very long contact of donor and recipient cells. Moreover, various methods and media are used. Trying to avoid the most serious errors arising in this way and to enable comparison of the results attained in different laboratories, we faced three main questions:

1. Which bacterial strain shall be used as recipient of R-factors?
2. What time of incubation of the conjugation mixture is necessary?
3. Which media and which concentrations of antibiotics are the most suitable for selection of R^+ transconjugants?

These questions are by no means new but we felt that certain aspects of them remain to be answered. The recipient strain shall receive an R-factor with high frequency, it must be easily distinguishable from the donor, produce no bacteriocines or enzymes which can alter the result, harbour no sex factor, no R-factor and, preferably, no prophage.

By comparing several strains of *E. coli* and one of *Citrob. freundii* it was found, as had been supposed, that the best recipient — at least for *E. coli* R-factors, are *E. coli* K 12 F^- strains. In order to easily distinguish transconjugants we used also *Citrobacter freundii* 0 7, 3b, 1c but this was not the best choice since it was later demonstrated that this strain, insensitive to colicines and, according to common diffusion tests, to col^- within 30 min. kills under mixed incubation 60–80% of the donor population by some bacteriocine previously not detected (Chaloupecký, in preparation). When the frequency of transfer is expressed as a number of R^+ transconjugant colonies in relation to the number of the donor colonies, an incorrect result, i.e. a higher frequency, is thus obtained. Another problem, the restriction of R-factors transmitted from various "wild" *E. coli* strains to *E. coli* K 12 seems to be at present at least partly elucidated. It was shown that at least some R factors can be restricted (Chaloupecký, prepared for publication) to a similar extend as F'lac (Glover and Colson, 1966).

On the basis of this we suppose that *E. coli* K 12 F^- r^- is the best recipient strain. Difficulties caused by colicines can be solved by addition of 10^{-3} (w/v) of trypsin both in conjugation and selective media.

E. coli C_b proved to be a poor recipient.

2. When donor and recipient strains are mixed, two events occur:
a) pairs or clusters of cells of opposite mating type are formed
b) R-factor is transferred

TABLE 1

The frequency of R-factor transmission in various E. coli strains

Donor strains	Recipient strains					
	B_{s-1}	C_b	JE 189	5K	C600	B
B_2 ($R_{1-19drd}$)	$3.9 . 10^{-6}$	$1.0 . 10^{-5}$	$1.0 . 10^{-4}$	$1.3 . 10^{-3}$	$2.8 . 10^{-4}$	—
W945 ($R_{1-19drd}$)	$5.7 . 10^{-4}$	$1.8 . 10^{-4}$	$1.1 . 10^{-1}$	$1.0 . 10^{-1}$	$1.8 . 10^{-1}$	$9.7 . 10^{-4}$

Characteristics of *E. coli* strains: *E. coli* K 12 C600 F^-, thr^-, leu^-, thi^-, $r_K^+m_K^+$; *E. coli* K 12 5K F^-, thr^-, leu^-, thi^-, $r_K^-m_K^+$; *E. coli* W945 F^-, pro^-, thi^-; *E. coli* K 12 JE 189 F^-, met^-, pro^-; *E. coli* B try^-; *E. coli* B_{s-1} prototroph^0 *E. coli* C_b prototroph.

a) The speed of pairing depends on a number of cell-collisions in the mating suspension, i.e. on cell concentration. When the number of cells in 1 ml of the mixture reaches or exceeds 10^8, the yield of R^+ transconjugants and consequently perhaps the number of collisions does not notably change by a further increase in the number of cells, if the strains are allowed to be in contact for 5 min. It seems that pairs are formed in the mixture within the first minute. In order to avoid any damaging of recipient (Skurray and Reeves, 1973), which might occur by conjugation with derepressed R-factor-donor, it is advisable to keep the donor recipient ratio at about 1 : 9.

b) R-factor transmission is a rapid process and has been so far impossible to interrupt the genom mechanically during transmission. On the other hand, the expression of markers introduced by R-factor takes some time. Cells of *E. coli* K 12 grown in glucose minimal medium supplied with aminoacids express chloramphenicol resistance in the 10th min. and streptomycin resistance in the 8th min. after mixing with the donor of R1–19rd (Chaloupecký, 1972). R-pili of transconjugants are unable to mediate the R-factor transfer earlier than 15–25 min. after acquiring the R-factor (Salzman, 1971) and the newly obtained R1–19rd factor can be again transferred from irradiated recipient cells only 20–30 min. after its entering the cell (Falkow et al., 1971).

The time of conjugation must therefore be short, so as to avoid transmission of the newly acquired R-factor from R^+ transconjugants, but sufficient to enable proper pairing and undisturbed R-factor transmission. According to the data mentioned above, this time shall not exceed 15 min. The short time also minimizes the possibility of a donor to be in contact with more recipient cells than one.

The process of pairing may be stopped by the destruction of the donors or by the prevention of further collisions. We prefer the prevention of collisions, as the destruction of donors is never immediate and always introduces additional factors into the system. The simplest way is the dilution of the donor-recipient mixture. But this method, too, has two disadvantages:

a) the sudden change of the environment of mating cells,

b) the necessary dilution (1000 times) can reduce the number of R^+ transconjugants in a unit volume under the detectable level when working with repressed R-factors transmitted with a frequency of about 10^{-4} (to a number of donor cells).

We therefore used another method. 0.1 ml of the conjugation mixture is after a relatively short time (3–5 min.) of conjugation applied by means of a pipet on a membrane filter placed on solid medium containing no antibiotics. After 20–30 min. of incubation the filter is transferred on a selective medium where only R^+ transconjugants are able to grow. This procedure has the following advantages:

a) only one colony of R^+ transconjugants will grow from one cell of the donor and one (or more) cell(s) of the recipient in contact, in spite of cell multiplying and contingent further mating with neighbouring cells;

b) all R-factor-controlled genetical markers can be expressed freely before the contact of the transconjugant cell with antibiotic.

The estimation of R-factor transmission frequency after several hours or even overnight growth in a conjugation mixture, when the culture very probably became a HFRT system (Watanabe, 1963) and transconjugants are multiplying, can be considered to be quite incorrect.

3. For the selection of transconjugants a solid medium with antibiotic(s) is mostly used. When testing resistance levels against chloramphenicol, streptomycin and tetracycline in R^+ *E. coli-strains* isolated in hospitals, we found that chloramphenicol and tetracycline R-mediated-resistance is sufficiently high (MIC 256 or more μg in 1 ml and 64 or more μg in 1 ml respectively) to enable selection of R^+ cells on a medium with 25 μg of antibiotic in 1 ml, but in the case of streptomycin the obtained resistance was in most cases rather low and, therefore, concentration of 5 μg in 1 ml only is advisable (Chaloupecký, 1974). A check of R^+ status of the colonies which have grown on this selective medium is then probably almost necessary.

We believe that testing of the R-factor transmissibility by using *E. coli* K 12 F^-r^- strains as recipients will be quite useful and that a method of estimating the R-factor transfer frequency on membrane filters, though not the simplest one, would increase the accuracy of estimation and can be used even by field laboratories.

REFERENCES

Chaloupecký, V. (1972). The kinetics of expression of antibiotic resistance in *E. coli* K 12. J. Hyg. Epidemiol. Microbiol. Imm. 16, 124.

Chaloupecký, V. (1974). R-factor-induced resistance levels. Zbl. Bakt. (in press).

Falkow, S., Tompkins, L. S., Silver, R. P., Guerry, P., Leblanc, D. J. The replication of R-factor DNA in *Escherichia coli* K-12 following conjugation. Ann. N.Y. Acad. Sci. 182, 153.

Glower, S. W., Colson, C. (1966). Stable and unstable alterations of the host-induced modification properties of *Escherichia coli* B, K and C. Genet. Res. 7, 223.

Salzman, T. C. (1971). Coordination of sex pili with their specifying R factors. Nature New Biology 230, 278.

Skurray, R. A., Reeves, P. (1973). Characterization of lethal zygosis associated with conjugation in *Escherichia coli* K-12. J. Bacteriol. 113, 58.

Watanabe, T. (1963). Infective heredity of multiple drug resistance in bacteria. Bacteriol. Rev. 27, 87.

EPIDEMIOLOGICAL AND CLINICAL PROBLEMS ATTENDING DRUG RESISTANCE

F. VÝMOLA

Institute of Hygiene and Epidemiology, Prague, Czechoslovakia

There is a constantly growing number of bacterial infections that resist chemotherapy; this applies both to acute forms and, much more so, to chronic processes. Since failure in treating these diseases with antibiotics need not be caused only by resistance of the ethiological agent, let us consider to what extent the situation is further complicated if the infection is caused by a multiple-resistant microbe.

Microbiologists need not be reminded that the cause of drug resistance developing in most bacterial species is the large-scale and indiscriminate use of antibiotics and other chemotherapeutics in medical and veterinary practice, whether to treat infections, for prophylactic purposes, or as additives to fodder in intensive livestock raising, as is still common in some places. On the other hand, the fact must be over and over again called to the attention of clinicians, agricultural workers and, where there is free sale of antibiotics, also the general public. We should not forget that antibiotics not only alter the original state of sensitivity in many obligatory pathogens, but to a considerable degree also influence the microbial ecology in man and animals. Moreover, they may unfavourably affect the general immunity in individuals and the whole population.

In considering the role of transferable resistance, we are worried most of all today by infections elicited by opportune bacteria, formerly regarded as saprophytes that carry so-called multiple-resistance patterns. Their share in the development of the infectious process is most often seen in health establishments, as a result of numerous medical procedures and operations that elicit microtraumas even if delicately performed, so that the infective agent easily metastasizes into such exposed sites notwithstanding even strict personal and environmental hygiene precautions. This may also explain the development of many autoinfections provoked by common commensals that colonize the alimentary, respiratory or urogenital tract, especially where the resistance of the macroorganism of the patient is reduced. It is obvious that successful control of these infections is considerably complicated if their agents are resistant to antibiotics.

If we want to cure resistant infections today, we cannot do without team work in which the clinician collaborates with the microbiologist, immunologist, biochemist, hygienist and epidemiologist. It is imperative for the near future to train quite a number of specialists who will determine the tactics and strategy of using antibiotics on a wide scale. What has to be done, therefore, is not merely to observe the rules of a rational chemotherapy, by which we understand the most advantageous exploitation of the therapeutic properties of antibiotics in relation to individual infectious diseases, but to establish general control over the employment of antibiotics, which means primarily the application of effective methods in combating the expansion of drug resistance. The present meeting should provide a stimulus for this problem to be taken up, in cooperation with the World Health Organization, on a global and not merely local scale.

A basic prerequisite for solving the complicated epidemiological and clinical problem of drug resistance is to keep the developments in and causes of drug resistance, under systematic surveillance, using unified methods. This relates especially to sensitivity tests and the adoption of sensitivity and resistance criteria, as well as a unified system of documentation. Modern surveillance of drug resistance cannot do without computerized data processing. Adequately drawn-up computer programmes are thus becoming an important aid not only in rational chemotherapy but also in antibiotic prophylaxis. They will help us determine priorities for the most useful preparations both for the present and the future and will thus also exert an influence on production trends. Drug resistance surveillance conceived in this way will help determine a useful antibiotic policy by signalizing in time which factors unfavourably influence the resistance trends, especially with reference to hospital facilities.

Most frequent RESISTANCE PATTERN
(in the hospitals 1973)

St. aureus	PNC, TC, Cl, Er, Ka	28%
Pseudomonas	CoT, TC, Cl, STM, Ne, Ca, Co	30%
Klebsiella	CoT, Cl, TC, STM, Ka, Fu	26%

Dangerous RESISTANCE PATTERN in

Ps. aeruginosa	Ca, Ne, Ge, Co, Po
Proteus mirabilis	Am, Ce, Ge, Fu, Na
Esch. coli (serotype 0111)	CoT, A, Ce, Ge, Co, Fu, Na
Klebsiella *Enterobacter*	Ce, Ge, Co, Na, CoT
St. aureus	r-PNC, Sp, Li, Ge, Va, Fc

In conclusion I should like to show on two slides, first, our findings of the most frequent resistance patterns in some bacterial species in Prague hospitals in 1973 and, second, some so-called dangerous resistance patterns in *Pseudomonas, Proteus mirabilis, Esch. coli, Klebsiella* and *Staphylococcus aureus.* The drawing up of these resistance patterns was made possible precisely by the use of computers. In our conditions this means, for instance, that the occurrence of these resistance patterns, even if extremely rare at present completely thwarts successful chemotherapy so that the life of the patients may be seriously endangered. If a bacterial agent bearing multiple drug resistance causes a nosocomial outbreak, tragedy is imminent. With this in view we have been insisting with constantly growing emphasis that today an integral part of a strict hygienic regime in hospital facilities is an adequate regime of handling and administering antibiotics.

IN VIVO EFFECT OF MACARBOMYCIN ON ESCHERICHIA COLI STRAINS CARRYING R FACTOR

Y. NAGAI*, T. FUKUYASU**, Y. YUSA**, Y. SHEKIZAWA**
and S. IYOBE***

*Central Institute for Feed and Livestock Research Laboratory, (ZEN-NOH)
Tsukuba-machi, Ibaragi, Japan

**Laboratory of Biology, Research Laboratories, Meiji Seika Kaisha, Ltd,
Yokohama, Japan

***Department of Microbiology, School of Medicine, Gunma University,
Maebashi, Japan

Macarbomycin is a new antibiotic containing phosphorus and inhibits the cell wall synthesis of *Staphylococcus aureus* (Takahashi et al., 1970). Mitsuhashi et al. (1970) reported that *Escherichia coli* strains harboring plasmid such as R or F factor were more sensitive to MM than their original strains without plasmids. It was also reported that the preferential antibacterial activity of MM was parallel to the sex pili formation of bacteria carrying plasmids (Iyobe et al., 1971).

We studied the R-factor transfer among the intestinal flora of specific pathogen-free (SPF) pigs and examined the *in vivo* effect of MM on *E. coli* R^+ strains in intestine.

MATERIALS AND METHODS

Animals. Eight SPF pigs were used. They were separated in 4 groups and each consisted of two pigs. They were fed a sterile feed which was free from antibiotics except macarbomycin (MM). The cages were kept in a clean and air-conditioned room.

Bacteria and phage. The strains used were *Escherichia coli* (S-14) NA^r, *E. coli* (S-14) $R(KM)^+$ of pig origin (NA^r, nalidixic acid resistant; KM, kanamycin sulfate), and *E. coli* K12 ML1410 NA^r. *E. coli* (S-14) NA^r was used for the recipient of R factor. *E. coli* $R(KM)^+$ was used for the donor of R factor. *E. coli* ML1410 (NA^r) was used for *in vitro* experiment as a recipient of R factor. The male bacteria, *E. coli* K12 W3630 F-*cml*$^+$, was used as the indicator strain for male phage infection. The F-*cml* factor is a recombinant between F and the chloramphenicol (*cml*) resistance determinant (Kondo and Mitsuhashi, 1966). Phage f1 was used as a male phage.

Media. DHL agar (Eiken) was used for the isolation of *E. coli* from the fecal specimens of pigs and for testing the *in vitro* transfer of R factor. Brain heart infusion broth (BHI, Difco) was used as a propagating medium for oral inoculating S-14 to pigs as well as for testing the *in vitro* transfer of R factor. L-broth and L-agar (Lennox, E. S. 1955) were used for the assay of sex pili formation.

Drugs. NA (50 mg/ml) and KM (25 mg/ml) were used as slective drugs for the isolation of *E. coli* strains from the fecal specimens of pigs and for the conjugal transfer.

Bacterial infection. Five ml of overnight broth culture (about 10^9 cells/ml) of (S-14) NA^r was administered orally to each pig once a day for four days. At ten days after oral administration of (S-14) NA^r, (S-14) $R(KM)^+$ was administered orally for five days by the same procedure as described above.

MM administration. The feeds which contained 100 mg of MM/kg or 300 mg of MM/kg were prepared and sterilized by ethylene oxide gas. The feed containing MM was administered every day to each pig from four days before or six days after oral inoculation of S-14 $R(KM)^+$.

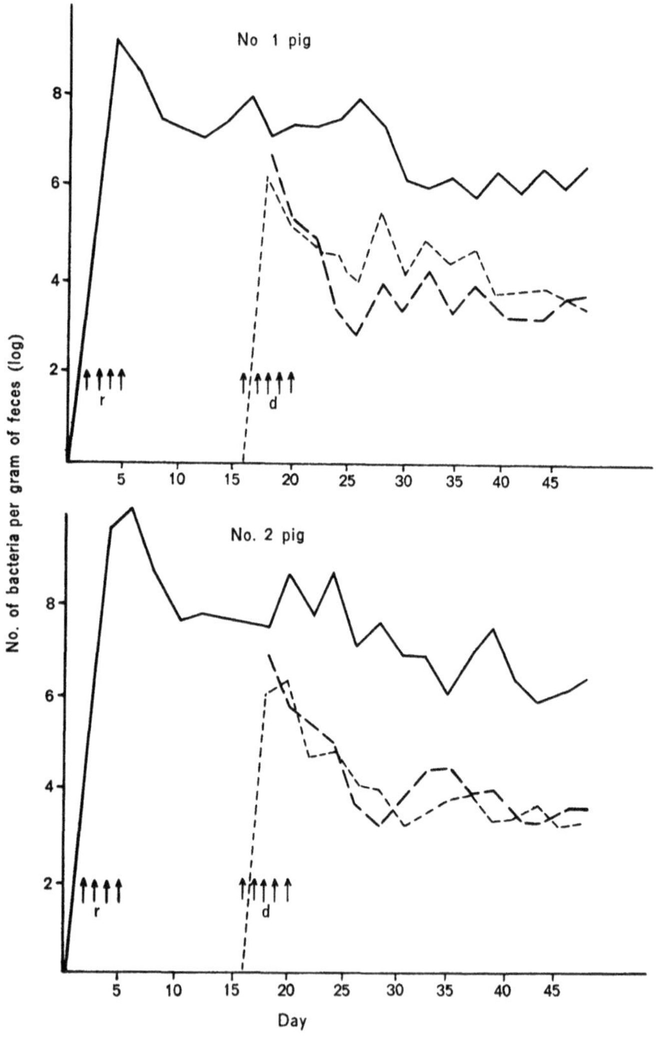

Fig. 1. Daily changes of the number of *E. coli* strains in the intestinal tract of pigs (Group 1, fed with MM-free feed). Five ml of overnight BHI culture of S-14 (NA^r) or S-14 $R(KM)^+$ was orally administered to each pig, and then the number of S-14 (NA^r) (▬▬▬), S-14 $R(KM)^+$ (▬ ▬ ▬ ▬), and exconjugant S-14 (NA^r) $R(KM)^+$ (-------) in faeces were counted on selective medium. Arrows r and d indicate the oral inoculation of S-14 (NA^r) and S-14 $R(KM)^+$, respectively.

Transfer of KM-resistance. Equal volumes (0.5 ml) of overnight broth cultures of (S-14) R(KM)+ and ML1410 NA^r were put into 1 ml of fresh BHI broth and incubated at 37 C. After overnight incubation, the ML1410 (NA^r) which had received KM-resistance was selected on DHL plate containing both NA and KM. Transfer frequency was expressed as a ratio of the number of exconjugants per that of donor cells.

Assay of sex pili formation. Two tenth ml of sample was withdrawn from overnight L-broth culture of bacteria and added to 0.2 ml of phage f1 lysate (10^{10} p.f.u./ml).

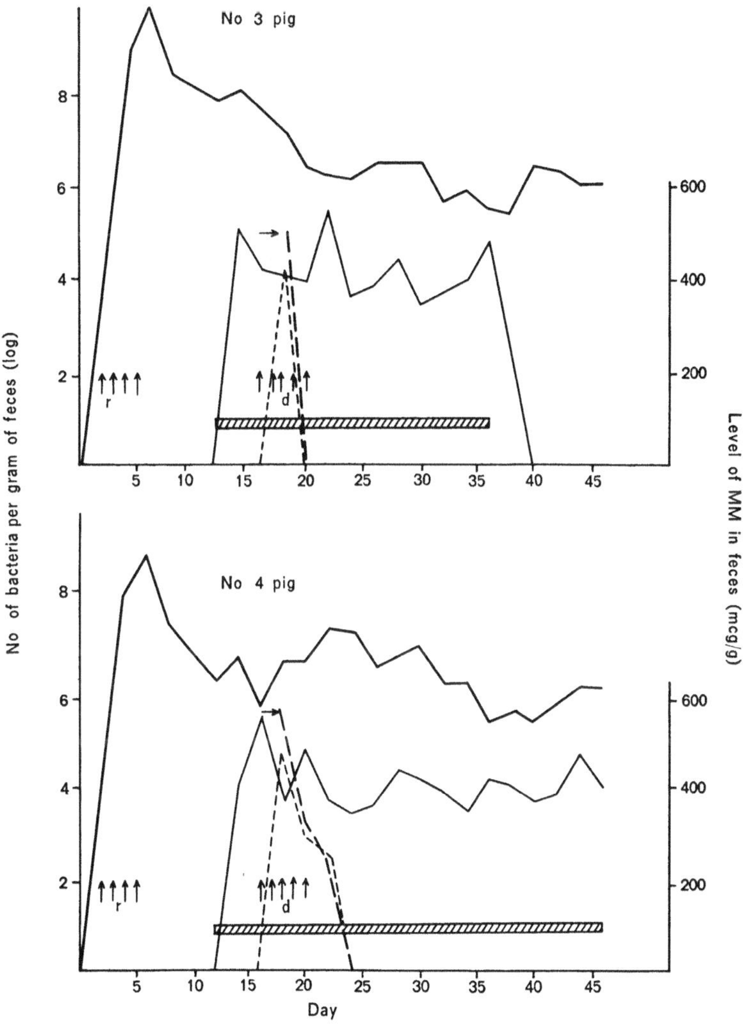

Fig. 2. Daily changes of the number of *E. coli* strains in the intestinal tract of pigs (Group *2*, fed with the feed containing 300 mg of MM/kg). The MM-containing feed (300 mg/kg) was administered from 4 days before the oral inoculation of S-14 R(KM)+, and thereafter every day during the experiment. [hatched bar], the period of the administration of MM-containing feed. The concentration of MM in feces (————) was measured by bioassay. (▬▬▬), (▬ ▬ ▬), (-------), see the footnote of Fig. 1.

After 5 min of adsorption at 37 C, free phages were inactivated with rabbit antiphage serum. The mixture was diluted 100- to 1,000-fold with cold L-broth, and the number of infective centers was counted on a lawn of the indicator strain W3630 F-*cml*+.

RESULTS

Administration of E. coli strains to SPF pigs and transfer of R factor in intestinal tract. About 10^9 cells of *E. coli* (S-14) NA^r was orally administered to pigs once a day for four days. As shown in Fig. 1, the number of NA-resistant bacterial cells increased to the level

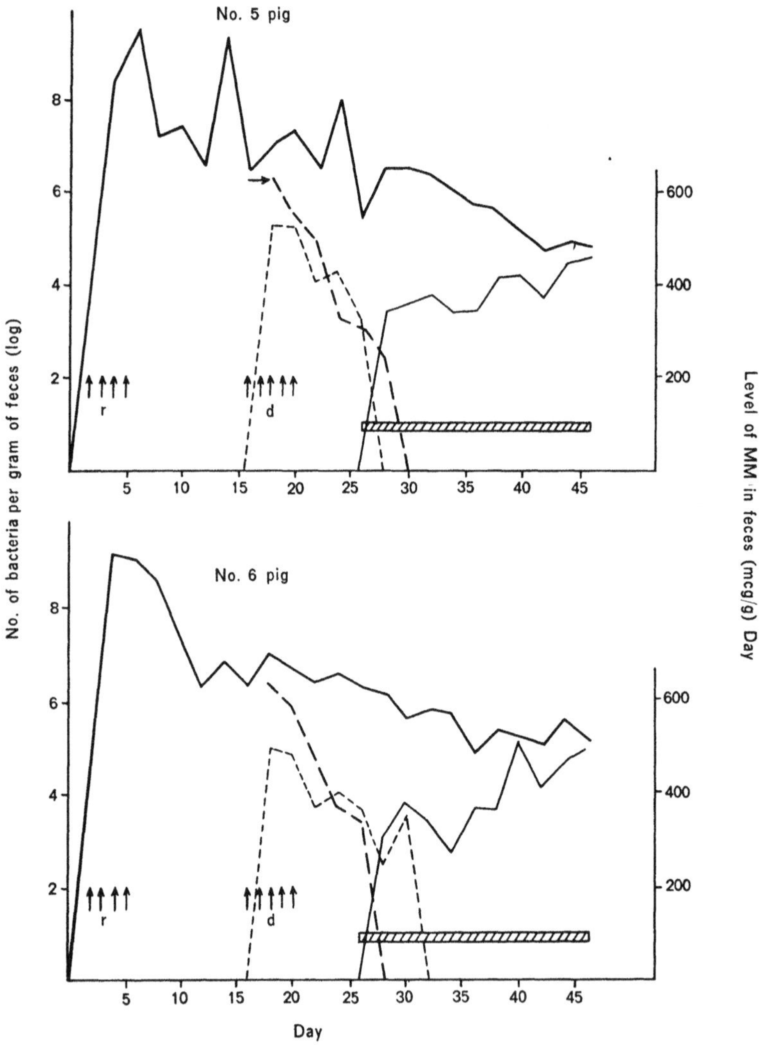

Fig. 3. Daily changes of the number of *E. coli* strains in the intestinal tract of pigs (Group 3, fed with the feed containing 300 mg of MM/kg). The MM-containing feed (300 mg/kg) was administrated from six days after oral inoculation of S-14 R (KM)+ and thereafter every day during the experiment. (▬▬▬), (▬ ▬ ▬), (-------), see the footnote of Fig. 1. and (————), [hatched bar], see the footnote of Fig. 2.

of about 10^{10} cells/g of faeces (wet weight), but thereafter immediately decreased to the level of about 10^6–10^7 cells/g and remainds constant during the period of experiment. Ten days after oral inoculation of S-14 (NA^r), the strain S-14 R(KM)$^+$ was orally administered to the pigs for five days. The number of S-14 R(KM)$^+$ cells was about 10^6 cells/g of faeces. Within 24 hours after administration of S-14 R(KM)$^+$, the exconjugants, S-14 (NA^r) R(KM)$^+$, were isolated from each fecal specimen of pigs, indicating that the transfer of R factor occurred in the intestinal tract of pigs. The number of exconjugants was slightly fewer than that of donor cells.

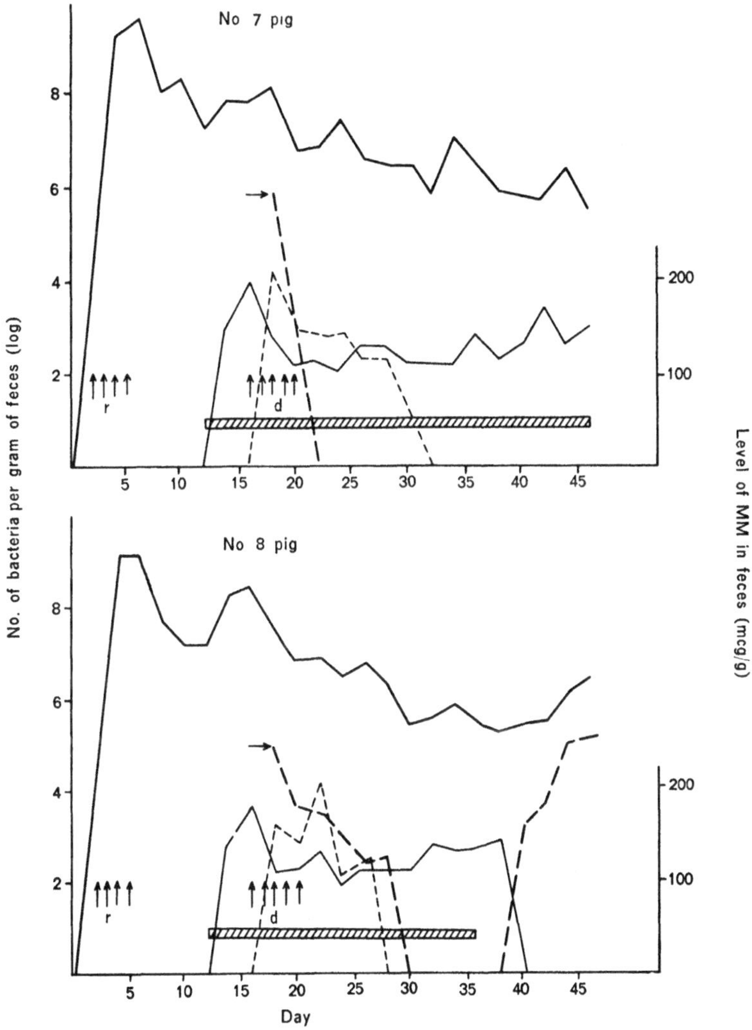

Fig. 4. Daily changes of the number of *E. coli* strains in the intestinal tract of pigs (Group 4, fed with the feed containing 100 mg of MM/kg). The MM-containing feed (100 mg/kg) was admistered from 4 days before oral administration of S-14 R(KM)$^+$, and thereafter every day during the experiment. (▬▬▬▬), (▬ ▬ ▬ ▬), (- - - - - - -), see the footnote of Fig. 1, and (————), and ▨, see the footnote of Fig. 2.

Decrease of E. coli strains carrying R factor by MM treatment. We divided pigs into two groups, one was a control group fed with feed without MM (group 1) and the others were the experimental groups fed with feed containing 100 mg of MM/kg or 300 mg of MM/kg before or after oral administration of (S-14) R(KM)$^+$ (groups 2, 3 and 4). Each group consisted of two pigs and the results are shown in Figs. 1, 2, 3 and 4. From the control pigs, two types R$^+$ strains, (S-14) R(KM)$^+$ and (S-14) NA^r R(KM)$^+$, were isolated at the level of 10^4 cells/g of faeces (Figs. 1). On the other hand, the numebr of R$^+$ strains decreased rapidly in the groups 2 and 4, to which the MM-containing feed was adminis-teerd four days before oral inoculation of S-14 R(KM)$^+$ and thereafter every day (Figs. 2 and 4). Even when the MM-containing feed was administered after oral inoculation of (S-14) R(KM)$^+$ (group 3), the number of R$^+$ strains decreased, but its decrease rate was not so rapid as seen in the group 2 (Figs. 3).

Even after interrupting the administration of MM-containing feed, the KM-resistant strains were not demonstrated from faeces of pigs (group 2, 3 and 4) except only one pig of group 4. The concentration of MM in faeces during the administration of MM-containing feed is shown in Figs. 2, 3 and 4, indicating almost the same concentration as that in feed.

Transfer of KM-resistance from KM-resistant strains of (S-14) isolated from faeces.

The KM-resistant strains of (S-14) were isolated from fecal specimens of a pig (No. 1 of group 1, control) and of a pig (No. 8 of group 4), and transferability of their KM-resistance was examined by mixed cultivation with (S-14) NA^r. All of the 52 KM-resistant (S-14) strains isolated from a pig (No. 1 of group 1, control) could transfer their KM-resistance. By contrast, only 4 out of 15 KM-resistant (S-14) strains isolated from a pig (No. 8 of group 4) could transfer their KM-resistance but not the remainging 11 strains. The nontransferability of 11 KM-resistant (S-14) strains was accountable for by the loss of ability to produce sex pili.

TABLE 1

Transferability of KM-resistance in KM-resistant (S-14) strains isolated from a pig (No. 1 of group 4)

	Organisms	Transfer frequency of KM-resistance	Sex pili formation
During MM-treatment	983	$<7.7 \times 10^{-8}$	—
	984	$<2.9 \times 10^{-8}$	—
	1120	3.2×10^{-1}	+
	1265	2.6×10^{-1}	+
	1266	5.0×10^{-1}	+
	1419	8.0×10^{-5}	+
	1420	$<2.0 \times 10^{-8}$	—
After MM-treatment	2225	$<4.0 \times 10^{-8}$	—
	2226	$<8.2 \times 10^{-9}$	—
	2348	$<1.7 \times 10^{-8}$	—
	2349	$<2.3 \times 10^{-8}$	—
	2488	$<2.5 \times 10^{-8}$	—
	2489	$<1.8 \times 10^{-8}$	—
	2619	$<2.9 \times 10^{-8}$	—
	2620	$<3.7 \times 10^{-8}$	—

Transfer of KM-resistance and sex pili formation, See Materials and Methods for details.

DISCUSSION

The R factors are conjugally transferable among various species of enteric bacteria and able to confer multiple resistance of their host. These properties of R factor are of great importance in practical medicine and hygiene of livestock. It is well known that the R factors are demonstrated from faeces of livestock at a high frequency, because they are fed with feed containing various antibiotics (Mitsuhashi, 1971).

Mitsuhashi and Umezawa started the screening studies of drugs capable of eliminating R factors from their host bacteria or of killing preferentially the bacterial strains carrying R factors. And they reported that *E. coli* strains carrying R factors were more sensitive to MM than their original strains without R factors (Mitsuhashi et al., 1970).

Then, we examined the *in vivo* effect of MM on *E. coli* strains carrying R factor in the intestinal tract of SPF pigs. From the results of experiments using SPF pigs, it was concluded that the R factors were conjugally transferable among the intestinal flora at a high frequency, even though the intestinal tract is known to be anaerobic.

The administration of MM-containing feed could eliminate preferentially *E. coli* strains carrying R factor from the intestinal flora of pigs. This fact is explainable by the results previously obtained by *in vitro* experiments, (1) MM preferentially killed *E. coli* R^+ strains in the mixed culture consisted of *E. coli* strains with or without R factor (Mitsuhashi, et al., 1970) and (2) the R factor mutants which could not produce sex pili and were nontransferable, were easily obtained by exposing *E. coli* R^+ strain to MM (Iyobe et al., 1973).

ACKNOWLEDGEMENT

The authors wish to express deepest appreciation to H. Umezawa and S. Mitsuhashi for their helpful advices to carry out this experiment. We wish to thank Dr. S. Suzuki, Director for his advice and encouragement to carry out this experiment.

REFERENCES

IYOBE, S., MITSUHASHI, S. and UMEZAWA, H. (1971). Relationship between sex-pili formation and macarbomycin sensitivity in *Escherichia coli*. *J. Bacteriol.* 108, 946.

IYOBE, S., MITSUHASHI, S. and SAITO, R. (1973). Sex pili mutants isolated by macarbomycin treatment. *Antimicrobial Agents* and *Chemotherapy*. 3, 614.

KONDO, E. and MITSUHASHI, S. (1966). Drug resistance of enteric bacteria. VI. Introduction of bacteriophage P1 *CM* into *Salmonella typhi* and formation of P1 *CM* and F-*CM* elements. *J. Bacteriol.* 91, 1794.

LENNOX, E. S. (1955). Transduction of linked genetic characters of the host by bacteriophage P1. *Virology*. 1, 190.

MITSUHASHI, S., IYOBE, S., HASHIMOTO, H. and UMEZAWA, H. (1970). Preferental inhibition of the growth of *Escherichia coli* strains carrying episomes. *J. Antibiot.* 23, 319.

MITSUHASHI, S. (1971). Epidemiology of R factors. In Transferable drug resistance factor R. University of Tokyo Press, pp. 25–32.

TAKAHASHI, S., OKANISHI, A., UTAHARA, R., NITTA, K., MAEDA, K. and UMEZAWA, H. Macarbomycin, a new antibiotic containing phosphorus. *J. Antibio.* 23, 48.

COLICINOGENY AND SPREAD OF R-FACTOR-CARRYING COLIFORMS

R. STEPHAN and E. BULLING
Institute of Veterinary Medicine, West Berlin

In antibiotic-feeding experiments involving pigs and calves we observed that the part of resistant coliforms in the faeces of the control group also increased if the antibiotic-feeding group had contact with the control group through animal caretakers. We therefore started some experiments: pigs with a high level of chloramphenicol-resistant coliforms in the faeces (after feeding of chloramphenicol) were reared together with pigs which had either a low level of chloramphenicol-resistant coliforms or non-resistant ones (<100 resistant coliforms per 1 g of faeces). In all the four experiments performed the proportion of chloramphenicol-resistant coliforms in the control group (= acceptor group) increased very quickly. After 24 hours all or nearly all coliforms were found to be chloramphenicol-resistant. We concluded that the resident sensitive coliform flora was displaced rather than R-infected by resistant strains from the donor group.

Then, new experiments were undertaken to establish whether colicinogeny was involved in the spread of R-factor-carrying coliforms.

MATERIAL AND METHODS

Eight piglets from one litter seven weeks old were reared together for five weeks. After this time they were separated into 2 groups. Four pigs were kept on the same diet as before, without antibiotics (acceptor group) and four with a dose of 100 ppm chloramphenicol (donor group) over a period of five weeks. At the same time, 17 chickens were included in the experiment. They were fed a diet without antibiotics (acceptor group). After five weeks of chloramphenicol feeding, the acceptor groups had contact with the donor group. 48 hours before contact, chloramphenicol feeding of the donor group was discontinued.

The faeces from each group of pigs and chickens were examined for their total count of coliforms and the proportion of chloramphenicol-resistant colonies once a week. After contact between the acceptor groups of pigs and chickens and the chloramphenicol-fed group the faeces were examined daily and, later on, twice a week. To avoid contamination, the chickens were slaughtered at these times (one or two animals each) and the contents of the caecum, colon and rectum were examined. Furthermore, both sensitive and resistant coliforms were examined for colicinogeny.

Coliform count: 25 g of faeces from each group of pigs and 15 g of faeces from chickens were first diluted at a ratio of 1 : 10 to form 2 parallel series with physiological saline. 1 ml of this suspension was diluted in 9 ml physiological saline up to 10^{-8}, two tubes for each level of dilution. 0.1 ml from each tube was plated on violet-red-bile agar (Difco), incubated at 37° for 18–24 hours, after which the colonies were counted. Then replica

Fig. 1

Chloramphenicol-fed pigs

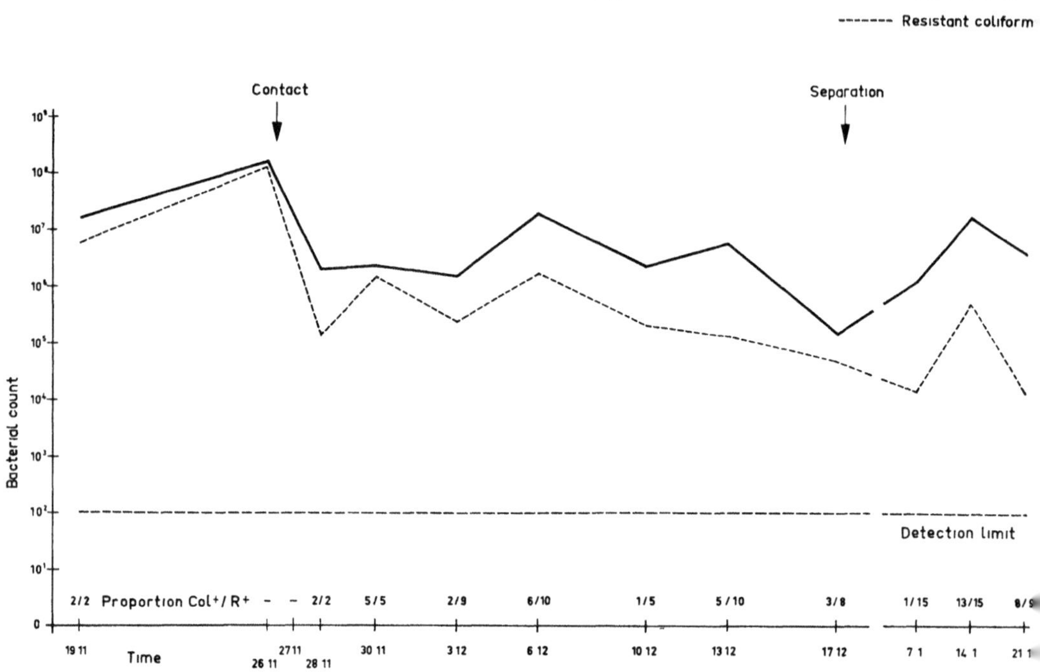

Fig. 2

Contact-pigs

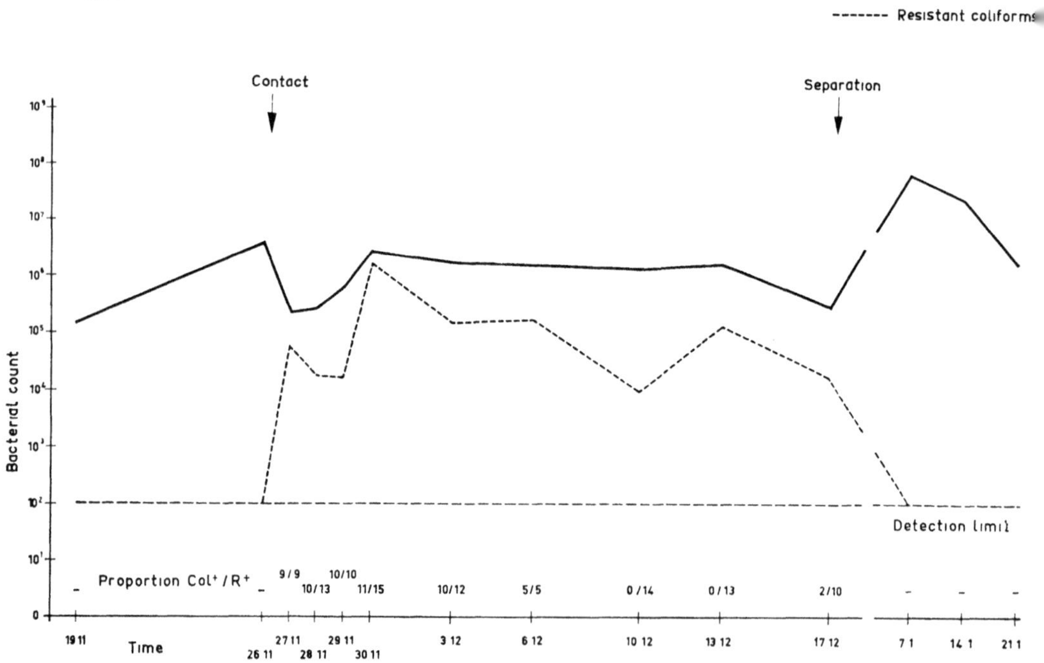

by Lederberg's method was performed on plates containing 30 γ chloramphenicol per 1 ml to establish the resistant proportion.

Counting of the colicinogenic strains: After producing replica on plates with chloramphenicol, the original plates were used to identify the colicinogenic colonies by Fredericq's method; *E. coli* 185-Nx served as indicator strain. Having counted the chloramphenicol-resistant colonies, a second replica was made on plates with chloramphenicol

Fig. 3

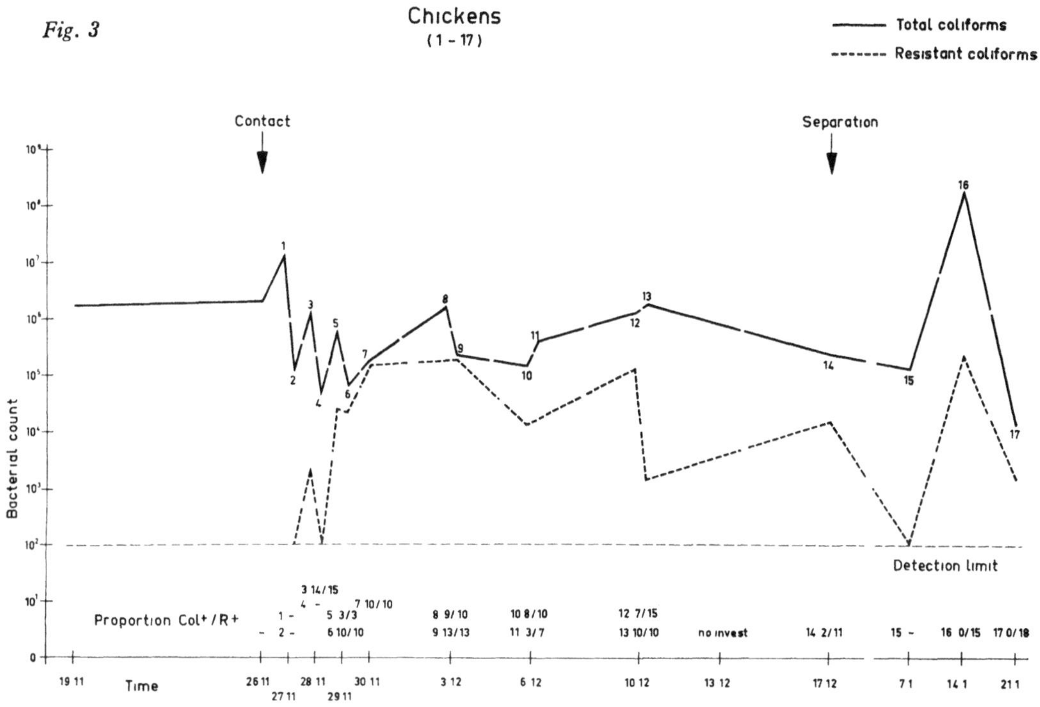

(30 γ/ml) to eliminate the inhibited sensitive bacteria. Then a third replica was made on plates without chloramphenicol. These plates were used to identify the colicinogenic colonies. Because of the insufficient exactness of this method, we isolated after contact, as far as possible, five colonies per level of dilution and examined them for colicinogeny.

RESULTS

The results are listed in Figures 1–3.

Fig. 1 contains the results for the chloramphenicol-fed group. Feeding of chloramphenicol caused a selection of resistant strains. The diffrence between the total number of coliforms and the resistant proportion was between zero and $\log_{10}$ 1, 5. Not all chloramphenicol-resistant coliforms were found to be colicinogenic.

Fig. 2 contains the results for the contact group of pigs. No chloramphenicol-resistant strains were found in the faeces before contact (detection limit 1.0×10^2). 24 hours after the first contact, resistant strains could be isolated. During three weeks when the two groups had contact, the difference between the total number of coliforms and the resistant proportion was about $\log_{10}$ 0.25–2. In the first days all or nearly all resistant strains were found to be colicinogenic. From the beginning of the 2nd week the resistant strains were no longer colicinogenic. After the 3rd week we separated the contact group from the donor group and interrupted the study for three weeks. Three weeks after the separation, no resistant strains could be isolated. We obtained identical results during the following three weeks.

Fig. 3 contains the results for the chickens. No chloramphenicol-resistant strains were found in the faeces before contact with the donor pigs.

24 hours after the first contact no chloramphenicol-resistant coliforms were found in the contents of the caecum, colon and rectum from chicken No. 1 and 2 either. 48 hours after contact, resistant coliforms could be isolated from chicken No. 3, but not from chicken No. 4. In the following investigations — over a period of eight weeks — resistant coliforms could be isolated in 12 out of 13 chickens. The difference between the total number of coliforms and the resistant proportion was about $\log_{10}$ 0.4–3. During the first week all or nearly all resistant strains were found to be colicinogenic. From the beginning of the 2nd week the number of noncolicinogenic resistant strains increased. Four weeks after separation no resistant strains were found to be colicinogenic.

CONCLUSIONS

These results confirm the results we obtained in the past. A spread of R-factor carrying coliforms was found to take place by infection without application of antibiotics. It was demonstrated that R-factor carrying coliforms could spread from pigs to chickens, too. The occurrence of colicinogenic R-factor carrying strains during the first phase of contact in the acceptor groups indicated that colicinogeny was involved in the spread of R-factor carrying coliforms. Further studies will be necessary to clarify this problem.

ACKNOWLEDGEMENT

The authors wish to thank Prof. P. Fredericq for *E. coli* 185-Nx and Mrs. A. Steinbeck. Mrs. E. Trommer for their technical assistance and Mr. Frank for his linguistic assistance in preparing the manuscript.

NEW MICROMETHOD SUITABLE FOR R PLASMID SCREENING IN ECOLOGICAL STUDIES

J. BOHUŠ

Dept. of Microbiology, District Station of Hygiene, Liptovský Mikuláš, Czechoslovakia

Although a lot of studies have appeared dealing with the ecology and epidemiology of R factors, many problems remain to be solved. Everyone dealing with the circulation of R factors in human and animal faces the task of testing a great number of strains for the presence of R plasmids. The classic test-tube method of Japanese authors (Watanabe and Fukasawa, 1971) though simple, consumes a lot of time and material, particularly if it is used on a wide scale. It therefore became necessary to find a simple, sensitive and economical method. We endeavoured to find one which would use time-saving mechanization elements and small volumes, saving material. In these efforts, we were inspired by Takátsy's micromethod employed in every virological laboratory (Takátsy, 1952). The result of our effort is the method termed MM (Bohuš, 1974) the principles and advantages of which we shall present here.

MATERIALS AND METHODS

Bacterial strains: We used, as a test series, 200 *E. coli* strains isolated in our laboratory from various clinical materials. They exhibited resistance at least to one of the following drugs: streptomycine (Sm), chloramphenicol (Cm), tetracycline (Tc), kanamycin (Kn), ampiciline (Am) and sulfonamides (Su). Later, we verified the usefulness of the MM method in the course of a one-year R-plasmids surveillance program which we shall present in our next lecture. We examined more than 6000 strains during a year, *E. coli* K-12 185 Nx strain kindly provided by Dr. Krčméry from the Research Institute of Hygiene, Bratislava, was generally used as recipient. We also isolated nalidixic acid resistant mutants from wild strains of *S. typhi* (267 Nx), *S. agona* (154 Nx), *Sh. flexneri* (60 Nx), *Sh. sonnei* (T Nx) and from a laboratory strain of *S. typhimurium* (SH3 Nx), and used them in the experiments in which the transfer of R plasmids from commensal *E. coli* strains to pathogenic recipients was studied.

Media: Tryptose broth Difco for conjugation and McConkey agar Difco for selection were used. Streptomycine, chloramphenicol, tetracycline, kanamycin, ampiciline and nalidixic acid were added to give final concentrations 10, 20, 30, 100 and 100 μg per 1 ml of media.

MM method: The basic technical equipment consists of stencils, two or more multi-inoculators and racks with microtubes (Figs. 1 and 2). The stencils have the shape of a lattice with 6 × 6 sectors, the side of one sector being 1 cm. They are numbered from 1 to 36 and from 37 to 72. This is the number of strains in one experiment. A multi-inoculator (Fig. 1) consists of 6 loops fixed in a plexiglass holder in a way enabling them to rotate freely and to shift along the longitudinal axis, the swing of the loops in the terminal point being minimal. The longitudinal movement of the loops in the holder is restricted by bits of a PVC tube fixed on the loops at corresponding distances. The loops are made of currently accessible material. Injection needles No. 10 are fixed on aluminium

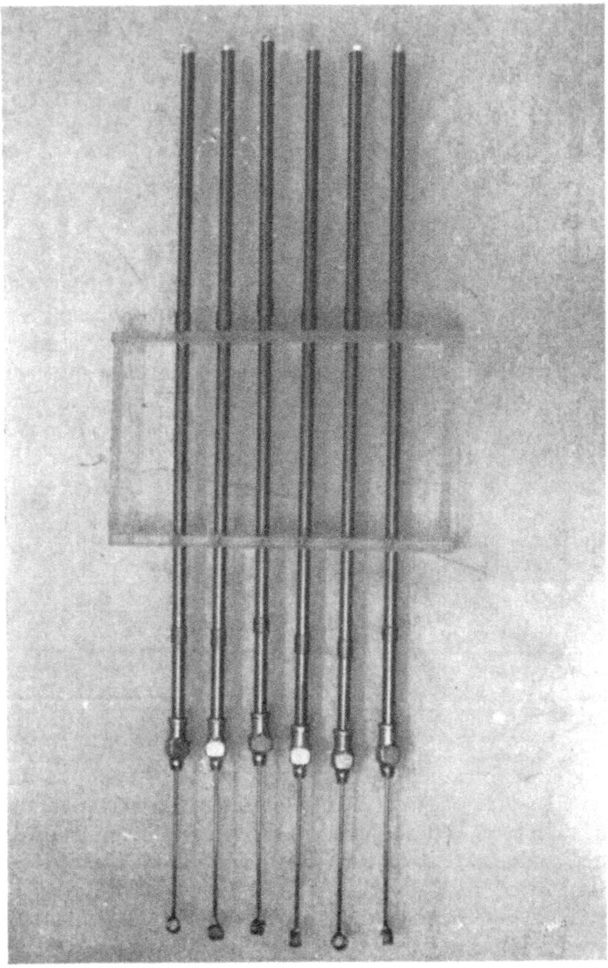

Fig. 1. Multi-inoculator.

rods (shortened aluminium knitting needles can be used with advantage). Bits of a slightly distended resistance spiral with four coils are inserted into the needles. The racks made of aluminium sheets have 6 × 12 holes, with an 8 mm diameter each. The microtubes, i.e. detection tubes commonly used for gas detection shortened to 2 cm are in the holes. The distances between the centres of the microtubes, loops and centres of sectors of the stencils are 1 cm.

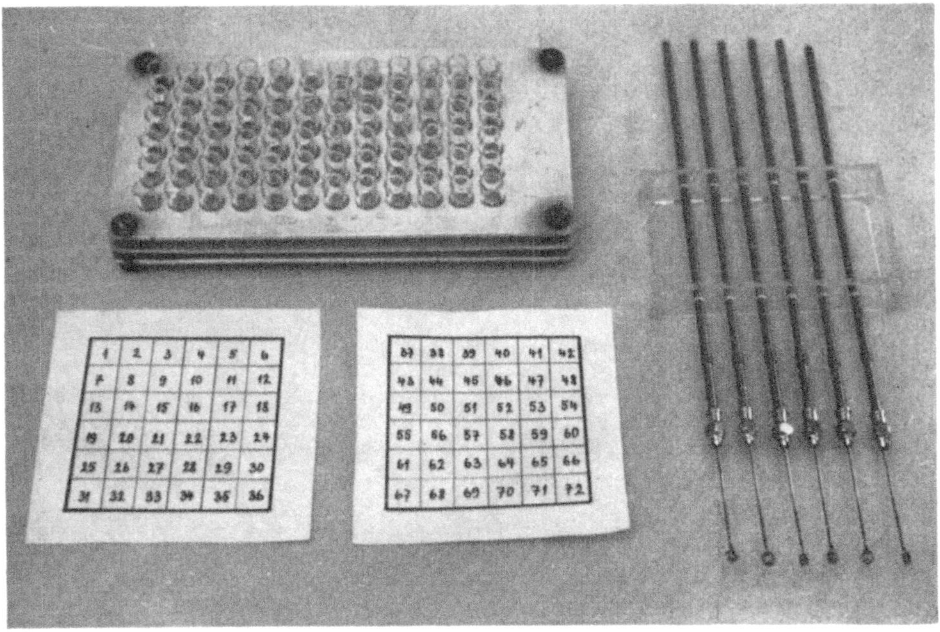

Fig. 2. Basic equipment for the MM method: a multi-inoculator a rack with microtubes and stencils.

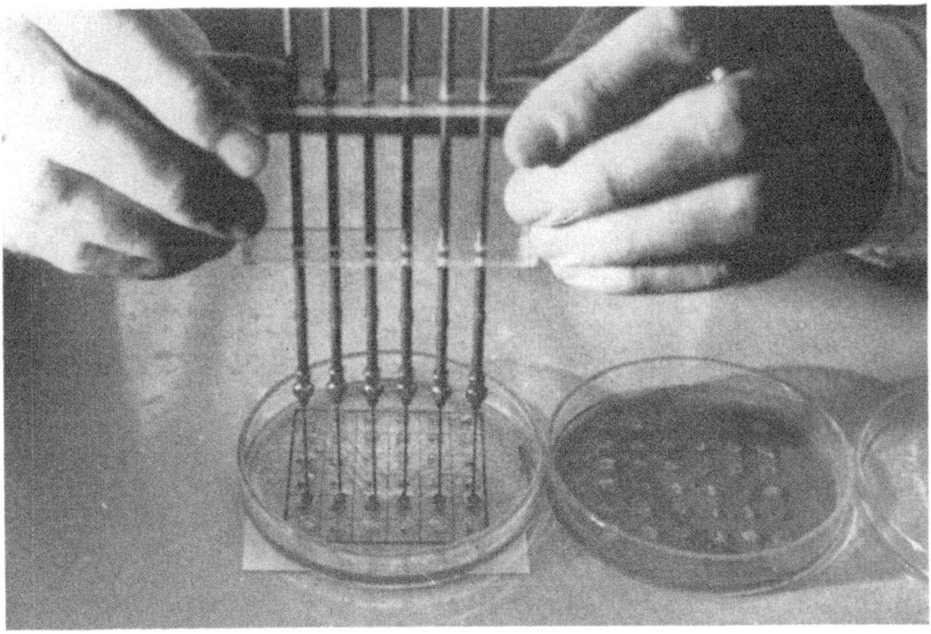

Fig. 3. Taking of donor strains from master plate.

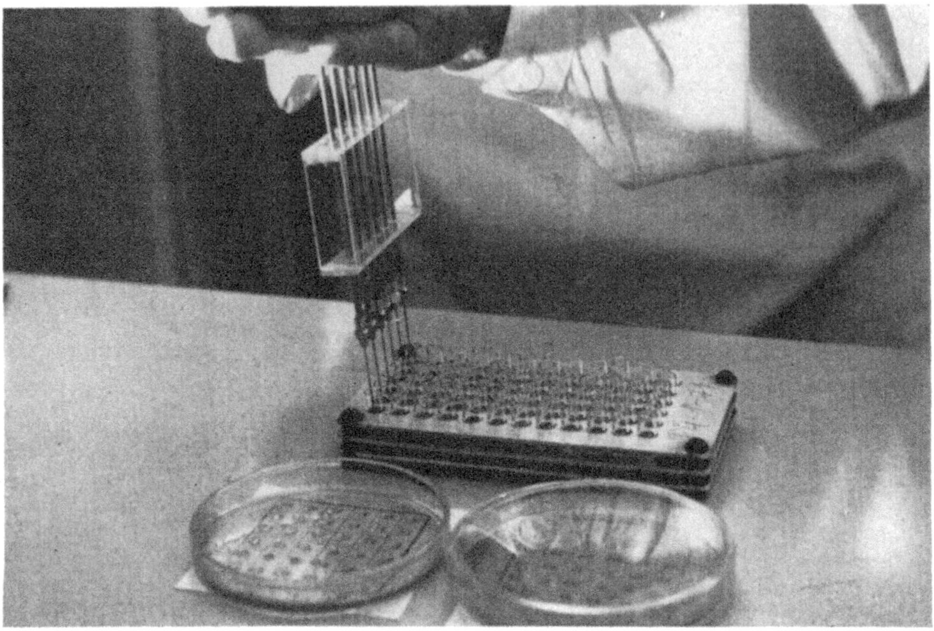

Fig. 4. Emulgation of donor strains in microtubes.

Fig. 5. Preparation of conjugation mixtures, Transfer of donor suspensions.

The procedure is as follows: The strains tested are inoculated by a needle on the master plates under the control of stencils. The first plate contains 36 strains and the second 35, the recipient strain is plated on the last sector as a control. The plates are incubated overnight at 37°C. Each microtube of rack No. 1 is filled with 0.2 ml of broth. Inocula of six trains of the first row of a master plate are simultaneously gathered by touch with the multi-inoculator (Fig. 3) and transferred to the first row of the microtubes of rack

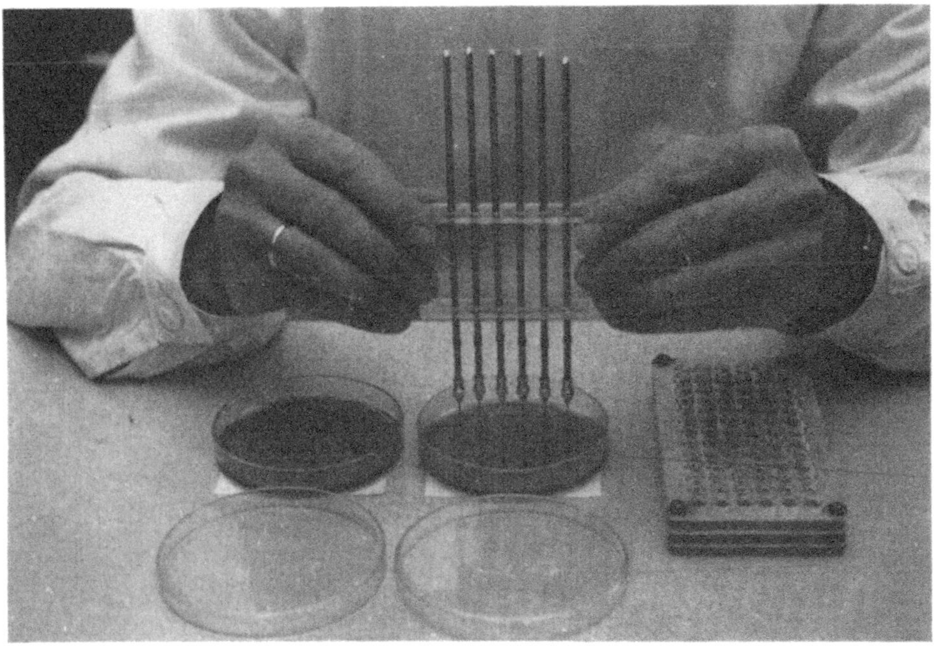

Fig. 6. Plating of conjugation mixtures on selection plates.

No. 1. The trains are emulgated by rotation of the loops between the palms of the hands (Fig. 4), the loops are then sucked by touch with cotton wool and flamed. It takes 10 minutes to prepare suspensions of the 72 strains. The microtubes are than covered with aluminium foil and incubated for 4 hrs at 37°C.

Four-hours broth cultures of the recipient strain are diluted in broth to the density of about 10^8 cells per 1 ml and 0.2 ml of this culture is pipetted into the microtubes of rack No. 2. Conjugation mixtures are prepared by transferring preincubated donor strains from rack No. 1 to rack No. 2 by means of a multi-inoculator (Fig. 5). This operation takes 10 minutes. Mixtures, together with donor strains are incubated overnight and then transferred (36 per plate) on selective plates by means of a multi-inoculator under the control of stencils. Six spots remain on the plate by one touch of the loops (Fig. 6). Donors from rack No. 1 are transferred on the control plates containing nalidixic acid only as contra-selective agent against donors. If selection for 5 antibiotics is performed, 12 plates are needed for 71 transfers (the 72nd microtube contains only a recipient as a control.)

The plates are incubated overnight, oriented on stencils and the results are rd. Either a continuous growth at a high frequency of transfer, or only sporadic colonies at the frequency 10^{-6} or less appear (Fig. 7). Donor strains growing on nalidixic acid control plates are excluded from the evaluation.

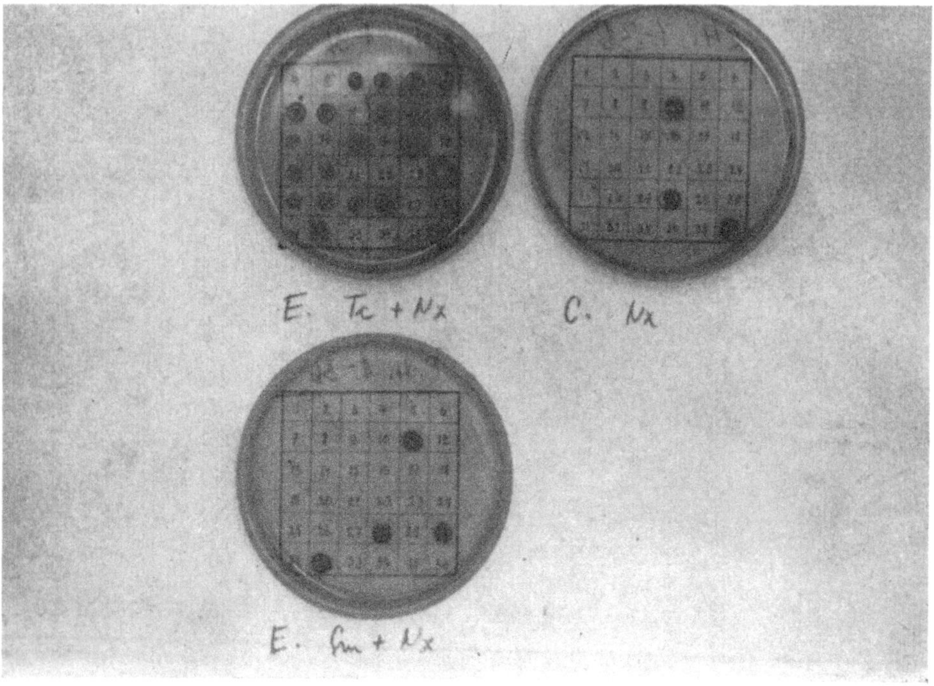

Fig. 7. Reading of results by means of stencils. E. Tc+Nx – Experiment, plate with tetracycline and nalidixic acid, Sm+Nx – with streptomycin and nalidixic acod, C. Nx – control plate with nalidixic acid.

To determine the inheritance of unselected resistance markers, the growth from selective plates is emulgated in saline to give it appropriate density and is tested for sensitivity by antibiotic paper strips having the concentration of standard sensitivity discs. Ten strains are tested in this way on two plates. If the inheritance of unselected markers or segregation by single exconjugant colonies are to be tested, positive microtubes are streaked additionally on selective plates to give isolated colonies.

RESULTS

Of 200 *E. coli* resistant strains, 91 (45.5%) transferred their resistance to *E. coli* 185 Nx recipient. Of these, in 22 only a portion of the R factor or one of the potentially present R-plasmids was transferred. In order to verify the reproducibility of the MM method, the test was repeated three times. Conforming results were obtained in 95–98% of strains. Discrepancies should be ascribed to the strains transferring their R plasmids beyond the limit of detectability by this method, which is 10^{-6} per one donor or recipient cell. The same strains were tested also by means of the classic test-tube method. The results differed only in those strains (2–5%) which transferred their R plasmids at a frequency lower than 10^{-6}.

TABLE 1

Transfer of R plasmids from 182 commensal E. coli strains to pathogenic recipients

Recipient	Number of transferring strains	%*
E. coli K-12 185 Nx	85	46
S. typhi 267 Nx	61	33
Sh. sonnei T Nx	47	25
Sh. flexneri 60 Nx	13	7
S. agona 154 Nx	7	3
S. typhimurium SH3 Nx	7	3

* 182 = 100%

TABLE 2

Materials needed for 72 transfers in the micromethod (MM) and the classic method (CM)

Material	MM	CM
Rack with microtubes	2	—
Multi-inoculator	2	—
Normal-size test-tubes	—	144
Two-row racks	—	6
Pipettes	—	72
Nutrient broth	30 ml	500 ml
Selective plates (selection for 5 drugs + controls)	12	45 *

* provided that 10 suspensions are plated as lines on one plate

TABLE 3

Time schedule for 72 transfers by means of the MM method and the classic method (CM)

Operation	Minutes MM	Minutes CM
Filling of 144 microtubes	5	40
Preparation of donor cultures	10	30
Preparation of conjugation mixtures	10	50
Plating of mixtures on selective plates (5 drugs)	15	50
Plating of donors	10	30
Cleaning of 144 microtubes	10	60
Total:	60	260

In view of its highly economical character, the MM method enabled us to carry out transfers of R plasmids from commensal *E. coli* strains to several pathogenic recipients simultaneously. It is surprising that *S. typhi* 267 Nx proved to be the best recipient among the pathogenic strains we used (Table 1).

Fig. 8. Material needed for 72 transfers in the MM method and the classic test-tube method.

In addition to the sensitivity and reproducibility of the MM method, the economic aspects play an important role, too. Materials needed for 72 transfers by means of the MM method and of the classic method are compared in Table 2 and Fig. 8. Although certain initial investments are needed for the former (racks with microtubes and multi-inoculators) they are very soon returned due to the saving of media, pipettes and stoppers. The time schedule of the individual procedures for 72 transfers using both methods is given in Table 3. The MM method is at least six times more economic than the classic one in this respect.

DISCUSSION

The technique by which donor strains are replicated on the surface of selective plates overlayed with recipient strain is simple and economic, but does not reach the sensitivity of that where the transfer of genetic material takes place in broth. In addition, some substances used as contraselective against donors may have a negative effect on the transfer of genetic material. Developing our method, we employed fluid medium for the transfer and adopted some mechanization elements. It should be pointed out that the MM method has a screening character and is useful mainly in ecological and epidemiological studies. It can be also used in other genetic experiments where hundreds of colonies or strains are to be tested in a short time.

Some additional tests and calculations made it possible to determine 10^{-6} per one donor or recipient cell as a value of the lowest frequency regularly detected by the MM method. On average, 0.06 ml of broth is taken by one loop of a multi-inoculator and about 0.02 ml are plated. The average densities of emulgated donors incubated for 4 hrs. are 5×10^8 cells per 1 ml. A loop contains 3×10^7 donor cells and if this amount is stransferred into 0.2 ml of a suspension of recipient strain with a density of 10^8 cells per 1 ml (i.e. 2×10^7 in 0.2 ml), one can state that conjugation mixture of the MM method contain an equal number of donor and recipient cells. The average density of conjugation mixtures after 20 hrs. incubation is $4–6 \times 10^8$ cells per 1 ml. The approximate calculation of the sensitivity of the MM method is based on the assumption that the density of the recipient or donor strain in a conjugation mixture after 20 hrs. of incubation is 3×10^8 cells per 1 ml. Provided that 6 R^+ recipient colonies are grown from 0.02 ml of a conjugation mixture applied on a selective plate, the number of R^+ cells in 1 ml of the mixture is 3×10^2. This corresponds to the transfer frequency 10^{-6} per one donor or recipient cell present in the conjugation mixture after 20 hrs. of incubation.

As to the economic aspects some factors are beyond numerical expression. Microtubes, for instance, need not be stoppered because the foil prevents their contamination. To verify this, 360 microtubes were filled with sterile broth and the aluminium foil was removed for 60 minutes. Only two tubes were contaminated with micrococci although no special sterile conditions were kept in the laboratory.

REFERENCES

BOHUŠ, J. (1974). A new screening method for proof of R factors transfer. Journal of Hygiene, Epidemiology, Microbiology and Immunology. In press.

TAKÁTSY, G. (1952). Kerszerü serologai mikroeljárások. Késerletes Orvostudomány, 4, 60.

WATANABE, T. and FUKASAWA, T. (1961). Episome mediated transfer of drug resistance in Enterobacteriaceae. I. Transfer of resistance factors by conjugation. Journal of Bacteriology. 81, 669.

ONE-YEAR SURVEILLANCE OF R-PLASMIDS IN CLINICAL SPECIMENS

J. BOHUŠ

Dept. of Microbiology, District Station of Hygiene, Liptovský Mikuláš, Czechoslovakia

Reports on R-plasmids content in bacterial strains published in our country up to the present dealt only with a limited number of strains collected in limited or undefined time intervals (Bohuš and Baláž, 1971; Horák, 1971; Janoušková and Krčméry, 1971). Discussions took place between "R-factorists", public health microbiologists and clinical workes on the real importance of transmissible resistance. These discussions induced us to realize a one-year R plasmid surveillance program which would help us to elucidate the significance of R-plasmids for human medicine. We decided to fulfil the following tasks: a) to carry out surveillance over a one-year period, b) to include all strains belonging to the *E. coli* – *Aerobacter* – *Klebsiella* group exhibiting resistance to "R-factor antibiotics", c) to cover any kinds of clinical materials, and finally, d) taking into consideration the uniform structure of the health service in our country, to do some calculations on the overall risk arising from the existence of R-plasmids for our population.

TABLE 1

Resistance and its transferability in strains isolated from various clinical materials. A survey of results.

Group No.	Kind – origin of material	Strains tested	Resistant	(%)[1]	Transferring	(%)[2]
1.	Urine	1471	972	(66)	480	(50)
2.	Sputa – upper respiratory tract	569	223	(39)	60	(26)
3.	Gynaecological materials	201	90	(44)	42	(46)
4.	Surgical materials	183	109	(59)	34	(31)
5.	Stools – New-born children's ward	555	298	(53)	112	(37)
6.	Stools – Pediatric ward I.	598	296	(49)	106	(35)
7.	Stools – Pediatric ward II.	1208	787	(65)	435	(55)
8.	Stools – Children in out-patient care	353	90	(25)	20	(22)
9.	Throats – Children generally	801	359	(44)	131	(36)
10.	Shigellas	169	71	(42)	29	(41)
	Total:	6108	3295		1449	

Note: [1]) calculated from number of strains in each group,
[2]) calculated from number of resistant strains in each group.

RESULTS

The method we used for R-plasmids screening is presented at this symposium and will be published (Bohuš, 1974). The survey of the results is presented in Table 1. It comprises the kind or origin of the material, the number of strains tested as well as the number of resistant and transferring strains. The strains are divided into 10 groups according to the material and origin. The column "Resistant" includes strains exhibiting resistance to at least one of the following drugs; streptomycine (Sm), chloramphenicol (Cm), tetracycline (Tc), kanamycin (Kn), ampicilline (Am) and sulfonamides (Su). The

TABLE 2

Prevalent resistance types and their transferability

Material	Number of types	Resistance types prevalent						Transferability (%)
Urine	36	Sm	Cm	Tc		Am	Su	60
		Sm	Cm	Tc	Kn	Am	Su	77
				Tc				30
Sputa – upper respiratory tract (adults)	18					Am		2
				Tc				31
		Sm	Cm	Tc	Kn	Am	Su	77
Gynaecological materials	13	Sm	Cm	Tc		Am	Su	66
				Tc				37
						Am		1
Surgical materials	22					Am		2
		Sm	Cm	Tc	Kn	Am	Su	80
		Sm	Cm	Tc		Am	Su	25
Stools – New-born children's ward	19	Sm	Cm	Tc	Kn	Am	Su	56
						Am		1
			Cm	Tc	Kn	Am		22
Stools – Pediatric ward I.	21					Am		7
		Sm	Cm	Tc		Am	Su	51
		Sm	Cm	Tc	Kn	Am	Su	58
Stools – Pediatric ward II.	31	Sm	Cm	Tc	Kn	Am	Su	72
		Sm	Cm	Tc		Am	Su	54
						Am		17
Stools – Children in out-patient care	15					Am		3
				Tc				41
		Sm	Cm	Tc		Am	Su	22
Throats – Children generally	15					Am		4
		Sm	Cm	Tc	Kn	Am	Su	70
		Sm	Cm	Tc		Am	Su	68
Shigellas	12	Sm	Cm	Tc		Am	Su	11
				Tc				75
		Sm	Cm	Tc			Su	60

Note: Percentage of transferability is calculated from number of strains in resistance type results in groups 1–4 indicate a high incidence of resistant and transferring strains in such materials as urine, sputa, gynaecological and surgical materials. Resistance and transferability of commensal strains in children differed according to wards or out-patient department (group 5–8). Generally, children hospitalized in new-born children's wards and in pediatric wards exhibited a higher percentage of resistant and transferring strains than those treated as out-patients. This proves a nosocomial aspect of R-plasmids spreading

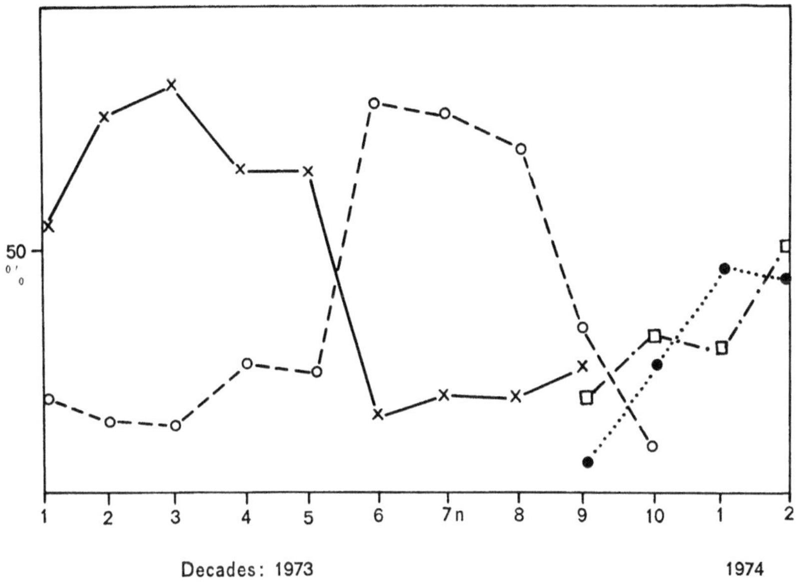

Fig. 1. Succesion of resistance types in new-born childrens' ward (Intestine). Legend: ○ - - ○ *E. coli* (Sm Cm Tc Kn Am Su), × — × *E. coli* (Am), □–·–□ *K. pneumoniae* (Cm Tc Kn Am), ●.....● *E. coli* 055 (Am)^{0} percentage calculated from the overall number of resistant strains

in hospital environment. The indidence of R-plasmids harbouring strains in new-born children is of special interest and will be analyzed in the next part of this communication.

Table 2 summarizes the number of resistance types recorded in materials as well as the prevalent types and their transferability. 12 to 36 resistance types were recorded in various materials and this number is proportional to the number of examined strains. Polyresistant strains with a high transferability were recorded among the most frequent types in all materials. Strains resistant to ampicilline only exhibited a significantly lower transferability than those with complex R plasmids.

We generally found that more complex types did not always transfer complex R-plasmid and, on the contrary, more simple ones were transferred from more complex types. For example, urinary strains belonging to the type Sm-Cm-Tc-Kn-Am-Su transferred 12 various R-plasmids, even though the original type was transferred most frequently. On the other hand, Am R plasmid was transferred from ten more complex types. This dispersity of transfer contributes to the wide scale of resistance types in all kinds of clinical materials.

We made the alarming finding that about 50% of new-born children and premature babies are colonized with resistant and transferring strains during the first days of their life. We evaluated the incidence of resistance types in every months of 1973 and found, that one or two types predominated at any time interval being then replaced by another. Figures 1 and 2 illustrate such a succession and identical colonization of the intestine and of the mouth cave. We also searched for sources of resistant strains and found only individual persons with resistant strains among members of the staff, but the types found did not agree with those which predominated in the children. This finding correlates

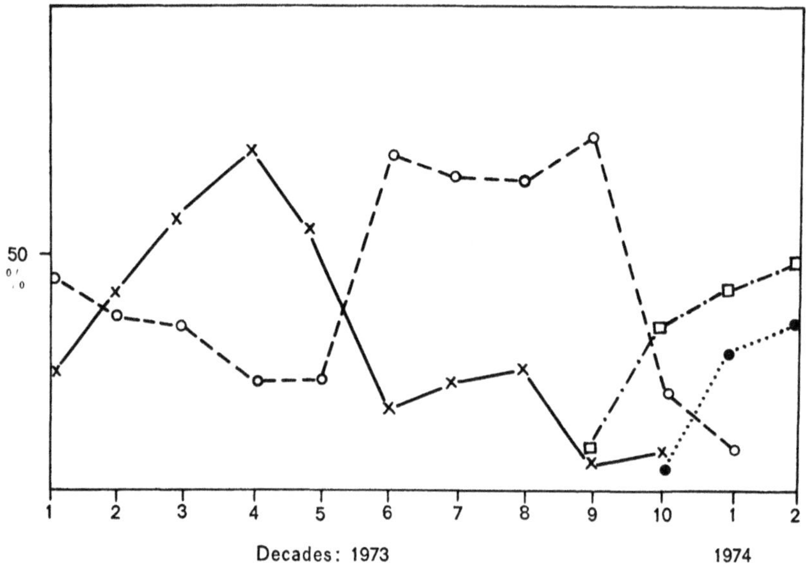

Fig. 2. Succession of resistance types in new-born children's ward (Mouth cave). Legend as for *Fig. 1.*

with our previous results (Bohuš and Baláž, 1971). We examined also 183 mother and child pairs and found only two of these pairs with conforming types of resistance We therefore excluded mothers as a source of polyresistant strains of their children. We suppose, in view of the above findings and of the fact that examinations of environment were not conclusive, that air is the main route by which the children are colonized with polyresistant strains. We base this hypothesis on three facts: 1. A new-born child inspires about 12 m^3 of air per day. 2. The mouth cave with persistent residues of a milk food create favourable conditions for the growth of bacteria. 3. Nearly in all the children the mouth cave and the intestine are colonized with identical strains.

Although air contamination is too low to be significantly proved by common methods, the large daily amount of air inspired by a baby, together with the favourable conditions for the growth of bacteria in the mouth cave are sufficient for colonization even during a few days. There is no hope of stopping the circulation of R plasmids harbouring strains in the wards by any hygienical measures. We assume that the only way of saving new-born children from contamination with R-plasmid strains is biological prevention based on artificial colonization by a suitable strain just after delivery. Such a strain should fulfil the following criteria: non-pathogenicity, ability to establish itself in macroorganism, rejection of the multiplication of other incoming strains and, finally, inability to accept any R-plasmid.

CONCLUSIONS

1. R-plasmids cause increasing trouble in antibiotic treatment of all kinds of bacterial diseases, not only of those having an infective character.
2. R-plasmids spreading in the hospital environment contribute to the general increase of the resistance of bacteria.
3. Special attention should be paid to the colonization of new-born and premature children.
4. The uniform structure of the Czechoslovak health service and bacteriological services permits us to calculate that, due to the existence of R plasmids, about half a million hospitalized persons, were in our country in 1973 restricted as regards treatment by antibacterial therapy.

REFERENCES

BOHUŠ, J. and BALÁŽ, M. (1971): Study of transmissible poly-resistance in routine laboratory practice. Journal of Hygiene, Epidemiology, Microbiology and Immunology, 15, 225.

BOHUŠ, J. (1974). A new screening method for proof of R factors transfer. Journal of Hygiene, Epidemiology, Microbiology and Immunology. In press.

HORÁK, V. (1971). R-factors of *Escherichia coli strains* causing urinary tract infections, their types of transfer factors and differences in their transmissibility to Citrobacter and *Salmonella typhimurium* recipient strains. Folia microbiologica, 16, 317.

JANOUŠKOVÁ, J. and KRČMÉRY, V. (1971). Výskyt faktorov prenosnej rezistencie na antibiotika a inych sex-faktorov u kmeňov *Escherichia coli* v ekologickom systéme hodonínských veľkovýkrmni. Časopis lékařů českých, 110, 806.

THE CHANGING PATTERN OF ANTIBIOTIC RESISTANCE IN SHIGELLA SONNEI STRAINS ISOLATED DURING OUTBREAKS OF DYSENTERY

E. SCHÖN and Z. MANDLÍKOVÁ

Department of Epidemiology, Charles University, Prague,
Department of Microbiology, Regional Station of Hygiene,
Prague, Czechoslovakia

During the past five years, monitoring of antibiotic sensitivity of Shigellae isolated in the microbiological laboratories of the Health Service of the Central Bohemian Region has revealed a variable annual incidence of the resistance to streptomycin (ST), tetracycline (TC), chloramphenicol (CM), and sulfonamides (SU). Because of the known emergence of explosive spreading of R-factors as observed by Watanabe (1963) in Japan and by Datta (1965) in England, a study was initiated to ascertain more precisely the prevalence of multiresistant Shigellae and the presence of transmissible R-factors in these isolates. Further, all strains isolated in the course of six outbreaks of dysentery were tested for sensitivity to study the possible changes in the resistance pattern during an epidemic.

MATERIAL AND METHODS

The majority of the strains of *Shigella sonnei* in this study were isolated either from out-patients suffering from dysentery or from symptomless excretors in the foci of dysentery. A small proportion of strains were isolated from patients on admission to the hospital. Cultures were made by plating rectal swabs on Endo agar and desoxycholate-citrate agar plates. The typical colonies were differentiated by selective biochemical tests and serotyped with typing sera. Disc sensitivity testing was performed on Mueller-Hinton agar by using LACHEMA discs.

Sensitivities to streptomycin (20μg), tetracycline (30μg), chloramphenicol (30μg), kanamycin (30μg), neomycin (30μg), gentamicin (20μg), colimycin (50μg), ampicillin (20μg), carbenicillin (100μg), cephaloridin (10μg), nitrofurantoin (100μg), nalidix acid (20μg), sulfisoxazol (200μg), and co-trimethoprim-sulphamethaxazol (25μg) were ascertained.

R-factor transfer was demonstrated *in vitro* by conjugating Shigella organisms as donor strains with Citrobacter Nx strain serving as a recipient (Horák, 1971).

Colicine testing was carried out by the method of Abbot and Shannon modified by Sekaninová (1974).

Phage typing was carried out by the method according to Kallings by dr. Aldová (1973) (Dysentery Reference Laboratory).

RESULTS

A five year (1969–1973) surveillance of Shigellae yielded 1664 isolates of *Sh. sonnei* strains. The resistance to streptomycin (ST), tetracycline (TC), chloramphenicol (CM), kanamycin (KA) and sulfisoxazol (SU) over the past five-year period is illustrated in Table 1.

TABLE 1
Drug Resistance of Shigella sonnei (1969–1973)

Year	Total strains	% resistant to				
		ST	TC	CM	KA	SU
1969	312	29	35	23	0,3	34
1970	346	29	26	15	1,0	35
1971	411	16	46	16	2,0	44
1972	395	20	23	32	0,5	53
1973	200	21	34	22	0	40

Annual changes were observed in the incidence of the resistance to ST ranging from 16% to 29%, to TC (23%–46%), CM (15%–32%), and SU (34%–53%).

In 1972–1973, sensitivities to ampicillin, carbenicillin, cephaloridin, neomycin, gentamicin, colimycin, nitrofurantoin, nalidixic acid, and co-trimethoprim – sulphamethaxazole (Septrin) were ascertained. (Table 2)

TABLE 2
Drug Resistance of Shigella sonnei

Year	Total strains	% resistant to								
		NE	GE	CO	AM	CA	CE	FU	NX	SE
1972	395	1,0	ND	2,0	6,0	ND	5,0	1,0	0,5	9,0
1973	200	ND	0	3,3	21,0	3,5	ND	0	1,0	5,0

NE – neomycin
GE – gentamicin
CO – colimycin
AM – ampicillin
CA – carbenicillin
CE – cephaloridin
FU – nitrofurantoin
NX – nalidixic acid
SE – septrin
ND – not done

A rapid increase in ampicillin resistance from 6% in 1972 to 21% in 1973 was found. Surprisingly, resistance to septrin was observed in 9% of strains isolated in 1972, i.e. before this drug had been introduced in Czechoslovakia. Neomycin and gentamicin remained uniformly inhibitory *in vitro* against all strains tested, as well as nalidixic acid and nitrofurantoin. Only 2%–3% strains were resistant to colimycin.

The resistance patterns with special reference to five drugs (ST, TC, CM, KA, SU) were studied in 637 strains isolated in 1971–1973 (Fig. 1).

About 50% of R-factors were single-resistant, the frequency of double- and tri-resistant strains was changing during the past three-year period. The frequency of the quadruple R-factor ST-TC-CM-SU increased from 6% in 1971 to 15% in 1972 with a subsequent decline to 11% in 1973.

With the use of the method previously described, conjugation of resistant Shigellae with the recipient organism Citrobacter Nx demonstrated the presence of a resistance transfer factor in 174 out of 220 resistant Shigellae (Table 3).

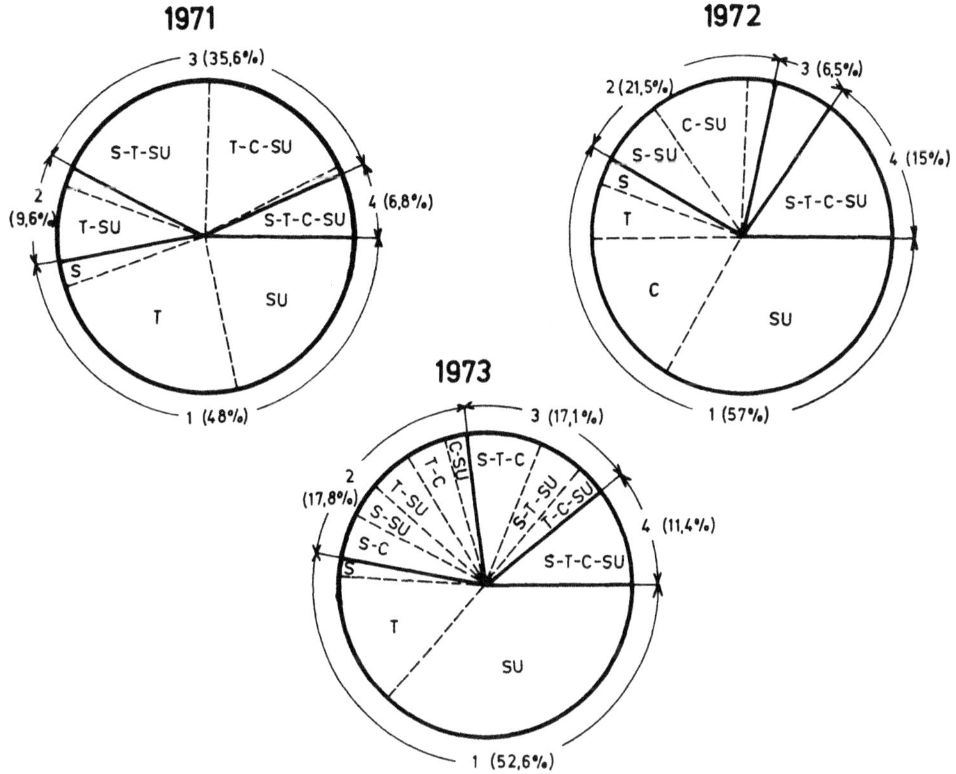

Fig. 1. Resistance patterns of R factors from *Shigella sonnei* strains isolated in 1971–1973 in the Central Bohemian Region

1971 – 250 R factors were demonstrated from 411 strains
1972 – 264 R factors were demonstrated from 395 strains
1973 – 123 R factors were demonstrated from 200 strains

Drug resistance was examined with special reference to streptomycin (S), tetracycline (T), chloramphenicol (C), and sulfonamides (SU).

The changing patterns of resistance indicated that the shigellae circulating in the population acquired R-factors *in vivo* on the one hand and, on the other hand, segregation of some resistance determinants occurred.

To confirm this hypothesis, Shigella strains isolated during six outbreaks of dysentery were studied in greater detail.

Two outbreaks (NY, ZL) noted in nurseries were caused by strains uniform in their resistance pattern, phage-type and colicine-type. In the third epidemic (MB), the first isolated strain carried the tetracycline R-factor, but the subsequent 12 isolations were resistant to TC-CM, three were ST-TC-CM and two ST-TC-CM-AM-KA resistant. In this case, the selection of multiresistant strains was probably caused by administration

TABLE 3

Transfer of R factors from Shigella sonnei strains

Resistance marker of R factor	Number	Characteristics transferred	Number	
ST	15	ST	4	
TC	47	TC	41	
CM	44	CM	34	
ST-TC	9	ST-TC	7	
ST-CM	25	ST-CM	21	
TC-CM	20	TC-CM	12	15
		CM	3	
TC-CO	1	TC-CO	1	
ST-TC-CM	59	ST-TC-CM	23	51
		ST-TC	7	
		ST-CM	7	
		TC-CM	6	
		ST	3	
		TC	2	
		CM	3	

TABLE 4

Resistance patterns in Shigella sonnei strains isolated in the course of outbreaks

Place	Number of strains		R factor	Number	Type	
	isolated	resistant			Phage-	Col-
NY	22	20	ST-TC-SU	20	66	12
ZL	5	5	TC-SU	5	12	2
MB	18	18	TC	1	ND	ND
			TC-CM	12		
			ST-TC-CM	3		
			ST-TC-CM-KA-AM	2		
VL	11	11	ST-TC-CM-SU-SE	11	65	6
VR	8	7	SU	7	65	6
KO	60	55	CM	45	21	0
			CM-SU	3		
			CM-TC	1		
			ST-TC-CM-SU	2		
			ST-CM	1		
			TC-SU	2		
			TC-CO	1		
ME	35	35	TC-CM-SU	27	ND	ND
			ST-TC-CM-SU	6		
			TC-CM-CO-SU	2		

of chloramphenicol. The fourth outbreak (VL-VR) affected two neighbouring villages. All strains were of phage-type 65 and colicine-type 6 but differred in the resistance patterns. Whereas 8 strains from the one village were only SU resistant, eleven strains isolated in the other village carried transferable ST-TS-CM-SU resistance. The fifth epidemic (KO) spreading through a village was caused by a Shigella strain of phage-type 21 and colicine-type 0. In the course of this outbreak 60 strains were isolated, 55 of them were resistant. The CM resistance was prevalent in 45 strains but strains with additional markers of TC, ST and SU were found, too. A similar situation occurred in the last epidemic (ME) where 27 out of 35 strains were resistant to TC-CM-SU, the resistance pattern of the remaining strains was ST-TC-CM-SU or TC-CM-CO-SU.

COMMENT

The annual changes in the incidence of R-factors in Shigella sonnei observed by us during the past five-year period in the Central Bohemian area as well as the changing patterns of resistance in the course of outbreaks of dysentery indicate that the transfer of R-factors occurs *in vivo* on the one hand and, on the other, resistance determinants may segregate. Data reported by Watanabe (1963), Davies (1970), Ross (1972) indicate that the frequencies and drug resistance markers of R-factor carrying bacteria in various countries are dependent on the amounts and kinds of antibiotics used in these countries.

Thus, the most frequent resistance pattern in Japan is sulfonamide-streptomycin-tetracycline-chloramphenicol, reflecting the widespread clinical use of these four anti-bacterial agents in that country. Ross (1972) found a rapid increase in ampicillin-resistant Shigellae in the metropolitan Washington area during 1968–1971 and a similar observation was reported by Davies (1970) in London covering a period from 1967 to 1969. This increase was connected with an increase of multiresistant Shigellae and with the precipitous rise in the frequency of R-factor transfer. The common resistance pattern in the United States may be sulfonamide-streptomycin-tetracycline-ampicillin which coincides with the extensive usage of ampicillin (Ross, 1972).

The findings reported here reflect the situation in the area of Central Bohemia in out-patients the majority of whom were treated symptomatically. Only a small proportion of them were treated with sulfonamides or chloramphenicol. Although the use of ampicillin is in our country limited only to hospitalized patients, an increase in the prevalence of ampicillin resistant Shigellae was observed. This finding points to a nosocomial origin of that R-factor and to its dissemination in the community.

Further monitoring of antibiotic sensitivities of Shigellae and other bacteria is desirable in order to ascertain the changes in the resistance pattern and the spread of R-factors in the community.

REFERENCES

ALDOVÁ, E., SUCHÁ, J. (1973). Shigella typing in Czechoslovakia. J. Hyg. Epidemiol. Microbiol. Immunol. 17: 184–201.

DATTA, N. (1965). Infectious drug resistance. Brit. Med. Bull., 21: 254–259.

DAVIES, J. R., FARRANT, W. N., UTTLEY, H. C. (1970). Antibiotic resistance of *Sh. sonnei*. Lancet, 2: 1157–59.

HORÁK, V. (1971). R factors of *Escherichia coli* strains. Folia Microbiol., 16: 317.

ROSS, S., CONTRONI, G., KHAN, W. (1972): Resistance of Shigellae to ampicillin and other antibiotics. Its clinical and epidemiological implications. J. Am. Med. Ass., 221: 45–47.

SEKANINOVÁ, G. et al. Kolicinptypizace v epidemiologii bacilarni dyzenterie. In press.

WATANABE, T. (1963). Infective heredity of multiple drug resistance in bacteria. Bact. Rev., 27: 87–115.

R-PLASMIDS IN HEALTHY CHILDREN: STUDIES IN NURSERIES, KINDERGARTENS AND OTHER CLOSED AND OPEN COMMUNITIES

V. LIPKOVÁ, A. SEČKÁROVÁ and V. KRČMÉRY
Research Institute of Hygiene, Bratislava, Czechoslovakia

From the ecological point of view, it is important to know how frequently transmissible antibiotic resistance plasmids occur in the gastrointestinal tract of a healthy city and country population. This reservoir of R-plasmids, together with R-plasmids in hospital and medical institutions and patients and animal farms, may play a significant role in influencing the frequency of infectious diseases caused by antibiotic-resistant bacteria. Datta (1969) reported that in 52% of patients R plasmid-bearing strains of *E. coli* occurred *before* admission to hospitals in London. Moorhouse (1969) showed in Dublin that 81% of clinically healthy babies carry R plasmid-bearing strains in their gut. Richmond (1972) pointed out the special sensitivity of the children population to the carriage of R plasmid-bearing commensal strains of *E. coli*. 73% of children and about 50% of adults in his materials were found to be carriers. Tschaepe et al. (1973) found similar data in extensive studies in the German Democratic Republic.

Undoubtedly, R plasmids represent even in non-pathogenic or conditionally pathogenic strains of human bacterial flora a serious threat to the therapeutic effectiveness of the antibiotics concerned: They may be transmitted to dangerous pathogenic species or even cause organic diseases, e.g. urogential infections of the same carrier.

In a long-term project of sanitary improvement of environmental conditions in children's collective institutions from the standpoint of hygiene, we undertook a prolonged study of the occurrence of R-plasmids in groups of children, living in or attending.

1. pre-school institutions,
   a) nursery schools (from 0–3 years of age),
   b) kindergartens (from 3–6 years of age),
2. living constantly in a family environment (same age),
3. living constantly in a collective of medical institution
   a) in a newborn ward (for many weeks),
   b) in a children's home (permanent inhabitation).

Two factors were especially followed:

a) the influence of age,
b) the influence of the living milieu on the carriage of R plasmid-bearing strains.

Rectal swabs of children and personnel and prints from environmental facilities were obtained by conventional techniques and processed on MacConkey Agar with 20 mcg/ml of tetracycline and without the drug (unselected clones). Transfers were performed using a nalidixic acid resistant recipient of *E. coli* K_{12}.

RESULTS

1. *R. plasmids in healthy children in nursery schools* (0–3 years), 5-days-a-week regimen. The incidence of R^+ strains of *E. coli* in the gut of the children in nursery schools was found to be particularly frequent, and a vast majority of such children was found to carry R-plasmids, mostly coding for T resistance, but spectra such as TA, TS, TSA, TCA were, also observed. Transferable colicinogenic factors were found to be also common in their *E. coli strains*. As they are obviously much more frequently treated with various antibiotics (Bakičová et al., 1972), this pressure may unfavourably combine with the fact of a frequent presence of fertility factors in children of a given age. Toilet facilities were found to be the main factor contributing to the spread of R^+ strains among the collectives of children, although personnel was also found to be a carrier factor at a significantly higher percentage than in institutions taking care of older children. An exchange of strains between adult personnel and children, and among the children themselves plus the contamination of the toilets should be suspected. Repeated swabbing of the children in certain periods of the year, e.g. at the beginning and at the end of the school-year, on admission to a new school and even to a new class, etc. was found to be important not only to reveal the dynamics of the spread of R-plasmids, but, as we shall demonstrate, to obtain exact data about their inter-school dissemination. The qualitative composition of the fecal flora of children of this age is significantly variable and unstable and is strongly influenced mainly by the composition of food, by living environment, persons of constant contact, drugs administered, etc. Another microbial factor was found to be of importance, i.e. the occurrence of *Proteus sp.* epidemies which time by time occurred in a non-manifested form among children in the institutions followed. During the *Proteus* infestation of the milieu we were always able to note a significant increase of R-plasmids in fecal isolates of *E. coli* in the children involved.

2. *R-plasmids in healthy children in kindergartens* (3–6 years). In these groups of children, the season factor was especially followed, i.e. the situation at the end of June when children were leaving the kindergarten for vacation, and again in September when new children from different nursery schools of the city pooled into the first class of the kindergartens studied. Each child was repeatedly examined at the very beginning of its attendance of the kindergarten, and each month after. Personnel and toilets as well as kitchen facilities were under surveillance, too.

In June 1972, at the end of the school-year, the nursery schools studied seemed to be in a very good condition as regards the occurrence of R-plasmids: very few children carried R-plasmids and almost all of them proceeded to regular elementary schools, so that the carriers in the ecological system of kindergartens were almost free of detectable R-plasmids. But the situation drastically changed in September. We followed several children possessing R-plasmids in their fecal *E. coli* flora going from a nursery school's last class to their respective kindergarten and found not only the prolonged occurrence of R-plasmids in the children involved but also a significant and previously never experienced contamination of the kindergarten: a large number of kindergarten children were found after vacation to have R-plasmids in their gut. This may be a consequence of the combination of two factors: new children coming, plus vacational pick-up of new microflora from steadily increasing reservoirs of R^+ bacteria. But this deterioration became in fact permanent and never improved to the original favourable R-plasmid-less situation which had existed before. Thus, no spontaneous loss or dilution of plasmids occurred in the current school-year 1972/73 and after, despite the fact that no significant antibiotic therapy of children was recorded. In contrast to the June 1972 results in adult members of personnel who were also found to be plasmid-less, we could find, in the September

1972 to June 1973 examinations, that many members of the nursing staff now carry R-plasmid-bearing *E. coli* strains. The last examinations in kindergartens in September 1973 showed that out of a total of 146 children swabbed, we could detect tetracycline-resistant *E. coli* in a clonal form in 104 children. *E. coli* strains from 27 children transferred

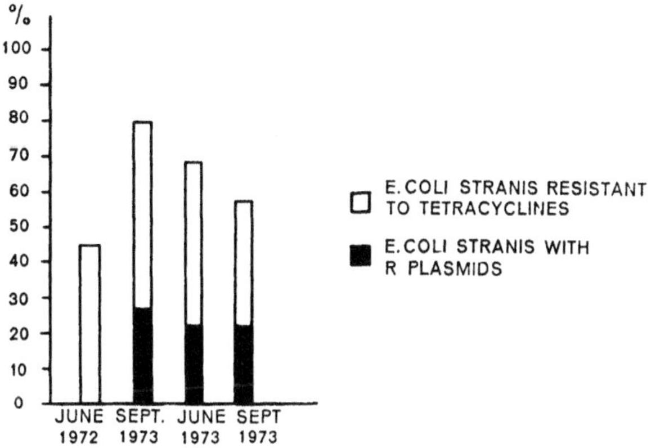

Fig. 1. Persentace of tetracyclinĕ-resistant strains of *E. coli* and strains vith R plasmids in the rectal swabs of children from the kindergardens

their resistance to tetracycline and other drugs, comparing to only one child in June 1972. Out of 24 persons of the staff, 15 shed *E. coli* clones resistant to the antibiotic and in six persons transferable resistance plasmids could be demonstrated, comparing with none in June 1972. The dynamics of R plasmids, in children from kindergartens can be seen in Fig. 1.

3. *R plasmids in gut flora of children in family care,*

a) *0–3 years of age, no school attendance,*

b) *9–10 years of age—attending school.*

The proportion of R-plasmid-bearing *E. coli* strains in the younger group of children was significantly higher, which can be perhaps ascribed to the fact that the majority of children in this age-group were younger than 1 year and as such experienced recent inhabitation in a hospital environment as well as a high frequency of domestic or institutional antibiotic treatment. In 37 out of 41 children examined as living at home, and aged under 3 years, antibiotic — resistant *E. coli* clones could be isolated and in 12 children also the presence of R-plasmids. In the group of home-living children, aged 9 to 10 years, altogether 28 children were repeatedly swabbed, but resistant *E. coli* could be scored only in three children, and none of them could be identified as bearing R-plasmid-containing *E. coli.* We shall see below, i.e. in older children from the childrens' homes that, despite their institutional inhabitation, their gut flora is almost plasmid-free; this suggests that the factor of age (representing the stability of gastrointestinal flora and a reduced frequency of antibiotics administration) is of considerable importance since, may be, the older children cannot be colonised by exogenous *E. coli* strains for a sufficient period of time.

4. *R-plasmids in children living permanently in collective institutions*

a) in a newborn ward (a type of all-week nursery, age 0–2 years).

This group of children significantly differed from both previously depicted groups of children of the same age: children living permanently at home and children attending nursery school. Despite their institutional inhabitation for a part of the day, the latter children interrupt their contacts which school-mates and personnel when taken home each day in the afternoon. In contrast to this, the children studied here, i.e. located in newborn wards, are permanently in the same environment. In addition, their morbidity frequency demands repeated antibiotic administration. This corresponds to the observed variety of multiple resistance patterns of a vast majority of bacteriological isolates from rectal swabs of children in this group.

32 children aged 0–6 months were periodically swabbed and *in all of them* antibiotic-resistant *E. coli* could be permanently isolated in a clonal form. It should be noted that babies under one month of age were as a rule nursed by their mothers. In 14 children, the occurrence of R-plasmid-bearing *E. coli* could be demonstrated.

In all 20 babies aged 6 to 24 months, antibiotic resistant, mostly multiple resistant *E. coli* could be also permanently demonstrated, and 5 babies shedded clones with transferable resistance factors. In the newborn wards studied, *Proteus sp.* frequently accompanied *E. coli* isolates from rectal swabs and the Proteus clones, as a rule, showed identical or very similar multiple drug resistance patterns, e.g. additional resistance to chloramphenicol, neomycin plus kanamycin, nalidix acid, etc. not observed in the previous groups of children investigated. It should be mentioned that the morbidity represented mostly upper respiratory infections, but not infrequently those of the gastrointestinal tract. For URI therapy, the children were given mostly tetracyclines and penicillins including oxacillin. Ampicillin resistant *E. coli* in the feces was a common observation in almost all children. Streptomycin and neomycin was given to diarrheatic babies with corresponding resistance findings in the gut of *E. coli* isolates.

b) in a children's home, aged 3 to 6 years.

All 28 children of this age, living in another home, were periodically swabbed: again in all of them antibiotic-resistant enteric bacteria could be permanently isolated. In 13 children of this group the presence of strains with transferable resistance could be demonstrated. As the multiple antibiotic resistance spectrum of *E. coli* strains from all children was found to be identical or very similar, maybe there occurs a permanent circulation of a single or very few *E. coli* strains among the children.

c) in a children's home, aged 6 to 15 years.

The regimen of the home involved was that children from 6 to 9 years of age were permanently living in the home, having their school classes under the same roof. In contrast, to this children aged 9 to 15 years attended an extramural school.

Out of the 28 children aged 6 to 9 years, 23 were found to possess antibiotic resistant enteric bacteria, but a transferable character of this resistance could be demonstrated in only 6 children.

The resistance spectra were confined mostly to tetracycline alone or tetracycline plus ampicillin and were significantly poorer than in younger children in the above homes.

31 older children, i.e. aged 9 to 15 years, were under study and in 29 we could isolate clonal forms of antibiotic-resistant enterobacteria. This is significantly higher than in children of the same age groups but living at home with their families (see above). However, none of the strains transferred resistance in our experiments and this con-

trasted with groups of younger children but was almost similar to children of the same home but aged 6 to 9 years. We cannot yet decide whether the low incidence of plasmids is a specifity of the home itself or reflects the higher age factor as in the above studies, and more institutions of this type are under study from the point of view of this problem. See Fig. 2 for comparison of results in different age groups.

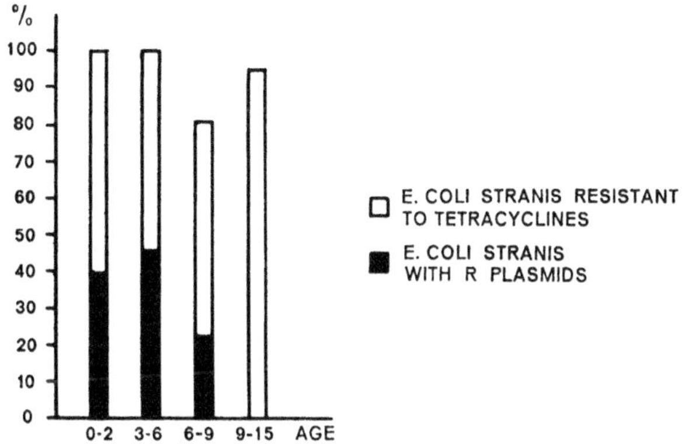

Fig. 2. Percentage of tetracycline-resistant strains of *E. coli* and strains with R plasmids ir rectal swabs of children liwing in childrens' homes

The wards, homes, institutions and families involved are continuously under study to reflect the dynamics of the feature of the emergence of resistance plasmid-bearing strains in the rectal swabs of children involved. Further studies are now in progress in newly included homes, etc., to obtain more precise correlations mainly of relations between antibiotic administration for therapy of children and the appearance, persistance, circulation and eventual dissapearance of (multiple) resistant strains of *E. coli* in fecal flora. Emphasis is also put on a more clear-cut evaluation of both age and institutional factors which, as we believe, are potent regulation elements of the ecological evaluation of R-plasmid occurrence in communities of healthy children.

REFERENCES

BAKIČOVÁ, Z. (1971). Long-term follow-up of antibiotic administration to children up to 3 years of age. Advances in antimicrobial and antineoplastic chemotherapy. Avicenum, p. 1167–1170.

DATTA, N. (1969). Drug resistance and R factors in the bowel of London patients before and after admission to hospital Brit. med. J. (2), 407–410.

LIPKOVÁ, V., SEČKÁROVÁ, A., KRČMÉRY, V. (1973). Incidence of R plasmids in children. J. Hyg. epidemiol. microbiol. immunol. 17, 368–374.

MOORHOUSE, E. C. (1969). Transferable drug resistance in enterobacteria isolated from urban infants. Brit. med. J. 2, 405–407.

RICHMOND, M. H. (1972). Some environmental consequences of the use of antibiotics. J. applied Bacteriol. 35, 155–176.

TSCHÄPE, H., KÜHN, H., RISCHE, H. (1973). R plasmide der *E. coli-flora* gesunder Kinder. J. Hyg. epid. microbiol. immunol., 17, 208–215.

INFLUENCE OF COLICINOGENIC STRAIN, GIVEN PERORALLY, ON THE OCCURRENCE OF VARIOUS PLASMIDS OF E. COLI IN PIGLETS FED WITH FLAVOMYCIN

A. SOKOL and F. FEDERIČ

Department of Microbiology and Zoohygiene, University of Veterinary Medicine, Košice, Czechoslovakia

The most widely spread and most studied plasmids of *E. coli* include: R. plasmids — conditioning resistance in the host strain to one and more often simultaneously; to several antibiotics. Col factors — determining the production of colicines; and the Hly factor — controlling the ability to produce alfa-hemolysine in the host strains (H. Williams Smith, 1969).

Domestic animals kept in large-scale breeding stations at present contain in their intestinal tract almost exclusively a tetracycline-resistant population of *E. coli*, in view of their being continuously fed additional chlortetracycline in nutritions doses.

Watanabe et al. (1972) described increased sensitivity of the *E. coli* cells carrying R plasmids, F factors, or Col factor Ib as well as the cells of *S. typhimurium* carrying R plasmids to flavomycin. The fact that derepressed types of the above mentioned plasmids lend to the host cells more pronounced sensitivity to flavomycin than the repressed types of the same plasmids has led the authors to the conclusion that the increased sensitivity to flavomycin is probably caused by the presence of sex pili on the host cells. The ability to produce sex pili is connected with the ability of the cells to function as a donor in the conjugation test. The presence of sex pili on the cell shows the presence of autotransferable plasmid in the cell (Meynell and al., 1968).

In the recent period data have been acumulating about papers connected with the mutual relationships of different categories of plasmids *in vitro* (Meynell and al., 1968; Datta, Hedges, 1971; Datta, Hedges, 1972). However, data about mutual relationships and possibilities of the interaction of different categories of plasmids *in vivo* are missing.

In the presented work we report about the rate of decrease of the three mentioned categories of plasmids in *E. coli* isolated from piglets additionally fed with flavomycin which were simultaneously given perorally a colicinogenic strain of *E. coli*.

MATERIALS AND METHODS

In pilot-plant conditions using 100 piglets over 28 days we examined the influence:

a) of flavomycin given at a dose of 10 mg/kg of food;

b) of flavomycin given at a dose of 10 mg/kg of food together with peroral application of a colicinogenic strain of *E. coli* (a weekly dose of 5×10^9 of *E. coli* Col Ia$^+$ for one piglet);

This investigation was carried out to establish the occurrence of *E. coli* markers (resistance to antibiotics, haemolytic activity and colicine production) which may be determined by plasmids. Each of the experimental and control groups consisted of 25 piglets.

Division of groups:

A) 1. The experimental group was for 28 days given additionally 10 mg of flavomycin per 1 kg of standard food mixture containing 28 mg of chlortetracycline (CTC)/kg. From the 28th day the group was fed a standard food mixture containing 28 mg CTC/kg mixture up to the conclusion of the experiment (56 days).

2. The control group was throughout the whole experiment (56 days) given a standard food mixture with a content of 28 mg of CTC/kg.

B) 1. The second experimental group had for 28 days added flavomycin at a dose of 10 mg flavomycin per 1 kg of standard food mixture with a content of 28 mg CTC/kg. In addition to this, the piglets received a colicinogenic *E. coli* strain at a dose of 5×10^9 for one pig per os in weekly intervals throughout the whole experiment (56 days).

2. The second control group was throughout the whole experiment (56 days) given a standard food mixture containing 28 mg CTC/kg. In addition, the piglets received weekly perorally 5×10^9 of *E. coli* Col$^+$ per one pig thoughout the whole experiment (56 days). The applied colicinogenic strain is a derivative of the *E. coli* strain 0–138, fsolated from a weanling, which originally did not carry a R factor, Col factor, or Hly factor, and to which the Col Ia factor was transferred by conjugation from the standard strain of *E. coli* CA53 (Kindly provided by J. Šmarda, from the Department of Biology, Faculty of Medicine, J. E. Purkyně University, Brno). A suspension of the *E. coli* germs 0–138 Col Ia$^+$ in physiologic solution at a concentration 5×10^9 of living germs/ml was used for peroral application. The suspension was applied after a thorough mix-up in the food for the whole group of animals (25 pieces) at once. One group of animals then received 25 ml of the suspension at a concentration 5×10^9 of *E. coli* 0–138 Col Ia$^+$/ml.

The experimental and control groups consisted of pigs of a weight of 21–27 kg so that each group included an equal number of animals. In case of possible cross contamination, the animals were subjected to double thorough disinfection (mechanical and chloramine) both before application of flavomycin — as well as after its discontinuation.

The intestinal flora of the experimental and control animals was examined before, during and after discontinuation of the application of flavomycin at 7-day intervals. Rectal swabs from the experimental and control animals were spread on the surface of a dish with MacConkey Agar DIFCO so that the individual colonies could grow after overnight incubation at 42–43°C. Five colonies of *E. coli* were taken from every tested sample and were submitted to a sensitivity test to tetracycline (method of monoantibiotic strips, Garrod, O'Grady, 1971); to a test for colicine production with an indicator strain of *E. coli* B1 (Šmarda, 1965), and to a test for haemolysis production (by spreading on blood agar prepared from defibrinated sheep blood).

All the colonies taken were at the same time submitted to a simple conjugation test by a derivative of *E. coli* strain K 12 — *E. coli* 185 Nx (Federič and al., 1972) for ability of T-resistance transfer.

RESULTS

The results obtained in this experiment are given in Tables 1, 2 and 3. All the percentual values shown in the Tables express the proportion of the single marker carrying *E. coli* strains in the total count of 125 strains (25×5 strains $=$ 125 strains at each taking in

TABLE 1

Dynamics of percentual occurrence of E. coli strains with mobile T-resistance

	Day		Feeding period of flavomycin				Discontinuation of flavomycin			
		0.	7.	14.	21.	28.	35.	42.	49.	56.
Colicino-genic strain not given	Group fed flavomycin	41.6	12.8	11.2	11.2	14.4	38.4	36.8	46.4	42.4
	Control group	38.4	41.6	35.2	37.6	42.4	46.4	41.6	36.8	36.0
Weekly perorally given 5 . 10^9 *E. coli* col$^+$ per piglet	Group fed flavomycin	36.8	11.2	8.0	8.0	6.4	17.6	34.4	40.8	38.4
	Control group	38.4	36.0	32.0	33.6	38.4	36.8	32.0	29.6	28.0

TABLE 2

Dynamics of percentual occurrence of haemolytic E. coli strains

	Day		Feeding period of flavomycin				Discontinuation of flavomycin			
		0.	7.	14.	21.	28.	35.	42.	49.	56.
Colicino-genic strain not given	Group fed flavomycin	19.2	16.0	4.0	4.8	4.0	8.0	22.2	18.2	24.0
	Control group	17.6	21.6	19.2	19.2	16.0	17.6	22.2	20.0	19.2
Weekly perorally given 5 . 10^9 *E. coli* col$^+$ per piglet	Group fed flavomycin	18.2	8.0	3.2	2.0	4.8	7.2	16.8	20.0	22.2
	Control group	20.0	16.0	14.4	18.2	16.8	12.8	20.0	23.2	22.2

TABLE 3

Dynamics of percentual occurrence of colicinogenic E. coli strains

	Day		Feeding period of flavomycin				Discontinuation of flavomycin			
		0.	7.	14.	21.	28.	35.	42.	49.	56.
Colicino-genic strain not given	Group fed flavomycin	34.4	22.2	23.2	16.0	14.4	25.6	36.4	33.6	30.4
	Control group	30.4	27.2	33.6	34.6	29.7	29.7	31.2	37.6	28.0
Weekly perorally given 5 . 10^9 *E. coli* col$^+$ per piglet	Group fed flavomycin	34.6	25.6	28.8	20.0	18.4	40.0	60.0	43.2	45.7
	Control group	28.0	49.6	46.4	44.0	48.0	51.2	42.4	45.7	43.2

one group of pigs). In the course of the whole experiment no tetracycline sensitive strain of *E. coli* was isolated.

I. During the period of additional feeding of flavomycin, the following results were obtained in the individual groups:

A) 1. In the experimental group, which was given flavomycin, the initial 41.6% occurrence of mobile R plasmids decreased to 11.2%. The incidence of haemolytic *E. coli* strains also significantly declined — from 19.2% to 4%. The initial 34.4% occurrence of colicinogenic strains decreased to 14.4%.

2. In the control group which was not administered flavomycin, the occurrence of tested markers varied in the same period within narrow limits — in mobile R plasmids between the initial occurrence of 38.4% to a minimum of 4%, in haemolytic activity between 17.6% to a minimum of 4%, and in colicinogeny between 30.4% to a minimum of 4.2%.

B) 1. In the experimental group which was additionally given flavomycin and which received simultaneously colicinogenic *E. coli* strain we observed a significant drop in mobile R plasmids (from 18.2% to 2%), and a less marked drop in the occurrence of colicinogeny (from 34.6% to 18.4%).

2. In the control group which received a colicinogenic strain of *E. coli* (without additional feeding of flavomycin), the occurrence of mobile R plasmids varies within narrow limits (from 38.4% to a minimum of 6.4%), the occurrence of haemolytic activity is decreasing (from 20.0% to 14.4%), but the occurrence of colicinogeny increases (from 28.0% to 49.6%).

II. In the period when flavomycin was discontinued (28–56 days), the following results were obtained in the individual groups:

A) 1. In the experimental group which had previously received flavomycine we observed a gradual return of the percentual occurrence of the mobile R plasmids, when the values obtained even exceeded the initial occurrence to 46.4%). Gradual return to the initial value, and even exceeding of the initial value can be seen also in the incidence of haemolytic *E. coli* strains (to 24.0%). Exceeding of the initial values can be observed also in the occurrence of colicinogenic strains (to 36.4%).

2. In the control group which did not previously receive flavomycin, the occurrence of markers examined varied in the same period within narrow limits — in mobile R plasmids the occurrence varied by a maximum of 8%, in haemolytic activity by 4.6%, and in colicinogeny by about 7.2%.

B) 1. In the experimental group which was previously added flavomycin, and which received simultaneously colicinogenic *E. coli* strain, we observed a gradual return of the values of mobile R plasmids to the original values, which in some cases were exceeded (to 40.8%). The initial values were also exceeded in the occurrence of haemolytic *E. coli* strains (to 22.2%). However, the occurrence of colicinogenic strains was in the same period markedly increased (to 60.0%), and later remained at relatively high values (43.2% and 45.7%).

2. In the control group which had received a colicinogenic strain of *E. coli* (without previous additional feeding of flavomycin), the occurrence of mobile R plasmids is gradually but less markedly decreasing to 28.0% which, in comparison with the initial values, represents a decrease by about 10.4%. The occurrence of haemolytic *E. coli* strains gradually returns to the initial value, and even exceeds it (to 23.2%). However, the occurrence of colicinogenic strains remains in the same period, at relatively high values to 51.2%).

CONCLUSION

Flavomycin applied at a dose of 10 mg/kg feed in combination with the currently used chlortetracycline at a dose of 28 mg/kg feed administered for 28 days induces a marked decrease in the incidence of *E. coli* strains with mobile tetracycline resistance in the intestines of animals as well as in the incidence of haemolytic and colicinogenic *E. coli* strains. This favourable depressive effect of flavomycin on the occurrence of *E. coli* strains carrying different plasmids is not significantly higher at 10 mg than at 5 mg flavomycin/kg of the ratio (Sokol et al., 1973); Federič et al., 1973).

Simultaneous peroral application of the colicinogenic *E. coli* strain (weekly 5×10^9 E. coli Col^+ per weanling) leads to a more pronounced decrease in the percentual occurrence of *E. coli* strains carrying different plasmids. The results obtained therefore suggest the possibility that *E. coli* strains in the intestines of animals carrying R plasmids and the Hly factor are replaced by the newly accepted *E. coli* strain carrying Col Ia factor.

In the field, the above-listed results are important from several points of view:

a) From the viewpoint of utilizing the growth stimulating effect of antibiotics under conditions of intensive large-scale animal husbandry, breeders should use exclusively such "nutritional" antibiotics (e.g. flavomycin) which cannot be used parenterally as therapeutics, which are not resorbed enterally, which do not leave any residues in the organism, and at the same time depress the appearance and circulation of mobile plasmids in the intestines of animals;

b) This procedure would fulfil, both in human and in veterinary medicine, the principal requirement of hygienists regarding the reduced transfer of multiresistance from *E. coli* to enteropathogenic germs.

REFERENCES

DATTA, NAOMI, HEDGES, R. E. (1972). "Host Ranges of R Factors." J. Gen. Microbiol., 70, p. 453–460.

FEDERIČ, F., SOKOL, A., KOPPEL, Z. (1972). "Frequency of Colicine Production and Occurrence of Transmissible Resistance in Haemolytic Strains of *E. coli*, Isolated from Diarrhoetic Pigs." Bacterial Plasmids and Antibiotic Resistance, Avicenum, Prague, p. 147–150.

FEDERIČ, F,, SOKOL, A., RAJTÁR, V., ŠIMANSKÝ, Š., HAMAROVÁ, Z., POLÁKOVÁ, A. (1973). "The Dynamics of Changes of Resistance to Antibiotics in Pigs Fed Flavomycin." Folia Microbiologica ASB, 18, 2, 176–177.

GARROD, L. P., O'GRADY, F. (1971). "Antibiotics and Chemotherapy." E. and S. Liwingstone, London.

HEDGES, R. W., DATTA, NAOMI. (1972). "R 124an fi^+ R Factor of a New Compatibility Class." J. Gen. Microbiol., 71, 403–405.

MEYNELL, E., MEYNELL, G. G., DATTA, NAOMI. (1968). "Phylogenetic Relationships of Drug-Resistance Factors and other Transmissible Bacterial Plasmids." Bacterial. Rev., 32, 55–83.

SMITH, H. WILLIAMS. (1969). "Veterinary Implications of Transfer Activity." Bacterial Episomes and Plasmids. A Ciba Foundation Symposium, J. and A. Churchill Ltd. London, 213–226.

SOKOL, A., KRČMÉRY, V., FEDERIČ, F., RAJTÁR, V., JANOUŠKOVÁ, A. (1973). "The Influence of Flavomycin on the Elimination of R Factors of *E. coli* in vivo." Folia microbiologica ASB, 18, 2, 176.

ŠMARDA, J. (1965). "Lysogenie a bakteriocinogenie." Medical Faculty, J. E. Purkyně University, Brno.

WATANABE, T., OGATA, Y., SUGAWARA, K., ODA, K. (1972). "Increase of Flavomycin Sensitivity of Bacteria by R Factors." Bacterial Plasmids and Antibiotic Resistance, Avicenum, Prague, 105–113.

OCCURRENCE OF VARIOUS CATEGORIES OF PLASMIDS IN HAEMOLYTIC E. COLI STRAINS ISOLATED FROM FREE-LIVING ANIMALS

F. FEDERIČ[1], A. SOKOL[1], M. ŠPENÍK[2] ard O. HARVAN[2]
1 Department of Microbiology and Zoohygiene, University of Veterinary Medicine, Košice;
2 Institute of Hunting Animals, University of Veterinary Medicine, Košice

World literature presents a great number of data about the occurrence of plasmid-determined markers in representatives of the family Enterobacteriaceae in humans as well as in domestic animals. In particular, there exist a great many data about the percentual occurrence of R plasmids in *E. coli* (H. Williams Smith, 1973) and in representatives of the genus Salmonella (C. E. Voogd et al., 1973). We have also data about the occurrence and the transmission of the Hly factor plasmid, determining the production of "alfa" haemolysin by the host strain in *E. coli* strains isolated from men and from domestic animals (H. Williams Smith, Sheila Halls, 1967; H. Williams Smith, 1969). There also exist papers dealing with the occurrence and the relationship of R plasmids, Col factors and Hly factors in *E. coli* strains isolated from farm animals (H. Williams Smith, C. L. Gyles, 1970; F. Federič et al., 1972).

However, we miss data about the occurrence of various above mentioned plasmid categories in haemolytic *E. coli* strains isolated from free-living animals.

MATERIALS AND METHODS

Ninety haemolytic strains of *E. coli* were examined which had been isolated from free-living animals from the school-grounds of Rozhanovce, University of Veterinary Medicine. A set of 90 examined strains was taken from a collection of haemolytic strains of *E. coli* isolated between 1971 and 1973 at the Institute of Hunting Animals, University of Veterinary Medicine, Košice. The strains were kept by periodical (6 month intervals) subcultivations on Dorset medium. The whole set of examined strains consisted of:

a) 30 haemolytic strains of *E. coli* isolated from game-birds (21 strains isolated from pheasants, four from partridges, three from quails and two from wild ducks); the animals of this group were given tetracycline (by additional feeding) in doses up to 100 mg of tetracycline/kg of food;

b) 30 haemolytic strains of *E. coli* isolated from birds of prey (eight strains isolated from owls, eight from falcons, eight from hancks and six from eagles); the animals of this group have never directly touched antibiotics;

c) 30 haemolytic strains of *E. coli* isolated from deer (six strains from roebucks, two from fallow deer and two from red deer) and small fur-bearing animals (15 strains isolated from hares and five from rabbits); the animals of this group, too, did not come into direct contact with antibiotics;

TABLE 1

Occurence of various plasmid categories in haemolytic E. coli strains isolated from different groups of free-living animals

Animals	T-resistance			Colicinogenic activity			Haemolytic activity			Sum of 3 markers examined		
	Total	Mobile	Mobilizable	Total	Mobile	Mobilizable	Total	Mobile	Mobilizable	Total	Mobile	Mobilizable
Game-birds	18	8	2	10	0	2	30	1	4	50/90	9	8
Birds of prey	8	5	2	9	1	2	30	2	2	47/90	8	6
Deer and small furbearing animals	10	6	2	5	1	2	30	4	4	45/90	11	8

The examined strains were tested:

1. for haemolytic activity. This test was carried out by injection of the strains on a blood agar plate, prepared from defibrinated sheep-blood;

2. for sensitivity to tetracycline; this test was carried out on blood agar by using antibiotic discs;

The examined strains were further submitted:

I. to simple conjugation tests (F. Federič and al., 1972) with *E. coli* 185 Nx a derivative of the *E. coli* K12 strain;

II. to the plasmids mobilization tests (F. Federič et al., 1972) with the use of the following strains:

a) *E. coli* 50 $Col^- Hly^+$; this strain was isolated from a weanling at the Department of Microbiology and Zoohygiene, University of Veterinary Medicine; in simple conjugation tests, this strain transmitted the Hly factor to the *E. coli* 185 Nx recipient with a relatively high frequency (10^{-2});

b) *E. coli* CA 53-type, colicin Ia producing strain, kindly provided by J. Šmarda from the Institute of Biology, J. E. Purkyně University, Brno;

c) *E. coli* dispar P14-type, colicin E1 producing strain, kindly provided by J. Šmarda.

All these strains fulfilled the role of transfer factor donors. 90 haemolytic strains of *E. coli* were used as intermediary recipients. The strain used as final recipient was again *E. coli* 185 Nx.

Hybrids from selective media were tested for the ability of haemolytic activity, for their resistance to tetracycline and for their colicinogenic ability.

RESULTS

The occurrence of various categories of plasmids in haemolytic *E. coli* strains isolated from different groups of free-living animals is given in Table 1.

I. a) The 90 strains which were examined included 36 strains resistant to tetracycline, i.e. 40.0% (36 : 90). The highest percentage of strains resistant to tetracycline — 60% (18 : 30) — occurs in strains isolated from game-birds, which can be explained by selective pressure of therapeutically used antibiotics. A high percentage of tetracycline-resistant strains can be also observed in strains isolated from birds of prey — 26.7% (8 : 30) and from deer and small fur-bearing animals — 33.3% (10 : 30).

b) Out of the 90 examined strains there were 24 strains i.e. 26.7% which displayed colicingenic activity; the highest percentage of them — 33.3% (10 : 30) were isolated from the group of game-birds. The occurrence of this marker in strains isolated from birds of prey was 30% (9 : 30) and in strains isolated from deer and small fur-bearing animals 16.7% (5 : 30).

c) All the strains examined showed haemolytic activity.

II. a) The 36 tetracycline-resistant strains included 19 strains (52.8%) able to transfer this marker in simple conjugation tests. Strains isolated from different groups of animals showed the following representation of strains carrying an autotransferable determinant of T-resistance: game-birds — 44.4% (8 : 18), birds of prey — 02.5% (5 : 8) and deer and small fur-bearing animals — 60% (6 : 10).

b) Out of the all 24 colicinogenic strains only 2 strains, i.e. 8.3%, transferred this marker in simple conjugation tests. Strains isolated from different groups of animals showed the following representation of strains carrying an autotransferable Col factor: game-birds — 0% (0 : 10), birds of prey — 11.1% (1 : 9) and deer and small fur-bearing animals — 20.0% (1 : 5).

c) Out of all the 90 haemolytic *E. coli* strains, haemolytic activity was transmitted in simple conjugation tests in 7 strains, i.e. 7.8%. The strains isolated from different groups of animals showed the following representation of strains carrying an independently transmissible Hly factor: game-birds — 3.3% (1 : 30), birds of prey — 6.7% (2 : 30) and deer and small furb-earing animals — 13.3% (4 : 30).

III. a) Out of the 36 tetracycline-resistant strains there were 6 strains i.e. 16.7% (6 : 36) able to be mobilized in at least one of the three applied systems of plasmid mobilization tests. The strains isolated from different groups of animals showed the following representation of a mobilizable determinant of T-resistance: game-birds — 11.1% (2 : 18), birds of prey — 25.0% (2 : 8), deer and small fur-bearing animals — 20.0% (2 : 10).

b) The 24 colicinogenic strains included 6 strains, i.e. 25% in which it was possible to mobilize the non-autotransferable Col factor in plasmid mobilization tests by using *E. coli* 50 Hly^+ Col^- strains. In strains isolated from different groups of animals the occurrence of a mobilizable Col factor was as follows: game-birds — 20.0% (2 : 10), birds of prey — 22.2% (2 : 9), deer and small fur-bearing animals — 40.0% (2 : 5).

c) The 90 haemolytic *E. coli* strains included 10 strains, i.e. 11.1% in which it was possible to mobilize the non-autotransferable Hly factor in plasmid mobilization tests by using *E. coli* CA53 and *E. coli* dispar P14 strains as donors of the transfer factor. The strains isolated from different groups of animals showed the following occurrence of strains with a mobilizable Hly factor: game-birds — 13.3% (4 : 30), birds of prey — 6.7% (2 : 30), deer and small fur-bearing animals — 13.3% (4 : 30).

The genetic properties of the examined strains are documented in Table 2, "Number, nature and relationships of the markers in 90 haemolytic strains of *E. coli* isolated from free-living animals". The results listed in this Table show that in *E. coli* strains isolated from free-living animals simultaneous transfer of plasmids is relatively rare. It can be observed only in two strains with genotype Hly, T (both markers are transferred) and in two strains with genotype Hly, T, Col (Hly and T are transferred). Much more frequently we can observe the presence of two or three plasmid-determined markers which cannot be transmitted simultaneously.

The results of our investigations show that the more plasmid-determined markers the host strains carry, the greater is the possibility that at least one of the carried plasmids is autotransferable.

Summing up, a total of 90 haemolytic *E. coli* strains were tested for the presence of autotransferable determinants of tetracycline resistance (T), Col factors and Hly factors by the simple conjugation test with *E. coli* 185 Nx. The strains carrying non-autotransferable plasmids were also subjected to plasmid mobilization tests using *E. coli* 50 Col^- Hly^+, E. coli CA 53 and *E. coli* dispar P14 strains as transfer factor donors.

a) Out of 30 strains isolated from *tetracycline- treated game-birds* 18 appeared to be T-resistant, 8 of these strains carrying an autotransferable T-resistance determinant and in 2 other strains the non-autotransferable T resistance determinant was mobilizable in at least one of the applied plasmid mobilization test systems, which means that the result obtained was 18/8/2. None of the 10 colicinogenic strains carried the autotransferable Col factor; mobilizable Col factors were found in 2 strains, with the result 10/0/2. Out of 30 haemolytic strains only one strain appeared to have an autotransferable Hly factor; in 4 strains the Hly factor was mobilizable — 30/1/4.

b) In 30 strains isolated from *birds of prey* (*no previous contact with antibiotics proved*) the following plasmid categories could be stated: T resistance — 8/5/2, Col factors — 9/1/1 and haemolytic activity — 30/2/2.

c) In 30 strains isolated from *deer and small fur-bearing animals* (*no previous contact*

TABLE 2

Number, nature and relationships of the markers in 90 haemolytic strains of E. coli isolated from free-living animals

Number of markers carried	Genotype	Number of strains isolated from game: Productive feathered	Harmful feathered	Deer and fur-bearing animals	Total
	HEM.	7	17	14	38
1	Hly	0	0	2	2
	Hly	1	1	2	4
	HEM., T	5	0	1	6
	HEM., T	3	1	3	7
	HEM., T	1	1	2	4
	Hly, T	1	0	0	1
	Hly, T	2	0	0	2
	Hly, T	0	1	1	2
	HEM., Col	3	3	1	7
	HEM., Col	1	0	0	1
	Hly, Col	0	0	1	1
	Hly, Col	0	1	0	1
	HEM., T, Col	2	0	0	2
	HEM., T, Col	2	1	0	3
	HEM., T, Col	0	1	0	1
	HEM., T, Col	0	1	0	1
	HEM., T, Col	1	1	1	3
	Hly, T, Col	0	1	1	2
	Hly, T, Col	1	0	0	1
	Hly, T, Col	0	0	1	1
	SUM	30	30	30	90

Legend: HEM. – Haemolytic activity
Hly – Plasmid of haemolytic activity
T – Tetracycline resistance determinant
Col – Determinant of colicinogenic activity
□ – Autotransferable (mobile) marker
– Non-autotransferable, but mobilizable marker

with antibiotics proved) various plasmid categories were present as follows: T resistance — 10/6/2, Col factors — 5/1/2, haemolytic activity — 30/4/4.

At the same time the number, nature and relationships of the markers were determined in all the 90 haemolytic *E. coli* strains examined.

The results obtained provide wider knowledge of the ecology — the occurrence and circulation of various plasmid categories in *E. coli* even without the selective pressure of antibiotics.

REFERENCES

FEDERIČ, F., SOKOL, A., KOPPEL, Z. (1972). "Frequency of Colicin Production and Occurrence of Transmissible Resistance in Haemolytic Strains of *E. coli* Isolated from Diarrhoetic Pigs". Bacterial Plasmids and Antibiotic Resistance, Avicenum, Prague, p. 147–150.

SMITH, H. WILLIAMS. (1969). "Veterinary Implications of Transfer Activity", Bacterial Episomes and Plasmids. A Ciba Foundation Symposium; J. and A. Churchill Ltd., London, p. 213–226.

SMITH, H. WILLIAMS. (1973). "Effect of Prohibition of the Use of Tetracycline in Animal Feeds on Tetracycline Resistance of Faecal *E. coli* of Pigs." Nature 243, p. 237–238.

SMITH, H. WILLIAMS, GYLES, C. L. (1970). "The Relationship between Different Transmissible Plasmids Introduced by F into the same Strain of *E. coli* K-12." J. Gen. Microbiol., 62, p. 277–285.

SMITH, H. WILLIAMS, HALLS, SHEILA. (1967). "The Transmissible Nature of the Genetic Factor in *E. coli* that Controls Haemolysin Production." J. Gen. Microbiol., 47, p. 153–161.

VOOGD, C. E., GUINÉE, P. A. M., MANTON, A., VALKENBURG, J. J. (1973). "Incidence of Resistance to Tetracycline, Chloramphenicol and Ampicillin among Salmonella Species Isolated in the Netherlands in 1969, 1970 and 1971". Antonie van Leeuwenhoek, 39, 2, p. 321–329.

A PRELIMINARY REPORT ON COLISTIN-POLYMYXIN RESISTANCE WITH SUSPECTED PLASMID DETERMINATION WHICH DEVELOPED IN ESCHERICHIA COLI AND KLEBSIELLA PNEUMONIAE UNDER COLISTIN THERAPY

J. BOHUŠ and V. KULTAN

Dept. of Microbiology, District Station of Hygiene, Liptovský Mikuláš, Czechoslovakia

Monitoring a new-born children's ward for the presence of R-plasmids we found that, by the end of March 1974, nearly all new-born and premature children were colonized with a *Klebsiella pneumoniae* strain resistant to chloramphenicol (Cm), tetracycline (Tc), kanamycine (Kn) and ampicilline (Am). Resistance to Cm, Kn and Tc was transmissible. An enteropathogenic *E. coli*, serotype 055 had spread and caused enteritidis in premature babies. Because of Am resistance (nontransmissible) of this pathogen, 14 premature children were treated with colistin perorally with a catheter. Of these, 7 children showed colistin-resistant variants of both microbes during or after therapy.

Fig. 1 demonstrates the incidence of these variants in dependence on the treatment. No colistin resistant strains had been isolated in this ward before the critical period of

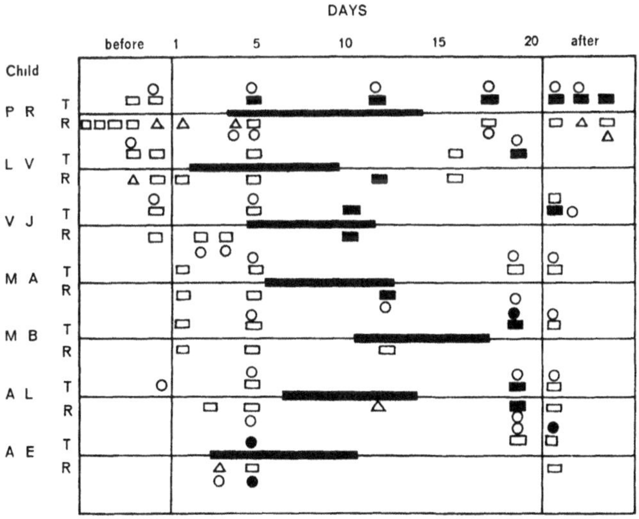

Fig. 1. Occurence of colistin resistant variants of *E. coli* 055 and *K. pneumoniae* in the dependence on colistin therapy of premature children. (*K. pneumoniae:* □ sensitive, ■ resistant; *E. coli* 055: ○ sensitive, ● resistant; △ other strains; T: throat, R: rectum; ▬▬: therapy)

20 days. This fact was reaffirmed by retesting of all the stored strains for colistin-resistance. This figure illustrates that colistin resistant variants of *K. pneumoniae* (black squares) and of *E. coli* 055 (black circles) were isolated from throat (T) and rectal swabs (R) during or after colistin therapy along with sensitive ones (empty squares and circles). In some children resistant variants emerged over the period of one month, or they disappeared after some shorter period, or the examination was stopped.

TABLE 1

Properties of colistin-resistant strains isolated from premature children

Child	T	Strain	Colistin resistance			Other determinants	
			MIC	TR	EL	present	transfered
A.E.	+	055	64	—	+	Am	
A.B.	+	055	64	—	+	Am	
		K.p.	320	—	+	Cm Tc Kn Am	Cm Kn (Tc)
M.A.	+	K.p.	320	—	+	Cm Tc Kn Am	Cm Kn (Tc)
V.J.	+	K.p.	320	—	+	Cm Tc Kn Am	Cm Kn (Tc)
A.L.	+	K.p.	128	—	—	Cm Tc Kn Am	Cm Kn (Tc)
P.R.	+	K.p.	128	—	—	Cm Tc Kn Am	Cm Kn (Tc)
L.V.	+	K.p.	128	—	—	Cm Tc Kn Am	Cm Kn (Tc)
M.I.	—	K.p.	128	—	—	Cm Tc Kn Am	Cm Kn (Tc)
C.H.	—	E.c.	320	—	?	Tc	Tc
O.A.	+	E.c.	256	—	—	Sm Cm Tc Kn Am Su	Sm Cm Tc Kn Am Su

T+—: treated or untreated with colistin
055 : *E. coli 055*
K.p. : *Klebsiella pneumoniae*
E.c. : *E. coli*
MIC : minimal inhibitory concentration (μg/1 ml)

TR : transfer
EL : elimination
Sm : streptomycine
Su : sulphonamides

Table 1 illustrates the properties of some representative colistin-resistant strains isolated from seven treated children. One resistant strain from an untreated child (M.I.) hospitalized in the same ward and two other strains isolated in other wards without any epidemiological relation to the above are added (C.H. and O.A.). Even though we presume that some children could be contaminated one from another by members of the staff, at least three independent colistin-resistances developed here. The first in *E. coli* 055 is expressed by minimal inhibitory concentration (MIC) 64 μg/ml and is eliminable by acridine orange (AO). The second that developed in *K. pneumoniae* is expressed by MIC 320 μg/ml and is also curable by AO. The third in *K. pneumoniae* is incurable and is expressed by MIC 128 μg/ml. All *Klebsiella* strains transfered resistance ot CM, Kn and irregularly to Tc as were, but we did not find co-transfer and co-elimination of colistin resistance with the above mentioned resistance determinants. Generally, we consider colistin resistance to be nontransmissible, even though we can presume it to be transmissible in future.

Fig. 2 demonstrates the course of colistin-resistance elimination in transfers in broth containing 20 μg/ml of AO. Elimination progressed very rapidly in *E. coli* 055, 963 strain when only sensitive cells were found in the third transfer. In two *K. pneumoniae* strains tested, one numbered as 53 and the second as 70, elimination began in the third and increased up to the fifth transfer. We observed spotaneous loss of colistin-resistance in *E. coli* 055 but not in *Klebsiellae*, in the fifth transfer in broth without AO.

As we had stored all strains isolated before colistin therapy, we could test them for colistin mutability under laboratory conditions. From all these sensitive strains of *E. coli* 055 and *K. pneumoniae* we isolated colistin-resistant mutants. We also found that this mutation is a one-step event. These laboratory mutants behave just as those isolated from

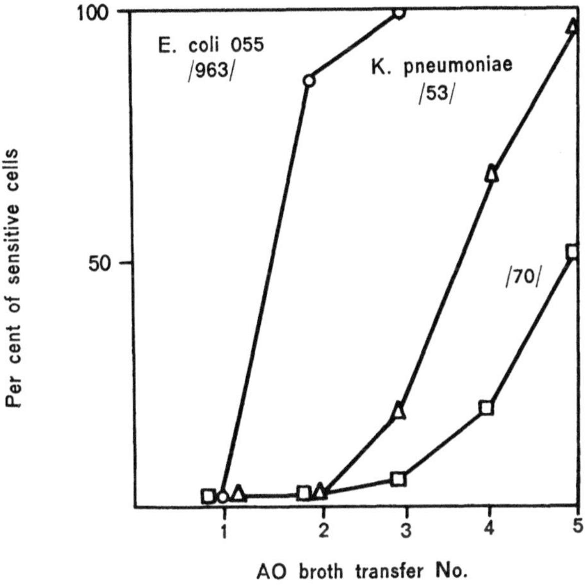

Fig. 2. Elimination of colistin resistance during AO broth transfers

children. This means that they have the same MIC and their colistin resistance is eliminable by AO. Another finding of some interest is the fact that "cured" strains can again mutate to colistin resistance but with a lower frequency than originally sensitive strains. This secondary colistin-resistance is also curable by AO, but with lower frequency. Preliminary estimations of mutation frequency showed that "primary sensitive" strains mutate with a frequency of 10^{-6} and those "cured" or secondary sensitive strains mutate with a frequency which is about ten times lower. Further experiments are in progress and will be reported in a special communication.

Sensitivity of colistin-resistance to AO, or its spontaneous loss, indicate that it may be of a plasmid nature. If we accept this hypothesis, the following explanations are possible: 1. A special kind of plasmid, if present, provides the cell with the possibility to mutate to colistin-resistance. 2. The cell contains many copies of this plasmid and only those mutated are selected. 3. In curing, only those cells are checked that had lost a resistant copy of the plasmid. The remaining "sensitive" copies provide the cell with a further possibility to mutate. 4. May be the non-cured copies are attached to the chromosome which protects them from elimination.

Chabbert, Baudens and Bouanchaud (1969) mention the colistin-polymyxin resistance among those having a plasmid nature. Other literary data are only very sporadic and will be listed in the final form of this communication. According to the work cited above and to the results obtained, colistin-resistance fulfils only one criterium for plasmid-determination — curing by acridins or spontaneous loss. Mobilization and studies of the molecular level will elucidate the true nature of this type of resistance.

REFERENCE

CHABBERT, Y. A., BAUDENS, J. G. and BOUANCHAUD, D. H. (1969). Medical aspects of transferable drug resistance. Bacterial Episomes and Plasmids, Ciba Foundation Symposium; J. and A. Churchill Ltd., London, p. 227.

ANTIMICROBIAL RESISTANCE AND R-FACTORS IN SALMONELLA ISOLATED FROM HUMANS AND ANIMALS

H. C. NEU, CH. E. CHERUBIN, E. LONGO

Division of Infectious Diseases, Columbia-Presbyterian Medical Center and The Division of Infectious Diseases, New York Medical College, New York, U.S.A.

The increasing resistance of members of the Enterobacteriaceae to antimicrobial agents has been noted on a world-wide basis (10, 11). Particular interest has focused on salmonellae species (3, 6, 7, 9, 12, 14, 19); and over the past decade a number of investigators have studied the antimicrobial resistance and presence of R-factors in isolates from both human and animal sources. We wished to compare the current antibiotic resistance of human salmonellae isolates to our previous results from 1968–69 (14, 19), and to determine if salmonellae isolated from various animal sources had different resistance patterns.

MATERIALS AND METHODS

The strains of salmonellae were those which had been submitted to the New York Salmonella Center at Beth Israel Hospital during 1973. The majority of strains were received between January and December 1973. Information on the age of the patient, sex, source of the isolation, therapy, and symptoms was available in the majority of instances. All duplicate cultures from the same individual were tested, but were not included in the statistical analysis unless the sensitivity pattern changed. There was a total of 718 salmonellae strains analyzed.

Salmonellae isolated from animals were obtained from The National Animal Disease Center at Ames, Iowa. These isolates represented strains collected from various animal species during the same period. The type of animal involved and source of the isolates permitted us to eliminate duplicate cultures. A total of 688 strains was examined.

Antimicrobial Susceptibility Tests: The disc method of Bauer et al. (5) was used to determine antimicrobial susceptibility. The following discs were used (μg size given): ampicillin 10; cephalothin 10; chloramphenicol 30; streptomycin 10; kanamycin 30; gentamicin 10; tetracycline 30; sulfisoxazole 300; nalidixic acid 30; polymyxin B 10; and trimethoprim 1.25. Zone sizes were interpreted according to standard recommendations, but organisms showing intermediate resistance were considered susceptible for purposes of analysis of the data. Agar plate dilutions of the antibiotics using a replicating apparatus were performed to check the disc method and excellent correlations were found in all instances.

Assay of Transferable Drug Resistance: The recipient organism was *Escherichia coli* K12 (W 1485, nalidixic acid resistant). Overnight cultures of donor and recipient organisms were mixed using 0.1 ml of donor, 0.9 of recipient in a final volume of 10 ml of Brain-Heart Infusion broth (Difco). This was incubated at 35C for 4 hours. A sample was removed and plated on eosin-methylene blue medium containing the appropriate antibiotics: ampicillin 100 μg/ml; chloramphenicol 30 μg/ml; streptomycin 30 μg/ml; kanamycin 30 μg/ml; tetracycline 30 μg/ml; cephalothin 50 μg/m; sulfisoxazole 300 μg/ml; all in combination with nalidixic acid 100 μg/ml. Selected transconjugant colonies were checked by replating on similar media.

RESULTS

Human Salmonellae

A total of 46 different serotypes were represented in the 718 human salmonellae strains. *Salmonella typhimurium*, including variety *copenhagen*, was the most frequently isolated serotype accounting for 34% of the isolates. The other common organisms were *S. enteritidis* (14%), *S. newport* (6.1%), *S. saint paul* (6.1%), *S. heidelberg* (4.2%), *S. agona* (3.3%), *S. derby* (2.4%) and *S. infantis* (4.1%). These serotypes were similar

TABLE 1

Antibiotic Resistance of Salmonella Isolated from Human Sources

Serotype	No. Tested	No. Resistant	Percent Resistant
S. typhimurium	241	139	57,6
S. newport	44	16	36.3
S. heidelberg	30	9	30.0
S. derby	17	5	29.4
S. st. paul	44	7	15.9
S. enteritidis	102	6	5.8
S. infantis	29	2	6.8
S. agona	24	1	4.1
S. blockley	19	1	5.2
Others	168	32	19.1
TOTAL	718	218	30.3

TABLE 2

Antimicrobial Resistance to Individual Antimicrobial Agents by Serotype Human Isolates

Antimicrobial Agent	No.	*S. typhimurium*	*S. heidelberg*	*S. newport*	*S. st. paul*	*S. enteritidis*	*Others* 257
	Tested	241	30	44	44	102	
				Percent Resistant			
Streptomycin		45.6	16.6	27.2	11.3	2.9	0.9
Chloramphenicol		2.9	0	0	0	0	0.4
Tetracycline		44.8	20.0	27.2	9.0	1.9	8.5
Sulfisoxazole		4.8	6.6	25	2.2	2.9	7.5
Ampicillin		36.9	30.0	31.8	13.6	0.9	4.2
Kanamycin		29.4	30.0	20.4	2.2	0.9	2.3
Cephalothin		2.1	3.3	0	0	0.9	0.8
Nalidixic acid		0	0	2.2	0	0	0
Trimethoprim		0	0	0	0	0	0
Polymyxin B		0	0	0	0	0	0

to those found in the 1972 salmonellae surveillance for the United States (8) of strains submitted to the Center for Disease Control. A total of 30.3% of the isolates were resistant to one or more antibiotic (Table 1.) Fifty-seven percent of *S. typhimurium*, 30% of *S. newport*, 30% of *S. heidelberg*, 29% *S. derby* and 15.9% of *S. saint paul* were resistant. In contrast, only 5.8% of *S. enteritidis*, 6.8% of *S. infantis*, 4.1% of *S. agona* and 5.2% of *S. blockley* were resistant. The resistance of other serotypes was 19%. A resistant strain was found in almost all of the serotypes except for *S. thompson* and *S. barielly*.

The resistance to specific antibiotics for different serotypes is shown in Table 2. Resistance to streptomycin was most common followed by resistance to tetracycline. *S. typhimurium* showed a significantly greater resistance to tetracycline (44.8%) and to streptomycin (45.6%) than did other serogroups in which the resistance ranged from 0.9% for other salmonellae to 27% in *S. newport*. Resistance to ampicillin was also common in *S. typhimurium*, *S. heidelberg* and *S. newport*. On the other hand, it was quite rare in *S. enteritidis* which is one of the commonly encountered causes of diarrhea in children. Resistance to cephalothin is very uncommon in all serogroups as is resistance to nalidixic acid. It was surprising to note the high resistance to kanamycin and a relatively lower resistance to sulfonamides. This is in contrast to the situation in shigellae (unpublished observations). None of the strains were resistant to gentamicin and the resistance to carbenicillin correlated exactly with resistance to ampicillin. None of the strains were resistant to polymyxin B or trimethoprim.

TABLE 3

Resistance of Salmonella Serotypes Isolated from Human Sources

Serotype	No. Resistant	Percent of Strains Resistant to the Number of Antibiotics Indicated					
		1	*2*	*3*	*4*	*5*	*6*
S. typhimurium	139	11.5	13.7	18	25.8	30.8	
S. newport	16	12.5	6.3	18.8	25	37.5	
S. heidelberg	9	22	55.5	11.2	11.2		
S. st. paul	7	28.5		57.2	14.3		
S. enteritidis	6	16.7	83.5		16.7		
S. derby	5	60	20	20			
Others	36	30.5	39	13.9	13.9	2.8	2.8

Table 3 shows the percent of strains resistant to the number of antibiotics tested. *S. typhimurium* and *S. newport* both had more than 50% of strains resistant to four and five agents. Both of these serotypes had in excess of 85% of the resistant strains resistant to more than one antibiotic. There are too few resistant strains in each of the other serotypes to comment on resistance to a single agent, but of all the other serotypes 34% were resistant to only one of the antibiotics tested.

Table 4 lists the antibiotic resistance in our current study in comparison with our previous studies (14, 19). It is notable that resistance to tetracycline has risen from 8.2% to 20.7%; resistance to kanamycin has risen from 3.3% to 12.5%, while resistance to ampicillin has not shown as great an increase, only 13.5% to 16%. Comparison of antimicrobial resistance by serotype for organisms isolated in the eastern United States shows that the resistance of *S. typhimurium* has risen from 18.5% to 57.6%, *S. newport* from 16.7% to 36.3%. In comparison with our own studies, resistance in *S. heidelberg*, *S. derby* has also increased, but *S. enteritidis* and other serogroups have not shown increased resistance.

TABLE 4

Comparison of Antimicrobial Resistance of Salmonella Isolated in 1968–69 and 1972–73

Antimicrobial Agent	1968–69	1972–73
	Percent	
Streptomycin	15.0	22.5
Ampicillin	13.5	16
Tetracycline	8.2	20.7
Chloramphenicol	0	0.8
Sulfisoxazole		20.0
Kanamycin	3.3	12.5
Cephalothin	3.2	1.2
Sulfamethoxazole-Trimethoprim		0.2
Nalidixic acid		0.1
Polymyxin	0	0

TABLE 5

Prevalence of R Factors in Resistant Strains of Salmonella Isolated from Humans

Serotype	No. Tested	No. with R Factor	Percent with R Factor
S. typhimurium	100	83	83
S. heidelberg	8	7	87
S. newport	15	12	80
S. st. paul	7	7	100
S. enteritidis	3	0	0
S. derby	4	3	75
S. infantis	2	1	50
Others	31	24	77
TOTAL	170	137	81

The prevalence of R-factors for the resistant strains is shown in Table 5. Not all strains were tested, but strains resistant to a different number and different antibiotics were tested for every serogroup. R-factor were common in *S. typhimurium*, *S. heidelberg*, *S. newport*, and *S. saint paul*. They were also common, in excess of 75%, in other serotypes except for the failure to demonstrate resistance transfer with *S. enteritidis* strains.

Resistance to ampicillin, streptomycin, sulfisoxazole, tetracycline, chloramphenicol and kanamycin was transferred, but we were unable to transfer resistance to cephalothin or nalidixic acid. The percent of transfer of resistance ranged from 73% for strains resistant to one antibiotic to 90–100% for those resistant to four or more antibiotics. There were 31 different antibiograms represented in resistant strains. The antibiograms as well as the resistance transferred were the same in all serotypes.

A seasonal variation in resistance to antibiotics was noted for *S. typhimurium*, but not other strains, Table 6. Resistance was highest in the winter months and lowest in the spring and summer for *S. typhimurium*. In contrast, resistance of the other serotypes did not show any regular pattern.

TABLE 6

Seasonal Variation in Antibiotic Resistance of Salmonellae

Months	No. of Isolates	Percent Resistant	Percent Resistant to Ampicillin
A. *S. typhimurium*			
Jan-Feb	21	81	86
Mar-Apr	33	39	61
May-June	39	23	33
July-Aug	43	23	44
Sept-Oct	45	44	67
Nov-Dec	42	50	64
B. *Other Salmonellae*			
Jan-Feb	24	12.5	20.8
Mar-Apr	61	8.2	18.0
May-June	79	16.5	25.3
July-Aug	90	4.4	11.1
Sept-Oct	77	3.9	15.6
Nov-Dec	96	5.2	8.3

TABLE 7

Resistance of Salmonellae Serotypes Isolated from Animal Sources

				Percent of Strains Resistant to Number of Antibiotics Indicated					
Serotype	No. Tested	No. Resistant	Percent Resistant	1	2	3	4	5	6
S. typhimurium	484	388	80	12	13	12	40	19	1
S. saint paul	77	47	61	25	17	25	27	4	
S. heidelberg	72	30	41	20	30	33	16		
S. san diego	26	16	61	43	18	18	18		
S. enteritidis	27	2	7		100				
Others	2	1	50	50					
TOTAL	688	488	70						

Animal Salmonellae

There were 7 serotypes represented in the 688 salmonellae from animal sources which were tested. *S. typhimurium* accounted for 70% of the isolates. Eighty percent of the *S. typhimurium*, 41% of *S. heidelberg*, 61% of *S. saint paul*, and *S. san deigo* were resistant, while only 7% of *S. enteritidis* were resistant to one or more antimicrobial agent, Table 7. The majority of *S. typhimurium*, *S. sait paul*, and *S. heidelberg* were resistant to three or more antibiotics, while 43% of *S. san deigo* were resistant to only one agent.

The resistance to specific antimicrobials is shown in Table 8. Resistance to streptomycin and tetracycline is high. Resistance to ampicillin is 31% in *S. typhimurium,* but lower in the other strains. Resistance to cephalothin is uncommon and only four *S. typhimurium* strains were resistant to chloramphenicol. Much higher resistance to sulfisoxazole (73%) was found for *S. typhimurium* than for other serotypes. Interestingly, resistance to kanamycin was 46% for *S. typhimurium* and 36% for *S. saint paul.*

TABLE 8

Antimicrobial Resistance to Individual Antimicrobial Agents by Serotype Animal Isolates

Antimicrobial Agent	*S. typhimurium*	*S. heidelberg*	*S. st. paul*	*S. enteritidis*	*S. san diego*
No. Tested	484	72	77	27	26
			Percent Resistant		
Streptomycin	64	32	45	0	34.5
Chloramphenicol	0.8	0	0	0	0
Tetracycline	61	18	35	0	37
Sulfisoxazole	73	22	27	0	22
Ampicillin	31	11	19.5	7.4	15.4
Cephalothin	1.6	0	1.3	0	0
Kanamycin	46	19	36	7.4	19.2
Polymyxin B	0	0	0	0	0
Nalidixic acid	0	0	0	0	0
Gentamicin	0	0	0	0	0
Trimethoprim	0	0	0	0	0

TABLE 9

Resistance of Salmonellae Isolated from Animals

Animal Source	No. of Strains	Amp.	Tetra.	Strep.	1 or more Antibiotic
A. *S. typhimurium*					
Chickens	72	12	33	33	36
Turkeys	54	28	72	74	81
Other fowl	34	9	9	9	12
Cattle	205	37	85	83	89
Pigs	53	43	72	70	76
Horses	23	35	65	70	74
B. *Other Salmonellae*					
Chickens	47	15	15	15	17
Turkeys	102	19	43	42	46
Other fowl	11	0	0	0	0
Cattle	13	38	46	46	46
Pigs	13	0	23	23	31
Horses	7	57	57	57	57

Table 9 gives a breakdown of the animal sources of the salmonellae. Resistance of *S. typhimurium* to tetracycline and streptomycin is similar from many different animal sources. However, resistance to ampicillin is lower in all instances. Resistance of other salmonellae serotypes to the three antibiotics is lower in strains from chickens, turkeys and pigs, but similar in strains from horses and cattle.

TABLE 10

Prevalence of R Factors in Resistant Strains of Salmonella Isolated from Animals

Serotype	No. Tested	No. with R Factor	Percent with R Factor
S. typhimurium	109	94	86
S. st. paul	45	36	80
S. heidelberg	28	26	92
S. san diego	14	8	57
TOTAL	196	164	84

R-factors were found in 86% of *S. typhimurium* tested, Table 10; and in 80% of *S. saint paul*, 92% of *S. heidelberg*, but only 57% of *S. san deigo*. Resistance to ampicillin, streptomycin, sulfisoxazole, tetracycline, kanamycin and chloramphenicol was transferred. The antibiograms represented were similar to those encountered in the salmonellae from humans. There was no seasonal variation in resistance of animal salmonellae.

DISCUSSION

Antimicrobial resistance of bacteria has become an increasingly important problem in the past decade (2, 10, 11). The contribution of R-factors to this resistance has been generally recognized, and studies have demonstrated the importance of R-factors as a mechanism of bacterial resistance in salmonellae from both human and animal sources (11). What has been less clear is whether there has been a significant increase in resistance of human salmonellae to specific antibiotics in the past few years and whether animal salmonellae have shown similar patterns of resistance.

In this study of salmonellae isolated from humans, *S. typhimurium* has continued to have the highest incidence of resistance, 58% in contrast to the second most commonly isolated salmonellae, *S. enteritidis* in which only 5.8% resistance was found. We had found a lower incidence of resistance in *S. typhimurium* in 1968 and 1969 in our previous two surveys (14, 19), and the current level of resistance is comparable to the 48% resistance found in a recent study from California (6). In that study resistance was not seen in *S. enteritidis*. We noted an increase in the antimicrobial resistance of *S. heidelberg* strains from 10.5% in 1968–69 to our current 30%, but it should be noted that seasonal variations in resistance do occur, since we had a 23% resistance in the spring and summer of 1969 in a repeat study (14). In general, there has been a gradually increasing antimicrobial resistance of most salmonellae serotypes isolated from humans over the past 5 years.

During this period there has been an increase in tetracycline resistance of *S. typhimurium* from 12.5% in 1968–69 to currently 44.8%. A similar increase in tetracycline resistance has been noted in *S. heidelberg*, *S. newport*, but not other strains. This increase is highly significant for the three strains at ($p < 0.001$), and is similar to the current tetracycline resistance seen in California (6). Of note is the low tetracycline resistance in strains such as *S. infantis*, *S. derby*, *S. saint paul*.

Ampicillin resistance has increased slightly from 13.5% to 16% in the 5 years, particularly in strains other than *S. typhimurium*. Resistance to ampicillin is similar in our study to that found in the California sample. Kanamycin resistance which was low in our previous studies has reached levels as high as the ampicillin resistance and is perhaps similar to the increased kanamycin resistance of *E. coli*. Chloramphenicol resistance has remained at a low level in the eastern United States as has cephalothin resistance. In contrast, in the western United States chloramphenicol resistance has been seen not only in *S. typhi*, but in other serotypes of salmonellae (6). The low resistance to cephalothin probably is because most of the R-factors of salmonellae specify beta-lactamases with a low affinity for cephalosporin antibiotics (13). It was encouraging to note that resistance to trimethoprim is not present, but since the use of that agent has been limited to some extent in this area, it will be necessary to re-evaluate this in a few years. Studies in Canada have also shown that resistance to trimethoprim is uncommon in contrast to the resistance to sulfonamides (12).

In this study, as in previous work (14), and from other investigators (16), there has been a high incidence of R-factors to account for the antimicrobial resistance. We were able to demonstrate R-factor transfer in 81% of the strains examined. Similar to other studies, transfer is highest in the multiply resistant strains. The high incidence of R-factor transfer suggests that most of the resistant strains seen today contain complete R-factors and that strains with loss of transfer factor are less common than formerly noted (2).

A number of studies have been performed to evaluate the resistance of salmonellae serotypes isolated from animals (3, 7, 15, 17, 18). Anderson and Lewis (3) in England in 1965 found a high incidence of resistance in *S. typhimurium*. In a similar study from England, H. Williams-Smith (18) in 1968, found only 4.2% of salmonellae from poultry in England and Wales were resistant and no resistance to ampicillin was found.

Salmonellae obtained from outbreaks of animal disease in the United States have been studied for antibacterial resistance and the presence of R-factors (1, 15). In one study 10.9% of *S. typhimurium* were multiply resistant and overall resistance to one of eleven drugs tested against all the salmonellae serotypes was 74% and 67% transferred resistance. In a large study (7, 17) in Germany in 1971 of 1,855 salmonellae strains examined, 12.1% were resistant to the antibiotics tested. The proportion of resistant strains was 32% for *S. typhimurium*, 66% for *S. panama*, 8.1% for *S. heidelberg*, and 6.3% for *S. enteritidis*. Resistance to tetracycline was the most common, followed by resistance to ampicillin. The overall resistance of our animal salmonellae is comparable to that seen in these other studies with a slightly higher level of multiple resistance, 61%. This study did show a higher incidence of kanamycin resistance than had been noted previously. Tetracycline resistance was highest in turkeys and ampicillin resistance highest in cattle (Table 9). This agrees with earlier data on beef *E. coli* resistance to ampicillin which ran 33%. These differences in resistance to specific antibiotics may reflect use of antimicrobial agents in the poultry and cattle industry.

Comparison of the resistance of the human and animal salmonellae of this study shows that tetracycline resistance is greater in the animal strains as is sulfonamide and streptomycin resistance. This may reflect the greater use of these agents in animals. The high incidence of kanamycin resistance in animals in this study is surprising in comparison with earlier studies which have shown lower kanamycin resistance in animal isolates of both salmonellae (15) and *E. coli* (4). The percent of strains containing R-factors was similar for the human and animal strains. Transfer of resistance for tetracycline, streptomycin and kanamycin was equivalent for strains from both sources, but ampicillin resistance was transferred in 36% of animal strains and 78% of human strains. This difference is of interest since it could reflect differences in the R-factors present in the strains. Studies are currently in progress to define the compatibility groups of the animal and human salmonellae strains which are ampicillin resistant.

Salmonellae of serotypes other than *S. typhimurium* that came from humans were consistently less resistant to all antibiotics than were animal isolates. For example, "other" human salmonellae had the following resistance: 7% ampicillin, 8.8% tetracycline, 7.5% sulfisoxazole, 10.9% streptomycin, and 4% kanamycin; whereas animal salmonellae had 14% ampicillin, 24% tetracycline, 21% sulfisoxazole, 33% streptomycin, and 24% kanamycin resistance.

This study confirms earlier data (15, 19) as well as recent observations from California (6) that resistance is low and R-factors uncommon in certain salmonellae; namely, *S. enteritidis* and *S. infantis*. There is no current explanation for this which could be due to cell wall composition, DNA restriction, or presence of incompatible plasmids conveying other properties on the strains.

The results of this survey clearly demonstrate that the resistance to antibiotics is increasing in salmonellae isolated from both humans and animals. Since there are great similarities in the resistance patterns of the human and animal isolates, it would be of great benefit to know if the R-factors are of a similar nature which might suggest that animal strains have contributed to the human pool of resistant organisms. However, it would seem that in the absence of knowledge of the compatability types of the R-factors we do not know if the resistant organisms are of common origin.

REFERENCES

1. Aden, D. P., Reed, N. D., Underdahl, N. R., and Mebus, C. A. (1969). Transferable drug resistance among Enterobacteriaceae isolated from cases of neonatal diarrhea in calves and piglets. Appl. Microbiol. 18, 961.
2. Anderson, E. S. (1968). The ecology of transferable drug resistance in the enterobacteria. Ann. Rev. Microbiol. 22, 131.
3. Anderson, E. S. and Lewis, M. J. (1965). Drug resistance and its transfer in *Salmonella typhimurium*. Nature 288, 579.
4. Babcock, G. F., Berryhill, D. L., and Marsh, D. H. (1973). R-factors of *Escherichia coli* from dressed beef and humans. Appl. Microbiol. 25, 21.
5. Bauer, A. W., Kirby, W. M. M., Sherris, J. C., and Turck, M. (1966). Antibiotic susceptibility testing by a standardized single disc method. Amer. J. Clin. Path. 45, 493.
6. Bissett, M. L., Abbott, S. L., and Wood, R. M. (1974). Antimicrobial resistance and R-factors in salmonella isolated in California (1971–1972). Antimicrob. Ag. Chemother. 5, 161.
7. Bulling, E., Stephan, R., and Sebeck, V. (1973). Die Entwicklung der Antibiotikaresistenz vonSalmonellabakterien tierischer Herkunft in der Bundesrepublik Deutschland einschl. Berlin Zbl. Bakt. Hyg. I Abt. Orig. A 225, 245.
8. Center for Disease Control (1973). Salmonella surveillance annual summary 1972. Center for Disease Control, Atlanta, Georgia.
9. Cherubin, C. E., Szmuness, M., and Winter, J. (1972). Antibiotic resistance of Salmonella, Northeastern United States – 1970. N.Y. State J. Med. 72, 369.
10. Davies, J. E. and Rownd, R. (1972). Transmissible multiple drug resistance in Enterobacteriaceae. Science 176, 758.
11. Dulaney, E. L. and Laskin, A. I. (ed) (1971). The problem of drug resistant pathogenic bacteria. Ann. N.Y. Acad. Sci. 182, 1–415.
12. Marks, M. I., Kazemi, M., and McKay, C. (1973). *In vitro* sensitivity of Salmonella to ten antimicrobial agents including sulfamethoxazole and trimethoprim, alone and in combination. Antimicrob. Ag. Chemother. 4, 555.
13. Neu, H. C. and Winshell, E. B. (1970). Purification and characterization of penicillinases from *Salmonella typhimurium* and *Escherichia coli*. Arch. Biochem. Biophys. 139, 278.

14. NEU, H. C., WINSHELL, E. B., WINTER, J., and CHERUBIN, C. (1971). Antibiotic resistance of Salmonella in Northeastern United States 1968–69. N.Y. State J. Med. 71, 1196.
15. POCURULL, D. W., GAINES, S. A., and MERCER, H. D. (1971). Surveys of infectious multiple drug resistance among Salmonella isolated from animals in the United States. Appl. Microbiol. 21, 358.
16. SCHROEDER, S. A., TERRY, P. M., and BENNETT, J.V. (1968). Antibiotic resistance and transfer factors in Salmonella, United States 1967. J. Amer. Med. Assoc. 205, 903.
17. STEPHAN, R. and BULLING, E. (1973). Die Entwicklung der Antibiotikaresistenz von Salmonellabakterien tierischer Herkunft in der Bundesrepublik Deutschland einschl. Berlin, Zbl. Bakt. Hyg. I Abt. Orig. A. 225, 257.
18. WILLIAMS-SMITH, H. (1970). The incidence of transmissible antibiotic resistance amongst Salmonellae isolated from poultry in England and Wales. J. Med. Microbiol. 3, 181.
19. WINSHELL, E. B., CHERUBIN, C., WINTER, J., and NEU, H. C. (1970). Antibiotic resistance in Salmonella in the Eastern United States. Antimicrob. Ag. Chemother. 1969, p. 86.

INDEX OF SUBJECTS

AUTHOR INDEX*

* Numbers in bold types refer to the first page of the authors communication.

THIRD INTERNATIONAL SYMPOSIUM ON ANTIBIOTIC RESISTANCE

will be held in the Castle of Smolenice, Czechoslovakia on June, 1—4, 1976

THEORETICAL AND MEDICAL PROBLEMS OF ANTIBIOTIC RESISTANCE

Chairman of the Symposium: Professor Susumu Mitsuhashi

I. Theoretical problems

1. Molecular Biology of Plasmid (Proposed chairman: R. C. Clowes)
2. Genetics of Plasmids (Proposed chairman: B. W. Holloway)
3. Envinonmental Consequences of Plasmids (Proposed chairmen: M. H. Richmond and L. Rosival)
4. Biochemistry of Plasmid Function (Proposed chairman: J. Davies)
5. Genetic engineering (Proposed chairman: R. Curtiss, III)

II. Medical problems

Proposed chairmen: S. Mitsuhashi, S. M. Navashin, H. Knothe

PROCEEDINGS WILL BE PUBLISHED BY AVICENUM, PRAGUE AND SPRINGER VERLAG, BERLIN - HEIDELBERG NEW YORK

GPSR Compliance

The European Union's (EU) General Product Safety Regulation (GPSR) is a set of rules that requires consumer products to be safe and our obligations to ensure this.

If you have any concerns about our products, you can contact us on ProductSafety@springernature.com

In case Publisher is established outside the EU, the EU authorized representative is:

Springer Nature Customer Service Center GmbH
Europaplatz 3
69115 Heidelberg, Germany

Batch number: 10844669

Printed by Printforce, the Netherlands